T0318559

Practical Handbook of Soybean Processing and Utilization

About the Editor

David R. Erickson has been involved in soybean processing and utilization for thirty-two years. He started his career in oilseed and edible oils research at Swift and Company, Chicago, Illinois, in 1963 and worked fifteen years in research and research management. In 1978, he became the technical director of the International Marketing Department of the American Soybean Association (ASA). In 1992 he became an independent consultant. During his tenure with ASA, he worked as a consultant engaged in technology transfer in over sixty countries; he continues to work with the ASA in that capacity and also as an independent consultant. Internationally he is recognized as knowledgeable in soybean processing and utilization, including product development and marketing. He has been very active in the American Oil Chemists' Society (AOCS): He served as its president in 1990 and received the Bailey Award in 1989. He has produced more than fifty publications, five patents, and has edited three AOCS monographs. He is also a professional member of the Institute of Food Technology and a member of the American Chemical Society, Alpha Zeta, Phi Kappa Phi, and Sigma Xi. He holds a Ph.D. in agricultural chemistry from the University of California–Davis (1963) and B.S. and M.S. degrees in dairy technology from Oregon State University.

Practical Handbook of Soybean Processing and Utilization

Editor
David R. Erickson
Consultant
St. Louis, Missouri

Published jointly by

AOCS Mission Statement

To be a global forum to promote the exchange of ideas, information, and experience, to enhance personal excellence, and to provide high standards of quality among those with a professional interest in the science and technology of fats, oils, surfactants, and related materials.

AOCS Press, Urbana, IL 61802

ISBN 978-0-935315-63-9

Library of Congress Cataloging-in-Publication Data

Practical handbook of syobean processing and utilization/editor,
David R. Erickson
p. cm.
Includes bibliographical references and index.
ISBN 0-935315-63-2 (alk. paper)
1. Soybean products—Handbooks, manuals, etc. I. Erickson, David R.

TP438.S6P76 1995
664'.805655—dc20

95-8610
CIP

Printed in the United States of America.
12 11 10 09 08 7 6 5 4 3

The paper used in this book is acid-free and falls within the guidelines established to ensure permanence and durability.

Preface

In 1980, the American Soybean Association (ASA) and the American Oil Chemists' Society (AOCS) jointly published the *Handbook of Soy Oil Processing and Utilization*. Its purpose was to serve as a single source for practically all of the published information on the subject available at that time. It has proved to be a valuable and useful reference and has been a best-seller for the AOCS.

Since its publication, it has been my privilege to work as a technical consultant in soybean processing in over sixty countries. This experience, plus that of other ASA consultants, has increased our knowledge and understanding of worldwide practices and concerns in soybean processing. Such experience has revealed a need for a new handbook that included all soybean products and was oriented more toward practical and actual industrial practice.

This new *Practical Handbook of Soybean Processing and Utilization* is intended to provide a single source of practical information on the subject to an international audience. It is designed not only to revise and update the content of the original "Handbook" on soy oil, but also to expand the content to include all soybean products.

This book is oriented toward inclusion of more practical "how to" information rather than being a reference on all published information; hence only key references are included. Authors have been selected, wherever possible, from those who have had the actual industrial experience required to provide information that is sometimes only gained through industry experience. Often, industrial practices stem from internal developments that are not published. Where such practices are generally known and not a trade secret, I have included them in this book.

I would like to thank the authors for their contributions and would also like to acknowledge the contributions of other ASA staff technical consultants, such as Lars Wiedermann, Dionisio Bazaldua, and Roger Leysen, as well as the contributions of other consultants who have worked for the ASA. Most especially, I extend my thanks to the many gracious people that I have met in my travels who with their questions, discussions, and input gave direction to what should be included in this book. I also thank the AOCS staff for their editorial advice and help in bringing this 2½-year project to fruition.

As a last and most important item I would like to dedicate this book to my wife Judy, who has been very patient with me during the putting together of this handbook and for the last 44 years.

David R. Erickson
St. Louis, Missouri

Contents

Chapter 1

Soybeans vs. Other Vegetable Oils as a Source of Edible Oil Products

David D. Asbridge

Director, Industry Information
American Soybean Association
St. Louis, MO

Introduction

Soybeans are the dominant oilseed in both U.S. and world markets. During a normal production year, soybeans make up well over one-half of all oilseeds produced worldwide. This domination has been documented (1) since the popularity of the crop skyrocketed in the United States and Brazil during the mid-1970s. Total world production of soybeans now accounts for almost 130 million metric tons (MMT) and is expected to continue to increase as world population increases and people continue to upgrade their diets as income increases to include more fats, oils, and livestock-based protein. This protein, in many cases, comes from livestock fed a nutritious diet of soybean meal and cereals.

Table 1.1 shows how important soybeans are in the world production of oilseeds. It also shows that the United States is the dominant source for soybeans, although Brazil and Argentina continue to increase production. China, a long-time producer of soybeans, has stagnated in its soybean production and will soon probably not even be able to meet its own internal needs as a rapidly growing economy continues to push demand ahead of supply. One growing producer of soybeans is India, where the oil is consumed and the meal is exported to generate hard currency.

Soybean dominance comes from a variety of factors, including favorable agronomic characteristics, reasonable returns to the farmer and processor, high-quality protein meal for animal feed, high-quality edible oil products, and the plentiful, dependable supply of soybeans available at a competitive price. The price factor is especially important, because many of the uses of the protein and oil are for "commodity" products that need to meet many different levels of uses. Soybeans have a good mix of amino acids and fatty acids that make the protein and oil fit into most uses. There are some "designer" soybeans for very specific uses, but most soy is raised to go into a generic market, where availability and price are the driving factors. A change in this situation may occur over time but will take a change in thinking of the entire industry.

TABLE 1.1 World Production of Major Oilseeds (1)

Oilseed and country	Production[a], million metric tons		
	1987/88–1991/92 average	1992/93[b]	*1993/94*[c]
Soybeans:			
United States	50.75	59.55	49.22
Brazil	19.40	22.30	24.40
China	11.01	10.30	13.00
Argentina	9.92	11.00	12.20
European Union	1.80	1.47	.78
Paraguay	1.38	1.75	1.80
Other	9.45	10.36	12.21
Total	103.70	116.72	113.61
Cottonseed:			
China	7.61	7.66	6.40
United States	5.33	5.65	5.69
USSR/C15[d]	4.76	3.68	3.88
India	3.84	4.67	4.24
Pakistan	3.27	3.08	2.61
Brazil	1.25	.73	.79
Other	6.82	6.06	6.12
Total	32.88	31.53	29.72
Peanuts (in shell):			
India	7.51	8.85	7.40
China	5.98	5.95	8.00
United States	1.83	1.94	1.51
Senegal	.78	.58	.63
Sudan	.40	.39	.39
Brazil	.16	.15	.15
Argentina	.42	.28	.30
South Africa	.14	.17	.18
Other	4.89	4.86	4.91
Total	22.10	23.17	23.46
Sunflower seed:			
USSR/C15[d]	6.31	5.69	5.41
Argentina	3.56	3.10	3.30
Eastern Europe	2.25	2.59	2.25
United States	1.09	1.18	1.18
China	1.18	1.18	1.25
European Union	4.00	4.06	3.57
Other	3.12	3.46	3.94
Total	21.51	21.25	20.91
Rapeseed:			
China	6.30	7.65	6.80
European Union	6.09	6.25	6.01
Canada	3.73	3.69	5.40
India	4.61	4.87	5.80
Eastern Europe	1.82	1.20	1.16
Other	1.64	1.70	1.84
Total	24.18	25.36	27.18

[a]Split year includes Northern Hemisphere crops harvested in the late months of the first year shown combined with Southern Hemisphere and certain Northern Hemisphere crops harvested in the early months of the following year.
[b]Preliminary.
[c]Forecast.
[d]Excluding Estonia, Latvia, and Lithuania.

Agronomic Characteristics

Soybean plants can be grouped into basically two main types, *determinant* and *indeterminant*, both grown mostly in temperate climates. The determinant varieties will flower at a certain time of the year, basically when the days begin to shorten. Indeterminant varieties will continue to flower and put on fruit until the weather dictates that it is time to curtail plant growth. There are many different varieties which allow soybeans to be produced in different maturity zones that stretch from North Dakota (latitude 49°N) to Louisiana (latitude 30°N) in the United States.

One of the most important agronomic characteristics of soybeans is that it can take nitrogen from the air and "fix" it to be used by the soybean plant. The symbiotic relationship between the soybean plant and its nodulating bacterium (*Rhizobium japonicum*) is responsible for the conversion of atmospheric nitrogen into plant-available nitrogen. This also makes soybeans a good rotational crop for use with high nitrogen-consuming crops such as corn. One can usually expect a 5 to 10% increase in the yields of both crops when rotated as opposed to either crop in a monoculture farming operation.

Another important benefit of this nitrogen fixation is that it helps to keep the production costs for soybeans relatively low compared to other crops that compete for the same land area. This is illustrated in Table 1.2 (1,2).

Soybean Meal and Protein

Probably the single most important factor in the soybean success story is the amount of high-quality protein meal produced—up to 1800 lbs per acre (2 MT/ha) average in the United States, a quantity greater than for any other commercial oilseed (Table 1.2). This means that the meal is usually the driving force in soybean prices, because it will account for two-thirds of the value of the bean on average.

Soybean meal is used extensively in high-protein commercial feeds for poultry and swine, and to a lesser extent for dairy and beef cattle (3). The use of soybean meal

TABLE 1.2 Crop Production Costs and Average Yield[a] per Acre (1,2,4)

Crop	1991 Production costs, U.S. dollars per acre	Average yields lbs. per acre		
		Total	Oil	Meal
Corn	183	7,000	300	—
Cottonseed	324	1,050	175	470
Oil palm (Malaysia)[b]	—	3,000	—	—
Peanuts	465	2,500	800	1050
Rapeseed (Canada)	65	1,200	490	720
Soybean	110	2,200	400	1800
Sunflower	80	1,200	480	600

[a]USDA, average 1990–1993.
[b]*Oil World Annual 1994.*

in U.S. feeds has just about doubled from 12.0 million tons in 1972 to 23.9 million tons in 1993—a growth of 98%. In 1993, soybean meal's share of the high-protein feed market was about 70% (3). As yet only about 2% of soy protein produced goes into edible protein products in the United States, but this amount is expected to increase in future years (4). However, rapid growth of vegetable protein utilization in human foods is not foreseen for the next decade unless a direct tie with chronic disease prevention or cure is more clearly defined. Some work currently points toward the conclusion that soy could have a major role in this area.

Soybean Oil

Soybean oil, as the other direct product of soybean processing, is also extremely important, especially in food. Soybean oil's share of total edible fats and oils consumption in the United States was 77% in 1992, up from 73% in 1986. The increase in domestic edible soybean oil consumption was 12.6% in the 1986–1992 period, while total edible fats and oils consumption increased only 7.2% over the same period (5). With increasing industrial uses for the oil, such as a base for a diesel substitute in high-pollution areas, soybean oil should continue to grow in the favor of the market. Its availability and versatility will continue to make it more and more popular for all types of uses in the future.

Soybean Oil vs. Other Oils

World and U.S. consumption (disappearance) figures for the major fats and oils are shown in Table 1.3 (2). In both cases, soybean oil is the leading oil, but palm oil, the

TABLE 1.3 World and U.S. Consumption (Disappearance) of Fats/Oils for Oct/Sept, 1993/94 (2)

	U.S.		World	
Fat/oil	Million Pounds	Million MT	Million MT	Rank
Soybean	13,114	5.95	18.19	(1)
Corn	1,230	0.56	1.68	(12)
Tallow/Grease	4,400	2.00	7.01	(5)
Canola/Rapeseed	1,168	0.53	9.00	(3)
Coconut	1,050	0.48	2.94	(9)
Cottonseed	992	0.45	3.63	(8)
Lard	850	0.39	5.60	(6)
Palm kernel	348	0.16	1.86	(11)
Palm	309	0.14	14.40	(2)
Olive	264	0.12	2.10	(10)
Sunflower	209	0.09	7.68	(4)
Peanut	187	0.08	4.16	(7)
Fish oil	77	0.04	1.12	(13)

second oil in the world, has little use in the United States because of the U.S. consumers' negative attitude (nutritionally) toward the so-called "tropical oils," not because of any technical deficiency of palm oil. Another difference between the United States and the rest of the world is that marine/fish oils have not until recently been considered as being edible in the United States.

An important factor in any emerging market for new uses of an oil will be its price, where soybean oil has a competitive advantage. Soybean oil is usually less expensive than corn, safflower, and sunflower oils, yet it has many of the desirable characteristics of these so-called premium vegetable oils (Table 1.4) (1). It has a high linoleic acid content and a low saturated fatty acid content, and thus it is more desirable nutritionally than the more saturated oils (see Chapter 23).

Soybean oil has a number of other advantages when compared to other vegetable oils:

1. A high level of unsaturation is present.
2. The oil remains liquid over a relatively wide temperature range.
3. The oil can be hydrogenated selectively for blending with semisolid or liquid oils (see Chapter 13).
4. The partially hydrogenated oil can be used as a pourable, semisolid oil because of the relatively low levels of palmitic acid, which in higher concentrations causes a plastic solid to form (see Chapters 13 and 20).
5. The oil can be processed readily to remove phosphatides, trace metals, and soaps, thereby improving stability (see Chapters 11, 12, and 14).
6. Naturally occurring antioxidants (tocopherols) are present and are not completely removed during processing (see Chapters 2 and 17).

There are only limited disadvantages:

1. Soybean oil has a relatively high content of phosphatides that must be removed in processing (see Chapters 10 and 11). The phosphatides can be removed and used as a source of commercial lecithin or returned to the soybean meal. If not removed by degumming, the phosphatides become a part of the soapstock (see Chapter 17).

TABLE 1.4 Composition and Price for Edible Vegetable Oils (1)

	Composition, g/100 g oil			
Vegetable oil	Linoleic acid	Total polyunsaturated	Total saturated	Crude oil, avg 1990–92 price[a], ¢/lb
Corn	57	58	13	22.9
Cottonseed	50	51	26	22.1
Palm	9	9	48	18.7
Peanuts	31	32	17	31.1
Canola	22	34	7	24.1
Soybean	51	58	15	20.4
Sunflower	64	64	10	23.3

[a]U.S. Department of Agriculture, 1994.

2. Soybean oil has a polyunsaturated fatty acid content of about 60%, which makes it susceptible to oxidation and the subsequent development of undesirable off-flavors. Modern optimal processing practices, as described in Chapters 11, 12 and 14, minimize this susceptibility. In addition, hydrogenation can be used to increase oxidative stability, as shown in Chapter 13.

Salad and cooking oils or frying fats can be made almost solely from soybean oil. Semisolid (i.e., plastic) baking and consumer shortenings generally contain a high percentage (80 to 93%) of partially hardened soybean oil and a minor proportion (up to 20%) of a completely hardened palm or cottonseed oil (Chapter 13). The latter two oils are added because of their high content of palmitic acid, which gives the β′ crystalline structure desired to ensure crystal stability and proper performance of these shortenings. Soybean oil is the primary oil used in margarine manufacture; currently about 90% of all margarine consumed in the United States contains soybean oil (5).

The Other Vegetable Oils

Corn Oil

On a moisture-free basis, the oil content of the corn kernel is about 5%, compared with about 20% for the soybean, and is concentrated in the germ, which is recovered both by wet corn millers in the production of starch and by dry corn millers in the production of grits, meal, and flour. Corn oil is a premium oil, because of its high polyunsaturated fatty acid content (Table 1.4) and its low content (<1%) of linolenic acid. Production of corn oil has increased sharply in recent years because of the increased production of ethanol for the fuel market and of high-fructose corn syrup. Corn oil exports have been gaining on U.S. domestic use in recent years; domestic consumption has shown only slight growth.

Cottonseed Oil

The availability of cottonseed oil has been relatively stable over the past few years as there has been virtually no growth in cotton acreage, which has leveled off at 13 to 14 million acres (5.5 Mha) annually, while whole seed feeding to dairy animals is growing. Cottonseed oil has become very popular in the United States based on its functionality and flavor. The increase in domestic usage has virtually eliminated it from the export market with only 13% of the 1993 production level of 1.15 billion lbs (0.52 MMT) going offshore (1). Cottonseed oil has a higher saturated fatty acid (mostly palmitic) content than soybean oil (Table 1.4) but contains <1% of linolenic acid. The high palmitic acid content favors the β′ crystal form, which is necessary for plastic shortenings.

Palm Oil

Imports of palm oil have declined from a maximum of 933 million lb (0.423 MMT) in 1975/76 to less that 300 million lb (0.14 MMT) (1). Palm oil now competes with soybean oil mainly in Asian and European markets. Over the past two years, palm oil prices have maintained about a $0.02/lb differential below soybean oil prices. The 42% of palmitic acid in the oil favors the β' crystal form. For this reason and because of its relatively low production costs, palm oil can be expected to continue its competitive position, especially in markets where perceived health concerns are not the driving force as they are in the United States.

Peanut Oil

Peanut oil has a lower linoleic acid content (31%) and a higher oleic acid content (46%) than soybean oil. There is a strong demand for whole peanuts in edible uses in the United States, and peanut oil production is currently only about 250 million lbs (0.11 MMT) per year in the United States (1). Exports of this premium-priced oil have been growing over the past five years, but are still only about 50 million lbs (23 kMT).

Rapeseed Oil

Rapeseed is agronomically suitable for northern climates and is grown extensively in Canada and Northern Europe. The so-called zero-erucic acid variety (also called *canola*) contains 53% oleic acid and 11% linolenic acid. This has been the fastest-growing oil for human consumption in the United States because of its perceived advantage in having a saturated fat level of only 7%. It is also usually priced above soybean oil due to this growth in popularity (Table 1.4).

Safflower Oil

Safflower is grown on both irrigated and semiarid lands in the southwestern United States. Safflower oil production, however, probably will not increase above about 100 million lb (45 kMT) per year (4). Safflower oil has a high linoleic acid content (73%) and a low linolenic acid content (<1%), and for these reasons it commands a premium price. A high-oleic safflower oil variety is also available, which contains 78% oleic and 12% linoleic acids.

Sunflower Seed Oil

Sunflower production increased rapidly in the late 1970s as the result of introduction of improved varieties but then just as rapidly declined because of increased disease damage. U.S. consumption of sunflower seed oil is expanding for use in cooking oil and margarine, but most is exported. This situation should change over the next few years as government assistance programs for sunflower oil are eliminated by the new General Agreement on Tariffs and Trade (GATT). This oil, when obtained from northern-grown seeds, has a high linoleic acid content (64%) and low linolenic acid content (<1%). Oil from southern-grown seeds contains 49% linoleic and a higher oleic acid content (34% vs. 21%). About 700 million lb (0.3 MMT) of oil will be produced in the United States in 1994 (1). It is usually priced at a 3 to 4 cent/lb premium to soybean oil.

Conclusion

Properly processed soybean oil has been established as a high-quality oil suitable for most edible oil uses. It is expected to continue its preeminence in edible oil uses as a result of its availability and relatively low cost. With the possible exception of canola (and possibly palm outside the United States), most other commercial oil sources probably will not expand to the same extent as soybean oil because of agronomic, compositional, or other characteristic limitations.

References

1. United States Department of Agriculture, various publications.
2. *Oil World Annual, 1994*. ISTA Mielke GmbH, Hamburg, Germany.
3. AUS Consultants, *Soybean Meal in the U.S. Swine and Poultry Industries*, a USB-sponsored publication, March, 1993.
4. American Soybean Association estimates/forecasts.
5. United States Department of Commerce, various publications.

Chapter 2

Composition of Soybeans and Soybean Products

Edward G. Perkins

Dept. of Food Science
University of Illinois, Urbana, IL 61801

Introduction

Mature soybeans are nearly spherical in shape and vary considerably depending on cultivar and growing conditions. The soybean seed consists of three major parts: the seed coat, or *hull*; *cotyledons*; and *germ*, or *hypocotyl*. The seed coat contains the *hilum*, which is the point of attachment to the pod. The basic structural features of the soybean seed are shown below in Fig. 2.1.

The soybean is a dicotyledon seed (two cotyledons) held together by the hull. On removal of the hull, the cotyledons separate and the germ is dislodged. The commercial soybean contains about 8% hull, 90% cotyledons, and 2% hypocotyl. The microscopic structure of the soybean is shown in Fig. 2.2. Within the seed coat is the *cuticle*, followed by the *palisade cells*, a layer of distinctively shaped *hourglass cells*, and several layers of *parenchyma cells*. The presence of the hourglass-shaped cells serves also to identify soy products in foods. The cotyledon is overlaid by a layer of parenchyma cells and *aleurone cells*. The cotyledon has an *epidermis* and the interior is filled with elongated *palisade-like cells* (not to be confused with the palisade cells in the hull) which contain most of the oil and protein in the form of oil and *protein bodies* (1).

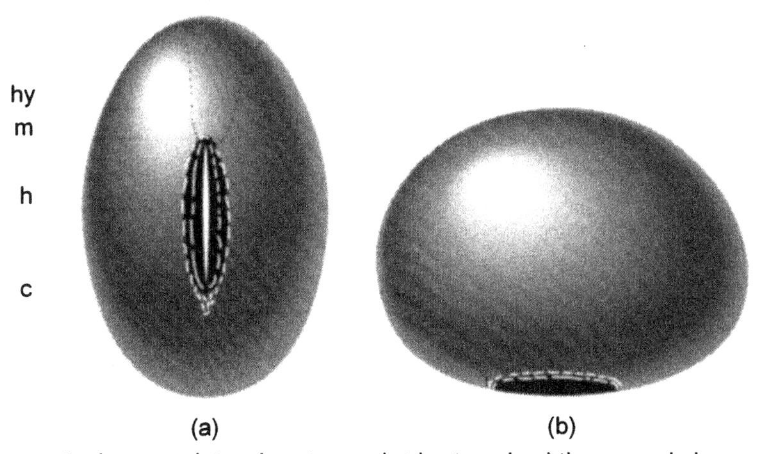

Fig. 2.1. Soybean seed, in edge view and side view; *h* = hilum; *c* = chalaza; *m* = micropyle; *hy* = outline of the hypocotyl under the sead coat. *Source:* Smith, A.K., and S.J. Circle, *Soybeans: Chemistry and Technology, Vol. 1, Proteins*, AVI Publishing Co., Westport, CT, 1972.

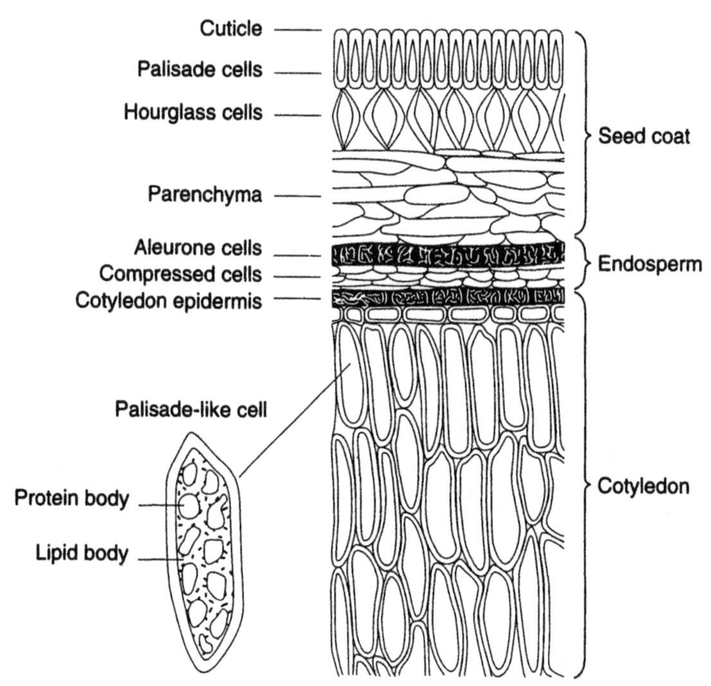

Fig. 2.2. Microscopic structure of soybean. *Source:* Smith, A.K., and S.J. Circle, *Soybeans: Chemistry and Technology, Vol. 1, Proteins,* AVI Publishing Co., Westport, CT, 1972.

Soybeans have been used in the Orient for hundreds of years. However, their use in the United States spans a much more recent history. The seeds were not commercially processed until 1922. Less than 108,000 metric tons (4 million bushels) were produced in 1922 (1). This has increased to about 54 million MT (2 billion bu) in 1992–93 (2) (see Chapter 1). The soybean and its structural components have the approximate composition shown in Table 2.1.

Modern soybean processing involves removal of the oil by solvent extraction; the oil is then further processed. The residual protein meal is of high quality and is used in food and animal feed (see Chapters 7 and 8). Soybean hulls may be either removed or left in the meal. The hulls have a high content of cellulose and hemicellulose, as well as acid-digestible fiber and a high content of zinc and iron (3).

TABLE 2.1 Chemical Composition of Soybeans and Their Components (Dry Weight Basis)

Components	Yield (%)	Protein (%)	Fat (%)	Ash (%)	Carbohydrates (%)
Whole soybeans	100.0	40.3	21.0	4.9	33.9
Cotyledon	90.3	42.8	22.8	5.0	29.4
Hull	7.3	8.8	1.0	4.3	85.9
Hypocotyl	2.4	40.8	11.4	4.4	43.4

TABLE 2.2 Typical Compositions for Crude and Refined Soybean Oil (4)

	Crude oil	Refined oil
Triglycerides	95–97	>99
Phosphatides	1.5–2.5	0.003–0.045
Unsaponifiable matter	1.6	0.3
Plant sterols	0.33	0.13
Tocopherols	0.15–0.21	0.11–0.18
Hydrocarbons (Squalene)	0.014	0.01
Free fatty acids	0.3–0.7	<0.05
Trace metals		
Iron, ppm	1–3	0.1–0.3
Copper, ppm	0.03–0.05	0.02–0.06

The oil removed by solvent extraction (see Chapter 6) is termed *crude* oil and contains many materials that must be removed by processing before the oil is ready for consumer use. Crude and refined soybean oil compositions are shown in Table 2.2 (4).

Neutral Lipid Composition

The neutral lipid composition of soybean oil is composed primarily of *triglycerides*, or *triacylglycerols*, with varying fatty acids as part of their structure, as shown in Fig. 2.3. The saturated and unsaturated fatty acids that occur in the glycerides of soybean oil are shown in Fig. 2.4. However, the fatty acids of soybean oil are primarily unsaturated (see Table 2.3) (5,6).

The three common unsaturated fatty acids—oleic, linoleic, and linolenic acid—can exist in more than one form. These are termed *geometric isomers* and *positional isomers*. They usually have different melting points and other properties. Naturally occurring fatty acids are found in the *cis* form, whereas those formed by catalytic hydrogenation are in the *trans* form as shown in Fig. 2.5.

The study of the attachment of fatty acids on the glycerol molecule has been the topic of much interest. In glycerides prepared in the laboratory or by interesterification (see Chapter 16), the fatty acids are placed on the glycerol molecule in a random fashion, unless the reactions are purposely directed, whereas in the glycerides prepared by nature, such as in soybeans, the acid residues are placed on the three hydroxyl groups of glycerol in a species-specific order if they are different.

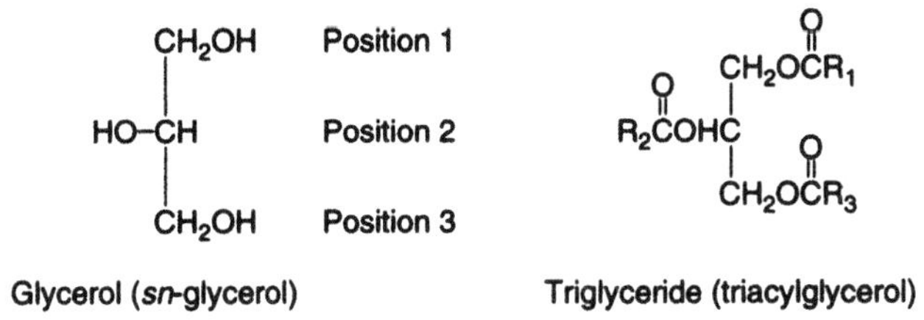

CH_2OH Position 1
HO–CH Position 2
CH_2OH Position 3

Glycerol (*sn*-glycerol)

$CH_2OC(=O)R_1$
$R_2C(=O)OHC$
$CH_2OC(=O)R_3$

Triglyceride (triacylglycerol)

Fig. 2.3. Structure of neutral lipid components.

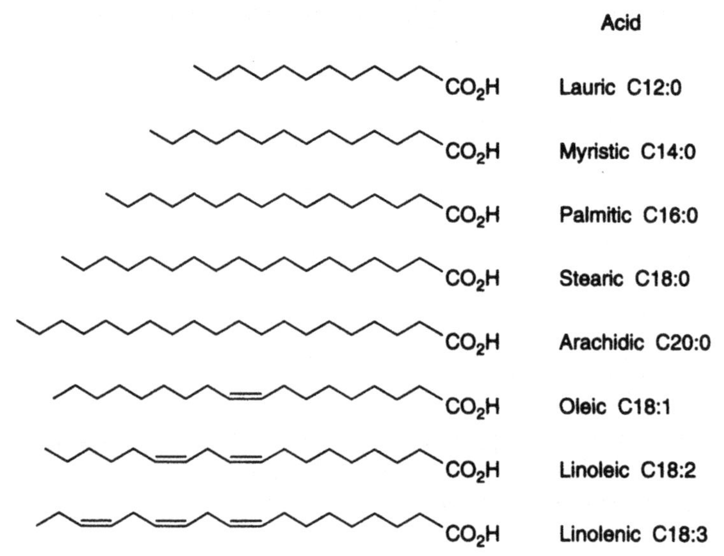

Fig. 2.4. Fatty acids represented in soybean oil triglycerides.

However, it is very difficult and time-consuming to carry out an analysis for the composition of each position esterified to the glycerol, and such techniques have not been generally used. The analysis requires enzymatic hydrolysis followed by thin-layer chromatography (TLC) and gas-liquid chromatography (GLC), chemical synthesis, enzymatic degradation by two different enzymes, and extensive chromatography. Significant data can, however, be obtained with a simple hydrolysis using pancreatic lipase, followed by TLC and GLC. With this data the compositions can be calculated (5,6). Considering all of the fatty acids in soybean oil only as saturated, S, or unsaturated, U, a simplified triglyceride composition can be calculated (7) to give the following values:

SSS = 0.07 USU = 0.7%
SSU = 5.2% UUS = 35.0%
SUS = 0.4% UUU = 58.4%

These calculations consider only the saturated and unsaturated categories of acids present in the fat; however, symbols for individual fatty acids can be substituted into the equations. An example of the triglyceride mixture obtained in this fashion is shown in Fig. 2.6.

Fatemi and Hammond (8) found that there was very little palmitic or stearic acid in the 2-position. The 1-position was richer in the saturated fatty acids and linolenic acid. The 3-position was enriched in oleic acid, whereas the linoleic acid was found in the 2-position. Generally speaking, the 2-position of the glycerol molecule contains the more unsaturated fatty acids. Some soybean varieties have somewhat differing distributions of fatty acids. Knowledge of the triglyceride structure of soybean oil is important, because it has been postulated that oxidative stability may be affected by the structure; however, studies have shown that this is not an important effect.

TABLE 2.3 Fatty Acid Composition of Soybean Oil

Component acid	Fatty acid content, wt%	
	Range	Average
Saturated		
Lauric	—	0.1
Myristic	<0.5	0.2
Palmitic	7–12	10.7
Stearic	2–5.5	3.9
Arachidic	1.0	0.2
Behenic	0.5	—
Total	10–19	15.0
Unsaturated		
Palmitoleic	<0.5	0.3
Oleic	20–50	22.8
Linoleic	35–60	50.8
Linolenic	2–13	6.8
Eicosenoic	1.0	—

Source: Pryde, E.H., in *Handbook of Soy Oil Processing and Utilization*, edited by D. Erickson, E. Pryde, O.L. Brekke, T. Mounts, and R.A. Falb, American Oil Chemists' Society, Champaign, IL, 1980, pp. 13–29.

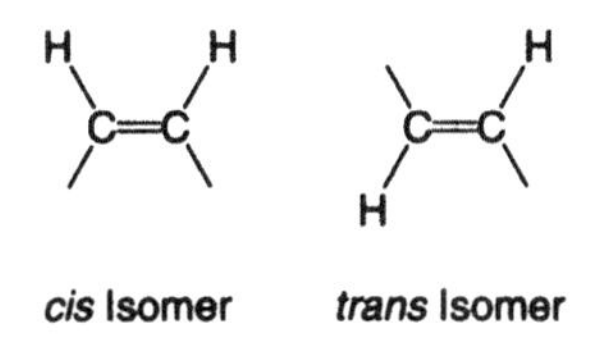

Fig. 2.5. Geometric isomers in an unsaturated fatty acid.

S	S	S	U	S	U	Position 1
S	S	U	U	U	S	Position 2
S	U	U	U	S	U	Position 3
	U	U		S	U	Positional Isomers
	S	U		U	S	
	S	S		S	U	

Fig. 2.6. Patterns of saturated and unsaturated fatty acids in soybean oil triglycerides.

Carbohydrates

Soybeans contain about 35% carbohydrates, and defatted dehulled soy grits and flour contain about 17% soluble and 21% insoluble carbohydrates (1). Soybeans have found limited use in the human diet because of the flatulence produced by

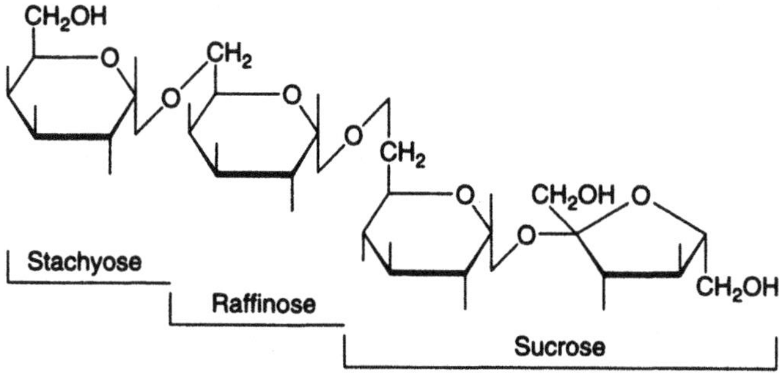

Fig. 2.7. Structure of carbohydrates found in soybeans.

raffinose and stachyose. Humans lack the enzymes to hydrolyze the galactosidic linkages of raffinose and stachyose (see Fig. 2.7) to simpler sugars, so the compounds enter the lower intestinal tract intact, where they are metabolized by bacteria to produce flatus.

The carbohydrate composition of dehulled, defatted soy meal is shown in Table 2.4 (10).

Inorganic Components

The primary inorganic components of the soybean are minerals, which vary in concentration according to the variety chosen. The seeds from twelve different varieties of soybeans were analyzed and found to be quite variable. On ash fractionation, mineral content was found to be (3) in the following ranges (10):

Potassium	1.5 to 1.92%	Magnesium	0.094 to 0.208%
Sodium	0.4 to 0.61%	Calcium	0.024 to 0.063%
Phosphorus	0.352 to 0.733%	Iron	0.0044 to 0.0163%

Typical mineral contents of soy protein products are given in Table 2.5.

TABLE 2.4 Carbohydrate Constituents of Dehulled Defatted Soybean Flakes

Carbohydrate	Source	% wt
Monosaccharides:		
Glucose	Cotyledons	trace
Oligosaccharides:		
Sucrose	Cotyledons	5.7
Raffinose	Cotyledons	4.1
Stachyose	Cotyledons	4.6
Polysaccharides:		
Arabinan	Cotyledons	1.0
Arabinogalactan	Cotyledons	8–10
Acidic polysaccharides	Cotyledons	5–7

Source: Smith, A.K., and S.J. Cirek, *Soybeans: Chemistry and Technology, Vol. 1. Proteins*, AVI Publishing Co., Westport, CT, 1972.

TABLE 2.5 Typical Mineral Content of Soy Protein Products

Element	Defatted soy flour	Soy protein concentrate	Isolated soy protein
Arsenic	0.1 ppm	0.2 ppm	<0.2 ppm
Cadmium	0.25 ppm	—	<0.2 ppm
Calcium	0.22%	0.22%	0.18%
Chlorine	0.132%	0.11%	0.15%
Chromium	0.9 ppm	<1.5 ppm	<2 ppm
Cobalt	0.5 ppm	—	<5 ppm
Copper	23 ppm	16 ppm	14 ppm
Fluorine	1.4 ppm	—	<10 ppm
Iodine	0.01 ppm	0.17 ppm	<2 ppm
Iron	110 ppm	100 ppm	160 ppm
Lead	0.2 ppm	—	<0.2 ppm
Magnesium	0.31%	0.25%	415 ppm
Manganese	28 ppm	30 ppm	11 ppm
Mercury	0.05 ppm	—	<0.5 ppm
Molybdenum	2.6 ppm	4.5 ppm	<0.3 ppm
Phosphorus	0.68%	0.70%	0.82%
Potassium	2.37%	2.1%	960 ppm
Selenium	0.6 ppm	—	0.2 ppm
Sodium	254 ppm	50 ppm	1.0%
Sulfur	0.25%	0.42%	0.63%
Zinc	61 ppm	46 ppm	36 ppm

Source: Ralston Purina Company, 1995, unpublished data.

Protein Components of the Soybean

Several protein products are available from soybeans, as indicated in Table 2.6 (see Chapters 7 and 8).

Soybean meals are material resulting from extracting the oil from the flakes in the solvent extraction process (see Chapter 6). The meals are available with or without the hulls, which serve to increase the fiber content of the meals. *Soy flours* are prepared by grinding and sieving defatted flakes. *Grits*, coarse ground flakes, are graded according to particle size.

TABLE 2.6 Approximate Composition of Soybeans and Soybean Products

Material	Moisture	Protein	Fat	Fiber	Ash
Soybeans	11.0	37.9	17.8	4.7	4.5
Defatted meal with hulls	10.4	44.0	0.5	7.0	6.0
Defatted meal dehulled	10.7	47.5	0.5	3.5	6.0
Full-fat soy flour	5.0	44.3	21.0	2.0	4.9
Defatted grits and soy flour	7.0	54.9	0.8	2.4	6.0
Lecithinated soy flour	5.5	49.9	15.5	2.1	5.0
Soy protein concentrate	7.5	66.6	—	3.5	5.5
Soy protein isolate	5.0	93.1	—	0.2	4.0

The amino acid composition of soy protein, soy flour, and concentrates is shown in Table 2.7 (12 and 13). The composition and nutrient content of soy protein concentrates, isolates, and other soy foods are given in Table 2.8 (2) (see Chapters 8 and 22).

Soybean protein contains proteinaceous substances known as *trypsin inhibitors* which inhibit the digestion of proteins, and *hæmagglutinins* (*lectins*) which are important nutritionally and must generally be heat-inactivated because they exert negative effects on the nutritional quality of soybean protein (Table 2.9) (14). Soybeans also contain *goitrogens*, as well as *antivitamins* D, E, and B_{12} (14). Furthermore, the requirement for certain metals, such as calcium, magnesium, zinc, copper, and iron, is increased in the presence of soybeans in diets (17). Heat-stable factors capable of causing an estrogenic response in animals have been isolated from soybeans. These are the *isoflavones*, which exist in the plant as glycosides. The structures of the major isoflavones isolated are shown in Fig. 2.8, and their estrogenic potency relative to diethylstilbestrol is shown in Table 2.10 (15).

Saponins are a type of glycoside widely distributed in plants. They consist of a sapogenin, which makes up the aglucon moiety, and a sugar. The sapogenin may be a steroid or a triterpene, and the sugar may be glucose, galactose, a pentose, or a methyl pentose. The saponins of soybeans are essentially innocuous on animal growth. Other nutritional and biochemical effects of the soybean, such as anticarcinogenesis and cholesterol-lowering ability are discussed in Chapter 23.

TABLE 2.7 Amino Acid Compositions of Soy Protein Concentrates, Soy Flours, and Soy Solubles[a]

Amino acid	Soy protein concentrate alcohol wash	Soy protein concentrate water wash	Soy flour	Soy solubles from alcohol wash
Alanine	4.5	4.5	4.0	3.9
Arginine	7.7	7.7	7.0	7.4
Aspartic acid	12.1	12.1	11.3	15.0
Cystine	1.5	1.4	1.6	4.1
Glutamic acid	18.9	18.9	17.2	20.7
Glycine	4.4	4.4	4.0	3.5
Histidine	2.7	2.7	2.7	2.5
Isoleucine	5.0	4.8	4.9	2.1
Leucine	8.2	8.1	8.0	3.2
Lysine	6.5	6.7	6.4	3.5
Methionine	1.4	1.4	1.4	3.6
Phenylalanine	5.2	5.1	5.3	5.6
Proline	5.6	5.6	4.7	3.5
Serine	5.3	5.3	5.0	3.4
Threonine	4.3	4.1	4.2	3.4
Tryptophane	1.4	1.4	1.2	1.3
Tyrosine	3.9	3.9	3.9	3.9
Valine	5.3	5.6	5.3	2.1

[a]Data expressed as grams amino acid per 100 grams protein. *Source:* Central Soya Company, Inc., Fort Wayne, IN.

TABLE 2.8 Soy Foods, Composition and Nutrient Content for 100 g Edible Portions (2)

Soy Food	Water, g	Kcal	Protein, (n x 5.71) g	Fat, g	Carbo-hydrate, g	Crude fiber, g	Calcium, mg	Iron, mg	Zinc, mg	Thiamin, mg	Ribo-flavin, mg	Niacin, mg	Vit. B6, mg	Folacin, mcg
Miso	41.5	206	11.8	6.1	28.0	2.5	66	2.74	3.32	0.10	0.25	0.86	0.22	33.0
Natto	55.0	212	17.7	11.0	14.4	1.6	217	8.60	3.03	0.16	0.19	0.00	n/a	n/a
Okara (fiber byproduct of soymilk/tofu)	81.6	77	3.2	1.7	12.5	4.1	80	1.30	n/a	0.02	0.02	0.10	n/a	n/a
Soy flour, defatted	7.3	329	47.0	1.2	38.4	4.3	241	9.24	2.46	0.70	0.25	2.61	0.57	305.4
Soy flour, full-fat, raw	5.2	436	34.5	20.6	35.2	4.7	206	6.37	3.92	0.58	1.16	4.32	0.46	345.0
Soy flour, full-fat, roasted	3.8	441	34.8	21.9	33.7	2.2	188	5.82	3.58	0.41	0.94	3.29	0.35	227.4
Soy flour, low-fat	2.7	326	46.5	6.7	38.0	4.2	188	5.99	1.18	0.38	0.29	2.16	0.52	410.0
Soy meal, defatted, raw	6.9	339	45.0	2.4	40.1	5.8	244	13.70	5.06	0.69	0.25	2.59	0.57	302.6
Soy protein concentrate	5.8	332	58.1	0.5	31.2	3.8	363	10.78	4.40	0.32	0.14	0.72	0.13	340.0
Soy protein isolate	5.0	338	80.7	3.4	7.4	0.3	178	14.50	4.03	0.18	0.10	1.44	n/a	176.0
Soy sauce, from HVP	75.7	41	2.4	0.1	7.7	0.0	5	1.49	0.31	0.04	0.11	2.83	0.14	13.0
Soy sauce, from soy (tamari)	66.0	60	10.5	0.1	5.6	0.0	20	2.38	0.43	0.06	0.15	3.95	0.20	18.2
Soy sauce, from soy and wheat (shoyu)	71.1	53	5.2	0.1	8.5	0.0	17	2.02	0.37	0.05	0.13	3.36	0.17	15.5
Soybeans, cooked, boiled	62.6	173	16.6	9.0	9.9	2.0	102	5.14	1.15	0.16	0.29	0.40	0.23	53.8
Soybeans, dry-roasted	0.8	450	39.6	21.6	32.7	5.4	270	3.95	4.77	0.43	0.76	1.06	0.23	204.6
Soybeans, raw	8.5	416	36.5	19.9	30.2	5.0	277	15.70	4.89	0.87	0.87	1.62	0.38	375.1
Soybeans, roasted	2.0	474	35.2	25.4	33.6	4.6	138	3.90	3.14	0.10	0.15	1.41	0.21	211.0
Soymilk, fluid	93.3	33	2.8	1.9	1.8	1.1	4	0.58	0.23	0.16	0.07	0.15	0.04	1.5
Tempeh	55.0	199	19.0	7.7	17.0	3.0	93	2.26	1.81	0.13	0.11	4.63	0.30	52.0
Tofu, raw, firm	69.8	145	15.8	8.7	4.3	0.2	205	10.47	1.57	0.16	0.10	0.38	0.09	29.3
Tofu, raw, regular	84.6	76	8.1	4.8	1.9	0.1	105	5.36	0.80	0.08	0.05	0.20	0.05	15.0

n/a = data not listed or available.

Source: 1994 Soya Bluebook, Soyatech Inc., Bar Harbor, ME, 1994.

TABLE 2.9 Trypsin Inhibitor Activities of Soybean Components (14)

	Antitrypsin activity (TIU[a]/g dry solids × 10^{-3})	Soy flour (%)
Soy flour (unheated)	86.4	100
Soybean isolate	25.5	30
Soybean fiber	12.3	14

[a]Trypsin inhibitor units.

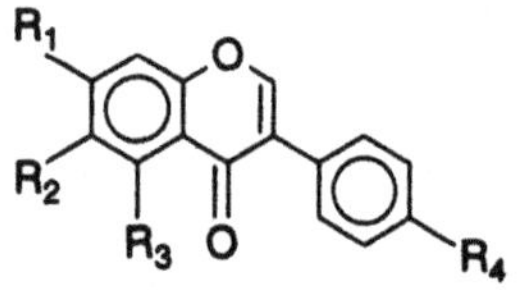

Isoflavones	R_1	R_2	R_4	R_5
Geinistein	OH	H	OH	OH
Genistin	O-glucosyl	H	OH	OH
Daidzein	OH	H	H	OH
Daidzin	O-glucosyl	H	H	OH
Glycitein	OH	OCH_3	H	OH
Glycitein 7-O-β-glucoside	O-glucosyl	OCH_3	H	OH

Fig. 2.8. Soybean isoflavones.

TABLE 2.10 Estrogenicity of Compounds Isolated from Soybeans (14)

Estrogen	Concentration (ppm)	Relative potency
Diethylstibestrol(DES)[a]	—	100,000
Genistin	1644 (15)	1.00
Daidzin	581 (15)	0.75
Glycitein 7-O-β-glucoside	338 (15)	—
Coumestrol	0.4 (16)	35

[a]DES included for comparison purposes.

Soybean Oil and Products (18)

Crudes

Crude soybean oil is oil extracted from soybean flakes with mixed hexanes. The solvent is removed to produce the initial crude oil. *Crude degummed soybean oil* is defined as pure soybean oil produced from fair to average quality crude soybean oil from which most of the natural gums (phospholipids) have been removed by hydration and mechanically separated. The analytical requirements for crude degummed soybean oil are shown in Table 2.11 (18).

Once-Refined Soybean Oil

Once-refined soybean oil is defined as pure soybean oil in which all of the free fatty acids and other nonoil material has been removed by chemical means and physical or mechanical separation with the analytical requirements shown in Table 2.12 (18). The analytical requirements for fully refined soybean oil are shown in Table 2.13 (18).

TABLE 2.11 Analytical Requirements for Crude Degummed Soybean Oil (18)

Test	Maximum	Minimum	AOCS Method
Unsaponifiable matter	1.5%		Cc 6a-40
Free fatty acids, as oleic	0.75%		Ca 5a-40
Moisture, volatile matter, and insoluble impurities	0.3%		Ca2d-25 (M & V) Ca 3-46 (I.I.)
Flash point		250°F	Cc 9b-55
Phosphorus	0.02%		Ca 12-55

Source: Yearbook and Trading Rules, 1993–1994, National Oilseed Processors Association (NOPA), Washington, DC.

TABLE 2.12 Once-Refined Soybean Oil Analytical Requirements (18)

Test	Value	AOCS Method
Clear and brilliant in appearance at	70 to 85°F (21 to 29°C)	
Free from settlings at	70 to 85°F (21 to 29°C)	
Moisture and volatile matter	0.10% maximum	Cs 2s-25
Free fatty acids	≤0.1% Maximum	Ca 5a-40
Color	<3,5 red and no green color	Ce 83-63
Flash point	250°F	Cc 9b-55
Unsaponifiable content	1.5% maximum	Ca 6a-40
Marine and marine animal oils	Negative	AOAC 28.121

Source: Yearbook and Trading Rules, 1993–1994, National Oilseed Processors Association (NOPA), Washington, DC.

TABLE 2.13 Fully Refined Soybean Oil Analytical Specifications (18)

Test	Maximum/Minimum	AOCS Method
Flavor Shall be bland		
Color (Lovibond)	Maximum 20Y/2.0R	Cc 13b-45
Free fatty acids	Maximum 0.05%	Ca 5a-40
Clear and brilliant in appearance at	70 to 85°F (21 to 29°C)	
Cold test	Minimum 5.5 hours	Cc 11-53
Moisture and volatile matter	No more than 0.1%	Ca 2d-25
Unsaponifiable content	No more than 1.5%	Ca 6a-40
Peroxide value (meq/kg)	No more than 2.0	Cd 8-53
Stability	Minimum 8 hours AOM or 35 Meq/kg PV	Cd 12-57
GRAS[a] preservatives are permitted		

[a]Generally recognized as safe. *Source: Yearbook and Trading Rules, 1993–1994*, National Oilseed Processors Association (NOPA), Washington, DC.

Soybean Lecithins

The generic term *soybean oil lecithins* is used to denote the gums formed by the initial degumming (hydration) of crude soybean oil. In addition to lecithin (phosphatidylcholine), the gums also contain cephalin (phosphatidylethanolamine) and phosphatidylinositol (see Fig. 2.9) as well as many other phospholipids in minor amounts. The fatty acid compositions of soybean phospholipds are predominately acids of the C_{16} and C_{18} series in various proportions, similar to the soybean oil triglycerides. The commercial lecithin byproduct is used as an emulsifier for foods and other products (see Chapter 10). Table 2.14 shows the NOPA specifications for six types of lecithin (18).

Hydrogenation of Soybean Oil

Commercial hydrogenation of soybean oil with conventional nickel catalysts produces fatty acids containing a variety of positional and geometrical isomers (see Chapter 13). The structures of linolenic acid and several of its common isomers is shown in Fig. 2.10.

Analysis of a production run of a lightly hydrogenated soybean salad oil showed the following distribution of linolenic acid and its isomers (19). The distributions of other isomers of unsaturated fatty acids are discussed later in this section (20).

c-9,*c*-12,*c*-15 68.60% *t*-9,*c*-12,*c*-15 10.10%
t-9,*t*-12,*c*-15 15.44% *c*-9,*c*-12,*t*-15 4.65%

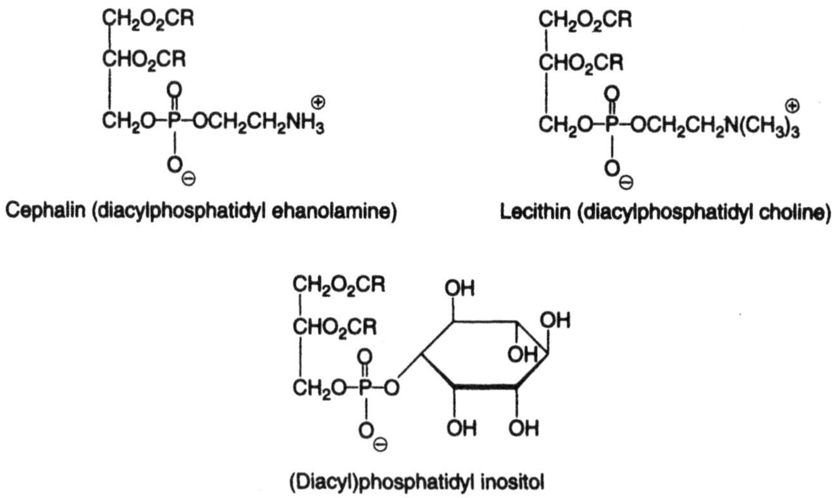

Cephalin (diacylphosphatidyl ehanolamine)

Lecithin (diacylphosphatidyl choline)

(Diacyl)phosphatidyl inositol

Fig. 2.9. Phospholipids occurring in crude soybean oil.

TABLE 2.14 Soybean Lecithin Specifications (18)

	Fluid lecithins			Plastic lecithins		
	Unbleached	Bleached	Double bleached	Unbleached	Bleached	Double bleached
Acetone-insoluble, min.	62%	62%	62%	65%	65%	65%
Moisture, max.[a]	1%	1%	1%	1%	1%	1%
Hexane-insoluble, max.		0.3%	0.3%	0.3%	0.3%	0.3%
Acid value, max.	32	32	32	30	30	30
Color, Gardner, max.[b]	18	14	12	18	14	12
Viscosity, Centiposes @77°F (25°C), max.[c]	15,000	15,000	15,000			
penetration, max.[d]				22 mm	22 mm	22 mm

[a]By Karl Fischer titration (AOCS Tb2-64)
[b]Undiluted basis.
[c]By any appropriate conventional viscosimeter, or by AOCS bubble time method Tq. 1A-64, assuming density to be unity. Fluid lecithin having a viscosity less than 7,500 cenipoises may be considered a premium grade.
[d]Using Precision Cone 73525, Penetrometer 73510; sample conditioned 24 h at 77°F (25°C).

Mossaba et al. (20) have also reported the presence of conjugated isomers, *cis-trans* and *trans-trans* linoleate, in hydrogenated samples of soybean oil. The iodine value decreases on hydrogenation, indicating a generalized loss of unsaturation. The lower the iodine value, the lower the amounts of dienoic and trienoic *trans* present in the fat (Table 2.15) (20). Various margarines have varying *trans* acid compositions, depending on how they are made (Table 2.16). The tub margarines are gener-

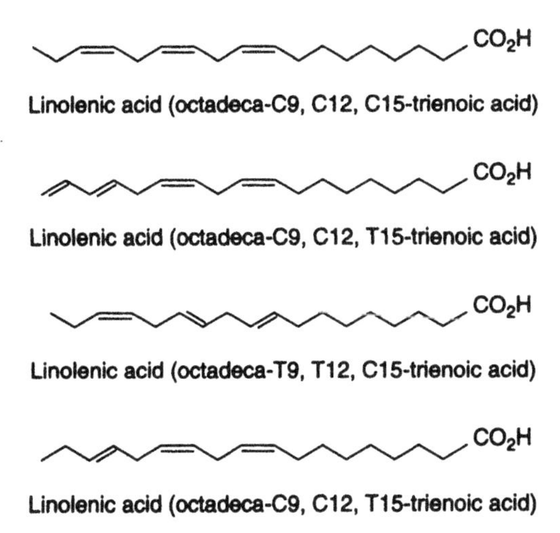

Fig. 2.10. Linolenic acid and its isomers.

TABLE 2.15 Isomeric Composition of Hydrogenated Soybean Oil (20) vs. Iodine Value

Iodine value:	91	123	111	96
Linolenic acid isomers:				
trans, trans, cis		0.3	0.5	—
cis, cis, cis	1.5	3.2	0.4	—
Linoleic acid isomers:				
conjugated *cis, trans*	0.2	0.4	0.4	—
conjugated *trans, cis*	0.2	0.3	0.3	—
conjugated *trans, trans*	0.2	0.4	1.1	0.3

Source: Mossaba, M.M., R.E. McDonald, D. Armstrong, and S.J. Page, *J. Chromatog. Sci. 29:* 324 (1991). Reproduced by permission of Preston Publications, a Division of Preston Industries, Inc.

TABLE 2.16 Detailed Analysis (wt %) of Representative Margarines

Fatty acid wt%	Hard cube (21)	Whipped cube (21)	Soft tub (21)	Unidentified (22)
12:0				0.2
14:0	<0.1	<0.1	0.1	0.1
*t*14:1				0.1
16:0	9.6	9.2	9.2	12.5
*t*16:1				<0.1
*c*16:1				0.2
18:0	5.5	5.7	5.6	7.5
*cc*18:1 + *t*18:1	67.6	62.6	55.8	53.1
*t*18:1				20.9
18:2-*tt.ct.cc*	11.0	15.3	24.2	24.9
*18:2 *xx*	1.6	1.2	0.8	—

*Tentatively identified as isomers of 18:2.

ally blends with polyunsaturated oils and, thus, have the most unsaturated and generally have the lowest *trans* content (Table 2.16). The analysis for dienoic and trienoic *trans* fatty acids is still a difficult and sophisticated procedure. However, the results from a study on the *trans* acid content of hydrogenated soybean oil, in which capillary and packed-column gas chromatography were compared, indicated that the coefficient of variation (CV) was greater for capillary analysis (23). The capillary analysis, however, allowed resolution of isomers that the packed column did not (Table 2.17).

Varietal Differences and Effects of Breeding

Manipulation of the fatty acid composition of soybean oil through plant breeding in order to obtain desirable characteristics in the oil has attracted much interest and is now commonplace (24–30). For many years it has been a goal to reduce the linolenic acid content of soybean oil. This would increase its oxidative stability and make it more desirable for use as a nonhydrogenated salad oil. This goal has finally been

TABLE 2.17 *Trans* Isomer Content of Hydrogenated Soybean Oil

Isomer	Capillary GC % composition	CV[a], %	Packed-column GC % composition	CV, %
trans-monoenes	38.8	1.7	38.9	0.5
cis-9, *trans*-12-18:2	4.6	4.5		
trans-9, *cis*-12-18:2	4.7	1.3		
Other *cis, trans*-18:2	7.0	2.9	13.6	0.5
trans-9, *trans*-12-18:2	6.2	1.1	9.5[b]	0.4
Total *trans*	61.3		62.0	
Total *trans*-dienes	22.5		23.1	

[a]Coefficient of variation
[b]Also includes *cis, trans* and *trans, cis* isomers
[c]Includes *cis*-9, *trans*-12 and *trans*-9, *cis*-12-18:2 methylene-interrupted isomers as well as *cis, trans* and *trans, cis* non–methylene-interrupted (NMI) isomers
Source: McDonald, R.E., D.J. Armstrong, and G. Kreishman, *J. Agric. Food Chem. 37:* 637 (1989).

reached. Low-linolenic acid oils have been produced at Iowa, Purdue, and North Carolina that have much lower levels of linolenic acid than the common Williams variety (Table 2.18). Furthermore, a variety of soy oil containing a considerably lower amount of linolenic acid is now becoming commercially available. This oil will be of increased stability for cooking, and because no hydrogenation is used in its production, it will be *trans*-acid-free. Other genetically defined oils that have high oleic, palmitic, and stearic acid content have now been developed, although they are not yet available for commercial use (Table 2.18) (24–30). Other varieties of the soybean with lower amounts or lipoxygenase and trypsin inhibitor have also been developed, but these seeds are not yet commercially available.

TABLE 2.18 Fatty Acid Composition of Various Soybean Varietals

Varietals	Fatty acid				
	16:0	18:0	18:1	18:2	18:3
C1640 (Purdue University)	10.0	3.1	26.5	56.2	4.11
N85-21224 (North Carolina)	10.0	3.5	24.5	57.2	3.7
A5-Iowa	9.4	4.3	46.6	36.3	3.4
A6-Iowa	6.7	29.4	29.3	27.1	4.9
Williams	10.2	3.9	20.9	57.0	7.7
8-26-3	4.9	3.2	54.3	33.1	4.5

TABLE 2.19 Sterol Content (mg/100 g) of Soybean Oil (31)

Sterol	Crude	Refined	Refined and hydrogenated
β-Sitosterol	183	123	76
Campesterol	68	47	26
Stigmasterol		47	30
Δ^5-Avenasterol	5	1	ND
Δ^7-Stigmasterol	5	1	ND
Δ^7-Avenasterol	2	<0.5	ND
Cholesterol	ND	ND	ND
Total	327	221	132

[b]ND = not detected
Source: Weihrauch, J.L., and J.M. Gardner, *J. Amer. Diet. Assoc. 73:* 39 (1978).

Maturity and Environmental Effects

During its various stages toward maturity, soybean oil exhibits a varied composition. Depending on the number of days after flowering, the soybean seed varies from 84.3% triglyceride content of the total lipid with only 8.1 μ-moles total glycerolipid per seed, at 30 days, to 94.7% triglycerides and 40.9 μ-moles of glycerolipid per seed, at 75 days (30). Soybean oil composition also varies with crop year, location, and weather.

Unsaponifiable Material

The unsaponifiable materials of soybean oil contain tocopherols, sterols, phytosterols, and hydrocarbons. The phytosterols (sitosterol) occur to the largest extent (31). The concentrations of these compounds are decreased during the typical oil processing steps; however, some still remain in the final finished oil. The typical composition is shown in Table 2.19, and typical structures of these components are shown in Fig. 2.11.

Tocopherols are minor components of most vegetable oils and are thought of as antioxidants with various degrees of effectiveness. Although the β and γ isomers may be present in larger amounts, the δ form is the most powerful and effective antioxidant. Such a compound is most useful when the oil is not to be used repeatedly or used for deep frying, because these processes develop colored compounds that darken the oil and contribute to off-flavor development. The tocopherol content and several isomers of soybean oil in various products are shown in Table 2.20.

Soybean Oil Byproducts (see Chapter 7)

Soapstock

Soapstock is a byproduct of caustic refining which is usually acidulated to recover the fatty acids (34) whose composition is essentially the same as in soybean oil.

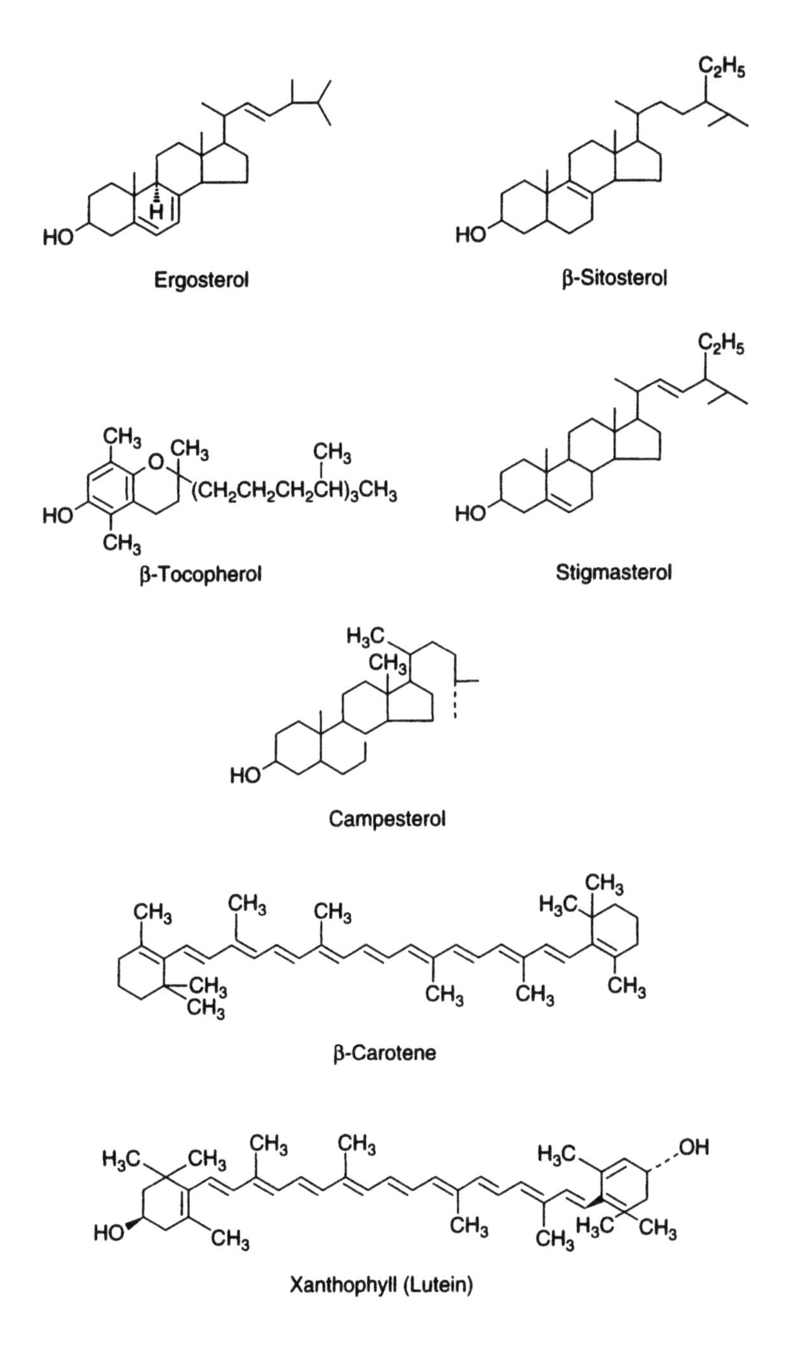

Fig. 2.11. Typical unsaponifiable components of soybean oil.

TABLE 2.20 Representative Tocopherol Content of Various Soybean Oil Products (32,33)

Soybean oil product	Tocopherol, mg/100 g			
	α	γ	δ	Total
Crude	9–12	74–102	24–30	113–145
Refined	6–9	45–50	19–22	73–77
Brand A	14	102	37	153
Brand B	10	80	22	112
Brand C	9	68	23	100
Brand D	5	42	11	58

TABLE 2.21 Deodorizer Distillate from Various Oils (34)

Item	Sunflower seed	Cotton	Soybean	Rapeseed
% Unsaponifiable	39	42	33	35
% Tocopherol	9.3	11.4	11.1	8.2
% alpha	5.7	6.3	0.9	1.4
% Sterol	18	20	18	14.8
% Stigmasterol	2.9	0.3	4.4	1.8

Spent Bleaching Earth

Increased awareness of the environment does not allow discard of spent earths; instead, the oil content must be reduced to low levels before discard. The fatty acid composition of the adsorbed oil reflects that of soybean oil.

Deodorizer Distillates

Deodorizer distillates are the products of steam distillation of heated soybean oil under high vacuum (see Chapter 14). The distillate contains all of the volatile compounds that result from autooxidation and tocopherols and other sterols present in the soybean oil. This process produces a bland oil. The distillates are valuable as a source of tocopherol and avenasterols. The composition of deodorizer distillates is shown in Table 2.21 (34).

References

1. Smith, A.K., and S.J. Circle, *Soybeans: Chemistry and Technology, Vol 1. Proteins*, AVI Publishing Co. Westport, CT, 1972.

2. *1994 Soya Bluebook*, Soyatech Inc., Bar Harbor, ME, 1994.

3. Korngay, E.T., *Feedstuffs 50:* 24 (1978).

4. Pryde, E.H., in *Handbook of Soy Oil Processing and Utilization*, edited by D. Erickson, E. Pryde, O.L. Brekke, T. Mounts, and R.A. Falb, American Oil Chemists' Society, Champaign, IL, 1980, pp. 13–29.

5. Conner, R.T., and S.F. Herb, *J. Amer. Oil Chem. Soc. 47:* 186A, 195A, 197A (1979).
6. Evans, C.D., D.G. McConnel, C.R. Schofield, and H.J. Dutton, *J. Amer. Oil. Chem. Soc. 43:* 345 (1966).
7. List, G.R., E. Emken, W.F. Kwoek, T.D. Simpson, and H.J. Dutton, *J. Amer. Oil. Chem. Soc. 54:* 408 (1977).
8. Fatemi, S.H., and E. Hammond, *Lipids 12:* 1032 (1977).
9. Kellor, R.L., *J. Amer. Oil Chem. Soc. 51:* 77A (1974).
10. Au-Kumar-Om, L.B. Sasikia, and S.B. Kannur, *J. Food Sci. Technol., Mysore 29:* 111–112 (1992).
11. Mustakis, C., L.D. Kirk, and E.L. Griffin, *J. Amer. Oil Chem. Soc. 39:* 222 (1962).
12. Allen, R.D., *Feedstuffs 56 (30):* 25–30 (1988).
13. *Soybean Processing for Food Uses*, edited by K. Tanteeratarm, University of Illinois, Urbana, IL, 1993.
14. Liener, I.E., *J. Amer. Oil Chem. Soc. 58:* 406 (1981).
15. Niam, M.B., Gestetner, S. Zikah, Y. Birk, and A. Bondi, *J. Agr. Food Chem. 22:* 806 (1974).
16. Knuckles, B.E., D. de Fremery, and G.O. Kohler, *J. Agr. Food Chem. 24:* 1177 (1976).
17. O'Dell, B.L., in *Soy Protein on Human Nutrition*, edited by H.L. Wilcle, D.T. Hopkins, and D.H. Waggle, Academic Press, New York, 1979, p. 187.
18. *Yearbook and Trading Rules, 1993–1994*, National Oilseed Processors Association (NOPA), Washington, DC.
19. Perkins, E.G., and C.S. Smick, *J. Amer. Oil Chem. Soc. 64:* 1150 (1987).
20. Mossaba, M.M., R.E. McDonald, D. Armstrong, and S.J. Page, *J. Chromatog. Sci. 29:* 324 (1991).
21. Carpenter, D.L., and H.T. Slover, *J. Amer. Oil Chem. Soc. 50:* 372 (1973).
22. Smith, L.M., W.L. Dunkley, A. Franke, T.J. Dairike, *J. Amer. Oil Chem. Soc. 55:* 257 (1973), 1978.
23. McDonald, R.E., D.J. Armstrong, and G. Kreishman, *J. Agric. Food Chem. 37:* 637 (1989).
24. Fehr, W.R., G.A. Welke, E.G. Hammond, D.N. Duvick, and S.R. Cianzio, *Crop Sci. 32:* 903 (1992).
25. Burton, J.W., R.F. Wilson, and C.A. Brim, *Crop Sci. 34:* 313 (1994).
26. Wilson, R.F., in *Designing Value Added Soybeans for Markets of the Future*, edited by R.F. Wilson, American Oil Chemists' Society, Champaign, IL, 1991.
27. Hammond, E., and W.J. Fehr, *J. Amer. Oil Chem. Soc. 61:* 1713 (1984).
28. Wilson, R.F., W.J. Burton, and C.A. Brim, *Crop Sci. 21:* 788 (1981).

29. Wilcox, J.R., J.F. Cavins, and N.C. Nielsen, *J. Amer. Oil Chem. Soc. 61:* 97 (1984).

30. Wilson, R.F., in *Soybeans, Improvement, Production and Uses*, Agronomy Monograph no. 16, 1987.

31. Weihrauch, J.L., and J.M. Gardner, *J. Amer. Diet Assoc. 73:* 39 (1978).

32. Gutfinger, T., and A. Letan, *J. Sci. Food Agr. 25:* 1143 (1974).

33. Carpenter, D.L., J. Lehman, B.S. Masson, and H.T. Slover, *J. Amer. Oil. Chem. Soc. 53:* 713 (1976).

34. Winters, R.L., in *World Conference on Edible Fats and Oils Processing: Basic Principles and Modern Practices*, American Oil Chemists' Society, Champaign, IL, 1990, pp. 402–411.

Chapter 3

Physical Properties of Soybeans and Soybean Products

Edward G. Perkins

Dept. of Food Science
University of Illinois, Urbana, IL

Introduction

Data concerning the physical properties of soybeans and its products is scarce and most of the information available is from the 1940s and early 1950s. The soybean and its constituents can vary in composition depending on such factors as the variety grown and climatic conditions. Furthermore, the soybean oil from such beans can also vary, and the physical properties will also change markedly depending on the processing conditions. Jefferson (1) and Swern (2) offer excellent reviews on the physical properties of soybean oil as well as fats and oils in general.

Density

Densities of vegetable oils will vary with temperature and are inversely related to the molecular weights of the oils. There is a direct relationship with the degree of unsaturation. Lund (3) has developed an equation to calculate the density of a vegetable oil based on its iodine value (IV) and saponification value.

$$\text{Density} = 0.8475 + 0.00030(\text{Saponification value}) + 0.00014(\text{IV})$$

Both of these values may be determined according to methods Cd 3-25 and Cd 1-25 of the American Oil Chemists' Society *Official and Tentative Methods* (4). Furthermore, an equation relating density at 20°C to the number of carbon atoms in the molecule has been developed (5):

$$d_n = d_1 b_n/(b_n + 1)$$

where d is the limiting density and b is a constant. Values of d_l and b are 0.877 and −4.486.

The densities of soybean oil at several temperatures have been published and appear in Table 3.1. Note the decrease in density with increasing temperature. The densities of soybean oil–solvent mixtures have been determined and are of practical value (6), because they are of importance in solvent extraction of soybeans.

Furthermore, Skau (8) has developed an equation for calculation of the density of glyceride oil–solvent mixtures:

TABLE 3.1 Density of Soybean Oil at Several Temperatures

Temperature	Density, g/mL	
	Magne and Skau (6)[a]	Johnstone et al. (7)[b]
–10.0	0.9410	—
0.0	0.9347	—
10.0	0.9276	—
25.0	0.9175	0.9171
37.8	—	0.9087
40.0	0.9075	—
50.0	—	0.9004

[a]Commercial edible soybean oil with an IV of 132.6 and containing 0.10% free fatty acid.
[b]Soybean oil that had been refined, bleached, and deodorized and had the following characteristics: iodine value, 130.1; peroxide number, 5.1; acid number, 0.11; thiocyanogen number, 80.0; and phosphorus content, 0.00095%.

TABLE 3.2 k Values for Commercial Hexane

Temperature °C	k, mL/g
10	0.0062
25	0.0080
40	0.0098

$$V = 1/D = a - (a - b + 4k)x + 4kx^2$$

where V = specific volume of the oil-solvent mixture, a = specific volume of solvent, D = density of oil-solvent mixture, b = specific volume of oil, x = weight fraction of oil in the mixture, $k = \Delta V$ when $x = 0.5$, $\Delta V = V_i - V$, ΔV_i = calculated specific volume

TABLE 3.3 Melting Points of Fatty Acids and Their Triglycerides Present in Soybean Oil and Partially Hydrogenated Soybean Oils (9)

Fatty acid triglyceride	Melting point (°C)	Name/composition	I (β)	Melting point II(β)
Name				
Palmitic	62.9	Tripalmitin	65.5	56.0
Tristearin	69.6	18:0-16:0-16:0	62.5	59.5
		16:0-18:0-16:0	68	65
		18:0-16:0-18:0	68	64
Oleic	16.3	Triolein	5.5	–12
		16:0-18:1-16:0	35.2	30.4
		18:0-18:1-18:0	41.6	37.6
		16:0-18:1-18:1	19.0	—
		18:0-18:1-18:1	23.5	—
Elaidic	43.7	Trielaidin	42	37
Linoleic		D-6.5 Trilinolein	–13.1	—
Linolenic		D-12.8 Trilinolenin	–24.2	—

of mixture, assuming ideal solutions. For commercial hexane the value of k is considered identical for fish and vegetable oils and is given in Table 3.2. Values of k for several other solvents have been given by Skau et al. (8). Such values are calculated from exact measurements of solution density at $x = 0.5$ and the exact densities of the oil and solvent. Using these values and the equation above, one can calculate the composition of solutions from the measured density. Detailed and comprehensive tables of oil–commercial hexane mixtures have been published (7).

The density of fat in the solid state as well as the specific volumes and subsequent changes in specific volume with temperature are rather important, because they are used to measure the changes in density or volume with temperature. This is termed dilatometry and is used to determine the amount of solid triglycerides in a fat at a specific temperature. Many fats produced for margarine or shortenings are complex mixtures of different-melting polymorphic forms in a liquid oil matrix (Table 3.3) (9). Although data is not readily available, the same applies to partially hydrogenated soybean oil. The relationship between the volume and melting point of a fat is the basis of the AOCS dilatometric method CDs 10-57 (4) for determining the solid fat content or solid fat index (SFI) of a partially hydrogenated oil. More recently, this value may be determined by low-resolution or wide-band nuclear magnetic resonance (nmr) spectrometr. (AOCS method Cd 16-8) (4).

Refractive Index

The refractive index is a basic value that depends on molecular weight, fatty acid chain length, degree of unsaturation, and degree of conjugation. The refractive index measurement is carried out in the liquid state or, in the case of a solid fat, at higher temperatures. Furthermore, the refractive index of glycerides is greater than that of the component fatty acids.

A mathematical relationship between refractive index and iodine value has been proposed by Zeleny et al. (10, 11).

$$n_D^{25} = 1.45765 + 0.0001164(\text{iodine value})$$

The refractive indices of a series of fractions from molecular distillation of soybean oil were determined by Detweiler et al. (12), who reported that there was good agreement between the observed and calculated refractive index over the narrow iodine value range of 128 to 138. The refractive index of refined, bleached soybean oil was first reported by Johnstone et al. (7) as 1.4377.

The reverse relationship, for calculation of the iodine value of crude soybean oil when the refractive index of the oil is known, is

$$\text{IV} = 8661.723(n_D^{25}) - 12{,}626.174$$

A relationship was also developed by Earle et al. (13) to calculate the refractive index at 40°C.

$$\text{IV} = 8555.559(n_D^{25}) - 12{,}425.928$$

Many fats, as well as those that have been partially hydrogenated, are solid or contain saturated triglycerides, which settle out. The refractive index of these materials is determined at 40°C. The effect of temperature on the refractive index is to increase it by 0.000385 units for each degree over 20°C. This can be used to estimate the refractive index at other temperatures [AOCS method Cc 7-25 (4)]. The various atomic and group refractions in a molecule may be considered additive and used to calculate the Lorentz–Lorentz molar refractions of fatty acid esters and glycerides.

A relation for the iodine value,saponification value, and specific rotation was developed:

$$r_{\mathrm{D}}^{26} = [(n^2 - 1) / (n^2 + 2)](1/d = 0.3307 + 1.68\ (10^{-5}\mathrm{IV} - 1.41(10^{-4})(\mathrm{SV})$$

where r = specific refraction, n = refractive index (at 20°C), d = density, IV = iodine value, and SV = saponification value (14).

Viscosity

Pryde has indicated that knowledge of viscosities is necessary for design calculations on pumps, piping, and heat transfer equipment (15).

Magne and Skau have reported the viscosities of soybean oil solutions (6), shown in Table 3.4. Viscosities of other fats and oils have also been published and are given in Table 3.5.

Melting Point

The melting point of a triglyceride reflects that of the component fatty acids. The melting point of a fatty acid depends on the chain length and the number and position of double bonds as well as the number and type of geometrical isomers. As one would expect, the melting point increases with increasing chain length. It decreases

TABLE 3.4 Viscosity of Refined Soybean Oil[a]–Hexane Solutions (6)

	Viscosity, centipoises			
Hexane[b], % oil/wt	0°C	10°C	25°C	40°C
0.00	172.9	99.7	50.09	28.86
11.45	49.03	31.78	18.61	11.95
20.69	21.88	15.40	9.88	6.68
30.80	10.34	7.83	.543	3.81
39.96	5.69	4.48	3.26	2.44
50.79	3.15	2.57	1.97	1.52
59.70	2.04	1.70	1.34	1.12
78.39	0.91	0.80	0.67	0.56
84.56	0.72	0.64	0.54	0.46

[a]Iodine value = 132.6.
[b]Skellysolve = B.

TABLE 3.5 Viscosity of fats and oils (16)

	Kinematic viscosity (cSt)				Saybolt viscosity (sec)	
Oil	Acid value	Specific gravity (20°/4°C)	100°F (37.8°C)	210°F (98.9°C)	100°F (37.8°C)	210°F (98.9°C)
Almond	2.85	0.9188	43.20	8.74	201	54.0
Olive	—	0.9158	46.68	9.09	216	55.2
Rapeseed	0.34	0.9114	50.64	10.32	234	59.4
Mustard	—	0.9237	45.13	9.46	209	56.9
Cottonseed	14.24	0.9187	35.88	8.39	181	52.7
Soybean	3.50	0.9228	28.49	7.60	134	50.1
Linseed	3.42	0.9297	29.60	7.33	139	49.2
Perilla, raw	1.36	0.9297	25.24	6.85	120	47.6
Sunflower	2.76	0.9207	33.31	7.68	156	50.3
Castor	0.81	0.9619	293.40	20.08	1368	97.7
Coconut	0.01	0.9226	29.79	6.06	140	45.2
Palm kernel	9.0	0.9190	30.92	6.50	145	46.5
Lard	3.39	0.9138	44.41	8.81	206	54.2
Neatsfoot	13.35	0.9158	43.15	8.50	200	53.1
Sardine	0.57	0.9384	27.86	7.06	131	48.3
Cod liver	—	0.9138	32.79	7.80	153	50.7
Refined whale	0.73	0.9227	31.47	7.48	147	49.7
Spermaceti	0.80	0.8829	22.99	5.70	110	44.1

with increasing *cis* unsaturation and rises with increasing concentrations of *trans* unsaturation. Monoacid triglycerides as well as their di- and triacid counterparts have increasing melting points as long as their corresponding fatty acids are saturated and are of the same or increasing chain length.

Polymorphism is one of the major factors affecting melting point. The formation of various polymorphic structures via partial hydrogenation is very important in controlling the melting points and physical properties of fats. This is illustrated in Table 3.3: the β polymorphic form generally has the higher melting point than the β′.

Thermal Properties

Specific Heat

The specific heat of soybean oil as well as other vegetable oils varies directly with chain length and temperature (see Table 3.6) and inversely with the degree of unsaturation.

Heat of Combustion

A general equation for the heat of combustion of vegetable oils has been developed by Bertram (18). This is a function of iodine value and the saponification values:

TABLE 3.6 Specific Heats for Soybean Oil with an Iodine Value of 128.3 (17)

Temperature		Specific heat, cal/g/°C
°C	°F	
1.2	34.2	0.448
19.7	67.5	0.458
38.6	101.5	0.469
60.9	141.6	0.479
70.5	158.9	0.490
80.4	176.7	0.493
90.4	194.7	0.504
100.4	212.7	0.508
120.8	249.4	0.527
141.3	286.4	0.531
161.9	323.4	0.550
182.7	360.9	0.567
200.1	393.6	0.594
250.5	483.8	0.621

$$-\Delta H_c \text{ (cal/g)} = 11.380 - (\text{IV}) - 9.15(\text{SV})$$

The calculated value for a soybean oil with an iodine value of 131.6 and a saponification value of 193.5 is 9,478 cal/g or 16,900 Btu/lb. The energy content of soybean oil is such that it can be used as diesel fuel (see chapter 21).

Smoke, Flash, and Fire Points

The smoke, flash, and fire points are functions of the free fatty acid content of oils, because the fatty acids are much more volatile than the triglycerides. All vegetable oils with comparable fatty acid composition have approximately the same smoke, flash, and fire points. The *smoke point* is the temperature at which smoke is first seen in a draft-free laboratory apparatus with special illumination. The *flash point* is the temperature at which volatile decomposition products are formed in amounts that they ignite but do not support a flame. The *fire point* is the temperature at which copious amounts of volatile decomposition products are formed and support a flame (AOCS Method Cc 9a-48 [4]). Such temperatures are, however, lower for oils that contain appreciable amounts of shorter-chain fatty acids, such as coconut oil. These

TABLE 3.7 Smoke, Flash, and Fire Points of Soybean Oil (19)

Soybean oil	Free fatty acid content, %	Smoke point, °F (°C)	Flash	Fire
Refined and bleached	0.01	453 (234)	623 (328)	685 (363)
Refined and bleached	0.01	443 (228)	625 (329)	685 (363)
Crude, expeller-pressed	0.51	365 (185)	565 (296)	660 (349)

TABLE 3.8 Dielectric Constants and Solubility Parameters of Solvents vs. Vegetable Oils (21)

Solvent	Dielectric constant	Solubility parameter[a]
Water	78.5	23.53
Furfural	41.9	10.09
Methanol	32.6	14.50
Ethanol	24.3	12.78
1-Propanol	20.1	12.18
2-Propanol	18.3	11.44
Acetone	21.4	9.62
Trichloroethylene	3.4	9.16
Vegetable oils	3.0–3.2	—
Cyclohexane	2.05	8.19
Hexane	1.89	7.27
2-Methyl pentane	—	7.03
Propane	1.61	5.77
Carbon dioxide	1.60	—

[a]Solubility parameter is defined as follows:
$\delta = (\Delta E/v)^{1/2} = [(\Delta H - P\Delta v)/v]^{1/2}$

temperatures for soybean oil are given in Table 3.7.

Solubility

Apolar and aprotic solvents dissolve soybean oil. In general, certain organic solvents such as hydrocarbons, esters, ethers, and ketones, as well as chlorinated solvents, all dissolve soybean oil. Several solvents have been evaluated for possible use as extraction solvents (20). The use of ethanol as solvent gave the best overall results. Solubilities of an oil in solvent can be predicted if the oil and the solvent are about the same polarity. This can be estimated from the dielectric constant and solubility parameter, shown in Table 3.8.

TABLE 3.9 Representative Values for Selected Physical Properties of Soybean Oil (15)

Property	Value
Specific gravity, 25°C	0.9175[a]
Refractive index, n_D^{25}	1.4728[b]
Specific refraction, r_D^{20}	0.3054
Viscosity, centipoises at 25°C	50.09[a]
Solidification point, °C	–10–16°C
Specific heat, cal/g at 19.7°C	0.458
Heat of combustion, cal/g	9,478[c]
Smoke point, °F (°C)	453 (234)
Flash point, °F (°C)	623 (328)
Fire point, °F (°C)	685 (363)

[a]IV = 132.6.
[b]IV = 130.2.
[c]IV = 131.6.

The solubility of oxygen in soybean oil is important, because it has detrimental effects on the quality and stability of the oil. The solubility of oxygen in soybean oil varies from 1.3 mL/100mL in refined oils to 3.2 mL/100mL in crude oil. The solubility of water in soybean oil is similar to the solubility in cottonseed oil, due to the similarity of the fatty acids that constitute the oils. The solubility of water in winterized cottonseed oil is 0.071% at 30°F and 0.141% at 90°F.

A summary of common values for soybean oil was constructed by Pryde (15) and is shown as Table 3.9.

Physical Properties of Soybean Lecithin

Lecithin has a multitude of food and nonfood uses. It is one of the most widely used emulsifiers for food products. The National Oilseed Processors Association (NOPA) has published specifications of various lecithin products; these have been given in Table 3.10. The percentage of acetone-insoluble materials is a crude indicator of the

TABLE 3.10 Soybean Lecithin Specifications (22)

Analysis	Fluid unbleached lecithin	Grade Fluid bleached lecithin	Fluid double bleached lecithin
Acetone-insolubles, min (%)	62	62	62
Moisture, max (%)[a]	1	1	1
Hexane insolubles, max (%)	0.3	0.3	0.3
Acid value, max	32	32	32
Color, Gardner, max[b]	18	19	12
Viscosity, centipoise, at 25°C (77°F max)[c]	15,000	15,000	15,000
	Plastic unbleached natural lecithin	Plastic bleached lecithin	Plastic double bleached lecithin
Analysis			
Acetone-insolubles, min (%)	65	65	65
Moisture, max (%)[a]	1	1	1
Hexane-insolubles, max (%)	0.3	0.3	0.3
Acid value, max	30	30	30
Color, Gardner, max[b]	18	14	12
Penetration, max (mm)[d]	22	22	22

[a]By Karl-Fischer Titration (AOCS Tb-2-64)
[b]Undiluted basis
[c]By any appropriate viscosimeter or by AOCS Bubble Time Method Tq 1A-64, assuming density to be unity. Fluid lecithin having a viscosity less than 7,500 centipoises may be considered a premium grade.
[d]Using Precision cone 73525, Penetrometer 73510: sample conditioned 24 hours at 77°F.
Source: Siemen, J.C., and H.J. Hirning, "Circular 1094," College of Agriculture, University of Illinois, 1974.

TABLE 3.11 Effect of Relative Humidity and Air Temperature on the Percent Moisture in Whole Soybeans (23)

Relative humidity of air (%)	0°C 32°F	5°C 41°F	10°C 50°F	15°C 59°F
40	9	8.5	8	7.5
50	10.5	10	9.5	9.0
60	12	11.5	10.5	10.0
70	14.0	13.0	12.0	11.5
80	16	15	14.5	13.5

amount of phospholipids present in a sample and is used as one of the specifications in commercial lecithin.

Miscellaneous Physical Properties of Soybeans

Several of the physical properties of soybeans are useful in designing drying and storage facilities. One of these is the angle of repose, which in the case of soybeans is 27°. Soybeans weigh 60 lbs per bushel (48 kg/ft^3 770 kg/m^3) and, in the case of beans with 17.2% moisture, have a specific heat of 0.47 to 0.49 over a temperature range of 75.2 to 129.2°F (24 to 54°C). Of major interest is the effect of relative humidity and air temperature on the percentage of moisture in the soybean shown in Table 3.1 (23). This type of data is used in determining prices and in the design of storage facilities.

References

1. Jefferson, M.E., in *Soybeans and Soybean Products, Vol. 1*, edited by K.S. Markley, Interscience, New York, 1950, pp. 247–273.
2. Swern, D., in *Bailey's Industrial Oil and Fat Products*, edited by D. Swern, Wiley-Interscience, New York, 1964, pp. 97–143.
3. Lund, J.Z., *Untersuch. Nahr. Genussmittelind. 44:* 113 (1922).
4. *Official and Tentative Methods of the AOCS*, American Oil Chemists' Society, Champaign, IL, 1989.
5. Lutskii, A.E., *Zhr. Obshch. Chim., 20:* 801 (1950).
6. Magne, F.C., and E.L. Skau, *Ind. Eng. Chem. 37:* 1097 (1945).
7. Johnstone, H.F., I.H. *Spoor, and W.H. Goss, Ind. Eng. Chem. 32:* 832, (1940).
8. Skau, E.L., F.C. Magne, and R.R. Mod, *ARS Bulletin No. 72-2 A.R.S.*, U.S. Department of Agriculture, Southern Regional Research Laboratory, New Orleans, Louisiana, 1955.
9. Singleton, W.S., *Oil and Soap 22:* 265 (1945).
10. Zeleny, L., and D.A. Coleman, U.S. Department of Agriculture, Technical Bulletin No. 554, 1937.
11. Zeleny, L., and M.H. Neustadt, U.S. Department of Agriculture, Technical Bulletin No. 748, 1940.
12. Detwiler, S.B., Jr., W.C. Bull, and D.H. Wheeler, *Oil and Soap 20:* 108 (1943).

13. Earle, F.R., T.A. McGuire, J. Mallan, M.O. Bagby, I.A. Wolff, and Q. Jones, *J. Amer. Oil Chem. Soc. 37:* 48 (1960).
14. Tels, M., A.J. Kruidenier, C. Boelhouwer, and H.I. Waterman, *J. Amer. Oil Chem. Soc. 35:* 163 (1958).
15. Pryde, E.H., in *Handbook of Soy Oil Processing and Utilization*, D. Erickson et al., American Oil Chemists' Society, Champaign, IL, 1980, pp. 37–47.
16. Rescoria, A.E., and F.L. Carnahan, *Ind. Eng. Chem. 28:* 1212 (1936).
17. Clark P.E., C.R. Waldeland, and R.P. Cross, *Ind. Eng. Chem. 38:* 350 (1946).
18. Bertram, S.H. *Chem. Technol.* (Dordrecht) *1*: 101 (1946) (in German).
19. Morgan, D.A., *Oil and Soap 19:* 193 (1942).
20. Beckel, A.C., P.A. Belter, and A.K. Smith, *J. Amer. Oil Chem. Soc. 25:* 7 (1948).
21. Hoy, K.L., *J. Paint Tech. 42:* 541 (1970).
22. *1993–1994 Yearbook and Trading Rules*, National Oilseed Producers Association, Washington, DC, p. 98.
23. Siemens, J.C., and H.J. Hirning, *Circular 1094*, College of Agriculture, University of Illinois, Urbana, IL, 1974.

Chapter 4

Harvest, Storage, Handling, and Trading of Soybeans

John B. Woerfel

Consultant
Tucson, AZ

Introduction

World production of soybeans in recent years has exceeded 100,000,000 metric tons. U.S. production accounts for approximately one-half, with the balance primarily produced by Brazil, Argentina, and the People's Republic of China (PRC), as shown in Fig. 4.1 (1).

Comprehensive figures on soybean production and utilization are compiled by the U.S. Department of Agriculture (USDA). Extensive information about soybeans and the soybean industry is reported in the *Soya Bluebook* (1) which is published yearly.

Harvest

Soybeans in the United States are primarily grown in the northern midwestern states from Ohio to Kansas and South Dakota, in states south along the Mississippi River, and in the southeastern states. In some growing areas, except for the extreme north, soybeans can be double-cropped with winter wheat, allowing farmers to get two crops per year from the same fields. Soybeans are grown as an alternative crop to corn and cotton and compete for acreage with these two crops.

Depending on the region, planting may be as early as May 1 or as late as July 15. Harvest may begin as early as September 15 and continue until December 20. The most active harvest period is October and November (1).

Once soybeans have matured and conditions are appropriate for harvesting, it is desirable to complete harvesting as quickly as possible to minimize losses from shattering or adverse weather conditions. Soybeans are directly loaded from combines into trucks and are delivered to storage bins on the farm, to an elevator, or to a nearby processing plant. In areas of intensive cultivation there may be several buyers competing for the supply. This stimulates price competition and quality of service.

The initial delivery point may be a country, subterminal, or terminal elevator or a processing plant, where sampling and grading are done on receipt and the price is paid based on grade. Receipts from farms will generally be "as harvested" and will vary in grade and moisture content, depending on growing and harvesting conditions.

Elevators may be individually owned, or they may be a part of an organization with interests in domestic processing and export. As country elevators fill up during harvest, soybeans will be moved to larger elevator facilities or processing plants to

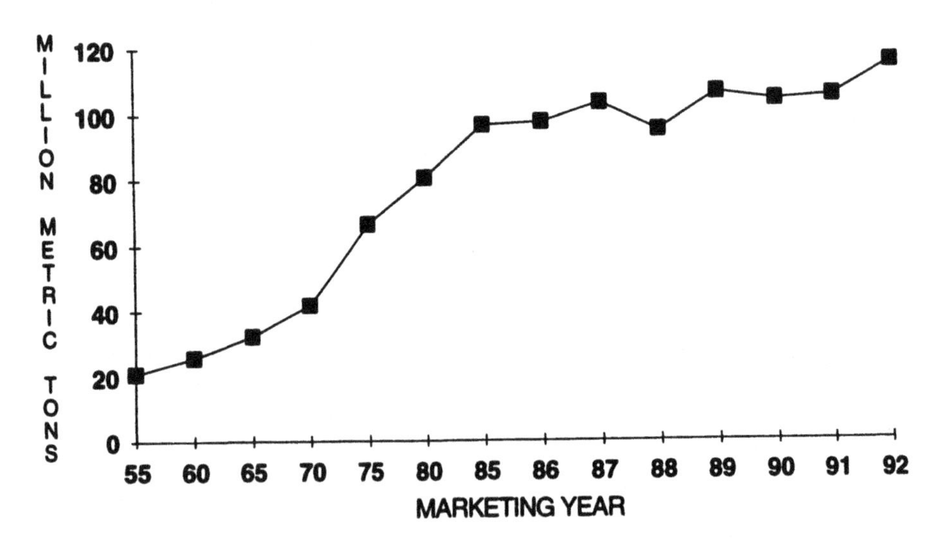

Fig. 4.1. World soybean production. (1) Total production, (b) 1992/93 forecast of market share. *Source: '93 Soya Bluebook*, Soyatech, Inc., Bay Harbor, ME, 1993, p. 203.

make room for additional receipts. When the crop is good, facilities may become filled, and storage will be at a premium.

Distribution

Fig. 4.2 shows the general flow of soybeans from the farm to the ultimate processor (2). Transportation may be by truck, rail, barge, or ship with brief or extended storage at each point.

Since processing is most efficient when carried out continuously, and since demand for product is fairly uniform, soybeans are handled on a year-round basis and may be in transit and stored for a year or more between harvest and processing.

Grading Soybeans

U.S. soybeans are commonly traded on standards established by the USDA, shown in Table 4.1 (3). Official inspection, supervised by the Federal Grain Inspection Service (FGIS), is mandatory on export shipments. This includes stowage examination, sampling, weighing, inspection, and certification. Domestic transactions do not require official inspection. However, official services are available on request and are frequently used in domestic trading.

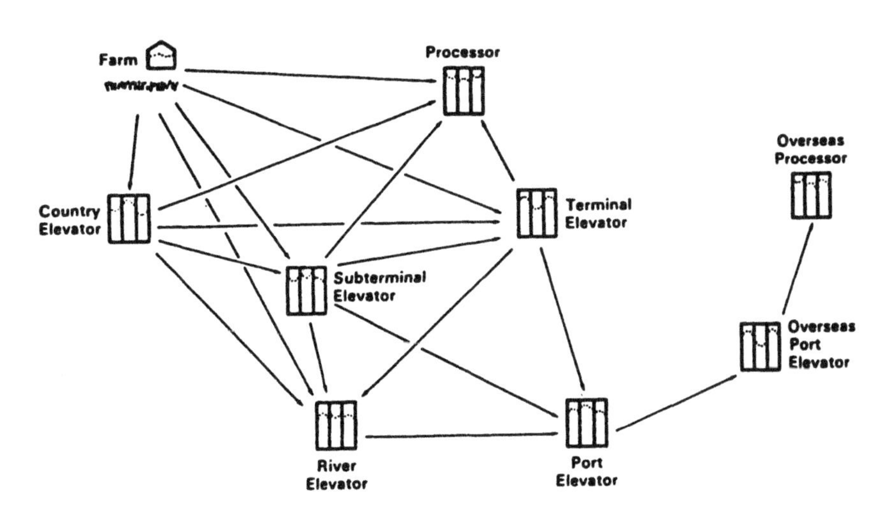

Fig. 4.2. The general flow of grain from the farm through the distribution system to the domestic and overseas processor. *Source:* Overview of Soybean Inspection, American Soybean Association, St. Louis, MO.

TABLE 4.1 Grades and Grade Requirements

		Maximum limits of:				
		Damaged kernels				
Grade	Min. test weight per bushel (lbs)	Heat damaged (%)	Total (%)	Foreign material (%)	Splits (%)	Soybeans of other colors (%)
U.S. No. 1	56.0	0.2	2.0	1.0	10.0	1.0
U.S. No. 2	54.0	0.5	3.0	2.0	20.0	2.0
U.S. No. 3[1]	52.0	1.0	5.0	3.0	30.0	5.0
U.S. No. 4[2]	49.0	3.0	8.0	5.0	40.0	10.0

U.S. Sample grade consists of soybeans that:

a. Do not meet the requirements for U.S. Nos. 1, 2, 3, or 4; or
b. Contain 8 or more stones having an aggregate weight in excess of 0.2 percent of the sample weight, 2 or more pieces of glass, 3 or more Crotalaria seeds (Crotalaria spp.), 2 or more castor beans (*Ricinus communis*), 4 or more particles of an unknown foreign substance(s) or a commonly recognized harmful or toxic substance(s), 10 or more rodent pellets, bird droppings, or equivalent quantity of other animal filth per 1,000 g of soybeans; or
c. Have a musty, sour, or commercially objectionable foreign odor (except garlic odor); or
d. Are heating or otherwise of distinctly low quality.

[1]Soybeans that are purple-mottled or stained are graded not higher than U.S. No. 3.
[2]Soybeans that are materially weathered are graded not higher than U.S. No. 4.
Source: Official U.S. Grading Standards for Grain, USDA, Washington, DC, June, 1993.

Voluntary grading of soybeans may be done by plant and commercial laboratories, not under United States Grain Inspection Service (USGIS) control, for controlling storage and processing practices at elevators and processing plants. Grading factors are directly affected by storage and handling practices, and oil quality is directly related to grade factors, especially oil from damaged soybeans.

As of September 9, 1985, moisture was eliminated as a grading factor, but it is required to be shown on the certificate. While moisture is the most critical factor in storage of soybeans, it is controllable and must be controlled to maintain quality.

Damaged Kernels

Damaged kernels are divided into two classifications: total and heat-damaged. Damage is identified by a visual system using interpretive line slides or line prints. Various types of damage are described, such as

Ground or weather damage	Mold damage
Frost damage	Sprout damage
Immature soybeans	Heat damage
Insect damage	

All of these, except heat damage, relate to growing conditions. Insects, mold, sprouting, and especially heat damage are related to storage conditions.

Foreign Material

Foreign material consists of all matter, including soybeans and pieces of soybeans, that will pass readily through a $^1/_8$ in (0.32 cm) round-hold sieve, as well as all material other than soybeans that remains on such sieve after sieving. It includes whole or parts of corn kernels or other grains; weed seeds; other vegetable matter such as pods, leaves, or stems; and dirt or other inorganic materials. Foreign material may result from field and harvest conditions, handling practices, residues left in transport vehicles or storage tanks.

Splits

Splits are soybeans with more than one-fourth of the bean removed and that are not otherwise damaged. Splits result from mechanical damage during handling of soybeans and can be controlled by proper design and operation of seed-handling equipment. Impact from soybeans free-falling onto hard surfaces can be especially damaging. To facilitate determination of splits, separation of the sample is made with a series of sieves.

Relation of Grading Factors to Crude Soybean Oil Quality

Foreign material, such as weed seed, green leaves, and immature soybeans, can promote oxidation and introduce pigments, especially chlorophyll. In addition, some of the materials may be high in moisture and cause heat damage in storage.

Nonhydratable phosphatides, free fatty acids, and metal content are increased by enzymatic and biological processes that are initiated when natural barriers present in sound soybeans are broken down by field damage, splitting, and breaking or by heat and other damage occurring in storage.

Economic effects include higher refining losses, increased use of refining materials, and poor-quality refined oil. There is a direct relationship between quality of soybeans as defined by these standards and their economic value. Although lower grades may be priced lower, preference is generally for No. 2 or better soybeans. There are strong economic incentives for farmers to produce high-quality soybeans and for elevators, shippers, and processors to use practices that enhance and preserve the quality.

Heat Damage

Control of heat damage is the most important factor in storage and transport. It is caused by chemical and biological activity that is promoted primarily by moisture and temperature. Certain other factors, such as foreign material or handling practices, also contribute to heat damage.

The moisture in soybeans normally is in equilibrium with the relative humidity (RH) of surrounding air. Table 4.2 shows this relationship at 25°C (4). Figure 4.3 provides comprehensive data on the moisture–RH equilibrium over a range of moisture contents and temperatures (5). Generally, 13% moisture is considered desirable for storage and handling of soybeans, and it has been shown that storage of up to one year

TABLE 4.2 Equilibrium Values for Soybeans at 25°C

Relative Humidity (%)	Moisture in Soybeans (%)
35	6.5
50	8.0
60	9.6
70	12.4
85	18.4

Source: Barger, W.M. *J. Amer. Oil Chem. Soc. 58:* 154 (1981).

is feasible at this level. A common contract specification is 14% maximum. Soybeans can be stored for shorter periods of time at higher moisture levels, but, as shown in Fig. 4.4, the allowable time decreases rapidly with increased temperatures (5).

Cleaning soybeans to remove foreign material before drying is desirable for quality and safety reasons. Ideally, cleaning is done at the earliest possible time after harvest.

Many U.S. farmers have drying and storage facilities to give them economic advantages in marketing their crops. Others deliver directly from the field to a commercial elevator, where they may be dried. During peak harvest, especially when moisture levels are high, drying capacity may not be sufficient to dry high-moisture soybeans as rapidly as received. In this situation, judgment must be made on how best to use available drying capacity.

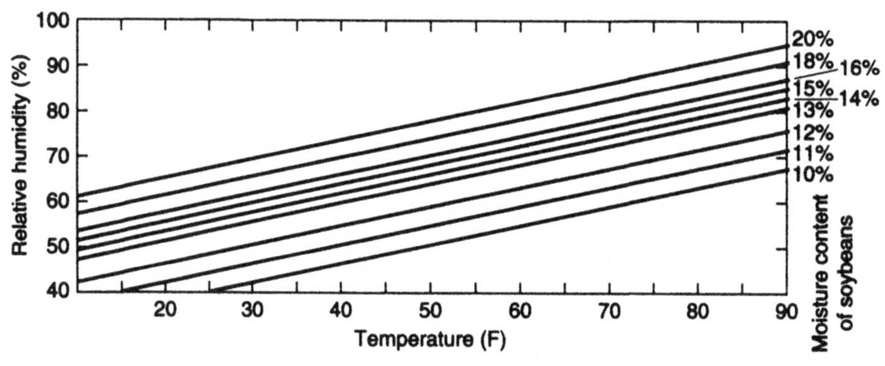

Fig. 4.3. Equilibrium level of soybeans with the temperature and relative humidity of the surrounding air. For example, an air temperature of 63°F (17°C) and a relative humidity of 70% is in equilibrium with 14% soybeans. With the temperature remaining constant, the moisture content of the soybeans could be reduced only with a decrease in the relative huumidity from 70%. There are other equilibrium levels for 14% soybeans, such as 50°F (10°C) and 65% relative humidity. Source: Spencer M.R., *J. Amer. Oil Chem. Soc. 53:* 238, (1976).

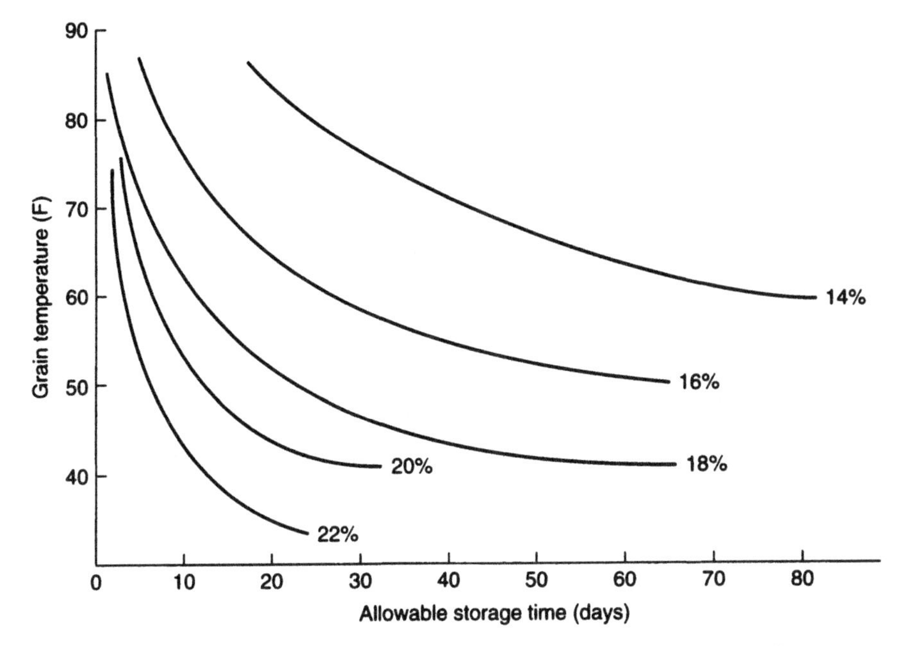

Fig. 4.4. Allowable storage time for soybeans (seeds). *Source:* Spencer, M.R., *J. Amer. Oil Chem. Soc. 53:* 238, (1976).

Drying Soybeans

Soybeans are usually dried in countercurrent *open-flame grain dryers* heated with natural gas or fuel oil. Uniform movement of the soybeans passing through the drying and cooling section is essential. Temperature of soybeans must be raised sufficiently to achieve the finished moisture content desired, but it must not exceed 76°C, since discoloration and protein denaturation will result.

Much improvement has been made in recent years to reduce fuel costs. Recirculation of exhaust air from the cooling system, as shown in Fig. 4.5 has been widely used to save fuel (6). Savings of 40% in fuel have been reported. Additional savings result from improvements in grain and air flow that reduce electrical usage. Dryers have also been built using low-cost steam generated from alternative fuels such as wood waste.

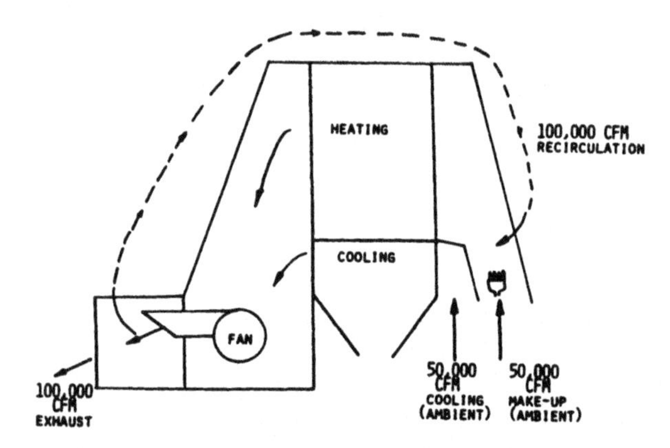

Fig. 4.5. Grain drying with recirculation of 50% of exhaust air. System volume = 200,000 CFM (94m^3/s).

Microwave vacuum dryers, as shown in Fig. 4.6, have been tested on soybeans. These represent an entirely new concept in the application of energy in that the soybeans are heated from the inside in combination with vacuum to remove water vapor. This results in a lower temperature, less temperature gradient, and more uniform drying. Benefits are summarized by the manufacturer as "quick, quiet, clean, efficient, quality, simple, safe and versatile" (7).

Fluid bed dryers are used for drying soybeans in the process of "hot dehulling."

Solar drying has been tested on a commercial scale (8).

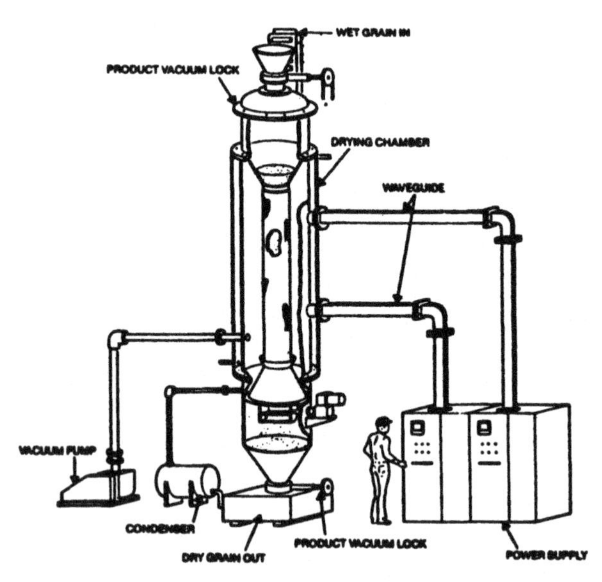

Fig. 4.6. Microwave vacuum drying. *Source:* Darla, S., *J. Amer. Oil Chem. Soc. 60:* 409 (1983).

Storage Facilities

Soybeans are usually stored in steel tanks or concrete silos. Each has certain advantages.

Steel tanks are generally less expensive. They may be of any size up to 60 m (200 ft) in diameter and hold up to 60,000 MT. The top is usually conical with a 27° pitch, which matches the angle of repose for soybeans.

Conveyors are provided to deliver soybeans to the center top, where they drop into the tank. Conveyor systems to remove soybeans are somewhat complex, because of the large flat bottom area of the tanks.

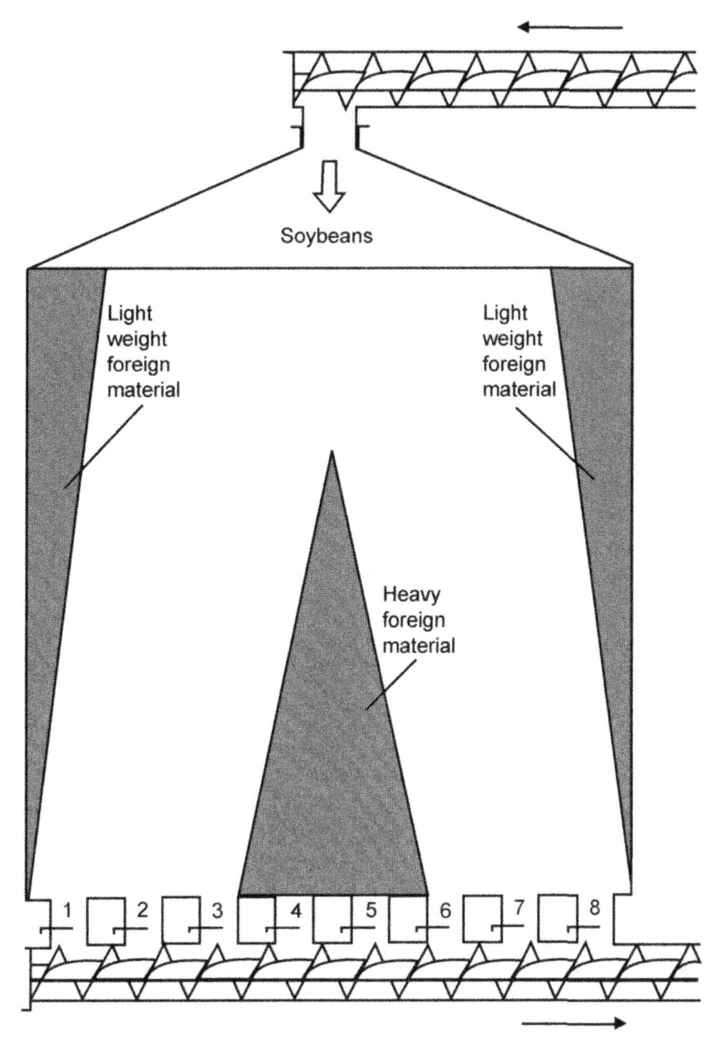

Fig. 4.7. Soybean storage tank. *Source:* Barger, W.M., *J. Amer. Oil Chem. Soc. 58:* 154 (1981).

Ventilating fans are provided to aerate the stored soybeans by blowing air into the bottom of the tank. An aeration rate of 0.1 m^3/min/MT (0.1 ft^3/min/bu) is typical. Aeration is used to cool and equalize temperatures within the tank. Soybeans can be dried in the tank by judicious use of aeration when ambient air temperature and relative humidity are favorable (see Fig. 4.3).

There is a tendency for foreign material to separate in tanks. Heavy foreign material tends to drop straight down and form a core in the center, whereas light foreign material tends to migrate to the outside, as shown in Fig. 4.7 (4). Heating often starts in the core of heavy foreign material. When this occurs, soybeans should be withdrawn from the center and sent directly to processing or dried and returned to storage.

Concrete silos are more expensive. They are usually built in groups. Individual silos and the interstices between them form units of relatively small cross-sectional area and capacity. This allows segregation of small lots of soybeans.

Filling and recovery are accomplished by using conveyors running the full length of the row of silos, and soybeans are readily moved or "turned" from one silo to another or from a silo to processing or drying. This is especially convenient in case soybeans are "heating" and need to be moved. Soybeans have less tendency to "bridge" than some other grains. As a result, they can be discharged from a silo by gravity at a higher rate.

Temperature monitoring of soybeans in storage is essential. Thermocouples in rigid conduits are placed throughout the tank or silo and monitored regularly. If any temperature rise is noted, the exact location can be identified and immediate corrective action taken.

"Flat" storage of soybeans is used in some installations. It is quite common, in plants that have been converted from cottonseed or process both cottonseed and soybeans, to use the typical "Muskogee" houses. These houses have short sidewalls and large capacity under the roof, which is built with a 45° pitch to match the angle of repose for cottonseed.

Because of the lower angle of repose and higher bulk density of soybeans, these houses cannot be filled to full volumetric capacity. Soybeans will exert excessive pressure on the sidewalls, so levels must be carefully restricted to less than the full height of the sidewalls, or the sidewalls must be reinforced.

Transportation and Handling

Soybeans are transported by truck, rail, barge, and ship. They are loaded, unloaded, and conveyed several times while on the way from the field to actual processing. Bulk handling is most common, although in some places with low labor costs they are handled in sacks.

Soybeans are free-flowing and have a density of about 670 kg/m^3 (42 lbs/ft^3). Bulk handling equipment for soybeans is similar to that used for handling other granular solids, grains, and oilseeds.

Concerns in transport and handling are to prevent splits and breakage, protect the soybeans from moisture, control dust, and avoid contamination or adulteration with foreign materials or other grains or oilseeds. Trucks are usually open flatbed trucks or semitrailers with side walls 1 to 2 m high and covered with a tarpaulin. Trucks are usually unloaded by putting them on a platform that is tilted by hydraulic power, allowing the soybeans to slide out into a hopper, from which they are conveyed to storage. Covered hopper cars are usually used for rail transit. Barges and ships are bulk cargo carriers, which are used generally for grain and oilseeds.

Screw conveyors, bucket elevators, and belt conveyors were standard for many years. In recent years en masse or chain conveyors have become widely used. En masse conveyors are highly flexible and can convey in many different configurations, both horizontally and vertically, with multiple feed and discharge points. Product is moved in a compact stream through a trough by flights attached to a chain. There is a minimum of turbulence, which can cause breakage.

Belts are an excellent choice where high capacity and long distances are involved, such as loading ships or barges. They can be reversed and used for both loading and unloading; they can transport on an incline or level; and there is minimum agitation of the soybeans on the belt.

Damage in Handling and Shipping

Mechanical damage to soybeans during shipping and handling causes splitting and breakage. This reduces the grade directly and promotes enzymatic activity and heat damage. Table 4.3 shows increase in splits between origin and destination of several export shipments (9).

A major cause of damage is free fall of soybeans, which occurs when soybeans are filled into tanks or silos or are loaded into the deep holds of ships. Table 4.4 shows breakage caused by dropping from different heights onto concrete or soybeans (10). It should be noted that the significant factor is the drop height. Whether soybeans drop onto concrete or onto soybeans, damage is reduced by 75% when height is reduced from 100 to 40 ft. Obviously, the condition of dropping onto concrete occurs for only a short time at the beginning of filling a tank.

TABLE 4.3 Splits in Samples of Soybean Shipments

Sample	Origin	Destination	% Increase
1	8.7	11.2	29
2	8.3	11.5	39
3	10.5	10.9	4
4	9.4	11.6	23
5	10.2	11.3	11
6	8.4	9.4	12
7	12.6	13.9	10
8	9.2	10.1	10
9	7.8	10.1	29
10	17.9	19.6	10

Source: USDA Marketing Research Report No. 1090 (1978).

TABLE 4.4 Breakage of Soybeans

Drop height (ft)	Onto Concrete (%)	Onto Soybeans (%)
100	4.5	3.2
70	2.1	1.4
40	1.1	0.7

Source: USDA Marketing Research Report No. 968 (1973).

Trading

Trading soybeans and soybean products is a complex and highly specialized activity. Effective trading requires knowledge of economics and experience in commodity markets. The large quantities of raw material processed and its high value and volatile prices make profits in soybean processing and handling very sensitive to trading decisions.

The following outlines a few basics of trading. References are included for those interested in more detail.

In the U.S., soybeans are traded in bushels, soybean oil in pounds, and soybean meal in short tons. Elsewhere in the world all three commodities are measured and traded in metric tons. While the term *bushel* implies volumetric measurement, a bushel (1 bu) of soybeans is legally 60 pounds avoirdupois (27.2 kg).

Table 4.5 lists conversion factors that are commonly used in statistical data concerning soybeans and soybean processing (1).

Cash Markets

Cash or spot markets are markets where a physical transfer of a commodity occurs. A cash market exists where buyers and sellers transact business, such as grain elevators, processing plants, and export markets. Cash market transactions are not standardized. Sales agreements for quality, quantity, and delivery terms vary from sale to sale as agreed upon by both buyers and sellers.

Trading rules issued by trade organizations, such as the National Oilseed Products Association (NOPA), the National Institute of Oilseed Producers (NIOP), (FOSFA), may govern such sales agreements.

Futures Markets

Futures markets, in contrast to cash markets, are centralized, regulated markets where soybeans, meal, and oil are not physically traded; instead, *futures contracts* are bought and sold. Futures contracts are legally binding agreements to deliver a commodity at a specified price, standardized according to quality, quantity, delivery time, and location.

Options to futures may also be traded on the floor of a regulated futures exchange. Options convey the right, but not the obligation, to buy or sell a particular futures contract at a certain price for a limited time.

Many commodity exchanges offer futures trading throughout the world. The Chicago Board of Trade (CBOT) is the world's largest commodity exchange and the largest marketplace for soybeans, meal, and oil. The CBOT contract provisions are briefly summarized in Table 4.6 (11).

Price is established by open auction of contracts by traders representing commercial members such as grain merchandisers, oilseed crushers, large banks and investment houses, or professional speculators.

Commercial members use the futures market and options contracts to protect themselves against price changes in soybeans, meal, and oil. Speculators trade to make profits from price fluctuations, and in doing so provide liquidity and assume risks from commercial traders.

Price changes during trading are immediately posted and transmitted around the world. Daily quotations of open, high, low, and close are published in the *Wall Street Journal* and other major newspapers.

Futures markets are used by soybean processors, feed manufacturers, and food processors to reduce risks that may occur from fluctuating raw material and product prices. This practice is known as *hedging*. Hedging is based on the fact that in general the cash prices are related to the futures prices and move up or down in a parallel pattern.

The difference between futures price and cash price at a specific location is known as *basis*. Fluctuations in basis tend to be smaller than fluctuations in either cash or futures prices, allowing hedging to work.

Transportation cost is the major factor in basis. Basis is generally lowest at the point of origin, such as in the U.S. Midwest, and increases as the product moves to processors, consumers, or export elevators in other locations. Other basis factors are cost of storage (which includes interest, insurance, and the expense of loading and unloading), sellers' profit margins, and local supply and demand.

TABLE 4.5 Statistical Conversions

Soya Conversions			
1 bushel of soybeans	=	60	pounds
	=	10.7	pounds of crude soy oil
	=	47.5	pounds of soybean meal
	=	39	pounds of soy flour
	=	20	pounds of concentrate
	=	11.8	pounds of isolated soy protein
1 metric ton of soybeans	=	36.74	bushels
1 short ton of soybeans	=	33.33	bushels
1 short ton of soybeans	=	.907	metric tons
1 long ton of soybeans	=	37.33	bushels
1 metric ton of soybean meal	=	46.39	bushels of soybeans
1 short ton of soybean meal	=	42.08	bushels of soybeans
1 long ton of soybean meal	=	47.13	bushels of soybeans
1 metric ton of soybean oil	=	206	bushels of soybeans
79.2% of a bushel of soybeans is manufactured into soybean meal.			
17.8% of abushel of soybeans is manufactuered into soybean oil.			
3.0% of a bushel of soybeans is manufacturing loss.			

Source: '93 Soya Bluebook, Soyatech, Inc., Bay Harbor, ME.

TABLE 4.6 Chicago Board of Trade (CBOT) Contracts

	Soybeans	Meal	Oil
Contract amount	5000 bu	100 tons	60,000 lb
Priced in	$/bu	$/ton	cents/lb
Price increments	1/4 cents	cents	1/100 cents
Contract months:			
Contract year begins	Sept	Oct	Oct
	Nov	Dec	Dec
	Jan	Jan	Jan
	Mar	Mar	Mar
	May	May	May
	July	Jully	July
		Aug	Aug
Contract year ends	Aug	Sept	Sept
Quality or grade	No. 2 Yellow Soybeans	Crude Soybean Oil CBOT Rules	48% Soybean Meal CBOT Rules
Delivery points[a]	[b]	Central Territory	Illinois Territory

Source: Soybean Complex, Futures and Options on Soybeans, Soybean Oil, Soybean Meal, Chicago Board of Trade, Education and Marketing Publications, Chicago, IL.
[a]Other delivery points at differential from contract price as determined by CBOT rules.
[b]Chicago Switching District and Burns Harbor, Indiana, Switching District.

Crushing margin is the difference between total value of products and soybean price. It is important to recognize that markets for soybeans, meal, and oil are separate and influenced by different economic factors. USDA publishes yearly statistics for margins between value of products and soybean price in dollars per bushel based on No. 1 yellow soybeans, Illinois points; 44% meal; and crude oil, tanks f.o.b. Decatur, Illinois. These statistics show wide variations in crushing margins from year to year.

Crushing margins influence the actions of processors and soybean prices. Favorable margins encourage processors to increase crushing and put upward pressures on soybean prices. Unfavorable margins discourage crushing and cause downward pressure on soybean prices or the introduction of marketing strategies to improve prices of meal or oil.

Hedging

Hedging can be a very complex process, and many types of hedges can be used by different hedgers for different purposes. Some of these are (from the Chicago Board of Trade, Introduction to Agricultural Hedging [12]).

1. Farmers, seeking protection against declining prices of stored grain or crops still in the field

2. Country elevator operators, seeking protection against changing prices between the time they purchase, or contract to purchase, grain from farmers and the time the grain is ultimately merchandised
3. Processors, seeking protection against rising raw material costs or decreases in inventory value
4. Exporters, seeking protection against an increase in the cost of commodities that have not yet been acquired but have been contracted for future delivery to importers
5. Livestock feeders or feedlot operators, seeking protection against declining prices of livestock or rising feed costs

Examples of various types of hedges are included in the publication referenced. The following example (12) is of one type of buying, or long, hedge that is used by processors or exporters wanting protection against rising prices.

Assume it is May and a processor anticipates the purchase of soybeans in August. The cash market price in May is $5.00/bu, but the processor is afraid that by the time he purchases soybeans in August, the price may be higher. To protect himself from this price increase, he goes long (buys) August soybean futures at $5.30/bu. The T-account below shows the outcome if soybean cash and futures prices increased by $1.00/bu:

Cash	*Futures*	
May 1		
Wants to lock in soybean price of $5.00/bu	Buys Aug soybean contract at $5.30/bu	
Aug 1		
Buys soybeans at $6.00/bu	Sells Aug soybean contract at $6.30/bu to offset initial long futures position	
	$1.00/bu gain	
Result:	Cash purchase price	$6.00/bu
	Less futures gain	–1.00/bu
	Net purchase price	$5.00/bu

In this example, the higher cost of soybeans in the cash market was offset by a gain in the futures market. The processor locked in the $5.00/bu price that he had wanted in May.

If the price decreased by $.50, the result would be as follows:

Cash	*Futures*	
May 1		
Wants to lock in soybean price of $5.00/bu	Buys Aug soybean contract at $5.30/bu	
Aug 1		
Buys soybeans at $4.50/bu	Sells Aug soybean contract at $4.80/bu to offset initial long futures position	
	$.50/bu loss	
Result:	Cash purchase price	$4.50/bu
	Plus futures loss	+ .50/bu
	Net purchase price	$5.00/bu

Although the processor could not take advantage of the decrease in cash soybean prices, he did lock in the $5.00/bu price for soybeans that he found acceptable in May.

Safety

Handling, storage, drying, and cleaning of soybeans involve safety hazards. Safety concerns are fire and explosion, powered machinery, entry into confined spaces, the risk of being smothered by sinking in a bin of soybeans, and transportation accidents.

Dust explosions can be catastrophic. Soybean dust, when mixed with air in proper proportions, will explode when ignited. Dust can also be an environmental problem. Dust control involves both design and operational factors. Uncovered conveyors or exposed loading and unloading can spread dust over a wide area. Cyclones and other types of dust collectors must be properly designed and maintained. Cleanup of dust should be a continuing program. Any excessive dust accumulation should be investigated and corrected at the source.

Dust can be controlled by treatment of soybeans with sprays of oil or water. The preferred material is soybean oil, since there is no question of compatibility. The use of water on grain is controlled by the USDA because of potential for "economic adulteration." Mineral oil use is subject to U.S. Food and Drug Administration (FDA) control as a food additive and is restricted to 200 ppm/application.

Fires may occur in soybean dryers. Paper or other readily ignitable foreign material must be removed in cleaning before drying. Dryers should be equipped with alarms and sprinklers.

Spontaneous combustion can result from heating of soybeans in storage and cause extensive damage.

Moving machinery can be a serious problem. A not uncommon occurrence in a soybean operation is injury to a worker trying to clear a plugged conveyor. All moving machinery should be protected by guards, which are not to be removed unless the machinery is shut off and locked out so that it cannot be accidentally started.

There are many other concerns to be covered in a comprehensive safety program, including not only the details of procedure but education and involvement of all employees. In the United States, such a program must conform with the U.S. Occupational Safety and Health Administration (OSHA) standards, which should be considered minimum, and attention should be given to specialized local and industry problems.

Pest Control

Soybeans, in common with other grains and oilseeds, are a food source, and hence highly attractive to a wide variety of pests. Large and complex storage and handling facilities provide isolated spaces in which pests can flourish. An effective control program requires professional organization and continuing attention.

Good housekeeping is a prerequisite. Spills and dust should be cleaned up immediately and meticulously. Good drainage, to avoid standing water, and careful inspection and maintenance of spray ponds, hot wells, drainage ditches, and surrounding areas will discourage colonization by rodents and breeding of insects.

Inspection by trained personnel of all areas from "roof to cellar" on a regular schedule, and immediate response to any evidence of infestation, are necessary. Close attention to the perimeter of the property helps to avoid invasion from neighboring areas.

When infestation is identified, response with appropriate control measures is necessary. Spraying surfaces, areas, and, in some cases, soybeans for insect control is appropriate, but it must be done with approved techniques and materials.

References

1. *'93 Soya Bluebook*, Soyatech, Inc., Bar Harbor, ME, 1993.
2. *Overview of Soybean Inspection*, American Soybean Association, St. Louis, MO, 1988.
3. *Official U.S. Grading Standards for Grain*, USDA, Washington, DC, June, 1993.
4. Barger, W.M., *J. Amer. Oil Chem. Soc. 58:* 154 (1981).
5. Spencer, Max R., *J. Amer. Oil Chem. Soc. 53:* 238 (1976).
6. Dada, S., *J. Amer. Oil Chem. Soc. 60:* 409 (1983).
7. *MIVAC Microwave Vacuum Dryer*, Aeroglide Corporation, Raleigh, NC, not dated.
8. Reisz, Al, ASME 79-WA/Sol-32, The American Society of Mechanical Engineers, New York, NY (1979).
9. USDA, Marketing Research Report No. 1090 (1978).
10. USDA, Marketing Research Report No. 968 (1973).
11. *Soybean Complex, Futures and Options Soybeans, Soybean Oil, Soybean Meal*, Chicago Board of Trade, Education and Marketing Publications, Chicago, IL, 1994.
12. *Introduction to Agricultural Hedging*, Home Study Course, Chicago Board of Trade, 1988.

Chapter 5

Overview of Modern Soybean Processing and Links Between Processes

David R. Erickson

Consultant
American Soybean Association
St. Louis, MO

Introduction

Modern soybean processing consists of a series of unit processes that in total determine the costs of processing and quality of final products. This handbook consists of a sequential discussion of each unit process, and this chapter will outline the processes employed and their interrelationships and interdependencies.

Soybean Quality Effects

The effect of soybean quality on processing and the quality of final products is a fundamental consideration and will direct the choice of processing conditions.

For trading purposes, soybeans are considered a fungible commodity, and any soybeans of one classification can replace soybeans of the same classification from another source. To an extent this convention is true; for example, No. 2 soybeans have to meet certain criteria to be classified as No. 2. However, such criteria are insufficient to guide processors in their choice of processing conditions or to predict quality of products without additional testing or experience (see Table 4.1).

The effects of harvest, storage, and handling on soybean quality are discussed in Chapter 4. At the time of harvest, soybeans are at their peak of quality; subsequent harvesting, storage, and handling reduce this quality. Table 11.1 shows some of the effects on oil quality of various factors that can be considered abusive (1). In that table, the extraction plant has control of the soybeans starting with the fourth item, whereas the refiner has control only over some portion of the last item of crude oil storage. A factor not specifically mentioned is the age of the soybeans, which may be a matter of weeks, months, or two years or more.

The foregoing discussion simply states that soybeans for processing are going to be of variable quality. Therefore, processors must be prepared to test for this variability and to adjust their processing practices accordingly. Wiedermann describes this in relation to degumming, refining, and bleaching as both "minimum and adjusted treatment practices" (1). The "minimum" is related mainly to costs of processing, and "adjusted" is related mainly to optimizing quality; however, in actual practice there is a natural overlap or balance between minimum costs and maximum quality. The foregoing applies to all soybean processing and products.

In general, the quality of soybeans will not affect the quality of soybean meal in terms of protein content and functionality as an animal feed. Soybean quality does affect the quality of soybean oil, as reflected in the National Oilseed Processors Association (NOPA) trading rules in terms of free fatty acid (FFA) content, bleachability, phosphorus content, and so forth (2).

Processing Overview and Links Between Processes

Modern emphasis in soybean processing has tended to concentrate on energy conservation and savings, increased production, and environmental considerations. These are real-world concerns and merit emphasis, but not at the expense of the basic technical goals in processing. First and foremost, the overall technical goal should be production of products of acceptable quality. Secondly, such production should be at the lowest possible cost that allows meeting the first goal.

Fortunately, the separate goals of lowest possible processing costs and highest quality are not incompatible for soybean oil. This has been demonstrated and reported by Charpentier (3) and Erickson (4).

Extraction Processes

Chapter 6 discusses the extraction processes for soybeans, which involve continuous screw presses, solvent extraction, or, in some cases, a combination of the two.

Mechanical Extraction

In the United States less than 1% of soybeans are processed by continuous screw presses ("expellers"); a flow diagram of the process, combined with solvent extraction, is shown in Fig. 5.1. The soybean meal (press cake) from pressing operations contains 4 to 5% residual oil, and the crude soybean oil is similar to that from solvent extraction, with generally a lower phosphatide content.

Solvent Extraction

Solvent extraction processes basically break down into the three steps of preparation, extraction, and desolventizing. The major differences in soybean solvent extraction processes occur in the preparation steps. A commonality in all the preparation processes is the choice of whether or not to dehull.

The conventional preparation sequence is shown in Fig. 5.2. One variation in the conventional system is the introduction of an expander after flaking or other method of size reduction (grinding), for the reasons discussed in Chapter 6. Another variation is the introduction of "hot dehulling," which replaces the tempering step in conventional preparation and has some other advantages, as discussed in Chapter 6. A final variation is the so-called Alcon process, which is a method of cooking the flakes prior to extraction. A flow diagram encompassing the existing variations in preparation is shown in Fig. 5.3.

In conventional solvent extraction, the thickness, moisture content (<10%), and integrity of the formed flakes are of key importance. It is also important that the time between flaking and extraction be a short as possible. Once flaked, enzyme action is at a maximum, and the FFA and nonhydrable phosphotides (NHP) increase in proportion to the time between flaking and extraction. The later introduction of the expander and the Alcon process reduced the importance of the flaking operation but also caused changes in the amount and type of phosphatides extracted into the crude oil.

A comparison of crude soybean oil properties for the different extraction processes are shown in Table 5.1. Although the expander and Alcon process have their advantages (see Chapter 6), they produce crude oils with higher FFAs, higher phosphatide content, and lower neutral oil content. This is especially true for the Alcon process, which practically doubles the gum content in the crude oil over that from conventional processes. In addition, the gums are more water-hydratable, which leads to lower phosphatide levels in the degummed oil. The drawbacks and the costs associated with the extra processing have to be balanced against the advantages.

One important precaution to take with the Alcon system is to ensure sufficient drying of the cooked flakes (<10% moisture) before extraction to avoid color rever-

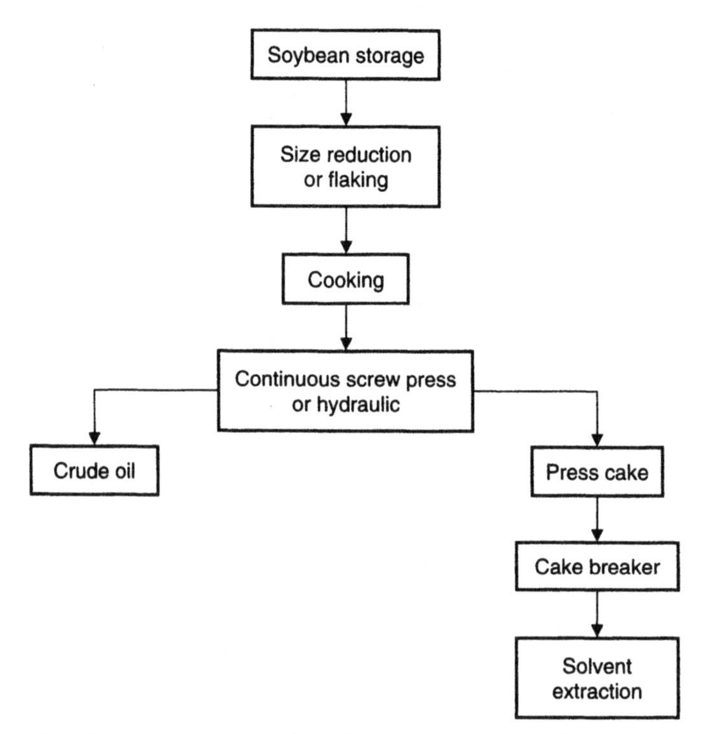

Fig. 5.1. Flow diagram of combined mechanical extraction and solvent extraction.

sion in the finished oil. Use of an expander will show effects proportional to those of the Alcon process as the temperatures and moistures in the expander approach those of the Alcon cooker. The collets from expansion should also be dried to <10% moisture to avoid color reversion.

Extraction

The preparation process affects both the efficiency of oil extraction and the quality of the oil. The efficiency of extraction is based on the residual extractables in the soybean meal in relation to production rates. Kock (5) has shown that there is a correlation between the temperature of extraction and undesirable enzymatic action. From this observation he patented the so-called Alcon process.

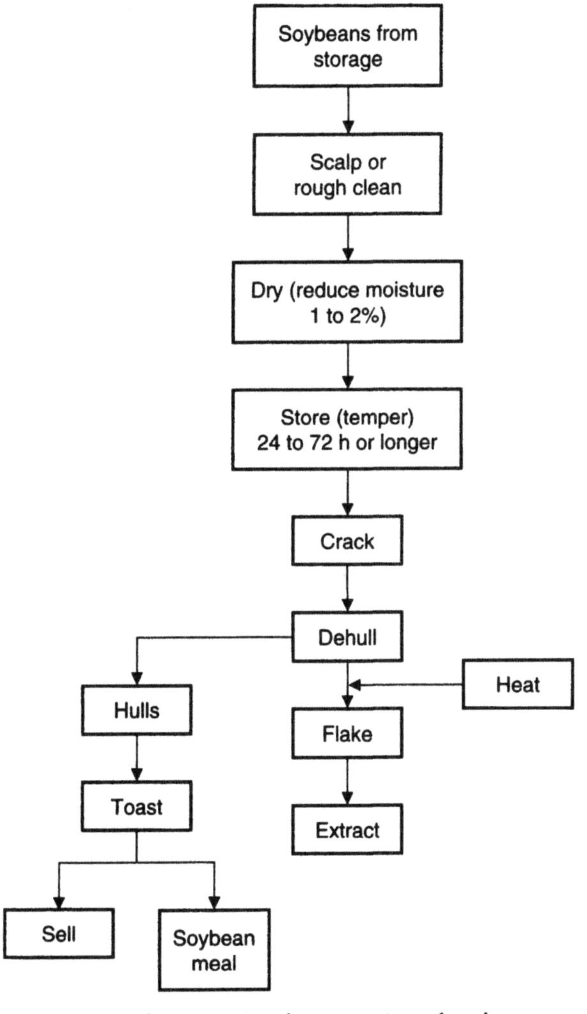

Fig. 5.2. Conventional preparation for extraction of soybeans.

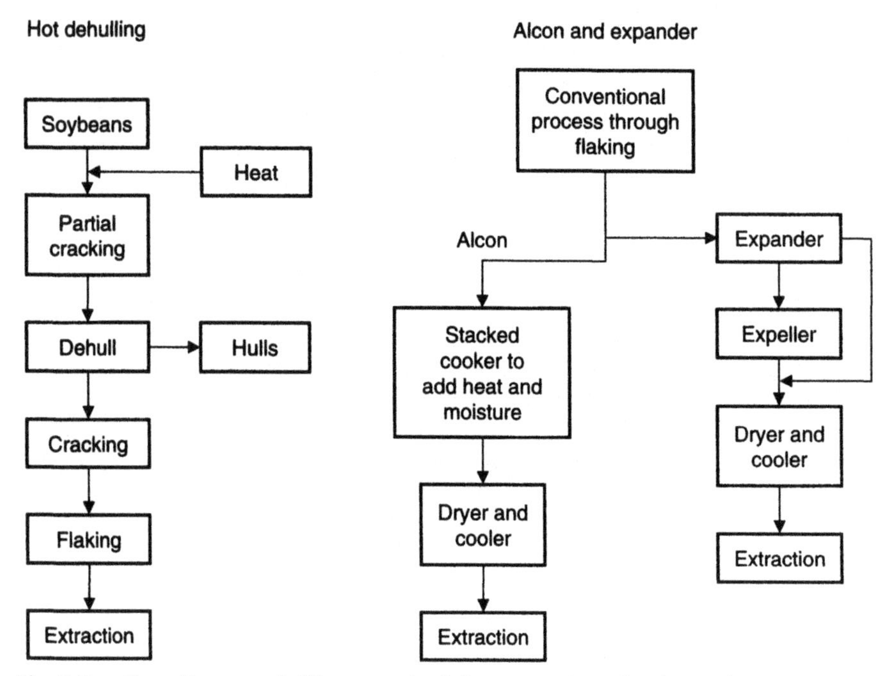

Fig. 5.3. Flow diagrams of different methods for preparation of soybeans for extraction.

TABLE 5.1 Composition of Crude Soybean Oils as Affected by Extraction Processes

	Crude composition		
Process variation	Phosphatides	FFA	NOL
Conventional extracton	2.0–3.0%	0.5–0.8%	2.5–3.8%
Expander	2.5–4.0%	0.8–1.0%	3.5–5.0%
Alcon	4.0–6.0%	1.0–2.0%	5.0–7.0%

[a]From good quality soybeans.

In normal operation, the flakes or collets entering the extractor will be cooled to a temperature below the boiling point of the extraction solvent. The efficiency of extraction increases with temperature, so normal operation is to have the temperature of extraction as high as is safe in the equipment being used.

Desolventizing

The desolventizing of soybean meal and soybean oil is the final step in solvent extraction. The advent of the expander process required some changes in the desolventizing process, because of less residual solvent in the collets from the extractor and more phosphatides in the oil. Use of the Alcon process also required changes in operation of the desolventizing equipment, especially for the much higher gum content in the crude oil.

The desolventizing of the miscella can have an adverse effect on the quality of the extracted oil. If the temperatures in desolventizing are too high (in excess of 115°C), the content of nonhydratable phosphatides may increase (see Chapter 4).

Preparation and Extraction for Soy Protein and Soy Food Production

The preparation and solvent extraction process for production of soy proteins is similar to that used for normal production. The major differences are the amount of cleaning of the soybeans, the need for dedicated equipment, and the desolventizing process. For soy protein production it is mandatory that the soybeans be thoroughly cleaned of soil and trash before preparation and that the preparation and extraction equipment is segregated from normal production. It would be a mistake to use the same equipment for both types of processing. The requirement for clean soybeans and equipment is to reduce the microbiological level as low as possible.

For production of textured products, some concentrates, and isolates, it is necessary to start with defatted soybean meals with higher protein solubilities (see Chapter 8). To produce "white flakes" with higher protein solubilities, a vapor desolventizer or flash desolventizer is used (see Chapter 7). The crude oil from the extraction processes used for production of soy protein products will be of better quality than that from normal extraction processes. Selection of soybeans for high protein solubilities will preclude use of damaged or aged soybeans and ensure good oil and lecithin quality. The extra cleaning will also reduce fines in the crude oil and hexane-insolubles in the lecithin.

For production of soy foods, the foregoing comments on the cleanliness of soybeans and processing equipment are equally applicable (see Chapter 22).

Usually, soybeans destined for use in soy foods are specially selected and marketed as identity-preserved as to either variety or growing area. In other cases, selection is made by removing a specially cleaned or sized fraction from bulk-purchased No. 1 or No. 2 soybeans. Work continues for identifying the compositional factors in soybeans that influence quality and yields of the various soyfoods prepared from them.

Crude Oil Quality

As just mentioned, the effect of the unit processes in solvent extraction have an effect on crude soybean oil quality, as shown in Table 11.1. In addition to these effects there are other factors involving crude oil storage, handling, and transport, as discussed in Chapter 9. NOPA trading rules for crude soybean oil are as follows (1):

1. Not more than 0.5% moisture and volatile matter
2. Color when green lighter than Standard B
3. Refined and bleached color not darker than 6.0 red
4. Neutral oil loss (NOL) not exceeding 7.5%
5. Not more than 1.5% unsaponifiable matter (exclusive of moisture and insoluble impurities)
6. A flash point not lower than 250°F

These NOPA specifications are not sufficiently detailed to give guidance to a refiner. This means that a refiner will have to develop a more detailed specification for purchase, which would include other tests such as FFA, insoluble impurities, calcium, and magnesium (see Chapter 24).

Degumming/Lecithin Production

The processes for degumming and lecithin production are described in Chapter 10. The increases in NHPs shown in Table 11.1 gives direction to the processor in determining the correct dosage of hydration water, whereas Ca and Mg determination would allow estimation of the nonhydratable phosphatide content and acid dosage.

The NOPA specifications for crude degummed soybean oil are more comprehensive than those for crude soybean oil. The analytical requirements are shown in Table 5.2. In addition to the NOPA specification for crude degummed soybean oil, a refiner might add additional analytical requirements for Ca and Mg content, NOL, bleachability, and other characteristics. Because phosphatide removal is of primary importance in soybean oil refining, a knowledge of the NHP content, as correlated with the Ca and Mg content, is particulary helpful to a refiner (see Chapters 11 and 12) for determining proper acid pretreatment dosage.

Lecithin quality depends on the crude oil quality and its content of fines, FFA, NHP, etc. If the crude oil has a low P content, a high proportion of NHPs, or both, then lecithin quality will be low.

TABLE 5.2 NOPA Specification for Crude Degummed Soybean Oil

	Maximum	Minimum
Unsaponifiable matter	1.5 %	
Free fatty acids	0.75%	
Moisture, volatile matter, and insoluble impurities	0.3 %	
Flash point		250°F (121°C)
Phosphorus	0.02%	

Neutralization

The soybean oil supply to the neutralization or caustic refining process can be crude oil, crude degummed soybean oil, or a mixture. The long-mix caustic refining system is recommended for soybean oil, as discussed in Chapter 11 and depicted in Fig. 11.2. Proper employment of a long-mix system will give minimum contents of phosphatides and soaps in the caustic-refined and water-washed soybean oil while keeping neutral oil losses also at a minimum. Phosphatide removal is more important for quality than is neutralization of FFA in soybean oil, allowing use of lower dosages of caustic as described by Charpentier (3). Less caustic use lowers neutral oil loss and lowers the cost of caustic and amount of soapstock to be handled. With use of two water washes, it is possible to produce a once-refined soybean oil with low P (1 to 2 ppm) and low soap content (<5 ppm), and if citric acid is added to the wash water, both P and soap values will be essentially nil.

Bleaching/Adsorption Treatment

Chapter 12 covers bleaching and adsorption theory and practice and shows the importance of a once-refined soybean oil's quality on the effectiveness and cost of bleaching. Any P or soap present will increase usage and cost of expensive bleaching earth and loss of neutral oil in proportion to the earth used. These costs are significant, as shown by the example in Chapter 12. The example reinforces the importance of having a clean feed stock for bleaching and the futility of depending on the use of extra bleaching to overcome deficiencies in the caustic refining process.

Properly bleached soybean oil should be further processed immediately or protected from oxidation (peroxide development). Bleached oil is very susceptible to oxidation, and storage or transport of it is not recommended.

If bleached oil becomes oxidized (peroxide value >1.0), then its exposure to the high temperatures of hydrogenation or deodorization will have similar results as an air leak in the equipment, causing fouling of equipment oil contact surfaces and reduce oxidative stability of the finished products. Similar undesirable effects will occur if any bleaching clay escapes the filtration process and remains in the bleached oil. All possible precautions must be taken to avoid this carryover.

Hydrogenation

Properly caustic-refined and bleached soybean oil provides a hydrogenation feedstock virtually devoid of catalyst poisons. The presence of soaps, bleaching earth, or phosphatides will interfere with both activity and control of selectivity and may require use of excess catalyst and extra time for hydrogenation, both of which increase costs. The importance of this has been demonstrated by Charpentier (3), who was able to reduce catalyst usage by nearly 60% and reduced postbleach of the hydrogenated oils (see Chapter 13).

In addition to the foregoing advantages of a clean feedstock, there is the additional advantage of much-reduced fouling of oil contact surfaces, leading to fewer cleanings of the hydrogenation equipment at longer intervals.

Deodorization

Deodorization is the last step in processing except for crystallization, tempering, and packaging. All of the foregoing statements relating to the importance of a clean feedstock apply equally to the deodorization step. Deficiencies in any of the previous unit processes cannot be overcome by deodorization. This is particularly true for oils with a peroxide value >2.0. Practically all deodorized oil products come out of the deodorizer at a peroxide value of zero and acceptable initial flavor, but for oils with a high initial peroxide value this is deceiving because such oils may exhibit a quicker deterioration of flavor or shorter shelf life than oils that go to the deodorizer with a peroxide value near zero.

For other details of the deodorization process, see Chapter 14. A final caution in deodorization is the caveat, "If it isn't sold, don't deodorize it." Deodorized oil is at its peak of quality as it exits the deodorizer, and quality subsequently decreases over time.

References

1. Wiedermann, L.H., *J. Amer. Oil Chem. Soc. 58:* 159 (1981).
2. *Yearbook and Trading Rules*, National Oilseed Processors Association, Washington, DC, 1993–1994.
3. Charpentier, R., *INFORM* 2: 208 (1991).
4. Erickson, D.R., and M.D. Erickson, in *Proceedings from the 4th Latin American Congress on Fats and Oils Processing*, Associación Argentina de Grasa y Aceites (ASAGA), Buenos Aires, Argentina, November, 1992.
5. Kock, M.J., *J. Amer. Oil Chem. Soc. 60:* 210 (1983).

Chapter 6

Extraction

John B. Woerfel

Consultant
Tucson, AZ

Introduction

Solvent extraction using hexane is the primary method for recovery of oil from soybeans, although some mechanical extraction is still used (see the section on "Mechanical Extraction"). The history of solvent extraction was reviewed by Langhurst (1), who detailed the process and equipment and its development in parallel with the early expansion of U.S. soybean production in the 1930s and 1940s.

Process improvement in extraction plants has continued with increasing emphasis on energy efficiency, cost reduction, reducing hexane loss, quality of meal and oil, and increased capacity, driven by a sixfold increase in soybean production since 1950 and by an increase in energy costs since 1970.

The conventional soybean extraction process that is in widespread use is illustrated by the flow diagram in Fig. 6.1 (2). The process consists of four steps; preparation, extraction, solvent recovery from miscella and desolventizing/toasting of meal.

Plant Location and Layout

Selection of a plant location is of primary economic importance. A number of factors must be considered based on current as well as projected future conditions.

Continuous operation at optimum capacity is both the safest and the most cost-efficient mode of operation. This requires a dependable supply of soybeans and effective distribution of meal and oil.

Inventories for the extraction industry are primarily as soybeans. U.S. soybeans are harvested in a period of about two months and may be stored for a year or more at various facilities in the distribution system (see Chapter 4). For example, an extraction plant located in a growing region may have elevator capacity for several months' operation at the plant site and additional capacity at nearby elevators. Soybean storage capacities at overseas plants will be based on the logistics of world trade and ocean shipping.

The market for products is primarily tied to the need for edible oil, as human food, and meal, as animal feed.

Bulk storage of meal presents problems of caking, and oil tends to deteriorate in quality. Extraction plants prefer to ship products as produced and not to build up inventories of finished products. The amount of product storage required will depend on the distribution pattern of meal to feed manufacturers, crude oil to refiners, and refined oil products to food processors.

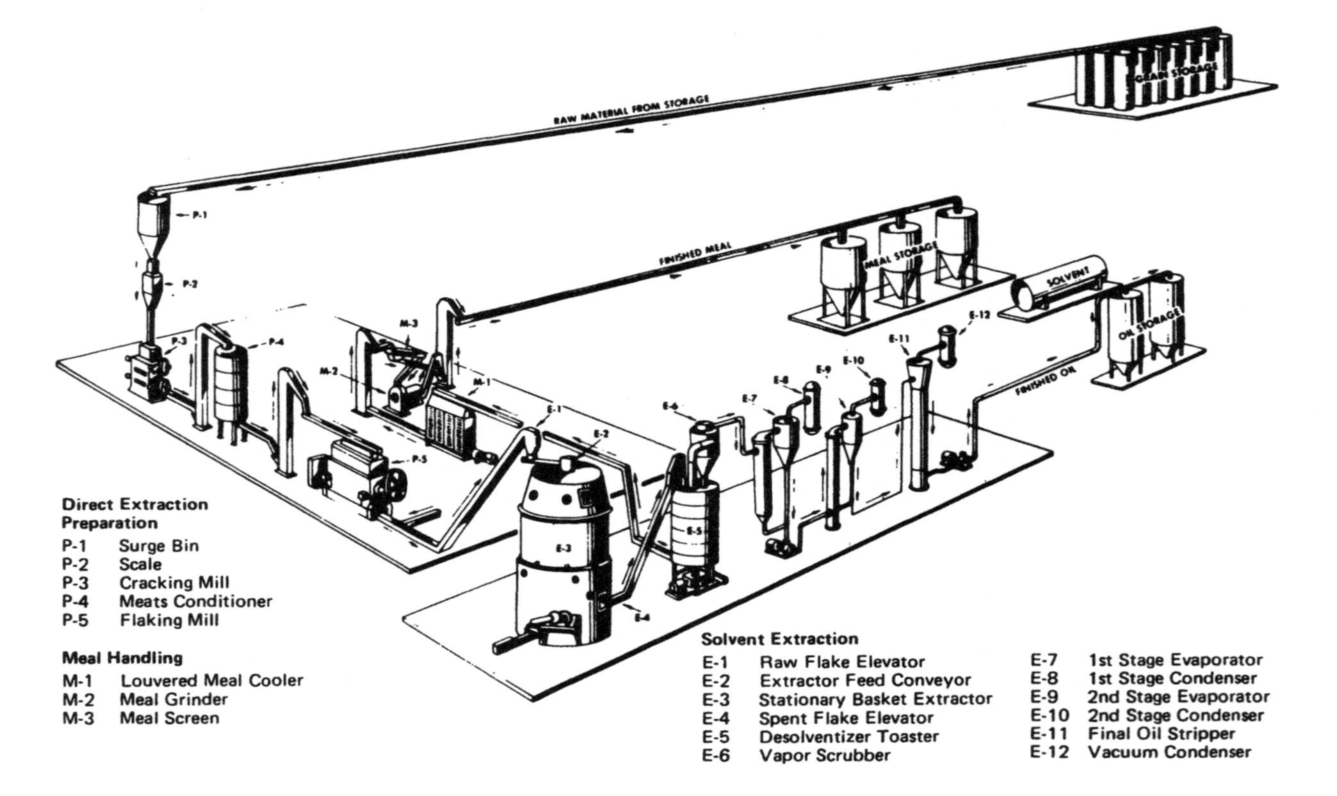

Fig. 6.1. Flowchart of a soybean extractor plant. *Source:* Courtesy of French Oil Mill Machinery Co., Piqua, OH.

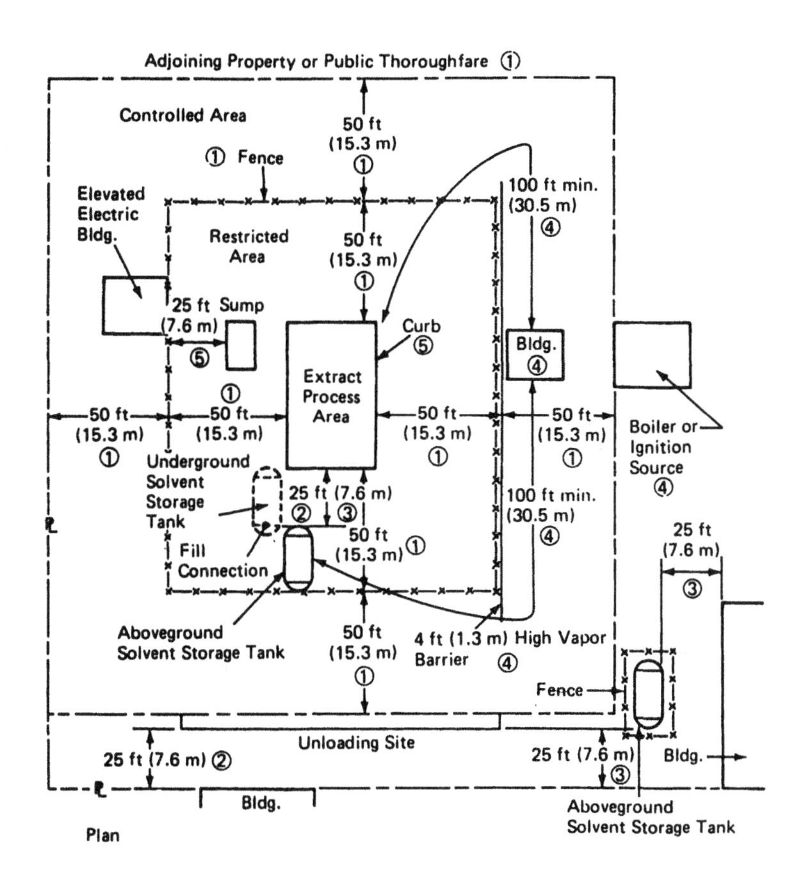

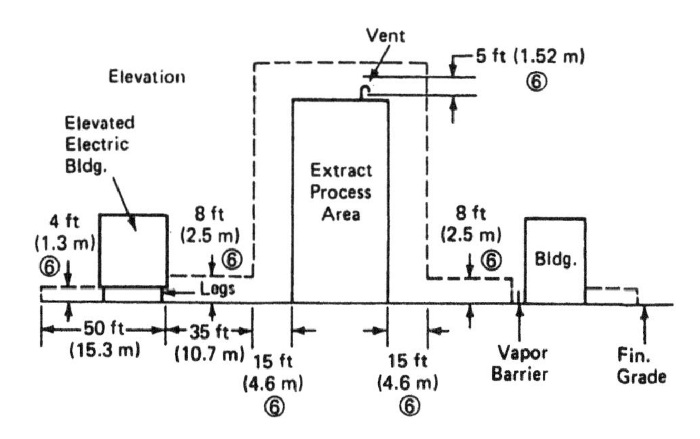

Fig. 6.2. Distance diagram for extraction process. Reprinted with permission from NFPA 36, *Solvent Extraction Plants*, Copyright © 1993, National Fire Protection Association, Quincy, MA, p. 36–11, Fig. 5.1. This reprinted material is not the complete and official position of the National Fire Protection Association on the referenced subject, which is represented only by the standard in its entirety.

Transportation is also a key point in plant location. Truck or rail transportation is necessary at all extraction plants. Suitable unloading, loading, weighing, and sampling facilities must be provided. Plants that are located in rural areas and receive soybeans directly from farmers, must provide traffic access for large numbers of trucks during the peak harvest period. Location on a navigable river is a significant advantage. This provides barge access to export markets for meal or oil. Also, soybeans may be delivered by barge from other producing regions, or imported soybeans may be shipped inland from ocean ports. Investment costs for docking, loading and unloading facilities, and additional storage may be substantial.

Utilities must be adequate, reliable, and competitive in cost because they represent the largest single cost factor in extraction. Natural gas, fuel oil, coal, or alternative fuels are necessary for drying and steam production. Where available, natural gas is the most convenient and cleanest-burning fuel. However, a variety of fuels and technologies has developed in recent years and provides some options. The optimum choice may be specific for a particular site. Examples of alternative fuels being used are a wood waste–burning boiler in a U.S. plant and a fluid-bed boiler, of U.S. design, that burns either lignite or rice hulls in a plant in India.

Electrical service must be reliable, because any interruption necessitates emergency shutdown of an extraction plant for safety reasons. Where electric service interruptions are frequent or likely, a standby generator or completely independent generation system should be considered.

Cogeneration, producing both steam and electricity, is a viable option. A natural gas turbine, running an electric generator, with the exhaust heating a boiler for process steam, has been used. Another option is to generate high-pressure steam to drive a steam turbine generator and use the lower-pressure exhaust steam for processing. In the United States, surplus cogenerated electricity is sold to the local electric utility company or power grid.

An adequate supply of cooling water is necessary. This may be provided by atmospheric cooling towers, spray ponds, or deep wells. Sea water has been used in some plants.

Treatment of waste water is required to a degree dictated by local conditions. Some plants in U.S. rural areas have complete tertiary treatment systems, including biological treatment in spray ponds or by an activated-sludge process (see Chapter 25). Other environmental concerns, which must be addressed in site selection and development as well as in plant design and operation, include hydrocarbon emissions, polluted storm water runoff, dust generation, accidental discharge of soybean oil and materials such as hexane, power plant emissions, and solid waste.

Soybean storage, fuel storage, crude oil tanks, railroad tracks, roads, power house, and waste treatment may require areas much greater than the actual extraction process equipment. Additionally, distance separation of the flammable solvent–processing area is essential for safety.

National Fire Protection Association (NFPA) 36, *Solvent Extraction Plants* (3) (Fig. 6.2) specifies a restricted area of 15.3 m (50 ft) plus an additional controlled zone of 15.3 m (50 ft) around the extraction process area. This standard covers many other factors of safety relating to construction and location of the plant, including the requirement that the extraction process must be 30.6 m (100 ft) from the property line or any ignition source, such as a boiler. These space requirements are minimal, and for practical reasons extra space, for possible expansion or as a buffer from neighboring industrial or residential areas, is highly desirable. In a rural area, space is usually easily available, but access to railroad or truck transportation may require construction of railroad spurs or roads. In contrast, space may be limited and expensive at a deep-water port or a main rail center.

If conventional refining is planned in association with the extraction plant, space should be provided adjacent to, but outside, the restricted/controlled area, because the two processes are not compatible, both for safety and production management reasons. In some instances, refining facilities, including chemical refining, bleaching, and deodorization, have been constructed within the restricted/controlled area of an extraction plant. There are several disadvantages to this practice, compared with the alternative of constructing refining facilities outside the controlled area. Construction costs for conventional refinery processes within the controlled area are greater because of the need to conform to the more restrictive standard for extraction plants, especially electrical codes. Refinery repairs and maintenance practices are complicated by the need to conform with extraction plant practices, such as purging, or doing needed electrical and other repairs when the extraction plant is operating or unpurged.

Spent bleaching clay can undergo spontaneous heating and combustion, so it must be handled with extreme care, and deodorizers and auxiliary high-temperature boilers employ temperatures in excess of 270°C (520°F).

Supplying fully refined bulk oils to food processors requires reliable and frequent shipments, usually based on a weekly cycle so that deodorized oil inventories are kept at a minimum to ensure shipment of fresh oil. Interruption of refinery production in the extraction plant because of unexpected or extended shutdowns can seriously affect customers' production schedules.

Miscella refining and degumming, appropriately integrated with the extraction process, become inherent parts of the extraction process, both by design and operation.

Preparation

Conventional preparation consists of scalping, drying, tempering, cleaning, cracking, optional dehulling, conditioning, and flaking as shown in Fig. 6.3 (4).

Drying

Drying has been discussed in Chapter 4 in connection with storage and handling. Similar dryers are used for process drying, and in practice the same dryers may be used for both purposes, depending on the needs of the plant.

Soybeans shipped in international trade or soybeans from storage elevators will typically have a moisture content of 13%, because they would have been dried to this level for protection against heating in storage or shipment. Further drying before cracking depends on their specific moisture content and whether or not they will be dehulled.

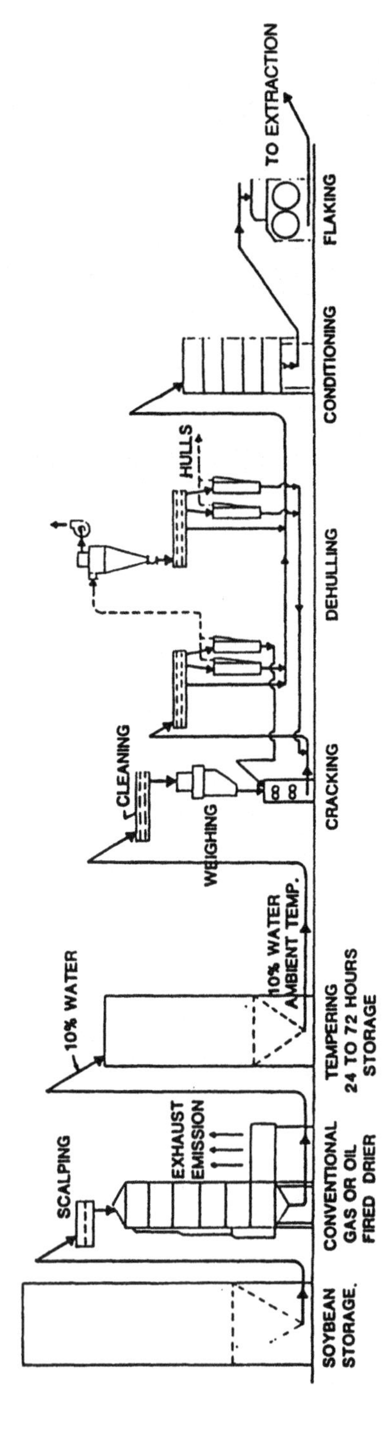

Fig. 6.3. Conventional preparation system. *Source:* Moore, N.H., *J. Amer. Oil Chem. Soc. 60:* 190, (1983).

For example, soybeans at 12% that are not to be dehulled might be processed without further drying; but, if they are to be dehulled, they will be dried to 10%.

Drying is energy-intensive. Reasonably efficient dryers will require 830 to 890 cal/kg (1500 to 1600 Btu/lb) of moisture removed. Therefore, it is imperative that the moisture content be considered along with other operating parameters to optimize overall efficiency and cost. After drying, soybeans are tempered for 24 to 72 h to allow moisture to equilibrate and the hulls to loosen to facilitate dehulling.

Cleaning

Soybeans are cleaned again after drying. Even though the beans may have been scalped before storage, additional cleaning is desirable. Cleaning is important for protection of equipment and for production of high-quality meal.

Soybeans are first passed over a magnetic separator as they are fed to a two-deck screen and aspirator as shown in Fig. 6.4 (4). Cleaned beans are weighed into further processing, usually with automatic hopper scales, which provide a means to control the rate of feed and the total amount of raw material for accounting purposes. The dried, cleaned beans are then ready for cracking.

Cracking

The cracking step is very important to traditional dehulling. With properly dried and cracked soybeans, the cotyledons should separate easily from the hulls. If the soybeans are too dry, they will pulverize in the cracking rolls and produce excessive fines, which could cause problems in the extractor. If too soft, or moist, they tend to mash and not release the hulls.

Proper operation and maintenance of rolls are critical. Conventional cracking rolls may be two or three high. A common size in the United States is a 25 cm (10 in) diameter roll that is 1.07 m (42 in) long. Rolls are corrugated in a sawtooth pattern, and opposing rolls run at different speeds, with corrugations arranged "sharp to sharp." Practice varies in the speed of the rolls and the size of corrugations. For example, the corrugations might be 2.4 mm (0.1 in), 3.6 mm (0.14 in) and 5 mm (0.2 in) from top to bottom for a 3-high stand, with six stands being needed for 1500 MT (1650 short tons)/day. More recently, there has been a trend for higher-capacity cracking mills in the 500 to 600 MT/day range. This is accomplished either by using larger-diameter rolls or by running the 25-cm (10-in) roll at a higher speed and using larger corrugations.

The objective of cracking is to break the soybean into suitable pieces for dehulling and flaking. There should be a minimum of fines and no mashed beans. The size of the flake will be in proportion to the size of the piece. Examination of a sample of the cracked bean shows the effectiveness of the crack. A screen test provides an objective measure. Table 6.1 shows a typical specification. The precise crack desired may vary from plant to plant. Typically, the bean will be broken into four to six pieces. The preparation operator must be careful to maintain proper adjustment of the feed and proper spacing of the rolls based on personal observations, and screen testing, of the cracks.

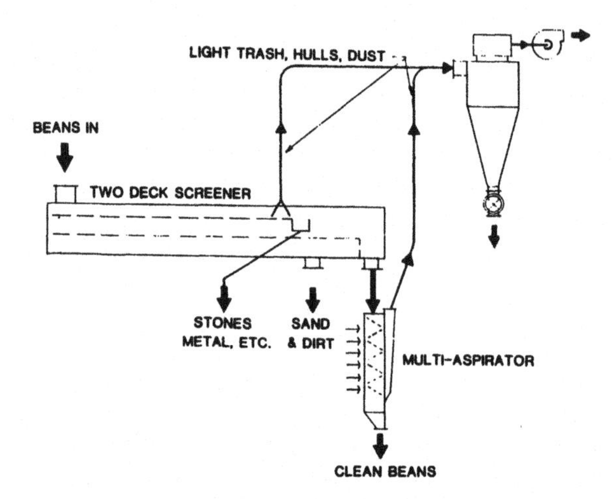

Fig. 6.4. Cleaner with multiaspirator. *Source:* Moore, N.H., *J. Amer. Oil Chem. Soc. 60*: 189, (1983).

A good maintenance program is essential. First, the beans should be well cleaned of any stones or other hard material, which will damage the corrugations. Magnets should be placed in the system just ahead of the rolls to remove any ferrous metal (tramp iron). The feed mechanism must provide a uniform feed across the entire width of the roll at all times so that the corrugations wear evenly. Cracks should be monitored daily and the rolls inspected frequently by the process and maintenance supervisors and a program established for change-out of the rolls and recutting of their corrugations. The objective is to keep the cracking operation at peak efficiency. Dehulling is not satisfactory without good cracking.

The cracked soybeans may go directly to conditioning, flaking, and extraction if 44% protein meal is to be produced. For high-protein meal the hulls are separated from the meats before conditioning.

Dehulling

Dehulling of soybeans is done to produce high-protein meal for animal feed or flour for human consumption. In particular, the poultry industry demands high-protein meal of high and consistent quality. For other feeds, low-protein meal is satisfactory. Meal that does not meet standards agreed on by shipper and buyer may be heavily discounted or rejected. State agricultural departments test feed ingredients and may fine the manufacturer if the analysis does not meet the guarantees that have

TABLE 6.1 Typical Screen Analysis Specification for Cracked Soybeans

On U.S. No. 6	30–40%
On U.S. No. 10	40–50%
Through No. 10	10% max.

been registered. Standards established by the National Oilseed Products Association (NOPA) are shown in Table 6.2 and are commonly used in trade.

Soybeans contain about 8% of hulls by weight. As a practical matter, it is not possible to get an absolute separation of hulls from the meats (cotyledons), but the objective is to get sufficient separation to meet the standard meal analyses required. Generally, if the fiber content is reduced to the desired value, the protein content will meet specification. There are exceptions to this, and at times the inherent protein content of the bean may be so low that it is difficult to produce a high protein level in the meal.

From an economic standpoint it is desirable that little of the meats be lost in the hulls. Obviously, small meat particles may adhere to hulls and be carried with them. This factor is monitored by analysis of the hulls. Since hulls contain almost no oil, the oil content of hulls is used as a dehulling control and optimally will be below 1.5%.

Separation of hulls is commonly done by screening and aspiration. A Rotex screen (Rotex Inc., Cincinnati, OH) with two decks, fitted with 4-mesh and 10-mesh screens is typical. Hulls are aspirated from the top screen. Discharge from the end of the 10-mesh screen goes to an aspirator for classification. The heavy material from the aspirator and the fines that went through the 10-mesh are mostly meats and go into a conveyor to the conditioner. The material aspirated from the top screen and from the multiaspirator is collected in a cyclone and discharged to a secondary dehulling system. The secondary dehulling may be a two-deck Rotex screen and aspirator similar to the first stage. In some cases other types of equipment, called "hull beaters", are used before the second Rotex screen.

Extraction is similar whether or not soybeans are dehulled. Dehulling increases the capacity of the extraction plant, because the hulls do not have to go through the extractor. The plant throughput can be increased by 10% or more by dehulling.

Some plants follow the practice of dehulling even when producing low-protein meal. The hulls are ground separately and added to the ground high-protein meal as it is being loaded, to adjust the protein content. Other mills bypass the dehulling when making low-protein meal. The decision as to which practice is best should be evaluated on the following factors:

Advantages of dehulling:	Increased extractor capacity
	Lower residual oil
Advantages of not dehulling:	Decreased energy for drying
	Decreased electricity for dehulling equipment
	Reduced wear and maintenance on dehulling equipment

TABLE 6.2 NOPA Standard Specifications for Soybean Flakes or Soybean Meal

	44% Protein	High-Protein
Protein, minimum	44.0%	47.5–49%*
Fat, minimum	0.5%	0.5%
Fibre maximum	7.0%	3.3–3.5%
Moisture	12.0%	12.0%

*As determined by buyer and seller at time of sale.

The installation of dehulling in a plant will affect the requirements for grinding. Meal from dehulled soybeans is generally easier to grind and takes less power than meal from nondehulled soybeans. Hulls are quite difficult to grind and require proportionately larger and higher-power grinders than meal does.

The systems described are typical of current dehulling practices in the United States. There is a considerable variation in detail in various plants, but the principles are similar. This is usually referred to as *front-end dehulling*.

An alternative system, called *tail-end dehulling*, is practiced in a few plants. In tail-end dehulling the separation of the hulls is done on the ground, extracted meal. This separation is done by a combination of sifting and separation of low-protein and high-protein fractions by gravity tables or aspirators. The disadvantage of tail-end dehulling in the United States is that only a portion of the production is high-protein, and a market for the lower-protein fraction must be found. One U.S. plant doing tail-end dehulling obtained about 30% high-protein and 70% low-protein meal. This plant was favorably located so that the 44% meal could be exported, while the high-protein meal could be sold to nearby poultry feeders.

Hot Dehulling. In recent years an innovation referred to as *hot dehulling* has been introduced, and variations are offered by several equipment suppliers. This process differs from conventional preparation in that the hulls are removed from split soybeans before cracking and flaking.

Cleaned whole soybeans with normal storage moisture are gently heated to approximately 60°C (140°F) in a period of 20 to 30 min, which allows moisture to migrate to the surface. They are then rapidly heated and dried to loosen the hulls. Surface temperatures may be as high as 85°C (185°F) and moisture reduction 1 to 3%.

The soybeans are then split into halves by corrugated rolls, and the hulls are released from the splits by mechanical impact or friction. Separated hulls are removed by aspiration. The hulls may be screened to recover fines, which are sent with the cracked soybeans going to the flaking rolls.

The split soybeans are cracked into four to eight pieces and may be *conditioned* before or after cracking. Conditioning consists of cooling and drying using ambient and recycled air to adjust the temperature and moisture for flaking; for example, 60°C (140°F) or 65°C (150°F) and 11% moisture.

Hot dehulling systems are offered by Crown Iron Works, Minneapolis, MN (5); French Oil Mill Machinery, Piqua, OH; Esher Wyss, Ravensburg, Germany; and Buhler, St. Gallen, Switzerland (6). Hot dehulling is more energy-efficient than conventional dehulling. The heat required to initially heat and slightly dry the beans provides heat for the entire process. In the conventional process, heat is needed to dry the soybeans to 10% moisture, and additional heat is needed to increase the temperature in the conditioner.

Conditioning

Cracked soybeans with, or without, hulls are conditioned in the same manner. Conditioners may be vertical stack cookers or rotary horizontal cookers, the latter being used in larger installations. In the conditioner the cracks are heated to approximately 71°C (161°F), and steam, or a water spray, is added to adjust moisture to approximately 11%, which makes the material plastic for flaking. Inadequate conditioning at too low a temperature may result in excessively fragile flakes.

Flaking

Flaking has traditionally been the final step in preparation of soybeans for extraction. Extractor design has developed using flakes as the primary feed.

Flaking mills consist of a pair of smooth-surfaced rolls of large diameter. The minimum diameter is 0.5 m (20 in) to provide an adequate angle of nip (1). Rolls are mounted with no clearance and driven at differential speeds. Typical early mills were 0.5 m (20 in) in diameter and 1.0 m (40 in) long.

Pressure between the rolls is generated by mechanical or hydraulic system, and flake thickness is controlled by adjustment of this pressure.

A feeder on each pair of rolls provides an even flow of cracked soybeans across the width of the roll to ensure uniform loading and wear. One system (4) permits overflow of cracks at the roll end to ensure even feeding to the very edges of the roll. Magnetic separators are provided with each roll to prevent damage from loose iron.

In recent years, the trend has been toward larger-capacity rolls. The additional capacity is provided by increasing the diameter, length, and peripheral speed. Larger-capacity rolls save space by reducing the number of rolls required. Currently, rolls with diameters of 60 to 80 cm (24 to 32 in) and lengths of 150 to 200 cm (60 to 80 in) are typical.

Flake thickness is generally in the range of 0.2 mm (0.008 in) to 0.5 mm (0.02 in) and, most commonly, from 0.25 mm (0.01 in) to 0.3 mm (0.012 in). Flake size and thickness are controlled by the size of the cracked soybeans, conditioning, and the adjustment of flaking rolls. Thickness is measured with a micrometer, and samples are taken on a regular basis at both ends and center of the roll and as frequently as needed, typically hourly, to obtain reliable control. All measurements should be recorded in an operating log.

A common problem is greater wear at the center than at the roll ends. As wear progresses, flakes in the center become thicker than those at the ends. Continued pressure adjustment to correct the center flake thickness may cause excessive pressure at the ends and chipping or pitting of the metal.

Flaking mills are supplied with roll grinding systems so that they can be evenly ground to compensate for wear without removal of the rolls.

Expanders

Expanders, also called extruders or Enhanser Press™, were originally developed in Brazil and were introduced in the U.S. soybean and cottonseed industries in the early 1980s. By 1989, 70% of the domestic tonnage of these crops was being processed by expanders (7).

The expander (Fig. 6.5) is added to the conventional preparation process after flaking and modifies the flakes to form collets, which are fed to the extractor.

Typical preparation conditions using the expander are as follows:

1. Dry soybeans to 10% moisture for cracking.
2. Crack into eight pieces.
3. Remove hulls.
4. Condition at 55 to 82°C (130 to 180°F) to 10 to 11% moisture.
5. Flake to 0.5 mm (0.02 in).
6. Process in expander with steam to exit die plate at 105 to 120°C (220 to 250°F).
7. Cool to 60°C (140°F) for extraction.

Watkins et al. (7) conclude, "The expander offers several advantages in processing oilseeds.

- The seed is finely macerated in the expander, freeing the oil for rapid extraction.
- Pellet-like collets are formed, which are more dense (weigh more per cubic foot).
- The collets are porous and do not restrict ("blind") percolation of solvent through the extractor bed, as may occur with the more fragile flakes.
- Slightly more oil is recovered by solvent extraction per ton of soybeans processed.
- The solvent drains more completely from the extracted collets, resulting in more complete removal of oil *and* use of less energy to desolventize collets compared to flakes.

All factors taken together, an extraction plant can expect to nearly double the throughput of its hexane extractor, to reduce the extraction costs per ton of soybeans processed, and to pay for purchase and installation of the expander in less than a year."

Alcon Process

Another practice in preparation of soybeans for extraction is the Alcon process, which was developed by Lurgi GmbH. of Germany (8). This process consists of an additional stage between the conventional preparation and extraction, aiming at eliminating enzyme activity by intensive moisture/temperature/time treatment. Added equipment, consisting of a conditioner, tempering equipment, and dryer-cooler, is shown in Fig. 6.6. Table 6.3 lists advantages and disadvantages claimed for the Alcon process. In particular, it has been reported that the water-degummed oil is low in phosphatides and, therefore, desirable for physical refining. However, physically refined soybean oil has not been acceptable in all markets (see Chapter 11).

Extractors

Solvent extraction of soybeans is a diffusion operation in which the solvent (hexane) selectively dissolves miscible components (oil) from other substances. The extractor

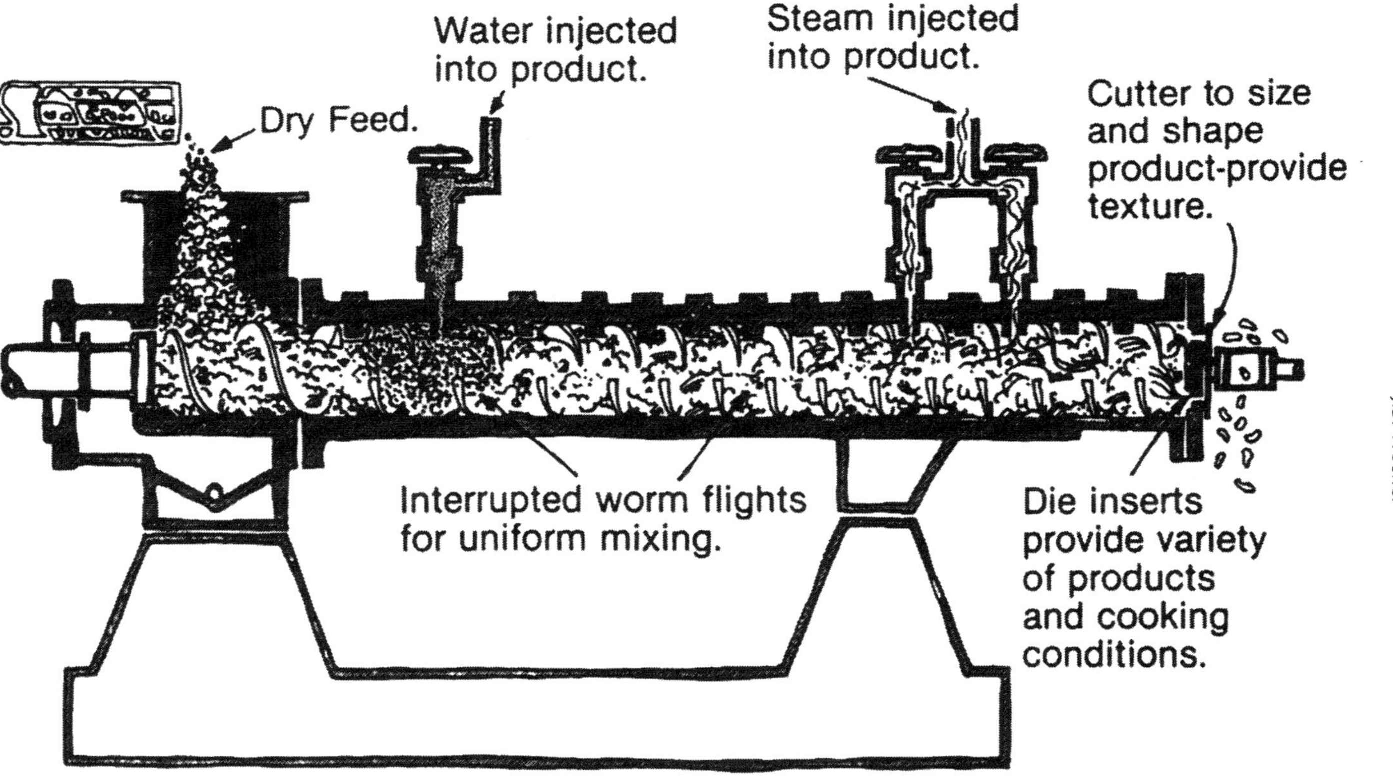

Fig. 6.5. Expander. Collets are discharged through die openings. *Source:* Watkins, L.R., E.W. Lusas, S.S. Koseoglu, S.C. Doty, and W.H. Johnson, "Developing the Potential Usefulness of Expanders in Soybean Processing," *Final Report of Research Sponsored by the American Soybean Association*, Dec. 15, 1989, p. 3.

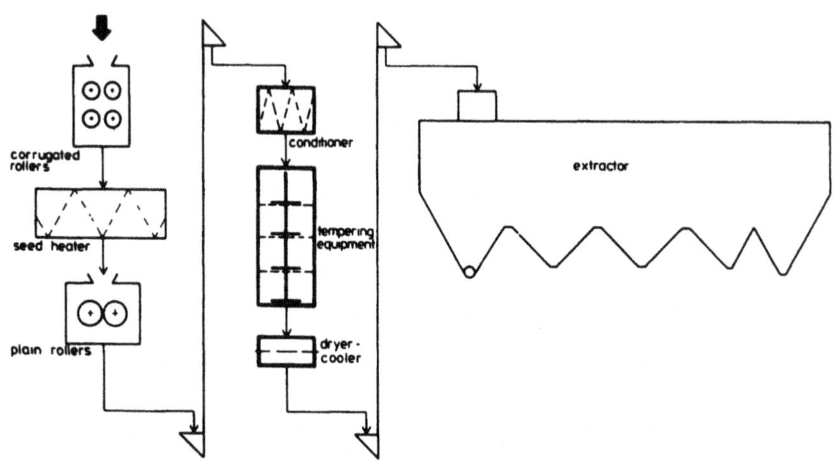

Fig. 6.6. Alcon process. *Source:* Penk, G., in *Proc. World Conf. on Emerging Technologies in the Fats and Oils Industry*, edited by A.R. Baldwin, American Oil Chemists' Society, Champaign, IL, 1985.

provides the physical means of contact between solvent and solids consisting of prepared soybeans. Contact can be achieved by immersion of the solids in the solvent, percolation of the solvent through a bed of solids, or a combination of the two. Although early extractors were primarily immersion types, later development favored percolation, utilizing a bed composed of soybean flakes and countercurrent flow of flakes and miscella (the mixture of oil and hexane). Successful designs have been developed over the years by a combination of theoretical considerations and practical experience. Many designs have come and gone, and those that have been successful have been consistently improved.

Good, in 1970 (9), discussed the theory of soybean extraction and the factors influencing extractor design. Karnofsky, in 1986 (10), derived equations for extractors based on laboratory data. Discussion of theory and subsequent design detail is

TABLE 6.3 Advantages and Disadvantages of the Alcon Process Extraction

Advantages	Disadvantages
Higher bulk density	Lower extractability
Larger extraction throughput capacity	Lower meal quantity
Higher percolation speed	
Lower hexane retention	
Larger lecithin quantity	
Better hydration of phosphatides	
Lower residual phosphatide content in crude oil	
Good meal qualtiy	

Source: Penk, G., in *World Conference on Emerging Technologies in the Fats and Oils Industry*, edited by A.R. Baldwin, American Oil Chemists' Society, Champaign, IL, 1985.

beyond the scope of this book, but general descriptions will be given, along with mention of features to be considered from an operations standpoint. The manufacturer of each type can advise on its design and modifications for a specific application and provide performance guarantees.

Four types of extractors are actively marketed at present.

Rotary or Deep-Bed

These extractors consist of a series of concentrically arranged cells, which are filled with oil-bearing material. Each cell is filled consecutively and brought into contact with decreasingly concentrated miscella, as shown in Fig. 6.7. The freshly charged solids are brought into contact first with the most concentrated miscella, and the fully extracted solids contact fresh solvent before discharge.

Two types of these extractors are built. In the first type the cells rotate under stationary solvent sprays and have a fixed position of loading and discharge. The original extractor of this type is the Rotocel®, sold by the Davy Corporation, Pittsburgh, PA (Fig. 6.8). The French Oil Mill Machinery Co. (Piqua, OH) Stationary Basket Extractor (Fig. 6.9) provides countercurrent extraction without moving cells. Fill spout, spray nozzles, bottom screens, and miscella collection pans rotate on a central shaft. The solids in each cell are washed by successively less concentrated miscella and finally with fresh solvent. The bottom screen has an opening the size of

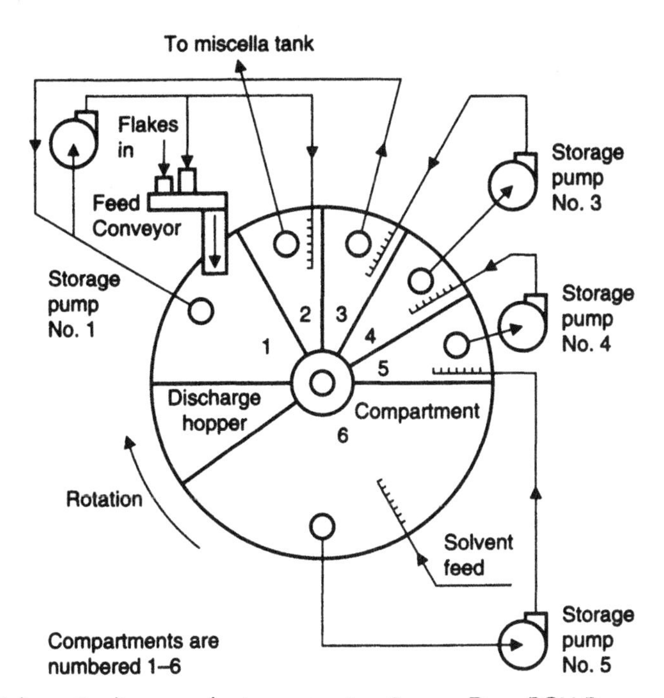

Fig. 6.7. Schematic diagram of rotary extractor. *Source:* Davy PGH Process, Pittsburgh, PA.

each basket, and when this opening rotates beneath the cell, the extracted solids drop into a discharge hopper and are conveyed from the extractor. Other versions of the rotary extractor are the Extraktions Technik "Carousel" and the Krupp (Hamburg, Germany) extractor.

Retention time in rotary extractors depends on rate of rotation and the capacity of each cell. Increased capacity is more economically achieved by increasing the depth of the cells rather than the diameter of the extractor. In larger capacities, depths are as great as 5 m; hence the designation "deep-bed."

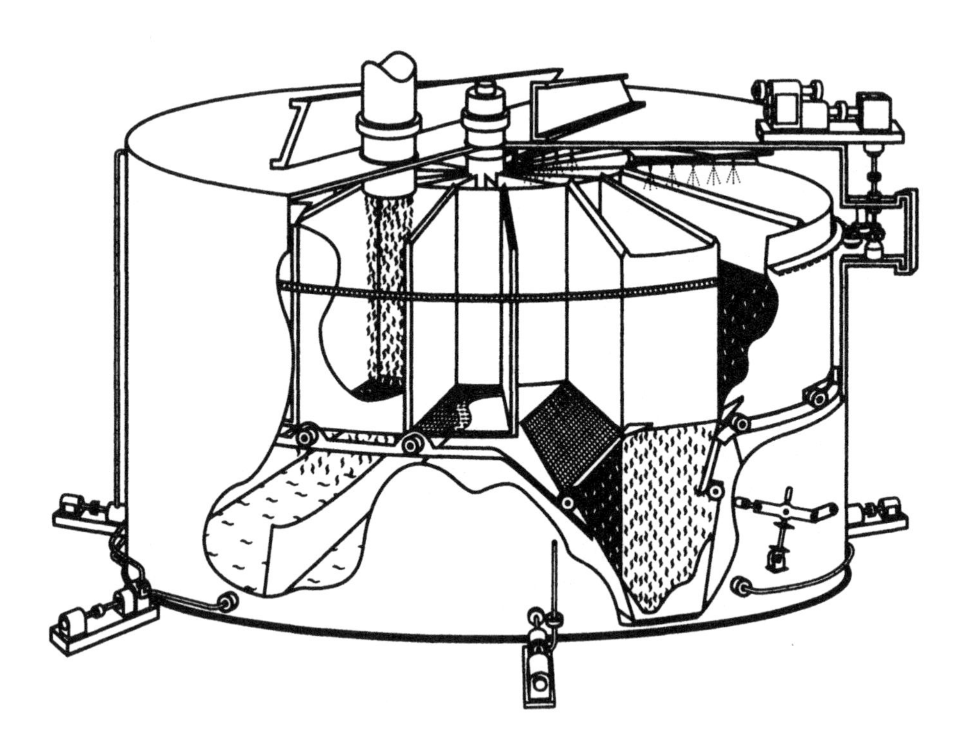

Fig. 6.8. Rotacel® rotary extractor. *Source:* Davy PGH Process, Pittsburgh, PA.

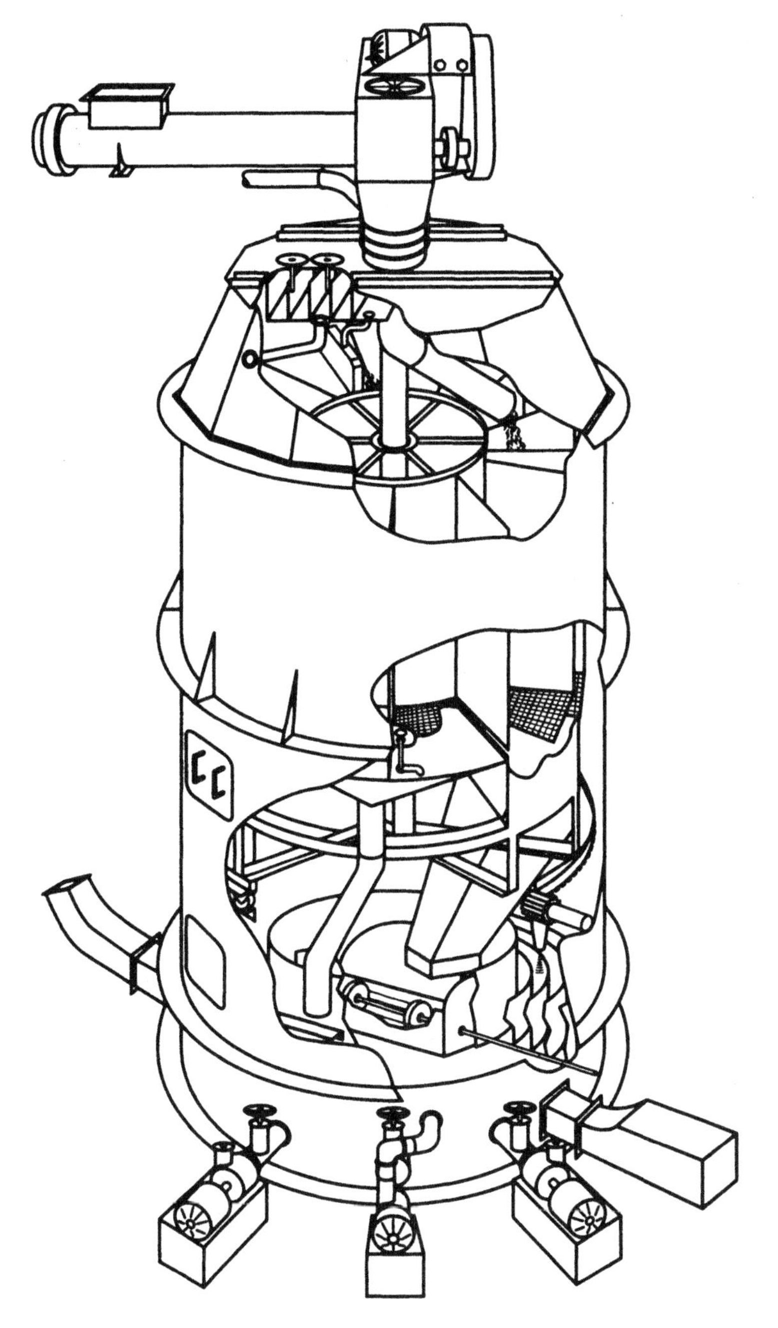

Fig. 6.9. Stationary basket-type solvent extractor. *Source:* Mustakas, G.C., in *Handbook of Soy Oil Processing and Utilization*, edited by D.R. Erickson et al., American Soybean Association and American Oil Chemists' Society, St. Louis, MO, and Champaign, IL, 1980, p. 59. Courtesy of French Oil Mill Machinery Co., Piqua, OH.

Horizontal Belt

The primary supplier of this type of extractor is Extraction De Smet, Antwerp, Belgium (11). This extractor, shown in Fig. 6.10, consists of a belt conveyor supported on rollers in a horizontal rectangular vessel. The oil-bearing material is carried on this belt, which has a slight upward slope, and is washed with a series of successive sprays operating countercurrently. After percolating through the bed, the miscella passes through a fine screen and is collected in a hopper under the conveyor. A pump under each hopper circulates the miscella to sprays installed above the same hopper or the next one. The speed of the conveyor can be varied over a wide range and the spray rate in each section adjusted. The capacity of the extractor depends on the width and depth of the bed as well as on the belt length and speed.

Continuous Loop Extractor

The loop extractor is a shallow-bed extractor in which the solid material is carried through an enclosed vertical loop, with a drag conveyor moving the solids through the loop, as shown in Fig. 6.11. Extractors of this type are made by Crown Iron Works, Minneapolis, MN (12). They are made with capacities of up to 4000 MT of soybeans per day. Feeding and discharge arrangements vary in different models. Extraction sections provide concurrent and countercurrent percolation and immersion. An interesting feature of the loop extractor is that the flake bed is completely turned over, permitting solvent to contact the flakes from both sides. The shallow-bed loop extractor is said to promote excellent contact, rapid draining, and more complete extraction with a wide variety of products and allows utilization of thinner, more fragile flakes with a higher fines content.

Other Extractors

A great variety of extractors is used worldwide, many of which are original types predating 1950, such as the Hildebrandt vertical U-tube immersion extractor and a vertical endless-chain basket extractor with both concurrent and countercurrent sections. Batch extractors are also still in use. In a batch extractor the entire sequence of multistage extraction and desolventizing can be carried out in the batch tanks.

One interesting adaptation is a semicontinuous plant using eight 3-MT batch extractors installed in a row and operated sequentially. Soybeans are prepared by cooking in a stacked cooker and prepressing to 8 to 10% oil in a screw press. The press cake is charged to each batch vessel, in turn, by an overhead conveyor, and the desolventized meal is dropped out on a conveyor that transfers it to a dryer. By appropriate piping and valves, solvent and miscella flow is countercurrent with the sequence of batches changed on a timed schedule. Solvent is recovered from the full miscella by a continuous multiple-effect evaporator.

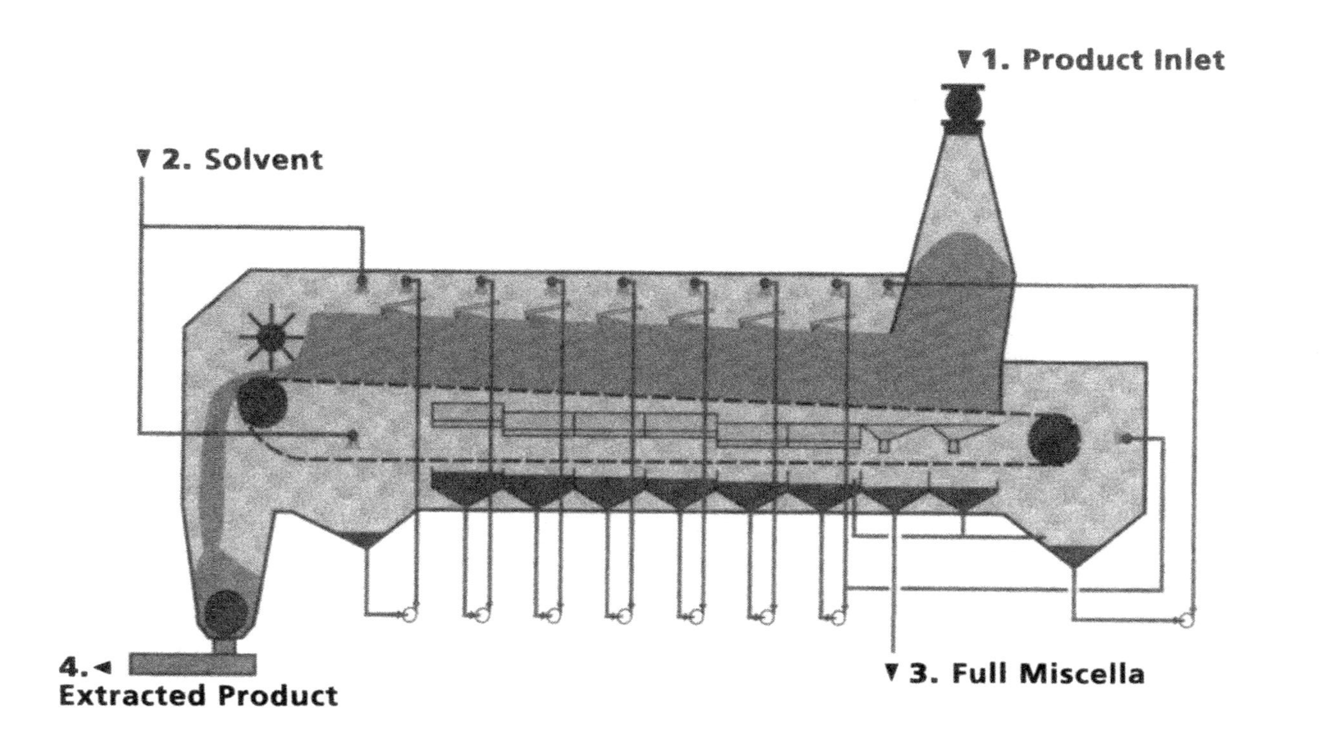

Fig. 6.10. De Smet belt extractor. *Source:* "Solvent Extraction Systems," Extraction De Smet, Edegem, Belgium, (1992).

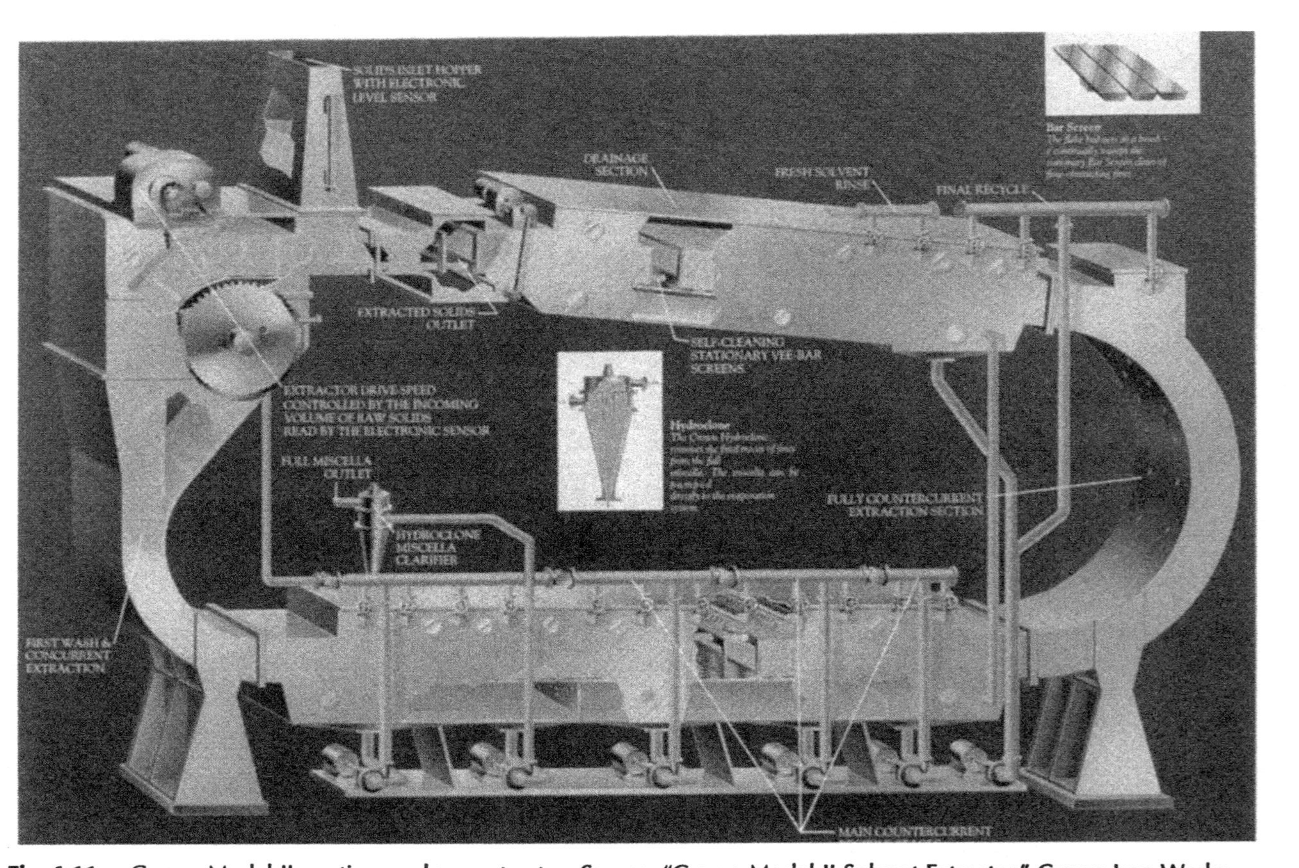

Fig. 6.11. Crown Model II continuous loop extractor. *Source:* "Crown Model II Solvent Extractor," Crown Iron Works Company, Minneapolis, MN.

Solvent

"Hexane" is the common name of the solvent used commercially in the extraction of soybeans. It is not pure *n*-hexane but a petroleum fraction consisting primarily of a mixture of 6-carbon-atom saturated hydrocarbons. The composition of extraction-grade "hexane" from different suppliers will vary. Typically the largest component is *n*-hexane, which can range from less than 50% to more than 90% by volume. Isohexane and methylcyclopentane may be present in appreciable amounts. Boiling points are carefully controlled, and extraction-grade "hexanes" in the United States fall within the limits of initial boiling point 65°C (149°F) minimum and dry point 70°C (158°F) maximum. Product from a single source may have a much smaller range. Table 6.4 shows specifications and analysis of hexane from a U.S. supplier. This lists the test methods and type of information commonly furnished by suppliers.

Sulfur is undesirable, because it may create problems with odors or with refining and hydrogenation of extracted oil. Sulfur is determined by analysis and by the Doctor test, which is a rapid test that was frequently used in the past to check each load of "hexane" delivered. This practice is no longer common in the United States, because improved sulfur removal practices at petroleum refineries are highly reliable. Benzene and other aromatic compounds are undesirable, because of concerns with toxicity, and are kept at low levels.

Lusas and coworkers (13) have reviewed possibilities for alternative extraction solvents in detail. They identify desirable characteristics for a solvent as plentiful supply, low toxicity, nonflammability, high solvency power, ease of separation from extracted material, desirable boiling point, low specific heat, low latent heat of vaporization, and high stability. Based on this, the principal disadvantages of "hexane" are flammability and dependence on petroleum supply. Both of these problems offer continuing challenges in plant design and operation. Meticulous safety management has proved effective, and required practices have been well documented (14,15,16,17,18). Solvent losses have been reduced significantly in the past 20 years and further reduction is a reasonable goal. A savings of 0.1 gallon (0.3 kg) per ton in a 1000-MT plant represents a savings of 100 gallons daily.

It is of interest to note that none of the alternatives meet all of the criteria for an ideal solvent. The organic solvents, with the exception of halogenated hydrocarbons, are flammable, whereas the halogenated hydrocarbons are generally toxic. Aqueous, or supercritical carbon dioxide, extraction if shown to be economically feasible, would not be adaptable to existing plants and would require large capital expenditures for new facilities.

Solvent Recovery

Solvent must be recovered and recycled with a minimum of loss. Solvent leaves the extractor as full miscella and as solvent in the white flakes transported to desolventizing. Removal of solvent from the meal in the desolventizer is discussed in Chapter 7. Solvent is removed from the miscella by double-effect evaporation and steam stripping, as shown in Fig. 6.12. Steam and solvent vapors from the desolventizer are used to heat the first-stage evaporator.

The first stage is usually a long-tube vertical evaporator with a vapor dome. Miscella passes upward through the tubes, where it is heated and the solvent evaporated, using vapors from the desolventizer-toaster (DT, see Chapter 7) for heat. This stage, sometimes called the *economizer*, is designed to make the most efficient use of the heat from the vapors and concentrate the miscella as much as possible (to 70 to 85%; see Chapter 7).

Concentrated miscella is then pumped to the second-stage evaporator for further concentration. This is a rising-film evaporator heated with steam. Vapors go to the condenser, and oil flows to the oil stripper. The second-stage evaporator operates under partial vacuum to save steam.

The oil stripper is typically a disk-and-donut type. It operates under vacuum provided by steam ejection. Oil flows downward in contact with stripping steam introduced at the bottom, and residual solvent is reduced sufficiently to achieve a flash point

TABLE 6.4 Specifications and Analysis of Hexane

Property	Typical value[a]	Specification	Test method
Distillation, Rec., °F at 760 mm Hg			ASTM D 1078
Initial boiling point	152.2	151.0 Min.	
10%	153.1		
50%	154.0		
90	154.9		
Dry point	155.9	156.9 Max	
API gravity at 60°F	76.8		ASTM D 287
Specific gravity, 60/60°F	0.676		ASTM D 1298
Density of liquid at 60°F, lbs/gal	5.73		ASTM D 1250
Vapor pressure at 100°F, psia	5.0		ASTM D 323
Flash point, approximate, °F	–15		Estimated
Composition, liquid volume %			Chromatography
Dimethylbutanes	1		
Normal hexane	62	60 Min.	
Methylcyclopentane	13		
Methylpentanes	24		
Sulfur content, ppm	<1	1 Max.	ASTM D 3120
Bromine no.	<0.001		ASTM D 1492
Nonvolatile matter, mg/100 ml	0.1	1 Max.	ASTM D 1353
Color, saybolt	+30	+30 Min.	ASTM D 1093
Acidity of distillation residue	Neutral		ASTM D 1093
Doctor test	Negative	Negative	STM D 235
Kauri-Butanol value	28.5		ASTM D 1133
Aniline point, °F	149		ASTM D 611
Benzene content, ppm	2	10 Max.	PPCo 6404-AA
Copper corrosion 2 hrs. at 212°F	1	1 Max.	ASTM D 130

Notice: Since the conditions of handling and use are beyond control, Phillips 66 makes no guarantee of results; nor is any of the above information to be taken as a license to operate under, or recommendation to infringe, any patent.

[a]The information contained herein is, to the best of our knowledge and belief, accurate, but Phillips 66 assumes no liability for damages or penalties resulting from use of or reliance on this information.

Source: Phillips 66 Co., Bartlesville, OK 74004, April, 1992.

of 150°C (300°F). Good vacuum is essential to avoid overheating the oil. Good practice is to operate at 50 mm (2 in) Hg absolute pressure and 115°C (240°F) maximum.

Excess vapors from heating the first-stage evaporator may go to a vapor con-

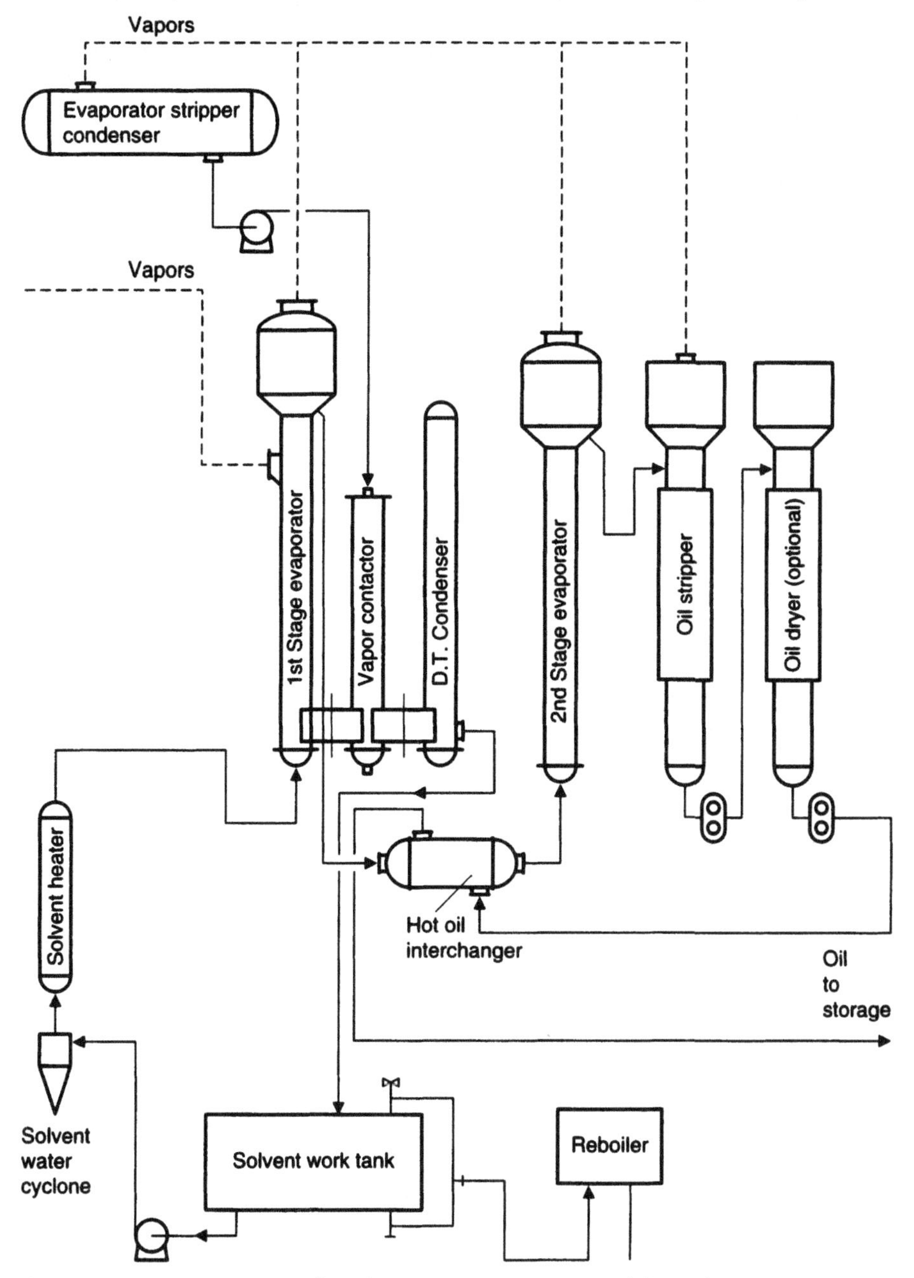

Fig. 6.12. Solvent recovery flowchart. *Source:* "Crown Model II Solvent Extractor," Crown Iron Works Co., Minneapolis, MN.

tactor, where the vapors are partially condensed by direct contact with condensate from the evaporator condenser and then flow to the solvent water separator. The vapor contactor reduces steam usage in heating the extraction solvent. Vapors that

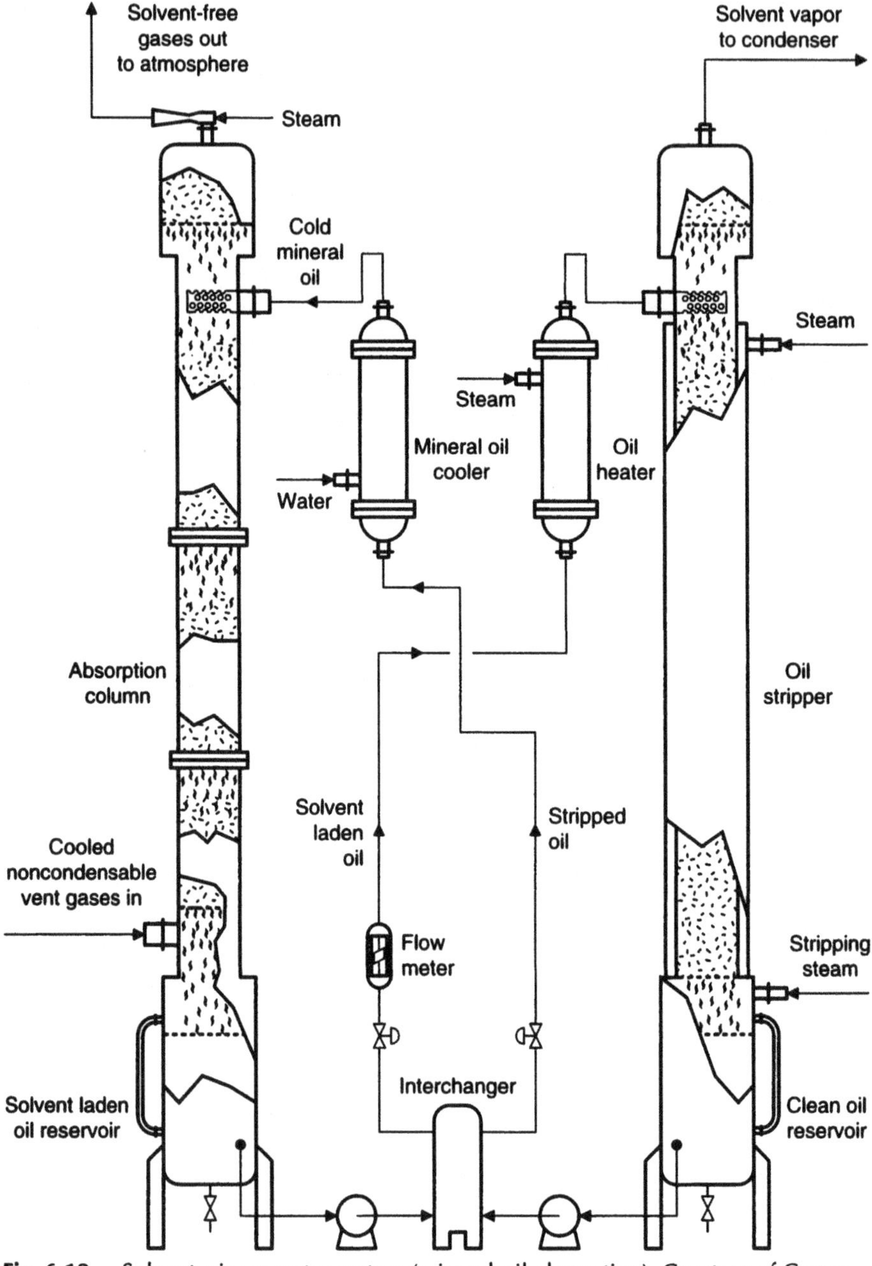

Fig. 6.13. Solvent–air separator system (mineral oil absorption). Courtesy of Crown Iron Works.

pass through the vapor contactor are condensed in a water-cooled condenser and flow by gravity to the solvent water separator.

Small amounts of air that have been introduced into the process must be vented safely without loss of solvent. Solvent is separated from the vented air most often by mineral oil absorption, as shown in Fig. 6.13. A vent condenser is provided for cooling vent gases before they enter the solvent–air separator system. Vent gases from the extractor, desolventizer, solvent–air separator, and waste water evaporator all pass through this condenser. A mineral oil absorption system removes solvent from the air before it discharges to the atmosphere. This consists of a packed absorption column, a packed steam-jacketed stripping column, and heat exchangers.

Vent gases pass upward though the stripping column, where they contact, and are absorbed by, cold mineral oil. The cold mineral oil is then heated and pumped to the top of the stripping column. Solvent is stripped from the mineral oil by contact with steam flowing upward through the column. Heat is exchanged between the cold mineral oil from the absorber and the hot mineral oil from the stripper for energy savings. Additional heat exchangers heat and cool streams to desired operating temperatures.

Steam from the stripper passes to a condenser for recovery of solvent. Solvent-free gases from the absorber are vented to the atmosphere through a steam ejector or fan, which provides a partial vacuum.

Solvent Loss

In the United States, solvent loss is expressed as gallons per short ton; in other countries it is expressed as kilograms per metric ton. For comparison, 2.8 kg/MT equals 1 gallon per short ton (specific gravity 0.674). For accounting purposes, the overall loss is calculated by dividing the total solvent disappearance by the weight of soybeans processed in a given period. In the 1970s, U.S. extraction plants typically used less than 1 gallon per short ton as an objective. Currently acceptable levels are less than one-third of this level. The disappearance, or overall loss, figure does not identify sources of loss, and operating practices to control loss must be more specific. Sources and control have been reviewed by Myers (19) and Venne (20) as meal from the desolventizer-toaster (DT) oil from the stripper; waste water from the evaporator; vent gasses; and mechanical losses.

Design and operation of the desolventizer-toaster or flash desolventizer controls the residual solvent in meal. This is covered in detail in Chapter 7. Values of 100 to 3000 ppm have been reported in the literature. Myers assumes levels of 500 to 1000 ppm for plants with losses of 1.4 to 2.8 kg per MT (0.5 to 1.0 gallons/short ton).

Oil from the stripper has a hexane content of 1000 ppm at a flash point of 120°C (250°F), the common trading standard in the United States, and 550 ppm at flash point 160°C (320°F). There is some evidence that 500 ppm may be the minimum that can be achieved. The stripper must be of proper design and capacity. Control of vacuum temperature and stripping steam is essential, and the stripper must be cleaned and maintained to ensure even flow for efficient stripping.

Vent loss is controlled primarily by operation of the mineral oil absorption. Venne (20) lists the importance of proper size, regulation of vapor flow and vapor volume, control of temperatures of oil and vent gas, flow of oil and stripping steam, vent gas temperature, and monitoring of vent gas to ensure low concentrations. In addition, mineral oil should be monitored and replaced if deteriorated or contaminated.

Mechanical losses may be the largest single factor. Leakage can occur from seals, packing glands, or gasketed joints. A continuing inspection and repair program is necessary. Pressure tests, instruments, and human senses can be used for detection. Repairs should be made promptly. A major source of solvent loss is shutdowns for repairs and maintenance. Frequent shutdown for emergency repairs is a problem, especially if equipment must be opened or the plant purged. Excessive downtime and unscheduled shutdowns are cause for management concern. Emphasis on planning the annual turnaround to anticipate and correct potential mechanical problems is desirable.

Safety

The most obvious and frequently discussed safety factors are those related to handling flammable solvent, with its potential for catastrophic accidents. Modern soybean processing involves large plants with many operations that must be considered. Transportation, material handling, drying, and milling are similar to other grain processing. Dust explosions and personal injuries from machinery, falls, burns, asphyxiation, and other causes must be considered in developing a comprehensive safety program.

NFPA 36 (1993) outlines safety considerations applicable to location, design, construction, maintenance, and operation of solvent extraction plants. Compliance with U.S. Occupational Safety and Health Administration regulations and local codes is mandatory. Several excellent papers by experts from various countries, published in the *Proceedings of World Conference on Emerging Technologies in the Fats and Oil Industry* (1985) and in *Proceedings of World Conference on Oilseed and Edible Oil Processing* (1983) as well as other sources provide comprehensive information on programs and practices.

Several of these experts point out the importance of a comprehensive safety program. Sandvig (21) lists the following basic contributors to total solvent plant safety, which involves all personnel (1) physical plant (process design, equipment design, critical vessel design, instrumentation/automation, and critical safety devices); (2) Plant operation (critical task, instruction/training, defined authority and responsibility, permitted work (system), incident investigation, and safety management).

Mechanical Pressing of Soybeans

An alternative to solvent extraction for extraction of oil from soybeans is the use of mechanical pressing. This is an old practice dating from the original lever and screw presses to more modern hydraulic and continuous screw presses. Continuous screw presses or "expellers" were the principal methods used on soybean oil extraction in the United States until about 1950 when solvent extraction became the preferred method (1). (Expeller is a trade mark for the continuous screw press patented by

Anderson in 1903 and has become somewhat of a generic term for all such types of equipment.) Mechanical pressing is still used in less develped countries and for small volume production of soybean oils for special purposes such as the production of so-called "cold pressed" oils and production of soybean meal (cake) with increased oil content as a premium feed ingredient.

The drawbacks to mechanical extraction are: (1) Low capacity (15 to 30 ton/day/press; (2) High residual oil in the press cake (4 to 7%); (3) High power requirements; and (4) High maintenance and operator skill. The advantages of mechanical pressing are: (1) Lower initial capital costs; (2) Suitable for up to 100 ton/day capacity; and (3) No use of solvents. Although there are some reports of a higher quality of soybean oil from pressing this is difficult to quantify because of the wide variation of operating conditions employed. Any economic gain in oil quality may be offset by a higher oil content in the press cake.

The usual sequence in pressing is to employ some type of size reduction and heating of the soybeans down to 4–6% moisture followed by pressing. A low moisture content and size redution are required to achieve maximum oil release from the soybeans during pressing.

Continuous screw presses are superficially similar in appearance and there are several manufacturers whose equipment varies in details of design, operation, and maintenance. Presses designed for one type of oilseed may require modification of use on other oilseeds. The press manufacturer should be consulted before making changes from one oilseed to another. The current major use of mechanical pressing is as a preparation or "prepress" for oilseeds with a higher oil content, such as cottonseed, rapeseed/canola, copra, sunflower, etc., prior to solvent extraction.

Soybeans with an oil content of 18–20% do not need prepressing and are directly solvent-extracted after suitable preparation. The prepress operation can be used in a multiseed extraction plant for soybeans. This is done where soybeans are a minor portion of the total oilseeds extracted, or where there are existing presses in older plants. This practice is neither particularly cost effective nor recommended but it does occur. More modern multiseed plants use a separate and conventional preparation process for soybeans.

In the United States there is an estimated total mechanical pressing capacity of about 1300 ton/day, which represents only 1 to 2% of the total crushing capacity (D.R. Erickson, Consultant, St. Louis, MO, private communication).

References

1. Langhurst, L.F., in *Soybeans and Soybean Products, Vol. 2*, edited by K.S. Markley, Interscience Publishers, New York, NY, 1951, pp. 541–567.
2. Mustakas, G.C., in *Handbook of Soy Oil Processing and Utilization*, edited by D.R. Erickson et al., American Soybean Association, St. Louis, MO, American Oil Chemists' Society, Champaign, IL, 1980, p. 54.
3. NFPA 36, *Solvent Extraction Plants*, National Fire Protection Association, Qunicy, MA, 1993.
4. Moore, N.H., *J. Amer. Oil Chem. Soc. 60:* 189 (1983).
5. "Hot Dehulling System," Crown Iron Works Co., Minneapolis, MN, 1992.

6. "Buhler Preparation for the Processing of Soyabeans with Hot Dehulling System 'Popping'," Buhler, Inc., Minneapolis, MN, undated, rec'd 1993.
7. Watkins, L.R., E.W. Lusas, S.S. Koseoglu, S.C. Doty, and W.H. Johnson, "Developing the Potential Usefulness of Expanders in Soybean Processing," *Final Report of Research Sponsored by the American Soybean Association*, Dec. 15, 1989.
8. Penk, G., in *Proceedings of the World Conference on Emerging Technologies in the Fats and Oils Industry*, edited by A.R. Baldwin, American Oil Chemists' Society, Champaign, IL, 1985, pp. 38–45.
9. Good, R.D., *Oil Mill Gazetteer*, Sept. 1970, pp. 14–17.
10. Karnofsky, G., *J. Amer. Oil Chem. Soc. 63:* 1011 (1986).
11. "Solvent Extraction Systems," Extraction de Smet, S.A./N.V., Prins Boudewijnlaan 265, B-2650 Edegem, Belgium (1992).
12. "Crown Model II Solvent Extractor," "Crown Model III Solvent Extractor," Crown Iron Works, undated, rec'd 1993.
13. Lusas, E.W., L.R. Watkins, and K.C. Rhee, in *World Conference Proceedings, Edible Fats and Oils Processing: Basic Principles and Modern Practices*, edited by D.R. Erickson, American Oil Chemists' Society, Champaign, IL, 1989, pp. 56–77.
14. Anderson, G., in *Proceedings of the World Conference on Emerging Technologies in the Fats and Oils Industry*, edited by A.R. Baldwin, American Oil Chemists' Society, Champaign, IL 1985, pp. 85–90.
15. Berteux, J.M., in *Proceedings of the World Conference on Emerging Technologies in the Fats and Oils Industry*, edited by A.R. Baldwin, American Oil Chemists' Society, Champaign, IL 1985, pp. 95–97.
16. Lujara, J., in *Proceedings of the World Conference on Emerging Technologies in the Fats and Oils Industry*, edited by A.R. Baldwin, American Oil Chemists' Society, Champaign, IL 1985, 84–88.
17. Lowry, F., in *Proc. of the World Conf. on Emerging Technologies in the Fats and Oils Industry*, edited by A.R. Baldwin, American Oil Chemists' Society, Champaign, IL 1985, pp. 75–83.
18. Kingsbaker, C.L., *J. Amer. Oil Chem. Soc., 60:* 245 (1983).
19. Myers, N.W., *J. Amer. Oil Chem. Soc., 60:* 224 (1983).
20. Venne, L., in *Proc. World Conf. on Emerging Technologies in the Fats and Oils Industry*, edited by A.R. Baldwin, American Oil Chemists' Society, Champaign, IL, 1985, pp. 135–136.
21. Sandvig, H.J., *J. Amer. Oil Chem. Soc., 60:* 243 (1983).
22. Erickson, D.R., Consultant, St. Louis, MO, private communication.

Chapter 7

Soybean Meal Processing and Utilization

Norman H. Witte
6631 Parsons Ct.
Fort Wayne, IN

Introduction

Extracted soybean flakes will normally exit a continuous extractor with a residual hexane content of 29 to 35% hexane on an "as is" basis. This chapter covers the removal, and recovery for reuse, of this residual hexane, and the cooking, drying, cooling, and sizing of the extracted meal as required for its principal end use in livestock feeds. Desolventizing processes for other uses of the extracted flakes, such as in edible products or industrial uses, are also briefly covered.

History

Effective desolventizing of extracted flakes from a hexane extraction process is relatively difficult, and some appreciation of the development of this technology will help in understanding the present process.

The first continuous solvent extraction plants for soybeans in the United States came from Germany and were erected between 1934 and 1937. In these plants, desolventizing of the extracted flakes was done in steam-jacketed tubes, often 500 mm (20 in) in diameter and 5.5 m (18 ft) long, arranged in vertical banks. The solids were conveyed through the tubes by means of a screw or paddle conveyor. The desolventized flakes were almost white in color and were very dry and fragile. Subsequent handling resulted in considerable breakage and problems with dust.

Nutritionists found that the white meal was not optimal for use in livestock feeds because of the presence of certain antinutritional factors. In 1939, Kruse and Soldner at Central Soya, Fort Wayne, IN, (1) obtained a patent on a method for toasting the meal under high-moisture conditions in vertically stacked kettles to produce a light golden brown product with improved nutritional value and better handling properties. In 1940 Hayward (2) at Archer Daniels Midland, Decatur, IL, described a toasting process using cookers with controlled live steam and temperatures. Toasting then became a standard part of the process when the meal was destined for livestock feed.

The term *toasting* usually implies treatment with heat under relatively dry conditions. In the case of soybean meal, the required procedure is more properly referred to as *cooking*, because the best results are obtained under conditions of high moisture and controlled temperature. Although not an accurate description of the process for soybean meal, the term *toasting* is so universally used that it will be used here also.

About 1949, Blaw Knox introduced a pressure toasting process in a rotary vessel using steam pressures of 0.1 to 0.3 atm. This process was usually operated under relatively dry conditions, and there were indications that this meal was nutritionally inferior to meal toasted under atmospheric high-moisture conditions.

An essential part of the toasting process patented by Kruse and Soldner (1) was wetting the flakes with water to a moisture content of 16 to 23%. It was recognized that the moisture played a protective role in preserving certain desirable nutritional traits while enhancing the destruction of antinutritional factors.

In 1948 Kruse (Central Soya) and Hutchins (French Oil Mill Machinery Co., Piqua, OH) (3) cooperated in testing a pilot-size *desolventizer-toaster* at Central Soya's Decatur, IN, plant. This apparatus furnished the heat for evaporation of the hexane by condensing live steam directly into the hexane-wet extracted flakes in the top chamber of an apparatus similar to the common meal toaster but made vapor-tight to retain the hexane vapors. The meal, wetted by the condensed steam, then flowed into the lower trays where the temperature was further raised by jacket heat and the meal held for the required toasting time. The pilot tests were very promising, and the first commercial desolventizer-toaster was installed at Central Soya's Decatur, IN, plant in 1949.

Both Kruse (4,5) and Hutchins (6) filed for patents on the new invention, and the United States Patent Office declared an interference proceeding. Kruse eventually won the proceeding because of earlier desolventizing tests utilizing live steam conducted several years earlier at the Link-Belt Co. Kruse's patent covered broad process aspects of the process and it was assigned to Central Soya, who proceeded to license manufacturers and users for a very nominal fee. Central Soya coined the abbreviation DT for the new machine, the first version of which is shown in Fig. 7.1.

All major extraction equipment manufacturers in the United States, and several in Europe, marketed the licensed DT machine, and within a few years the DT became the standard process for soybean meal desolventizing and toasting because of its very significant advantages of better solvent recovery, lower energy consumption, greater reliability, and consistently better meal quality.

Many design improvements were developed in the next few years; one such improvement is shown in Fig. 7.2 (10). In 1972, Ing. Heinz Schumacher developed the *desolventizer-toaster-dryer-cooler* (DTDC) concept, in which the desolventizing-toasting and drying-cooling functions are combined in a single vessel, as depicted in Fig. 7.3.

In 1981 Schumacher (7) developed a DT design (now referred to as the Schumacher DT), using multiple desolventizing-toasting trays in which the live steam was injected into the bottom toasting tray, thus providing more effective steam stripping of the final traces of hexane from the meal. The bottom of each toasting tray was provided with hollow *staybolts*, through which the steam from the kettle below could pass into the meal above. More recently the Schumacher DT concept has been combined with predesolventizing and drying and cooling into a single vessel.

Process Requirements

The process for producing toasted meal for livestock feeding is logically divided into the following steps. Process requirements for each of the steps are discussed below.

Desolventizing-Toasting

This step encompasses removal of the hexane from the flakes, followed by cooking at 100 to 105°C (212 to 220°F) at elevated moisture levels ranging from 16 to 24% for times in the range of 15 to 30 min.

The hexane-wet extracted flakes leave the extractor at a temperature of about 57°C (135°F) and enter the top chamber of the DT or DTDC apparatus. Live steam is injected below the level of the meal at the bottom of one or more sections of the apparatus; the steam that condenses furnishes the latent heat required for hexane evaporation, and the condensed steam raises the moisture level to facilitate the "toasting" operation.

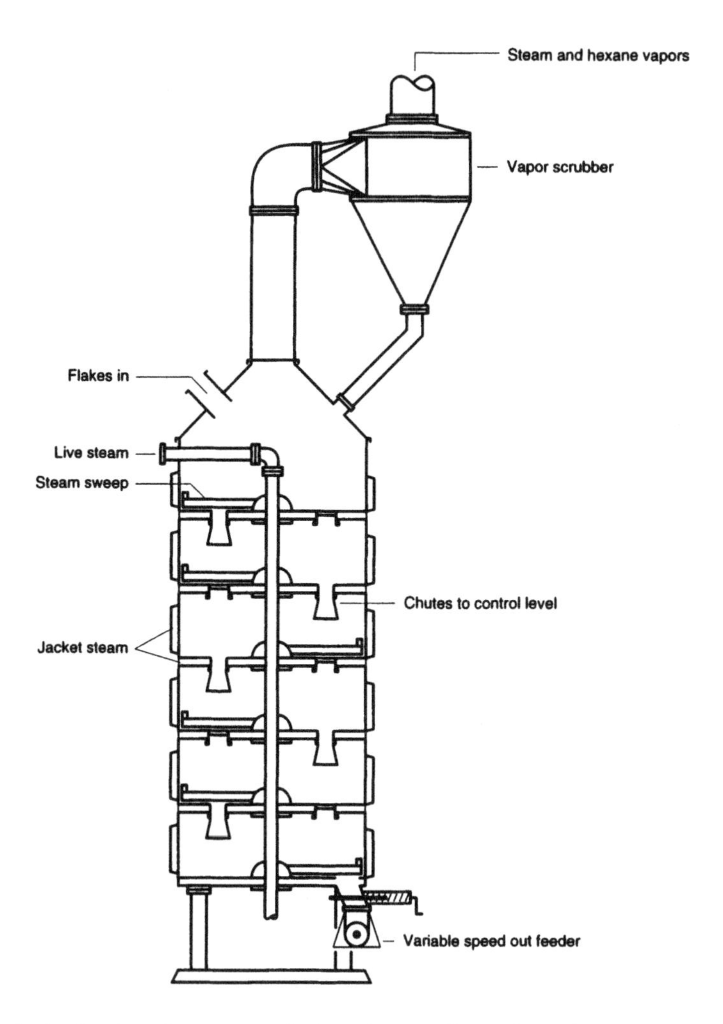

Fig. 7.1. Early desolventizer-toaster design. *Source:* Sipos, E., and N.H. Witte, *J. Amer. Oil Chem. Soc., 38:* 11 (1961).

The meal leaving the DT section will normally have a residual solvent content of less than 500 ppm; with the Schumacher DT, values below 300 ppm are attainable. The urease activity should be reduced so that the pH rise in the standard urease test is less than 0.2 pH units (8). This test is the normal process control for toasting, because it is fast and easy to run. In actuality, this test is only an indicator of the efficacy of destruction, or inactivation, of the many antinutritional factors present in raw soybeans. A test more directly related to nutritional values, but more time-consuming, is the measurement of the remaining antitrypsin activity (9).

Toasting time, temperature, and moisture conditions affect the degree of cooking and the nutritional properties of the resulting meal. However, optimization of these process variables is difficult in practice, because nutritional testing is a slow process and using such a test for direct process control is not practical.

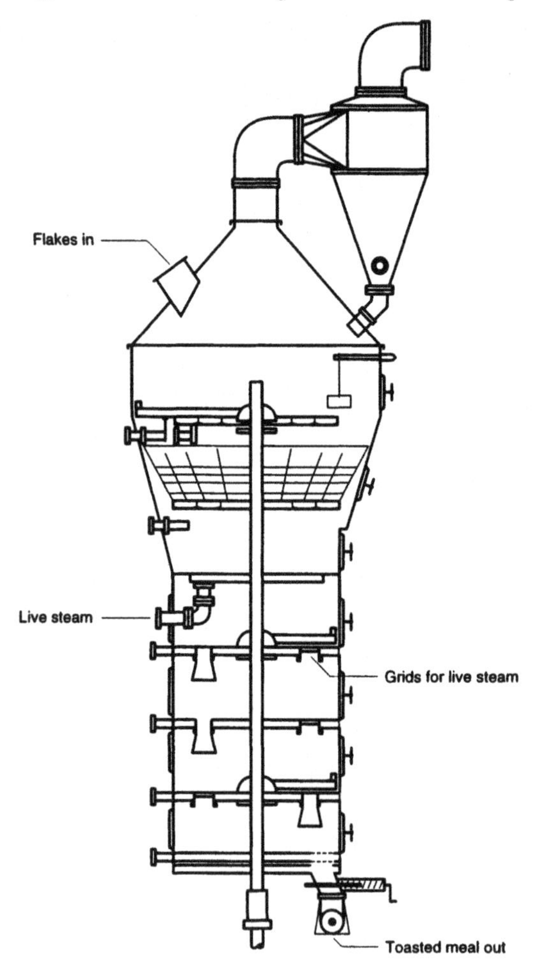

Fig. 7.2. Improved desolventizer-toaster. *Source:* Sipos, E., and N.H. Witte, *J. Amer. Oil Chem. Soc. 38:* 11, (1961).

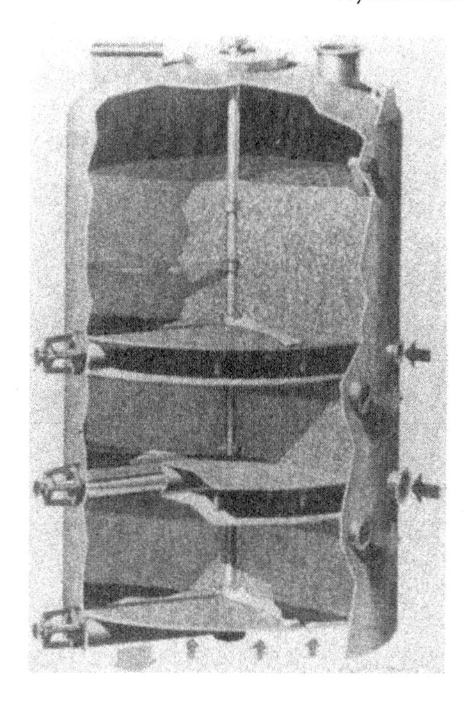

Fig. 7.3. Original DTDC. Courtesy of Crown Iron Works, Co., Minneapolis, MN.

Drying and Cooling

The meal leaving the DT section may be dried and cooled in drying/cooling (DC) trays immediately below the DT section; in a separate dryer/cooler of similar design; or in rotary or other types of solids-drying/cooling equipment.

In this step of the process, the toasted meal is dried from the 16 to 22% moisture at which it exits the DT section to near the 12% level required for finished meal and is then cooled to a safe level for storage and handling. Meal should exit the cooler at a temperature less than 32°C (90°F) or within 6°C (10°F) of the ambient air temperature, whichever temperature is higher.

The drying step usually occurs at temperatures well below that of toasting; thus there is no concern for possible nutritional damage to the meal. However, certain types of dryers, which use indirect steam, could result in overheating of the meal and have a deleterious effect on nutritional quality.

Meal Grinding and Sizing

This step is normally the final unit process in continuous extraction. This step reduces the particle size of the meal to that required by the customer. Some plants outside the United States omit this step because grinding is done in the feed mill after formulation. Because the meal is handled in bulk conveying and transport systems, it is desirable to produce a minimum of flourlike dust (<100 mesh in size) in the grinding process.

Storage and Shipping

Storage and shipping of the finished meal is the last step at the processing plant. Storage capacity may contain only a few hours of production, or it may be designed to hold meal for months to wait out surges in market demand. Storage must be able to keep the meal without spoilage and must be so designed that the meal does not cake during storage and become difficult, or impossible, to discharge from the bins.

Desolventizing-Toasting

As noted previously, the desolventizer-toaster process is almost universally used as the first step in processing flakes from the extractor when the meal is to be used for livestock feed. The original form of the DT, shown in Fig. 7.1, comprised a single chamber or tray at the top, where the live steam was injected, followed by one or more lower trays with jacketed bottoms and sides, where the toasting took place. Meal levels were maintained either by means of a variable-speed screw feeder at the bottom discharge or by level control gates in each tray. In the first case, the meal was held back to maintain the level in the top tray, and each intermediate bottom was fitted with a spout to prevent overfilling the tray below.

Each intermediate tray bottom was fitted with one or more slotted grids that permitted steam generated in the toasting kettles to pass upwards and into the desolventizing section, where some of its heat value could be recovered.

In the original DTs the solvent vapor, together with the equilibrium quantity of water vapor, left the desolventizing section at a temperature in the range of 74 to 80°C (165 to 176°F). Reducing the live steam to give a lower vapor temperature resulted in incomplete removal of the solvent in the desolventizing section; higher vapor temperatures gave no useful results and only wasted steam.

Many modifications were made to the original DT design to increase capacity and to incorporate a more countercurrent design to improve efficiency. Fig. 7.2 (10) shows a widely used concept wherein the live steam was injected in the third tray from the top and flowed countercurrently to the meal, finally exiting at the top. Such a design often permitted the reduction of the live steam usage, resulting in a DT vapor temperature below 71°C (160°F) (10).

In the multitray design, toasting time presented no problem, because as many kettles as required could be added; the limiting capacity of the apparatus was determined by the cross section of the top tray, where the solvent vapor disengaged from the meal. A soybean-crushing capacity in excess of 110 MT/day/m^2 of top flakes surface (11 short tons/day/ft^2) is likely to result in significant carryover of flakes in the vapor stream; a wet vapor scrubber, using hexane or water, is required in such cases to prevent fouling of the following heat exchange surfaces with dust.

Initially, the live steam was injected into the meal via perforated pipes that trailed the sweeps and were fed from the hollow drive shaft. Later, one or two bottoms in the upper sections of the machine were drilled with many small holes, and the steam was injected into the meal through these holes.

In 1972, Schumacher introduced the DTDC wherein the DT section is contained in a single very deep [1.2 to 2.0 m (4.0 to 6.6 ft) deep] flake bed in the top tray. In this design, the requirement for more surface area in the drying stages resulted in reduced vapor velocities in the top section and, in many cases, eliminated the need for vapor scrubbing. However, the single-bed desolventizing-toasting zone proved to be very sensitive to attempts to reduce the live steam flow and could readily discharge flakes containing hexane into the drying stage below.

About 1981, Heinz Schumacher (7) returned to the multitray desolventizing-toasting section and developed the idea of the hollow-staybolt design for the steam-jacketed bottoms, as shown in Fig. 7.4. This design gave more open area in the bottoms for upward movement of steam than did the grids used in the original DT and permitted all of the live steam to be injected in the bottom chamber of the toasting section. Additionally, the tendency of the grids to blank over with sticky meal was obviated, because the hole areas remained hot and prevented sticking. This gave the ultimate in countercurrent operation, permitting minimum steam consumption and the most efficient stripping of the solvent from the meal. A typical Schumacher DT is shown in Fig. 7.5.

The live steam used in the DT or in the DT section of the DTDC is the single largest user of steam in the extraction process and will usually be in the range of 35 to 55% of the total steam consumption of the extraction plant. The steam required depends primarily on three factors:

1. The hexane content of the extracted flakes leaving the extractor
2. The temperature of the vapor leaving the DT dome
3. The temperature of the extracted flakes entering the unit (to a smaller extent)

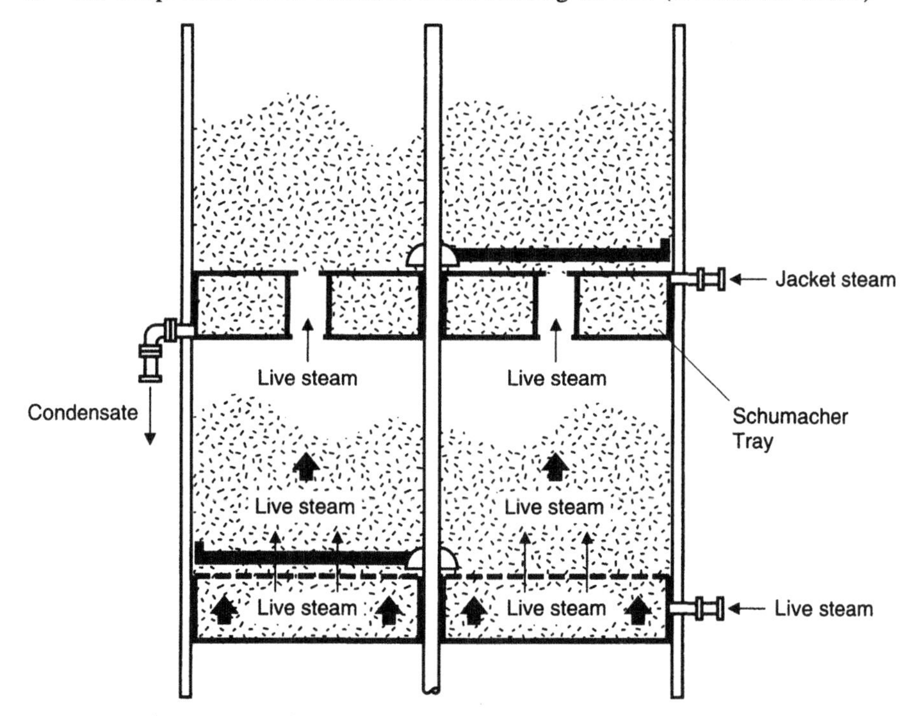

Fig. 7.4. Schumacher DT basics. Courtesy of Crown Iron Works Co., Minneapolis, MN.

The hexane content of the flakes from the extractor can vary from about 29% to as high as 35% depending on the type of extractor and the drainage conditions existing in the flake bed. The live steam requirement is directly proportional to the hexane content.

Not all of the live steam injected into the meal condenses on the meal. The solvent vapor stream leaving the DT contains water equilibrium vapor in a concentration such that the partial pressure of the water is equal to the equilibrium vapor pressure of pure water at the temperature of the DT vapors. Thus, as the vapor temperature rises, the vapor stream will contain a higher quantity of uncondensed steam. Fig. 7.6 shows the relationship between the mass of water and the mass of solvent in the vapor stream as a function of the temperature of the vapors. Note that the amount of steam that does not condense rises rapidly at the higher vapor temperatures.

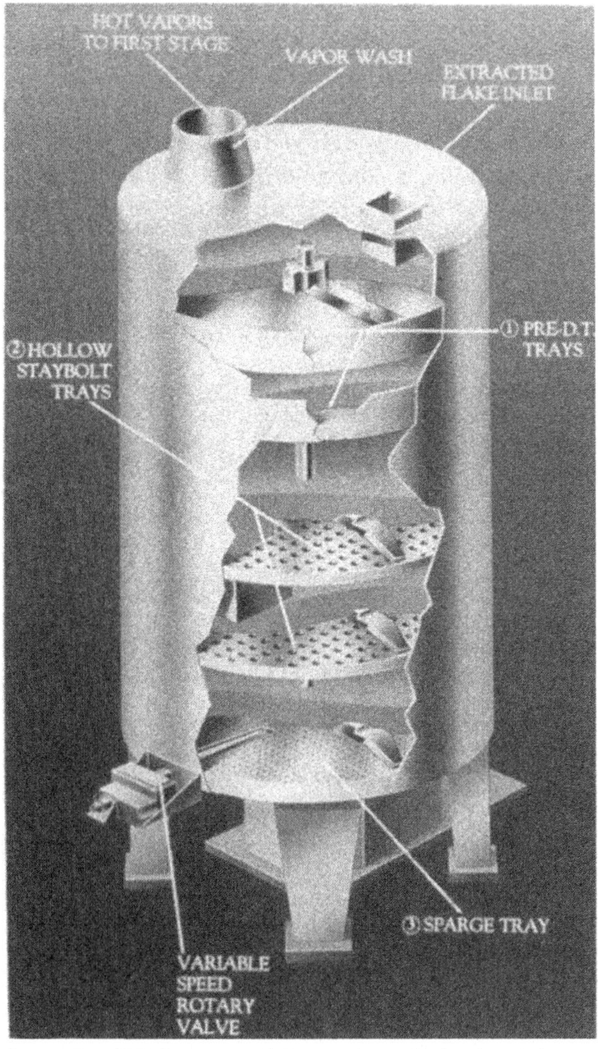

Fig. 7.5. Schumacher DT. Courtesy of Crown Iron Works Co., Minneapolis, MN.

Table 7.1 shows typical steam consumption quantities for the various stages of the DTDC process under several process conditions. Columns 1 and 2 show that increasing the live steam flow so that the temperature of the DT vapors increases

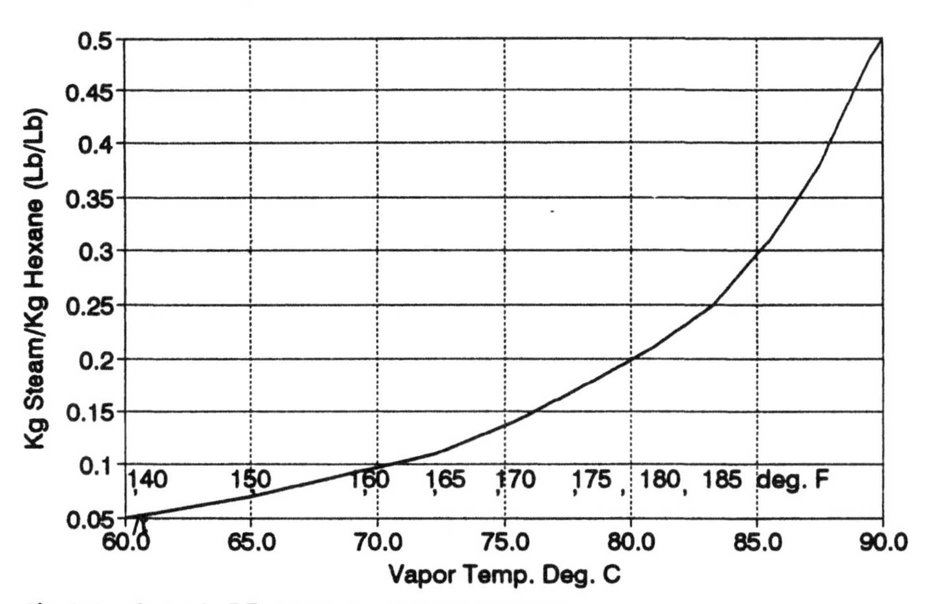

Fig. 7.6. Steam in DT vapors vs. vapor temperatures.

TABLE 7.1 Effect of Process Conditions on Meal Desolventizing/Toasting/Drying Operations

	1	2	3	4	5	6
Process condition	Normal DTDC	High vapor temperature	Better drainage	Expander	Predesolv-entizer	Expander and predesolvenizer
% Hexane in flakes	33		30	25	33	25
DT vapor temp °C (°F)	72 (163)	79.4 (175)	72 (163)			
Waste steam in vapor	43.1	68.3	37.5	29.2	43.1	29.2
Steam condensed	76	76.3	69.3	59.4	52	3.52
Predesolventizer steam	0				29.2	29.2
Total DT Steam	119.2	144.6	106.9	88.6	124.3	93.6
% H_2O meal in toasting section	20.6	20.7	20	19	18.3	16.5
Dryer steam	45	45.7	37	25.2	17.2	–3
Total system steam	164.1	190.3	144	113.8	141.5	90.6

Notes: All steam consumptions given in kg/MT beans. For lb/short ton multiply indicated values by 2.0. Temperatures are °C (°F).
Blank columns repeat conditions in previous column.

from 72°C (163°F) to 79.4°C (175°F) results in an increase of waste steam in the DT vapor from 43.1 kg/MT (86.2 lb/short ton) to 68.3 kg/MT(136.6 lb/short ton), or a waste of some 25.2 kg/MT (50.4 lb/short ton) of steam.

Most plants will use the vapors from the DT section as the source of heat for the first-stage miscella exaporator (see Fig. 6.12; also referred to as the No. 1 evaporator or the economizer). Much of the excess steam present in the DT vapor stream is condensed and used in this evaporator. With the usual contents of hexane in the full miscella and in the extracted flakes, the heat content of the solvent alone in the extracted flakes is not sufficient to evaporate all of the solvent available in the miscella stream (up to the 75 to 85% level possible with conventional vacuum) so at least some of the excess steam in the vapor stream is usefully recovered. Generally, DT vapors at a temperature of 70°C (158°F) will contain sufficient waste steam to furnish the energy required in the No. 1 evaporator.

The amount of latent heat that is finally wasted to the condensers in the discharge from the shell side of the No. 1 evaporator or any subsequent heat recovery steps is determined by many factors in the plant, including the solvent concentration of the full miscella, the solvent concentration in the extracted flakes, the sizing of the No. 1 evaporator and its vacuum condenser, and the DT vapor temperature.

Predesolventizing

A predesolventizing process step, ahead of the DT section, is often used to reduce the amount of hexane going to the DT section and, therefore, the amount of live steam used in the DT. The predesolventizer uses jacket steam heat and, therefore, evaporates hexane without condensing live steam on the flakes.

Predesolventizing may be done in large screw or paddle conveyors with steam jackets on the conveyor shell, a steam-heated shaft and paddles, or both. Alternatively, a predesolventizer section can be incorporated into the top section of the DT by means of one or more steam-jacketed bottoms, with no provisions for live steam contact with the meal.

Predesolventizing reduces the amount of water condensed on the flakes in the DT section and, consequently, the moisture level during the toasting step. This results in a corresponding reduction in the drier steam requirement. However, the reduced moisture present during the toasting step will give less protection against overheating and possible destruction of nutritional value in this step. Columns 1 and 5 in Table 7.1 show the effect of predesolventizing on steam consumption in the DTDC process. A predesolventizing step using 29.2 kg/MT (58.5 lb/short ton)(column 5) of steam will reduce total system steam by 22.6 kg/MT (45.2 lb/short ton).

Effect of Expanders in Preparation on Meal Desolventizing

The use of expanders in the preparation of soybeans for extraction can have a marked effect on desolventizing operations. Expanded flakes (called collets) have a higher bulk density than unexpanded flakes and will usually drain to a lower residual solvent content in the extractor. These effects combine to raise the effective capacity of the DT stage by increasing the holding time in the DT and reducing the live steam requirement for desolventizing.

Expanded flakes are claimed by some (11) to leave the extractor with as low as 25% residual hexane, as compared with the normal value of about 33% from a deep bed extractor. Columns 1 and 4 in Table 7.1 show the results of such a reduction in retained solvent on DTDC operations. Note that the DT steam usage is reduced from 119.2 kg/MT (238.4 lb/short ton) to 88.6 kg/MT (177.2 lb/short ton). Likewise, there is a reduced requirement for drying the flakes, resulting in a total system steam usage reduction from 164.1 kg/MT (328.3 lb/short ton) to 113.8 kg/MT (227.5 lb/short ton).

Column 6 of Table 7.1 shows the effect of a combination of expander and predesolventizer operation. This case results in the lowest desolventizing/drying steam consumption. Note, however, that the moisture of the meal in the toasting section is only 16.5%, as compared with a more normal value of 20% or higher (columns 1 through 3); such a reduction in the toasting moisture level could result in lowered nutritional value in the meal product.

Preparation processes for soybeans utilizing expanders will generally have a positive effect on DT desolventizing operations; however, the effect on overall plant operations (total steam consumption, extraction efficiency, maintenance costs) is not as clear and probably depends on the type of extractor the plant is using, whether the plant capacity is being "pushed," and where the actual capacity bottlenecks are (12).

Drying and Cooling

The meal leaving the DT section will contain 16 to 20% moisture and be at a temperature of about 105°C (221°F). Drying and cooling can be accomplished in separate processing vessels or can be located directly below the desolventizing-toasting section in a single apparatus (the DTDC design).

Twenty years ago the most common arrangement involved a steam tube rotary drier for drying and a rotary cooler with air as the cooling medium. Today, the DTDC design is probably the most popular although for very large processing lines, separated drying and cooling equipment is more common. This can consist of the rotary units just noted or a separate DC unit. Fig. 7.7 shows a freestanding DC unit; the same process described below for this equipment also pertains to the DC unit when it is incorporated into the same shell as the DT unit.

In the drier unit, hot air is injected into the meal bed via holes in the bottom, which forms the top of the air plenum. The energy efficiency of the drier rises as the temperature of the drying air is raised; however, temperatures above 150°C (300°F) should be avoided because of the danger of fire. The volume of drying air is usually in the range of 375 m^3/MT (12,000 ft^3/short ton) of beans; the cross-sectional area of the drier is designed for a superficial air velocity through the bed of less than 0.4 m/s (1.3 ft/s). Air velocities greater than this will result in excessive dust carryover with the exit air. If the meal cannot be dried in a single tray without going to excess air velocities or temperatures, a second drying tray is placed below the first.

Depending on the temperature of the entering drying air, the meal will leave the dryer section at a temperature of 45°C (113°F) to 75°C (167°F). The cooling section utilizes the same construction as the drying section(s), except that the entering air is at ambient temperature. The cooled meal leaving the unit should be at a temperature of 32°C (90°F) or within 6°C (10°F) of the ambient temperature, whichever is higher.

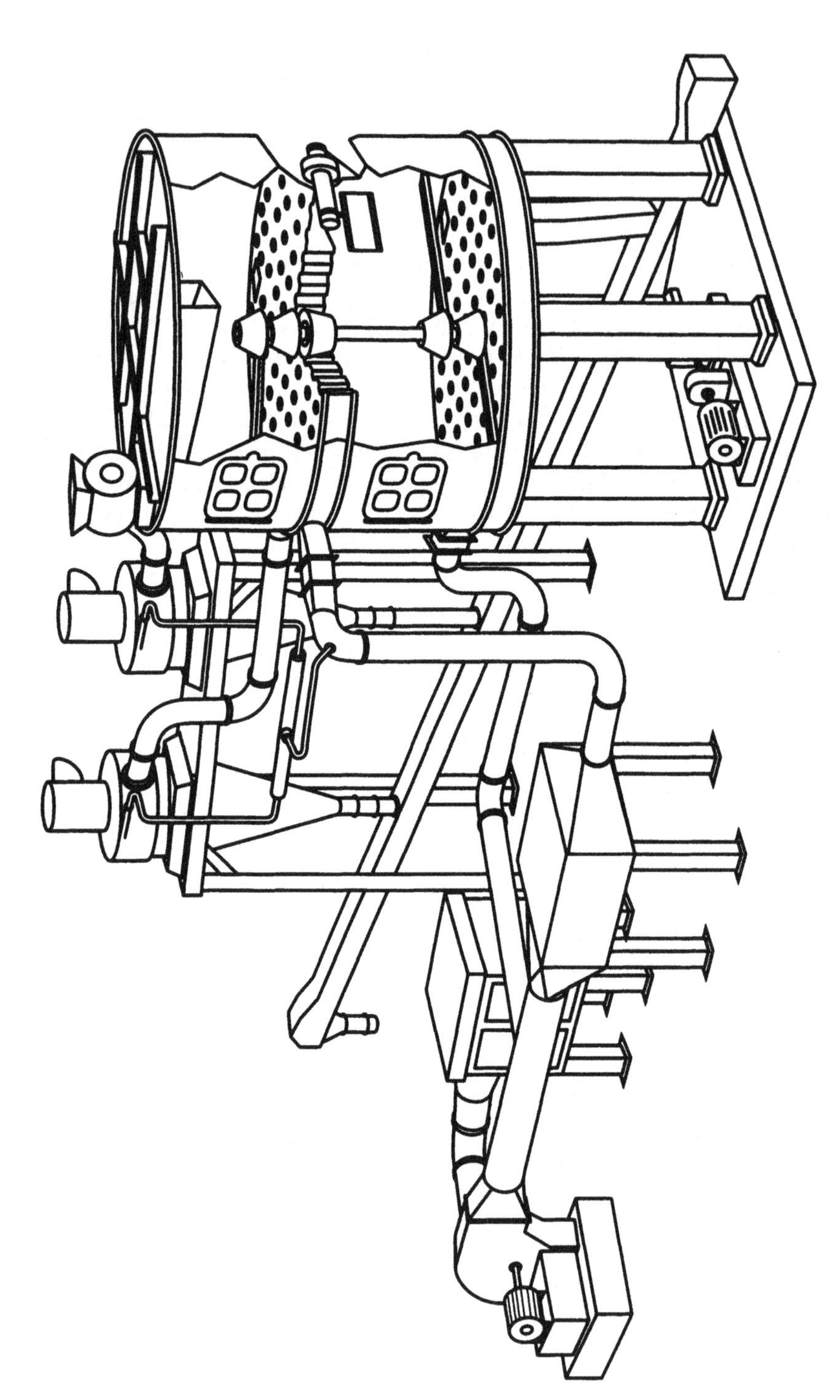

Fig. 7.7. Separate DC unit. Courtesy French Oil Mill Machinery Co., Piqua, OH.

Combined Desolventizer/Toaster/Dryer/Cooler (DTDC)

The combined functions of desolventizing, toasting, drying, and cooling are often incorporated into a single vessel. Fig. 7.8 shows such a design, also including the enhancements of predesolventizing trays and the Schumacher-type toasting section.

Figure 7.9 shows another modern design with the same process steps (except no predesolventizing) but with a larger diameter for the drying and cooling section of the apparatus, to optimize the vapor and air velocities in the various sections. This design is mechanically more complex and expensive. Both units incorporate the following sequence of process steps:

1. Predesolventizing with steam-jacketed bottoms in the top trays (only in Fig. 7.8)
2. Desolventizing with live steam and toasting in the Schumacher DT design, where the live steam is injected in the bottom kettle of this part and flows upward through several trays of meal, passing from one tray to the next through the hollow staybolt openings in the bottom; spouts maintain levels in the individual trays
3. A rotary valve to seal the DT section from the following drying section
4. One or more drying sections, in which hot air is injected into the meal bed
5. A cooling section, where ambient-temperature air is injected into the meal bed
6. Rotary valves in each drying and cooling section to maintain levels

Solvent Recovery

The mixed hexane–water vapor stream from the DTDC is usually routed through the shell side of the first-stage miscella evaporator (called the No. 1 Evaporator or the Economizer) to furnish heat for concentrating the miscella from the 25% oil at which miscella typically leaves the extractor to the range of 70 to 85%.

Vapors not condensed in this evaporator may be routed to other smaller types of heat recovery apparatus; final condensation of the vapors takes place in a water-cooled condenser. The condensate from all units in the desolventizing vapor flow contains a mixture of hexane and water and is routed to the plant hexane–water separator.

Vapors from meal desolventizing will contain some meal dust, and care must be taken to minimize the deposit of this dust in the heat exchangers and ductwork through which this vapor is routed. Evaporators and heat exchangers usually will have the vapor routed through the shell side, which should have minimum baffling and amply sized flow passages.

Safety and Environmental Concerns in the DTDC Process

The desolventizing section of the extraction process is one of the more dangerous spots in the process, because here the solids leave the enclosed solvent-handling area and enter a process area not designed for solvent vapors. In the case of the DTDC this passage takes place through a rotary lock, which operates at variable speed to maintain the proper levels of meal in the DT sections above it.

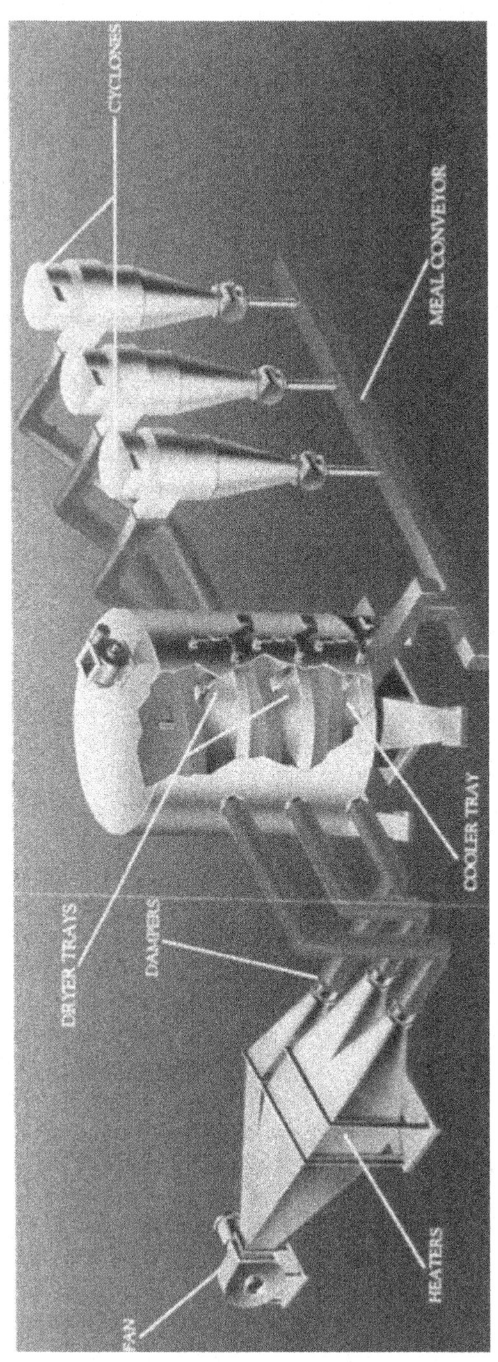

Fig. 7.8. DTDC. Courtesy Crown Iron Works Co., Minneapolis, MN.

When process conditions are correct, the rotary lock is a very effective seal. However, if the temperature of the meal leaving the tray above the lock is allowed to drop much below 90°C (195°F), significant quantities of hexane will be contained in the meal, and a hazardous mixture of air and hexane could occur in the dryer sec-

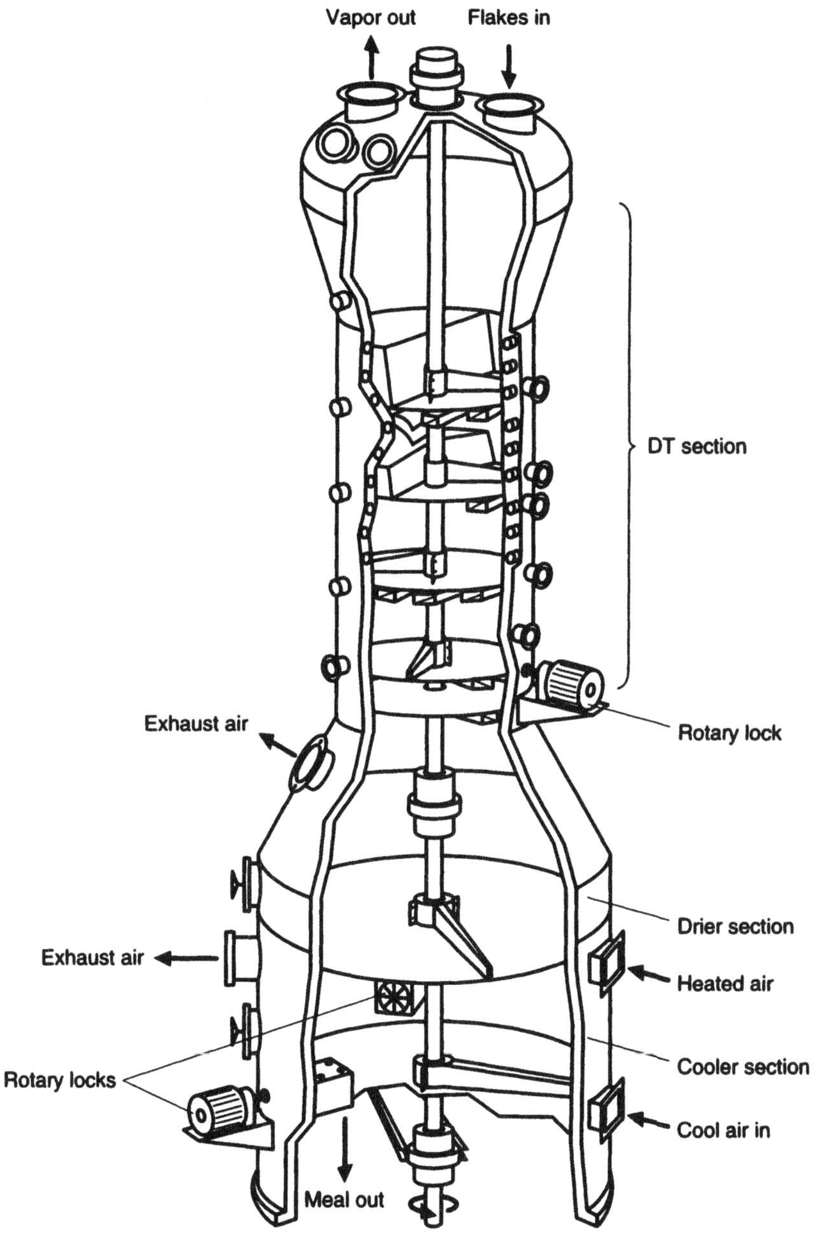

Fig. 7.9. DTDC with variable diameter vessel. Courtesy of Krupp Maschinentechnik GmbH, Division Extraktionstechnik, Hamburg, Germany.

tion or in the process building. Hexane vapors that escape from the desolventizing apparatus may, if present in sufficient quantity, represent a fire and explosion hazard. Even at levels much lower than this, though, the vapors may represent a health hazard.

Fires in the drying sections of the DTDC are another important safety concern. The combination of hot meal and hot air can result in a very damaging fire. The fires seem to start in areas where there is stagnant dust, which dries out and, if in contact with the hot air entering the dryer section, will eventually ignite. Such fires present an especially hazardous situation because of the proximity of the hexane-wet flakes in the preceding desolventizing-toasting section. A fire in the dryer is usually controlled by stopping the process flow, shutting off the fans, and injecting live steam into the air ducts in order to shut off or reduce the available supply of oxygen.

The exhaust air from the drying and cooling sections will contain a significant quantity of dust. Efficient cyclone collectors are the minimum requirement for cleaning up this air; in some situations, more efficient collectors may be required. The exhaust air from the first drying stage is very high in humidity and presents an especially difficult collection situation.

The water effluent from the solvent extraction process contains small quantities of dust carried over to the condensing system with the DT vapor. This can represent a significant effluent problem in some situations. Weber (13) has described a closed system wherein the water from the solvent water separator is boiled in an evaporator heated with plant steam; the steam so produced is used in the DT, and a concentrated solids suspension from the process side of the boiler is added to the DT meal stream.

In some cases, the air exhaust from the meal dryer(s) can present an odor problem. In one such case in Europe the air is passed through a biological filter consisting of a shallow bed of biologically active soil (13); Kratz (14) describes such a system used by an oleochemical manufacturer. In less demanding areas the dryer exhaust can be adequately diffused from a tall stack.

Extraction plants recover and reuse almost all of the solvent used in the extractor, but some is lost through leaks in the equipment, through process seals, and through retention by the meal and oil. The total loss in well-operated plants can be as low as 0.5 kg of solvent per metric ton of beans processed (0.2 gal/short ton); about half of this loss is probably accounted for by solvent that remains in the meal that leaves the DT.

Desolventizing Processes for Producing Untoasted Flakes

The DT process produces a toasted product that is nutritionally ideal for livestock feeding. Other uses of extracted soybean flakes, including human food products and products for further processing (into protein concentrates or isolates) or products for industrial uses, require that the flakes be desolventized in a manner that minimizes the amount of cooking (toasting) in order to preserve a high content of water-soluble protein. These products are generally referred to as "white flakes" (see Chapter 8).

Desolventizing for the production of white flakes has been done in three different types of processes:

1. In steam-jacketed conveyor tubes (originally called "schneckens") arranged in a vertical stack so that the extracted flakes move downward from one tube to the next. This system is similar to the desolventizing system used for conventional meal before the introduction of the DT.
2. By using superheated circulating solvent vapor to furnish the required heat energy. The flakes are contained in a horizontal drum equipped with an agitator/conveyor to facilitate contact with the superheated vapor. Figure 7.10 shows this type of apparatus, usually referred to as a *superheated vapor desolventizer* (15).
3. By using superheated solvent vapor as in system 2, except that the vapor and the flakes are brought into contact in a conveyor pipe, where the conveying action results from the velocity of the superheated gas. This apparatus (Fig. 7.11) is generally called a *flash desolventizer* because of the relatively short residence time of the flakes in the apparatus (16).

In either system system 2 or system 3 the temperature of the solvent vapor leaving the circulating heater and contacting the flakes is controlled by the heat required but will usually be in the range of 135 to 175°C (275 to 347°F).

All of these systems are usually followed by a deodorizing drum where live steam contacts the meal to remove additional residual solvent. Conditions in the deodorizing drum can be adjusted to produce flakes of varying protein solubility depending on the end-use requirement.

Hexane losses in white flake desolventizing systems are generally higher than in DT systems, because more equipment is involved in the process and the amount of live steam used for stripping residual hexane vapors is limited.

Meal Grinding and Sizing

In the United States and Canada, meal leaving the drying/cooling section of the process is normally ground and sized to meet a size specification such as the following:

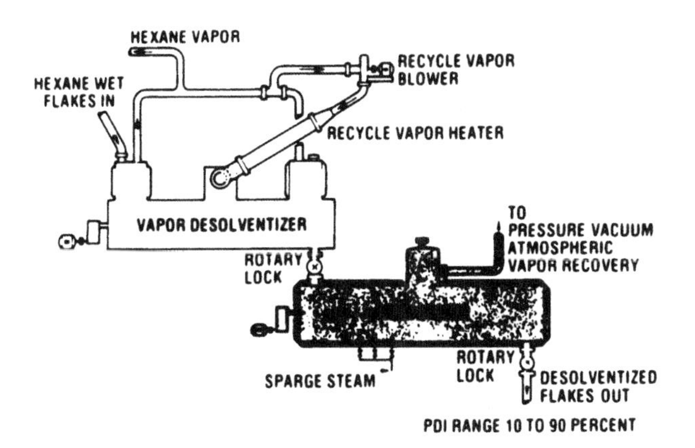

Fig. 7.10. Vapor desolventizing. *Source:* Becker, K.W., *J. Amer. Oil Chem. Soc. 60,32:* 1702 (1983).

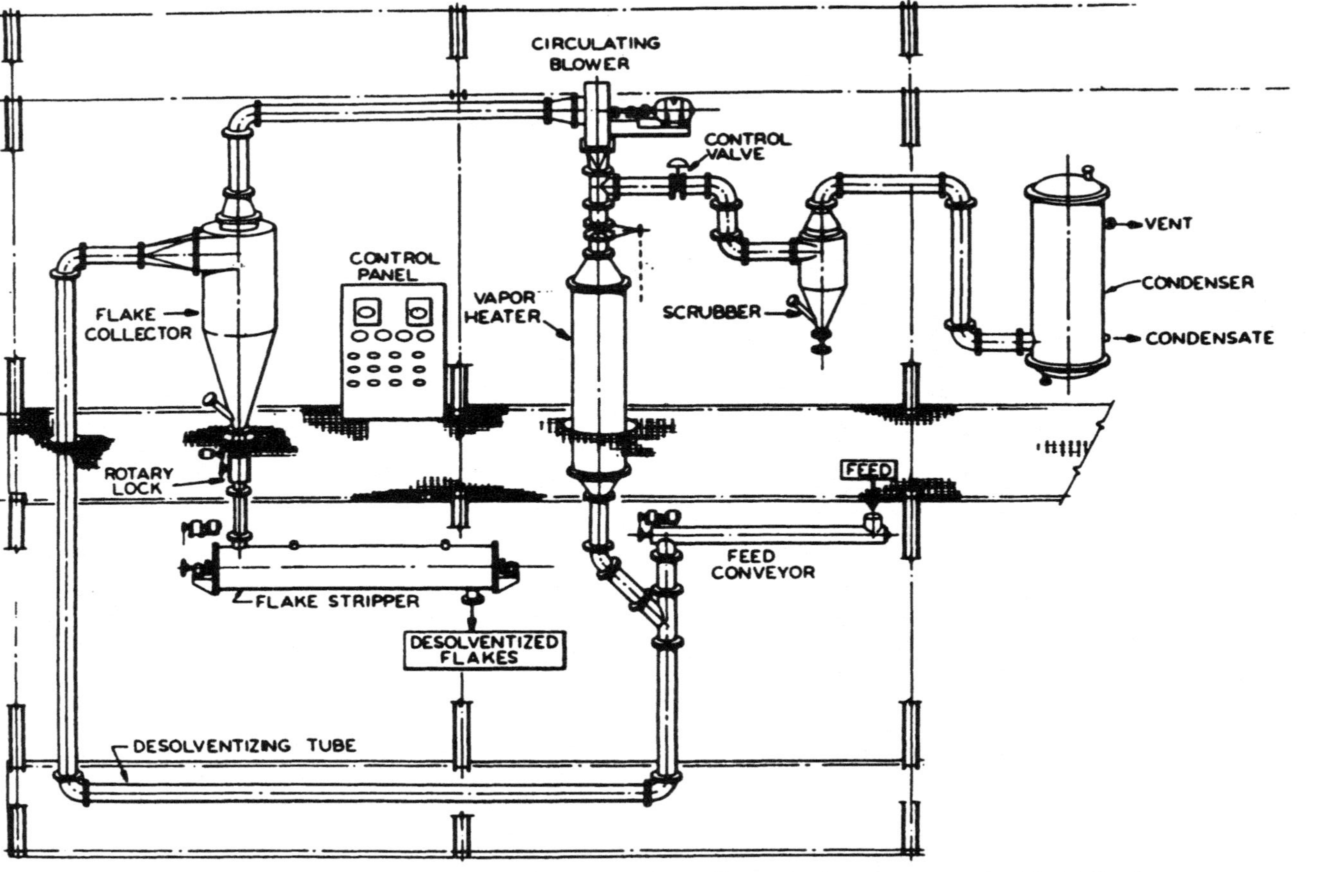

Fig. 7.11. Flash desolventizing system. *Source:* Vivlitis, A., and E. Milligan, in *Proceedings of the World Conference on Oilseed Technology and Utilization,* edited by T.H. Applewhite, American Oil Chemists' Society, Champaign, IL, 1993, p. 287.

Through U.S. No. 10 mesh	95% minimum
Through U.S. No. 20 mesh	40% minimum
	60% maximum
Through U.S. No. 80 mesh	6% maximum

Meal meeting these standards can be incorporated into livestock feeds without additional grinding.

In the grinding system the size reduction can be achieved by hammermills or by roller mills. Screening is almost universally used to minimize overgrinding of the product and can be accomplished with shaker screens or cylindrical centrifugal screeners.

The flow of the product through the grinding system may be closed-loop, where oversized meal from the screener is recycled back to the grinding mill, or open-loop, where the grinder configuration is adjusted so that no oversize particles can pass through. Closed-loop grinding was the traditional system until product standards became less stringent, about 1979, and open-loop systems came into use because they required less power. Flow diagrams for closed-loop and open-loop grinding with hammermills are shown in Fig. 7.12. A roller mill grinding flow is shown in Fig. 7.13.

The principal advantage of roller mill grinding is the reduction in power made possible by the more efficient grinding action in the roller mill. A comparison of grinding power for a 1350 MT/d (1480 short ton/day) soybean plant is shown in Table 7.2.

In addition to a reduced cost for power, the roller mill system produces a more uniformly sized meal with a minimum of dust. The major disadvantage of the system is that the bulk density of the meal is 5 to 10% lower than conventionally ground meal, resulting in less meal in the shipping container. Processing changes to eliminate this deficiency will add additional complications and costs and tend to shrink the advantage of the roller mill system.

Soybean Meal Storage and Shipping

Surge storage, to even out shipping cycles, is provided at processing plants in vertical bins varying in individual capacity from 100 to 600 short tons (90 to 540 MT).

TABLE 7.2 Power Consumption for 1350 MT/d Plant Capacity (45 MT/hr meal) (17)

Hammermill system	Connected HP	Roller mill system	Connected HP
2 mills at 100 HP each	200	1 roller mill at 75HP	75
Air assist 2100 CFM each			
Mill at 5 HP each	10	Air assist not required	0
2 deck screeners at 7.5 HP	15	1 centrifugal screener	25
3 conveyers at 3 HP each	9	2 conveyors at 3 HP	6
Prebreaker not required	0	Prebreaker at 50 HP	50
Total connected power	234 HP (174.56 kw)	Total connected power	156 HP (116.38 kw)
At 85% usage, 350 days/yr, and $0.05 /kw:			
Annual power cost:	$62,319	Annual power cost:	$41,546

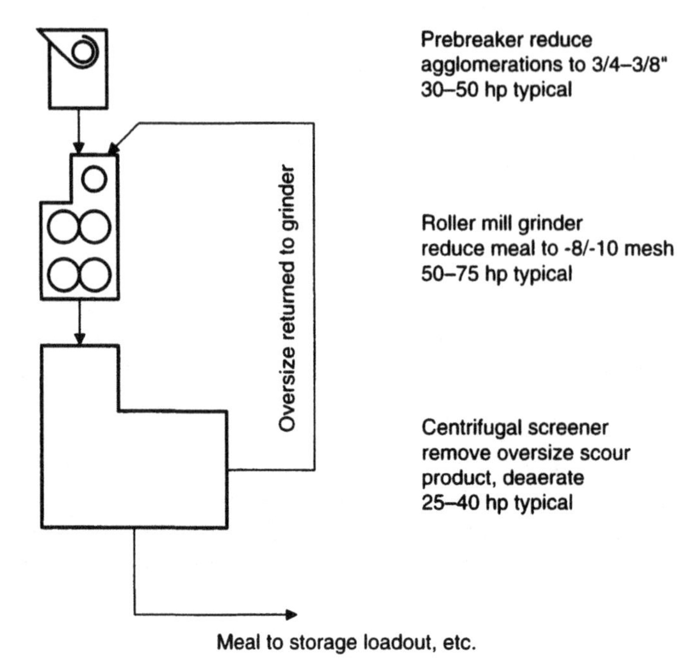

Fig. 7.12. Closed and open loop meal grinding. Courtesy Roskamp Champion.

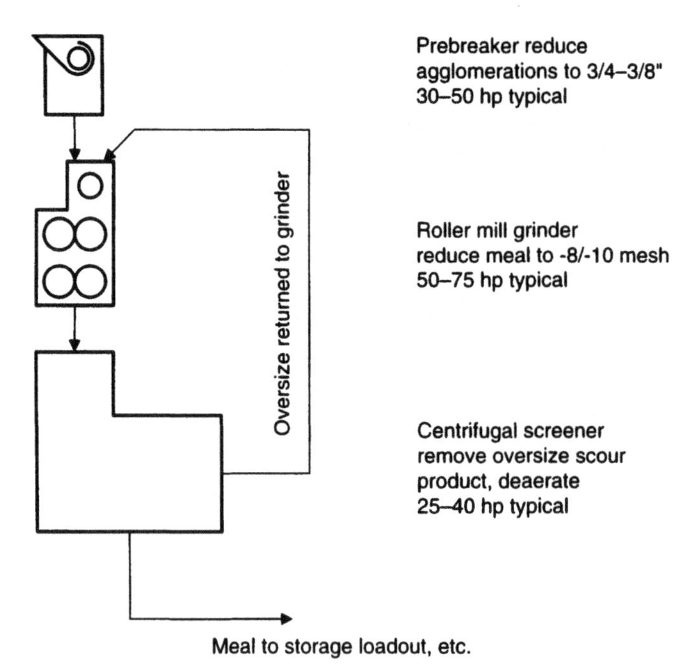

Fig. 7.13. Roller mill grinder system. Courtesy Roskamp Champion.

TABLE 7.3 Effect of Soybean Meal Cooking Time on Broiler Feed Efficiency (24)

Cooking time (min)	Urease (ph rise)	T inhibitor (mg/g)	Mean body wt	Feed/gain
0	0.19	12.12	605	1.61
5	0.11	7.84	625	1.53
10	0.02	1.77	643	1.51
15	0	0	626	1.54
20	0	0	596	1.59
25	0	0	565	1.68

Longer-term storage, to accommodate the demands of the market, may be done in conventional grain silos, holding on the order of 1000 tons each, or in flat storages, which are discharged by front end loaders or similar solids-handling vehicles.

Soybean meal at the normal commercial moisture of 12%, particularly at summer temperatures, will cake in storage and interfere with good discharging of the bin (18, 19). Flow coating agents to minimize the sticking are often added to the meal to improve storability. Reducing the moisture of meal being put into storage by about 2% will greatly improve storability, but this is only done where extreme storage conditions exist. Carson (18) also stresses the importance of cooling.

Caking of the meal in storage becomes progressively worse with time. If the meal is *turned* (moved from one bin to another) before caking has progressed to the point where the meal will not discharge, flowability can be maintained. Milligan (19) describes a storage system specifically designed to turn meal on a continuous basis and thus maintain flowability.

The best bin design for minimum problems should include steep hoppers, wide discharge openings and relatively small diameter bins (18). Unground meal will generally store better than normally ground meal.

In the United States, and other developed countries, most of the soybean meal is shipped in bulk containers from the processing plant to the feed mill end user. Overland shipments are hauled in trucks, railroad cars, or barges. Overseas movements are in bulk cargo ships. Loading and unloading stations must be designed to contain the dust that is generated or released in these operations.

Soybean Meal Utilization

The principal market for toasted soybean oil meal is incorporation into livestock feeds as the principal source of protein. Volumewise, soybean oil meal amounts to

TABLE 7.4 Performance-Influencing Variables in Soybean Meals (26)

Meal	Raw	Severely undercooked	Under-cooked	Quality processed	Over-cooked	Severely overcooked
Amino acid digestibility (%)	40	57	75	100	80	45
Usable energy (%)	49	80	85	100	92	86
Trypsin inhibitor activity (mg/g)	58	>21	12–20	6–7	3–4	0
Urease activity ΔpH	2.0	0.51–2	0.21–0.50	0.05–0.20	0.04	0

55% (65 million MT) of the total world production of defatted protein meals (119 million MT) (20). The dominance of soybean oil meal is due to its high nutritional value and favorable cost of production.

Because it is the principal source of protein in many formulas, the maintenance of the optimal nutritional value in the toasted meal is very important. Additional emphasis on this point arises from the current interest, especially in Europe, in more controlled feeding so as to lower the impact of unused nitrogen (protein) and phosphorus on the environment. Intensive livestock farming is being criticized severely by environmentalists for its excessive nitrogen and phosphorus output (22), which causes highly eutrophic effluents and the elevation of nitrate levels in the groundwater. Feeding programs that decrease the protein level require the highest possible protein quality to maintain optimal growth of the animal.

It has long been known that soybean meal must be cooked (toasted) (21, 27) to eliminate antinutritional factors that interfere with absorption and utilization of the protein. It has also been known that meal can be cooked for too long a time with a resulting decrease of nutritional value. Cooking at high moisture levels (15 to 25%) provides some protection against the effects of too long a cooking time; additionally, high moisture speeds the destruction of trypsin inhibitor and other antinutrients, thus reducing the amount of cooking required.

Recognizing that there is an optimal level of cooking for maximum nutritional value then leads to the need for an analytical method with which to maintain production control and to provide a basis for measuring quality in trading.

The method used today for rapid production control is the measurement of remaining activity of the urease enzyme by means of the Caskey-Knapp urease procedure (8), which expresses urease activity as the rise in pH of a meal sample solution treated with urea under standardized conditions. A pH rise of less than 0.2 units is generally regarded as indicating a satisfactorily cooked meal. Urease itself is not nutritionally significant, but a low level is important in dairy feeds that contain urea; levels of 0.2 pH rise or above may generate ammonia in a urea-containing feed from the action of the enzyme.

Another analytical method which offers reasonably rapid response but is considerably more complicated than the urease test, is the measurement of residual trypsin inhibitor (9). It has been generally accepted that whatever is an antinutrient in meal can be considered as eliminated provided that the trypsin inhibitor activity has been destroyed by heat treatment to the proper extent (22), generally meaning a level below about 6 mg/g of residual activity.

A disadvantage of both the urease test and the trypsin inhibitor test is that they provide a poor measurement of the amount of overheating. Color of the meal is as good an indication of overheating as is available, with the exception of actual feeding tests.

Many investigators have documented the role of moisture in the toasting step (5,21,23,24,25). All show that moisture increases the rate of urease and trypsin inhibitor destruction. The same levels of destruction can usually be attained at lower moisture levels with increased time or temperature of cooking; however, at moisture levels below 12%, satisfactorily low levels of urease and trypsin activity cannot be attained even with excessively long autoclaving times at 120°C.

Undercooked meal has lower efficiency, due to the effect of the antinutritive factors (trypsin inhibitors, etc.), reduced palatability, and poorer amino acid and energy digestibility. Overcooking results in the loss of methionine and lysine and, in severe cases, a reduction in digestibility.

The effect of proper cooking is illustrated by data from McNaughton (24) listed in Table 7.3. An undercooked commercial soybean meal was given additional cooking in incremental steps and then fed to broilers.

In this experiment the best growth was attained after 10 min of additional cooking. Further cooking resulted in decreased growth, but the urease and trypsin inhibitor measurements would not be able to give a good indication of the optimal point.

The improvement from undertoasted to properly toasted was about 7%. Another study (23) reported a range of 15% in the nutritive values of four samples of commercial meal fed to rats. Table 7.4 (taken from a bulletin from a feed manufacturer's research group) contains data compounded from various sources.

Many broiler growers in the United States believe that soybean meal with a urease rise of 0.1 unit is approaching overtoasted and does not give optimal results with broilers.

On the other hand, Matrai (22) "would stress caution regarding the existing allowances for heat treatment; much more steaming action may be needed for products to be used in especially sensitive animal groups, such as infants". His data show a significant level of antigen (especially important to young animals) remaining at a trypsin inhibitor level of 3.1.

In pigs, underprocessing of soybean meal can reduce gains up to 30% and increase conversion (feed/gain) up to 40%. Field experiences indicate that undercooked meal requires 14 to 21 days longer to market (454 to 200 lb) and 50 to 75 (23 to 35 kg) more feed per hog (26); overcooked meal can be nearly as costly in terms of reduced performance as undercooked meal. Hayward (21) cites instances of improvement in value with properly cooked over improperly cooked meals varying from $37.50 to $58.00/short ton ($41 to $64/MT). Sipos (27) cites extensive references showing the value of proper cooking.

Feed rations for livestock and poultry are formulated on the basis of the least cost of ingredients to fit the nutritional needs of the animal at its particular stage of the life cycle. Soybean meal contains a well-balanced mix of amino acids in concentrated form, is normally available in good supply, and is competitively priced. These factors account for its position as a preferred ingredient in feeds.

References

1. Kruse, N.F. and W.L. Soldner, U.S. Patent 2,260,254 (1941).
2. Hayward, J.W., *Flour and Feed 41:* 24 (1940).
3. Staff article, *J. Amer. Oil Chem. Soc. 54:* 202A (1977).
4. Kruse, N.F., U.S. Patent 2,585,793 (1952).
5. Kruse, N.F., U.S. Patent 2,710,258 (1955).
6. Hutchins, R.P., U.S. Patent 2,695,459 (1954).

7. Schumacher, H., U.S. Patent 4,503,627 (1985).
8. AOCS Official Method Ba 9-39 (updated 1993).
9. AOCS Official Method Ba 2-75 (updated 1993).
10. Sipos, E., and N.H. Witte, *J. Amer. Oil Chem. Soc. 38:* 11 (1961).
11. Pavlik, R.P., and T.G. Kemper, *INFORM 1:* 200 (1990).
12. Brueske, G.D., in *Proceedings of the World Conference on Oilseed Technology and Utilization, Budapest,* 1992, edited by T.H. Applewhite, p. 126, American Oil Chemists' Society, Champaign, IL, 1993.
13. Weber, K., *INFORM 4:* 499 (1993) (abstract).
14. Kratz, G. and L. Jeromin, in *Proceedings of the World Conference on Emerging Technologies in the Fats and Oils Industry, Cannes, France,* edited by American Oil Chemists' Society, Champaign, IL, 1986, p. 141.
15. Becker, K.W., *J. Amer. Oil Chem. Soc. 60:* 168a (1983).
16. Brueske, G.D., in *Proceedings of the World Conference on Oilseed Technology and Utilization, Budapest,* edited by T.H. Applewhite, p. 126, American Oil Chemists' Society, Champaign, IL, 1993.
17. Anonymous, *Efficient Processing of Solvent Extracted Meals,* Roskamp Champion, 1992.
18. Carson, et al. *Solving Soybean Meal Flow Problems,* Jenike & Johanson Inc., Westford, MA, May 1957.
19. Milligan, E.D., and J.F. Suriano, *J. Amer. Oil Chem. Soc. 51:* 150 1974.
20. Kohlmeier, R.H., in *Proceedings of the World Conference on Fats and Oils Processing,* Maastricht 1989, American Oil Chemists' Society, Champaign, IL, 1990, p. 390.
21. Hayward, J.W., *Feedstuffs,* April 28, 1975.
22. Matrai, T., in *Proceedings of the World Conference on Oilseed Technology and Utilization, Budapest,* edited by T.H. Applewhite, p. 126, American Oil Chemists' Society, Champaign, IL, 1993.
23. Wright, K.N., *J. Amer. Oil Chem. Soc. 58:* 294 (1981).
24. McNaughton, J.L. *J. Amer. Oil Chem. Soc. 58:* 321 (1981).
25. Mustakas, G.C., et al., *J. Amer. Oil Chem. Soc. 58:* 300 (1981).
26. Fabor, J.L., and J.C. Russett, *Soybean Meal Quality: Is It Constant?,* Master Mix Feeds, Decatur, IN, June 1984.
27. Sipos, E., Biological Value of Soy Proteins in Animal Nutrition and the Quality of Commercial Soy Protein Products, in *International Symposium on Amino Acids in Livestock Nutrition, Brno, Czechoslovakia,* 1988.

Chapter 8

Soy Protein Processing and Utilization

Edmund W. Lusas and Khee Choon Rhee

Food Protein Research and Development Center
Texas A&M University System
College Station, Texas

Introduction and Definitions

Types of Products

The term *soy proteins* typically refers to processed, edible dry soybean products other than animal feed meals. Many types are produced for use in human and pet foods and milk replacers and starter feeds for young animals.

Full-fat soy flours and grits are made by milling dehulled cotyledons to specific sizes and typically contain 40% protein (N × 6.25) on an "as is" basis. *Defatted* soy flours and grits are prepared by milling solvent-extracted flakes of dehulled soybeans and contain 52 to 54% protein as is. At least 97% of a *flour* must pass through a U.S. Standard No. 100 sieve. *Grits* are milled to specific particle size ranges to pass through sieves between U.S. Nos. 8 and 80, according to the manufacturer's or buyer's specifications. Full-fat and defatted flours are available in enzyme-active forms or in various degrees of water solubility, expressed as Protein Dispersibility Index (PDI) or Nitrogen Solubility Index (NSI). *Refatted* or *lecithinated* (0.5 to 30%) flours are made for applications in which a crude soybean oil flavor is not acceptable, dustiness must be minimized, fat must be provided in the formulation, or rapid dispersibility of powders is desired.

Soy protein concentrates contain 65% minimum protein, mfb (moisture-free basis), and essentially are flours from which the water- or alcohol-soluble components, including flatulence-promoting sugars and strong flavor compounds, have been leached before drying. Fiber is additionally removed in making *soy protein isolates*, which contain 90% protein minimum, mfb. Functionalities of soy concentrates and isolates may be modified by adjustment of pH with sodium or calcium bases, application of mechanical stress, and hydrolysis by proteolytic enzymes before drying (1). Approximate composition ranges of soy protein products as is and mfb, reported by the Soy Protein Council, are shown in Table 8.1.

Other soy protein products include dried soy milks and tofus, as well as mixtures of soy flours, concentrates, or isolates with milk or egg fractions, gelatin, or other components for specific functional applications. Extruder-texturized flours and concentrates and "spun" fiber isolates (which resemble muscle meat in appearance) may be made at the point of use but usually are supplied in bulk from strategically located production facilities. Edible coproducts of soy protein manufacture include hulls and the fiber coproduct from production of soy protein isolates.

TABLE 8.1 Typical Compositions (%) of Soy Protein Products

	Defatted Flours and Grits		Protein Concentrates		Protein Isolates	
Constituent	as is	mfb[a]	as is	mfb[a]	as is	mfb[a]
Crude protein (N x 6.25)	52–54	56–59	62–69	65–72	86–87	90–92
Crude free lipid (pet. ether)	0.5–1.0	0.5–1.1	0.5–1.0	0.5–1.0	0.5–1.0	0.5–1.0
Crude Fiber	2.5–3.5	2.7–3.8	3.4–4.8	3.5–5.0	0.1–0.2	0.1–0.2
Ash	5.0–6.0	5.4–6.5	3.8–6.2	4.0–6.5	3.8–4.8	4.0–5.0
Moisture	6–8	0	4–6	0	4–6	0
Carbohydrates (by difference)	30–32	32–34	19–21	20–22	3–4	3–4

[a]mfb: moisture-free basis.
Source: Soy Protein Products, Soy Protein Council, Washington, DC.

Most soy proteins are sold in bulk as ingredients for commercial meat processing, baking and for remanufacture into grocery store, fast food shop, institutional, and restaurant convenience foods. Except for imitation bacon bits, most forms of soy proteins are seldom seen directly by the general public. Limited quantities of soy flours, protein concentrates and isolates, texturized chunks, and processed organically grown soybean products are sold through health food stores. Products with pareve and kosher certification are available.

History

The development of soy protein processes and uses may be followed through publications including the handbooks by Circle, 1950 (2), and by Smith and Circle, 1972 (3); patent abstracts (4,5); and proceedings of the American Oil Chemists' Society World Conferences at Munich, 1973 (6), Amsterdam, 1978 (7), Acapulco, 1980 (8), Singapore, 1988 (9), and Budapest, 1992 (10). Notable bibliographies were prepared by Wolf and Cowan (11) and by Shurtleff and Aoyagi (12,13,14,15). In January 1994, the *Food Science and Technology Abstracts* global database, established in 1969, contained approximately 4,700 abstracts on preparation, utilization, and nutritional characteristics of soy proteins and foods or on uses of soy proteins as reference standards for comparing other foods. The major research concentrations for soybeans have been on soy milk and tofu, on soybeans as aids in meat processing, and on extrusion processing. Since global enforcement of 200-mile-offshore commercial fishing limits and consequent interest in possible substitutes for fish products, Japanese scientists have become the most prominent national group in soy protein research publications, with South Korea and the U.S. also well represented.

The preparation of soy proteins is described in this chapter, from the simpler to the more complex processes, but the technology did not develop in such an orderly fashion. A crude soy flour was sold in the United States as early as 1926, primarily as "health flour" (3). Flours were initially made from whole soybeans, from hydraulic press or expeller cakes, and later from solvent-defatted soybean meals.

The strong beany flavor limited market growth of soy flours, and considerable research effort was put into developing "debittering" processes to remove the objectionable compounds. In the presence of moisture and heat the natural lioxygenase enzymes act rapidly to develop beany flavor in disrupted cotyledon tissue during processing, and they continue to foster oxidative rancidity during product storage unless deactivated. Recognition of this fact led to improved processes and products. In 1966, Mustakas et al. (16) reported an extruder cooking process for making toasted, full-fat soy flour for use in developing countries. Heating soybeans to deactivate lipoxygenases and other enzymes before milling, and rushing soybean cotyledons into heat treatment after cracking and dehulling, have resulted in modern soy flours remarkably low in beany flavor.

Soy isolates were developed next and introduced to the market in 1937 by the Glidden Company as replacements for bovine casein as pigment binders in paper coatings and for production of blanketing foam for fire fighting. The availability of isolate also led to research on spinning soy protein into textile fibers, which did not materialize as an industry. However, protein isolate fibers were commercially spun as tows and worked, flavored, and colored to mimic meat products during the late 1960s and early 1970s (3). Short spun fibers are produced today for use as texturizing agents in fabricated food products.

Research in debittering soy flour showed that extraction with aqueous ethanol was the preferred means for removing strong flavor components and flatulence sugars. Soy concentrates were first produced commercially in the early 1950s, also for industrial applications (17). A patent issued to Morse in 1943 for a process to remove sugars, salts, and other soluble materials by leaching soy flours at the isoelectric point is one of the first descriptions of commercial making of soy protein concentrate (18).

In 1959, Sair (19) developed the first process specifically for making soy concentrates for food use. Defatted soy flakes were acid-leached, then neutralized before drying. With this development, concentrates were rapidly accepted as food ingredients, intermediate between flour and isolates in cost, protein content, and flavor. Cooking-texturizing extruders also were introduced at about this time and enabled the production of texturized soy flour or concentrate meat analogs at appreciably lower costs than those produced by the protein isolate spinning process.

Water solubility is considered an index of a soy protein's functional physical properties and biological and enzymatic activities. Hexane extraction of soy flakes is essentially an anhydrous process, with very little water present to react with the protein. However, approximately 40% of the weight of the *marc* (extracted drained flakes) is "hold up" solvent, and energy must be supplied for its vaporization. Energy for vaporizing hexane in desolventizer-toasters (DTs) is provided by contact heating and direct injection of steam. Hexane and water form a low-boiling azeotrope at about 94.5:5.6 weight:weight ratio. Complete evaporation of the hexane is favored by an excess of steam condensate in the extracted flakes. Both moisture and heat exposure are detrimental to protein solubility. Soybeans seldom reach Nitrogen Solubility Index (NSI) values above 70 in meals processed in contact-heated DTs, nor above 50 when steam is injected. The development of flash desolventizing systems (FDSs), a technique for desolventizing marc without the addition of water (see

Chapter 7), enabled the production of "white flakes" (WFs) with high NSIs. In this process, the marc is transported by superheated solvent vapors through a desolventizing tube. As heat is surrendered, the adhering solvent is evaporated and the vapors are swept away to a condenser.

Analysis

The soy proteins industry uses the applicable Official Methods and Recommended Practices of the American Oil Chemists' Society (AOCS) for analyses and for trading rules and litigation. Only part of the nitrogen in soy proteins is of protein origin. The AOCS conversion factor for soybean protein is $N \times 5.71$; however, industry practice is to label protein as "Protein ($N \times 6.25$)" (20).

Two methods are broadly used to evaluate protein solubility in soy proteins. The Protein Dispersibility Index (PDI; AOCS Official Method Ba-10-65, 1993) "rapid stir" method uses a blender to disperse the sample, and the Nitrogen Solubility Index (NSI; AOCS Official Method Ba 11-65, 1993) "slow stir" method uses a laboratory stirrer. In both methods, the protein or nitrogen that is leached into the liquid phase is compared with total protein or with nitrogen in the sample, determined by Kjeldahl analysis. The NSI method gives lower values and has been related to PDI by the formula (21)

$$PDI = 1.07(NSI) + 1$$

One objective in heating soy protein is to inactivate (primarily Kunitz-type) trypsin inhibitors, which act as protease inhibitors and antigrowth factors by restricting protein digestion in monogastric animals. At least 80% reduction of the approximately 85 to 95 trypsin inhibitor units (TIU)/mg of solids, normally present in raw soy flour, is sought. The rationale is that test animals have well demonstrated the ability to tolerate low levels of inhibitor, and the soy ingredient will receive additional heat treatment before consumption of the product (22).

A relationship between trypsin inhibitor activity, protein efficiency ratio (PER), and steaming of soybean meal is shown in Fig. 8.1 (23). PER values, determined by rat tests, are no longer used to assess quality (the essential-amino-acids balance) of proteins intended for food use, because of rats' and fur-bearing animals' extraordinarily high requirements for sulfur amino acids. However, these values are cited in this review as the best indexes of protein quality in earlier research literature. The current technique for evaluating protein quality for adults and children over one year of age is the Protein Digestibility-Corrected Amino Acid Score (PDCAAS) (24,25,26).

Only 40% of growth inhibition in test animals has been related to trypsin inhibitor activity. Heating also at least partially inactivates heat-labile hemagglutinins (lectins), goitrogens, antivitamins, and phytates, but not the heat-stable saponins, estrogens, flatulence factors, and allergens. Further, heat denaturation in itself increases digestibility of bean and soy proteins (27). It should be remembered that many of the antinutritional factors and enzymes that affect product quality in soybeans are rich in the essential amino acids and beneficial to the diet once inactivated.

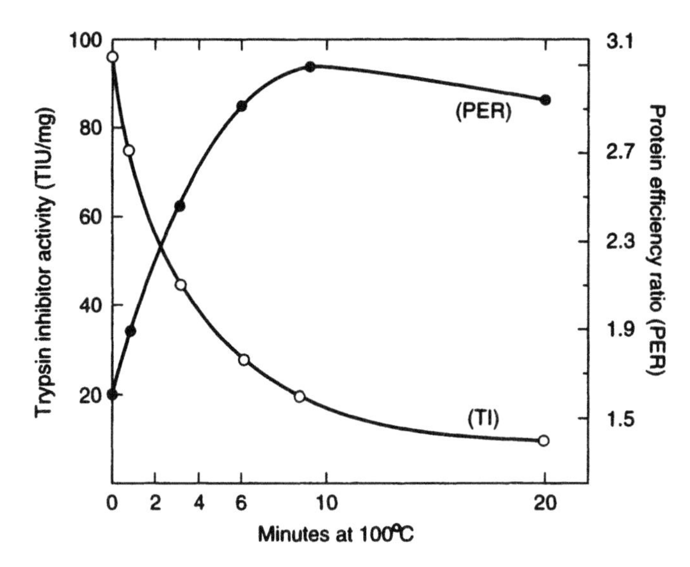

Fig. 8.1. Effect of atmospheric steaming on trypsin inhibitor activity and protein efficiency ratios of soybean meal fed to rats. *Source:* Rackis, J. J., *J. Amer. Oil Chem. Soc. 51(1):* 161A, 1974.

Urease activity has come to be used as an index for trypsin inhibition activity and is easier to analyze. Albrecht et al. (28) have shown that trypsin inhibitor activity is destroyed at approximately the same rate as urease (Fig. 8.2). High initial moisture promotes rapid decrease in both NSI and urease. Small particle size influences the reduction of urease activity but has little effect on the rate of NSI reduction. By steaming soybean fractions of small particle size (under 20-mesh) and low moisture (8%), it is practical to destroy urease activity and retain high NSI. Figure 8.3 shows the relationship between urease activity and NSI under conditions of a typical atmospheric steaming process (29).

Handling of Soybeans and Soy Protein Products

Considerable secrecy exists about processing of commercial soy proteins. The combinations of options for making different products are limitless. Even in making the same type of product, techniques and equipment differ between manufacturers and result in slightly different products. In processing all soy proteins, it is essential to start with thoroughly cleaned, sound, mature, yellow soybeans sorted to uniform size. They should be dried and subsequently handled at low moisture (9 to 10%) and mild heat if high-PDI (85+) enzyme-active products are to be made. Typically, food-grade soy proteins are produced on different lines than are used for oil and feed meals, with split and rejected soybeans diverted to animal feed meal extraction operations (30,31). Some manufacturers have washed soybeans to remove dirt and small stones (32). Specially designed extractors (with self-cleaning, no-flake-breakage features) and use of a narrow-boiling-range (66 to 68°C, 151 to 154°F) hexane are recommended for producing WFs (30).

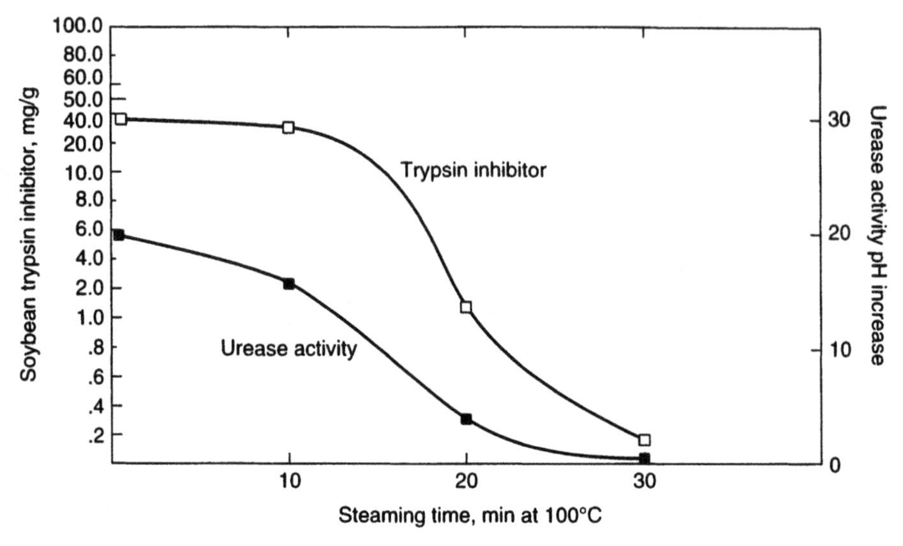

Fig. 8.2. Relationship of urease activity to trypsin inhibitor. *Source:* Wright, K.N., *J. Amer. Oil Chem. Soc. 58:* 294, 1981.

McDonald (33) reported that dehulling of soybeans is enhanced by drying on receipt—first rapidly at 79°C (174°F) to lower the moisture content to a range of 12.5 to 13.0%, and then at a lower temperature of 65°C (150°F) to a range of 9.0 to 10.0% moisture. The soybeans should be stored in this condition for 15 to 30 days before processing.

Although not as extensively studied as in dry field beans *(Phaseolus* species), loss of protein solubility on storage is well known in whole soybeans and in soy proteins (3,34,35). Changes in dried field beans can be slowed by lowering storage temperature and relative humidity (RH). Thomas et al. (36) reported a decrease of 14% in protein extracted into soy milk from soybeans stored for 8 months at 30°C (86°F) compared to those stored at 20°C (68°F) at RH of 85% and 65%, respectively. Tofu from soybeans stored at 85% RH became less uniform in microstructure toward the end of the storage period. Earlier research by others had shown that extractability of the 11s (glycinin) protein declines more rapidly than that of the 7S (conglycinin) (37). Storage at higher RH is more deleterious than higher temperature. Further, at high storage humidities and temperatures of soybeans, the color of extracted soy milk darkened, its pH decreased, and electron spin resonance spectra of stored soybeans suggested that the combination of phosphatidylcholine with the proteins in soy milk was strengthened (38). Some soy protein manufacturers recommend storage of their products, already in multiwall bags, at below 24°C (75°F) and 60% RH.

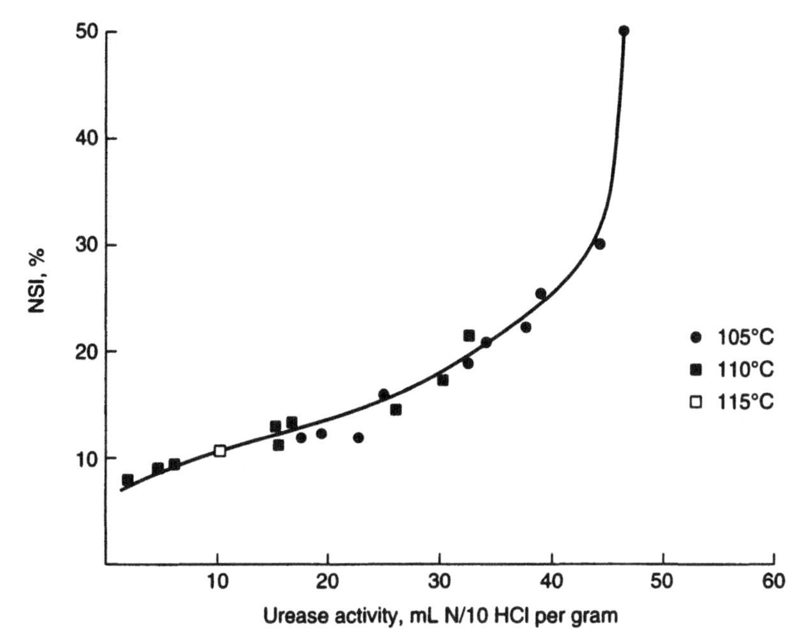

Fig. 8.3. Relationship of urease activity to Nitrogen Solubility Index. *Source:* Wright, K.N., *J. Amer. Oil Chem. Soc. 58:* 294, (1981).

Full-Fat Soy Flours and Grits

Three types of full-fat soy flours are produced: (1) enzyme-active; (2) toasted; and (3) extruder-processed. Refatted flours sometimes are referred to as "fully-fatted" or "full-fat", but they are described in this chapter in the section on defatted soy flour.

Enzyme-Active Soy Flours

Enzyme-active flours are used for the action of their lipoxygenases in bleaching wheat flours and conditioning doughs in Western-type breads. Soy β-amylases are more heat-stable than those of wheat or barley and remain active longer in the early stages of baking, contributing to improved texture. Enzyme-active soy flours are available in full-fat and defatted forms, with the former more popular in Europe and the latter in the United States (31).

The cleaned soybeans are cracked into 6 to 8 pieces, and the hulls removed by aspiration. The hulls may be loosened by moisture adjustment and mild heating before cracking or by passing the cracked pieces through corrugated rolls revolving at different speeds (see Chapter 6). Hulls are removed by shaker screen and aspiration (30,31). The dehulled pieces then are ground by a hammer mill or by an impact pin mill like the Alpine Contraplex (Alpine American Corp., Natick, MA) or Entoleter (Entoleter, Inc., New Haven, CT) into flours with desired particle sizes. Full-fat products are very difficult to pulverize or sieve. Customarily, they are not screened but are milled in two steps with separation of the coarse from the fine particles by air classification between grindings (30,31,39). One U.S. manufacturer's specifications for a commercial full-fat enzyme active soy product are

Protein (mfb): 42.0 ± 1%
Moisture: 10.0% maximum
Fat: 21.0 ± 0.5%
Ash: 4.7 ± 0.2%
Granulation: less than 1% on U.S. No. 45.

In 1991 a processing plant was established in Iowa (Mycal Group, Niichi Corp., Jefferson, IA) to make full-fat flakes from white hilum soybeans for export to the Far East for making soy milk and tofu.

Toasted Full-Fat Soy Flours and Grits

Toasted full-fat soy flours are also called "heat-treated full-fat soy flours." In order to minimize development of beany flavor by lipoxygenase, the cleaned whole soybeans often are steamed under slight pressure for 20 or 30 minutes, then cooled, dried, cracked, usually passed over a shaker screen and aspirated to remove hulls, and milled with sieving to produce full-fat grits or flour. When properly processed, the product is yellow to slightly tan in appearance, with a nutty flavor and aroma. The undesirable enzymes have been destroyed, and product PDI is in the range of 20 to 35. Toasted soy flours, ground to U.S. No. 100 or 200 mesh, are available domestically with special granulations possible. Specifications for grits vary with the manufacturer, for example (30,31,39):

Coarse: through No. 10 screen on No. 20
Medium: through No. 20 on No. 40
Fine: through No. 40 on No. 80

Heating of whole soybeans by infrared (IR) radiation before cracking has also been reported an effective means for deactivating lipoxygenase (40), but is not practiced commercially.

Optionally, the dehulled soybean chips may be screw-pressed to partially remove the oil and the press cake cracked, ground, and screened to produce low-fat soy flour or grits (39).

Extruder-Prepared Full-Fat Soy Flours

Extruded full-fat soy flours were pioneered by Mustakas and coworkers at the USDA Northern Regional Research Center, Peoria, IL (16,41). An extruder consists of a rotating screw in a barrel, both designed to compress relatively low(<18%)-moisture powders into a flowing mass, which can be sheared, cooked, cooled, and shaped into continuous extrudates that then can be cut into pieces. The product will expand ("puff") at the die if released from a pressurized zone into the atmosphere, and it will retain an expanded volume if sufficient active protein or starch is present

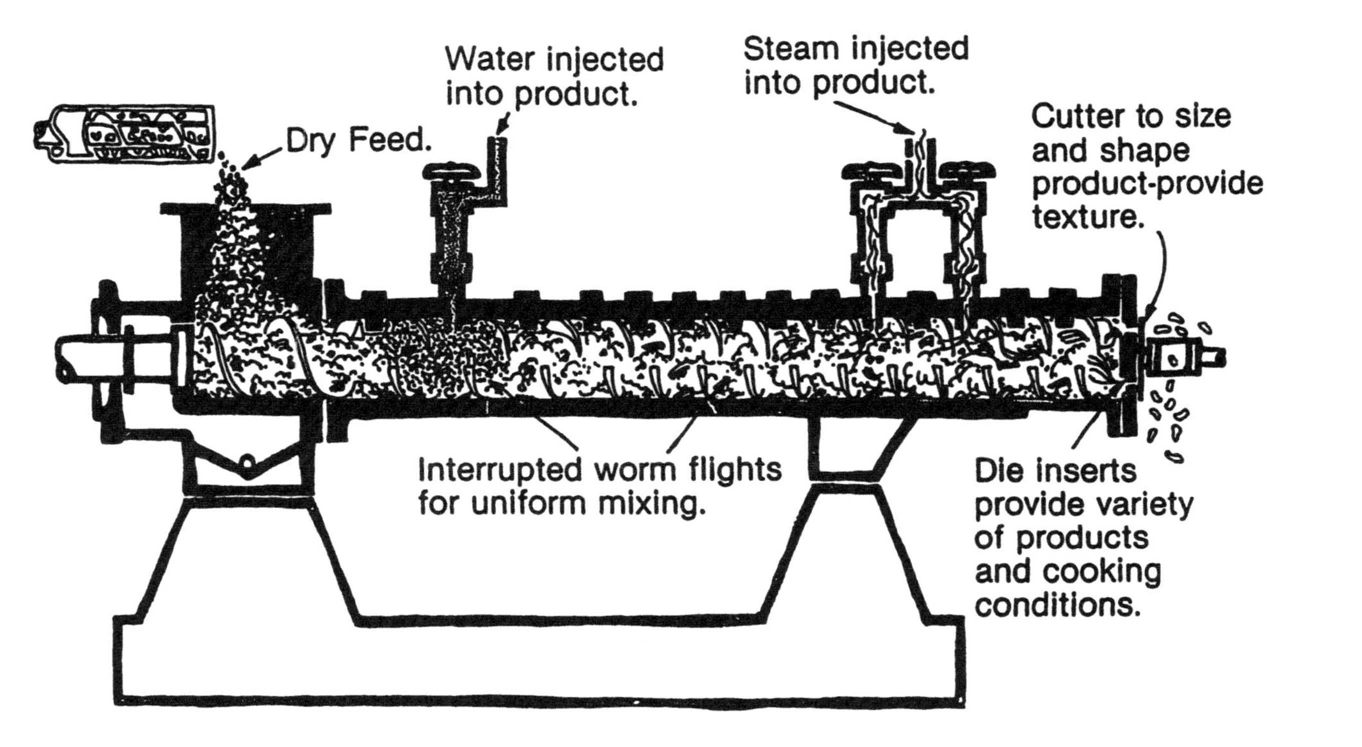

Fig. 8.4. Cross-section of interrupted-flight extruder used for production of toasted full-fat soy flour. Courtesy of Anderson International Corporation, Cleveland, OH.

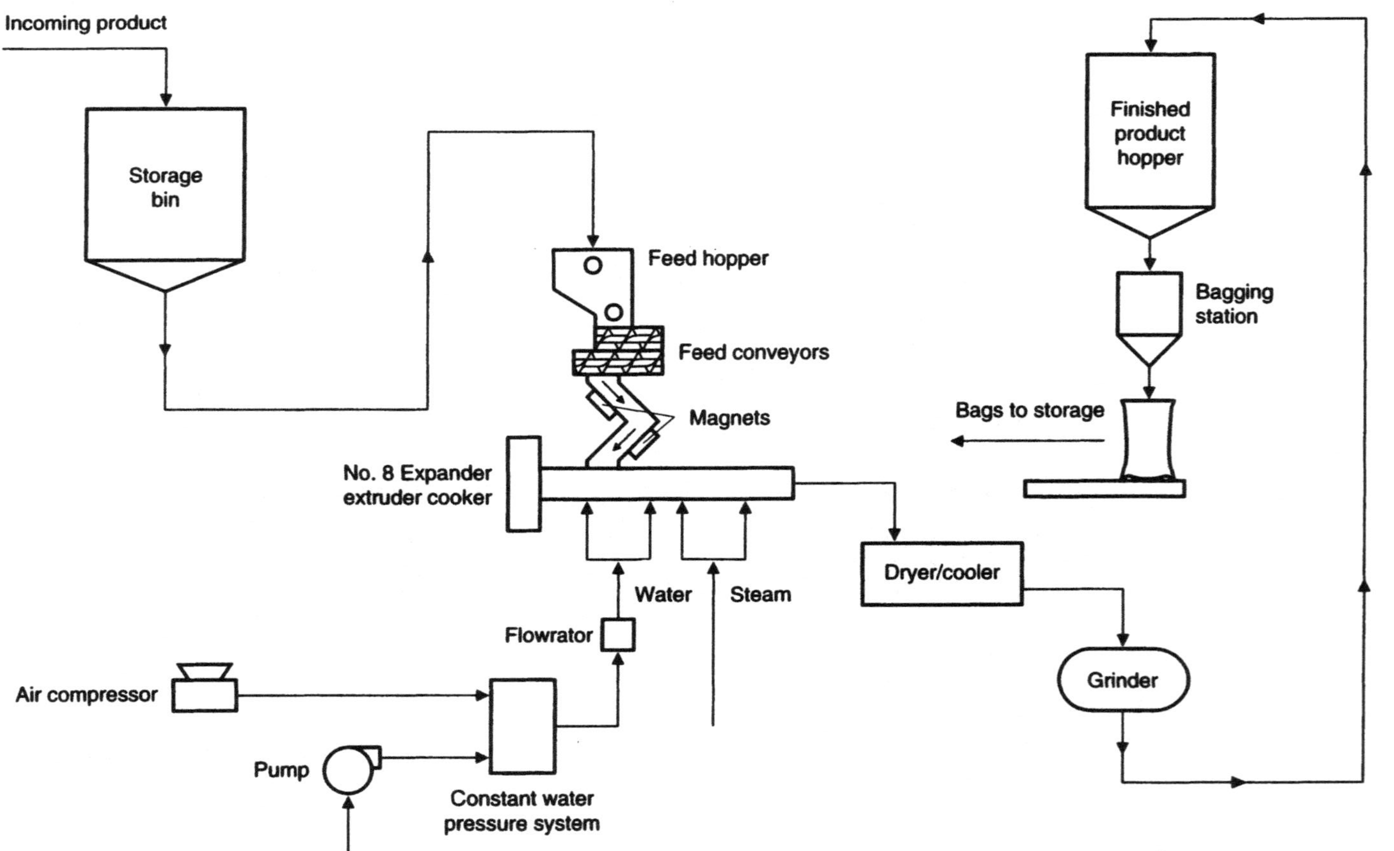

Fig. 8.5. Flow diagram for making extrusion-cooked full-fat soy flour. Courtesy of Anderson International Corporation.

for forming a matrix. If discrete shapes are not desired, the face plate, dies, and cutter often are replaced with one of several types of adjustable-cone discharges to produce granulated products. Designs of the screws and barrels vary with the process and manufacturer. Depending on capital limitations, extruders may be heated by passing steam or hot oil through the barrel jacket or by direct steam injection. Low cost "autogenous" machines, which create their heat by friction and do not require local steam-generating capabilities, also are used.

Mustakas and coworkers (41) obtained good results by cracking soybeans in 6-inch (15 cm) corrugated rolls and dehulling by shaker screen and aspirator to obtain grits in the 12 to 30 screen range. These were conditioned to inactivate the lipoxygenase by dry heating for 6 to 8 minutes and attained a discharge temperature 103 to 104°C (218 to 220°F). Tempering to 20% moisture and extruding at 135°C (275°F) with a 2 min retention resulted in a maximum dietary PER of 2.15, trypsin inhibitor inactivation of 89%, urease activity of 0.1 pH change, and a product NSI of 21%.

Fig. 8.6. Dry extruder used for preparation of infant and child foods in developing countries. Courtesy of Insta-Pro® International, Des Moines, IA.

Several types of extruders are used for preparing full-fat soy flours, including the interrupted "cut" flight extruder shown in Fig. 8.4. A flow diagram for making extruded full-fat flour is shown in Fig. 8.5.

Extruder-processed full-fat soy flours have been used for preparation of high-protein content foods and beverage bases for worldwide child and infant feeding programs (42). Studies also have shown that fats are stable in flours cooked to NSIs of 30, but antioxidants are required to ensure stability in flours processed to 11 and 19 NSI values (43).

Considerable research in producing weaning foods for famine relief and improved nutrition in developing countries was conducted by Harper and coworkers at Colorado State University in the late 1970s to mid-1980s using low-cost extrusion cookers (LECs) (44,45,46). These were autogenous machines with simple screw and barrel configurations for "dry extrusion" (moisture <15%) and included the Brady Crop Cooker, the InstaPro Extruder (Fig. 8.6), the Anderson Grain Expander, and local designs. Approximately a dozen LEC plants were brought to production stage, or designed, for developing countries. Full-fat soy flours and coprocessed mixtures of dehulled soy grits with corn, wheat, and rice were produced.

Nutritionists have found that mixtures of approximately 50:50 oilseed or legume protein and cereal protein are complementary in producing the essential-amino-acids profiles desired in infant and child foods. Because of differences in protein content, this has meant the use of soybeans with cereal grains in weight:weight ratios of about 30:70 (47,48). Because of limitations in screw designs, coarsely ground ingredients have worked better than flours in LEC machines. The protein efficiency ratio of full-fat soy flour was optimized by inactivation of antigrowth factors by dry extrusion at 143°C (289°F) (44). Other workers have found that lipoxygenase is completely inactivated when extruding 10% moisture content soybean and corn mixtures in an InstaPro extruder at temperature ranges of 127 to 160°C (260 to 320°F). In the same trials, trypsin inhibitor activity was reduced by 48.9 to 98.8%, and residual lipase activity was reduced from 63.7 to 2.7 μmol H^+/min/g (49).

Some low-cost extruders have the ability to grind whole soybeans and corn kernels into flours during processing. As an example, whole soybeans can be pulverized and heated in InstaPro extruders to make trypsin inhibitor–inactivated full-fat soybean meals for feeding poultry and swine. However, doing so bypasses the opportunity for reducing the fiber content through a separate dehulling step. In large installations, often it is more profitable to select specialized machines for grinding, preheating, extrusion, and cooling if capital is available.

Extracted Flake Products

By far the majority of soy protein products are made from WFs (White Flakes, hexane-defatted flakes of dehulled soybeans graded for food use). The production of WFs has been reviewed by Witte in Chapter 7 of this book.

Johnson (31), Fulmer (50), and Kanzamar et al. (30) have also described the making of WFs. The Soy Protein Council has defined the following flake–NSI categories (1):

"white" 85+
"cooked" 20–60
"toasted" NSI below 20

However, these terms are used loosely in the industry. For example, "toasting" means steam cooking rather than dry heat, and "white flakes" can mean enzyme-active flake or simply defatted flakes or meal used for making soy flours, concentrates, or isolates.

The majority of soybean extraction plants produce soybean oil and feed-grade meal and utilize a DT equipped for direct steam injection and sometimes an additional dryer to remove the condensate. Expanders, to shear soybean flakes and produce collets with enhanced oil extractability, also are common. The resulting extracted products have NSIs of 50 or less, because of denaturation of soy protein by moisture at high temperatures.

In contrast, the production of flakes with high PDIs/NSIs typically utilizes a flash desolventizing system (FDS), also sometimes called a "white flake system". Superheated gaseous hexane [boiling point 70°C (160°F)] under pressure at 116 to 138°C (240 to 280°F) is used to transport the extracted drained flakes pneumatically in a desolventizing tube and evaporates the hold-up solvent in 2 to 5 seconds. Moisture in the meal is also reduced by 3 to 5% during the process. The remaining hexane (about 0.3 to 0.5%) is then removed in a flake stripper using superheated steam under vacuum to obtain high-PDI flakes. Care must be taken to avoid condensation of steam into water on flake surfaces (30,39,51). Once a FDS with solvent vapor circulation is installed, it can be operated to produce flakes with a PDI range of 10 to 85%, depending on how much steam is applied.

White Flakes

WFs are an item of commerce in their own right. Some are produced by soybean processors who supply soy protein manufacturers but do not market soy flours, concentrates, or isolates. The major difference between WFs, grits, and flours is the granulation. It is critical that white flours be made carefully, since they limit the properties and yields of subsequent products. One domestic supplier of high-PDI WFs offers a product with 86 to 88 PDI, 80% lipoxygenase (bleaching activity), 2.2 minimum pH rise urease activity, and the following granulation:

On U.S. No. 20: 35%
Through U.S. No. 20 on U.S. No. 100: 45%
Through U.S. No. 100: 15%

Defatted Soy Flours and Grits

A general flow sheet for the manufacture of full-fat and defatted soy flours is shown in Fig. 8.7. Grinding of WFs into grits typically is done by hammer mills and sifters. Grinding into flours may be done by hammer mills, pin mills, or classifier mills. Particle size distribution can be controlled by air classifiers, with narrower particle

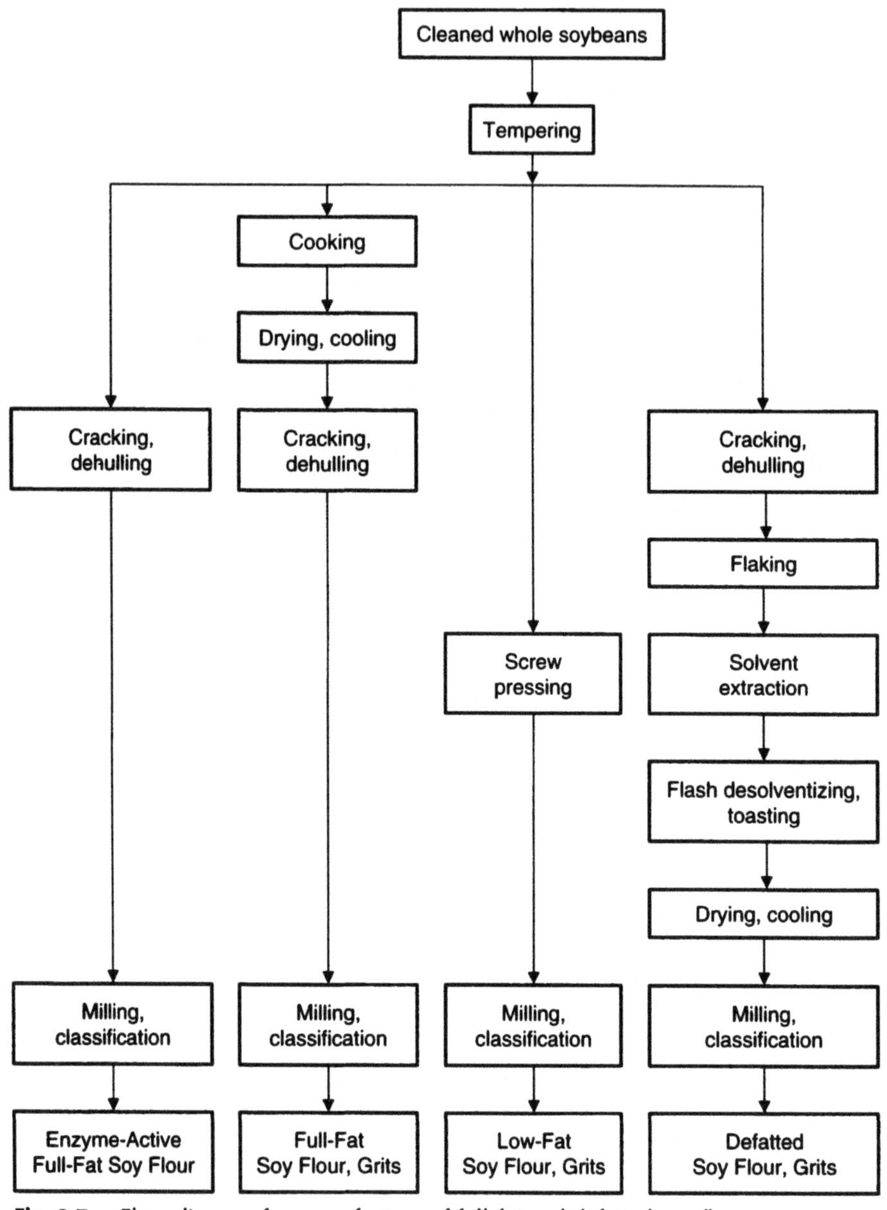

Fig. 8.7. Flow diagram for manufacture of full-fat and defatted soy flours.

size distributions possible when using sifters. Grinding capacities are dictated by flake PDI value and product mesh size. Commercial lines for grinding with a mill and air classifier are shown in Fig. 8.8, and for a classifier mill in Fig. 8.9 (30).

During processing, protein content increases from 41.1% mfb, for soybeans, to 49.4%, for defatted, nondehulled soybean meal, to 53.9% for dehulled soybean meal. Defatted soy flour (Table 8.2) contains about 38% total carbohydrates, including 15% soluble mono- and oligosaccharides, and 13% polysaccharides that are later removed if soy protein concentrates or isolates are made. Maximum total bacterial count specifications range up to 50,000/g depending on supplier and product. However, counts of known disease-producing microorganisms, such as *Salmonella*, coagulase-positive staphylococci, and *E. coli*, must be zero (50).

Relationships between heat processing and nutritional indicators of soy flours are shown in Table 8.3. Increased heating reduces trypsin inhibitor activity, which is

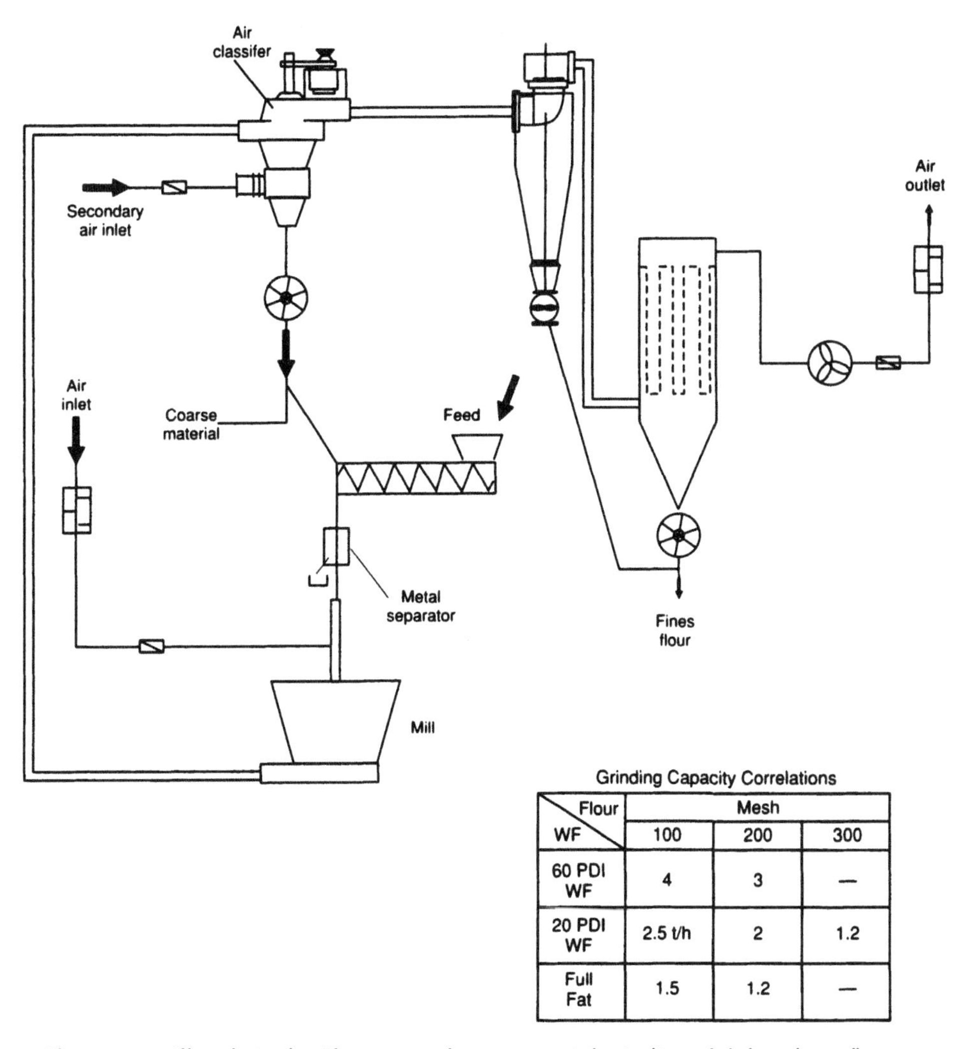

Grinding Capacity Correlations

WF \ Flour	Mesh 100	200	300
60 PDI WF	4	3	—
20 PDI WF	2.5 t/h	2	1.2
Full Fat	1.5	1.2	—

Fig. 8.8. Mill and air classifier system for commercial grinding of defatted soy flour. *Source:* Kanzamar, G.J. et al., in *Proceedings of the World Conference on Oilseed Technology and Utilization*, edited by T.H. Applewhite, American Oil Chemists' Society, Champaign, IL, 1993, pp. 226–240.

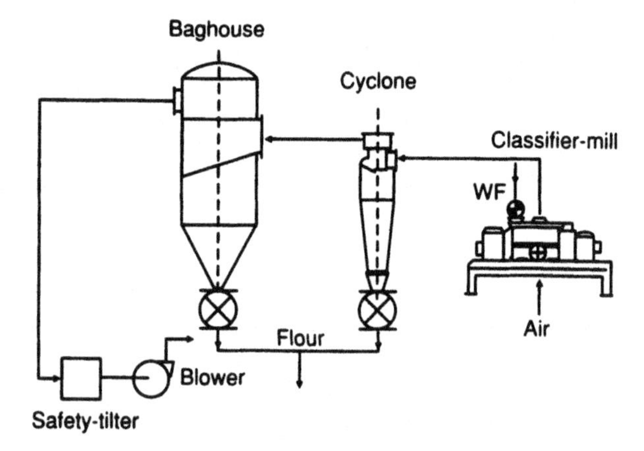

Fig. 8.9. Classifier mill system for commercial grinding of defatted soy flakes. *Source:* Kanzamar, G.J., et al., in *Proceedings of the World Conference on Oilseed Technology and Utilization*, edited by T.H. Applewhite, American Oil Chemists' Society, Champaign, IL, 1993, pp. 226–240.

reflected in lessened enlargement of the pancreas in test rats. Although minimum protein denaturation may be desired, to maximize soy flour solubility and functionality, the fabricated product must be adequately cooked before consumption (50).

Recommended soy flour PDI values for various applications are shown in Table 8.4 (50,52). Domestic manufacturers generally offer soy flours with PDIs of 90

TABLE 8.2 Carbohydrate Constituents of Dehulled Defatted Soybean Flakes

Carbohydrate	Source	Percent
Monosaccharides		
Glucose	Cotyledons	0.3
Arabinose	Hulls	Trace–0.1
Ribose	Nucleic Acids	Trace–0.1
Oligosaccharides		
Sucrose	Cotyledons	8.1
Maltose	Cotyledons	0.6
Raffinose	Cotyledons	1.1
Stachyose	Cotyledons	4.9
Verbascose	Cotyledons	Trace
Polysaccharides		
Arabinan	Cotyledons	15.0
Arabinogalactan	Cotyledons	5
Xylan (hemicellulose)	Hulls	3.5
Galactomannans	Hulls	Trace
Cellulose	Hulls	1–2

Source: Fulmer, R.W., in *Proceedings of the World Conference on Vegetable Protein Utilization in Human Foods and Animal Feedstuffs*, edited by T.H. Applewhite, American Oil Chemists' Society, Champaign, IL, 1989, pp. 55–65.

TABLE 8.3 Processing and Nutritional Parameters of Heat-Treated Soy Flours

Heat,[a] min	NSI[b]	TI, TIU/mg[c]	PER[d]	Pancreas wt, g/100 g body wt
0	97.2	96.9	1.13	0.68
1	78.2	74.9	1.35	0.58
3	69.6	45.0	1.75	0.51
6	56.5	28.0	2.07	0.52
9	51.3	20.5	2.19	0.48
20	37.9	10.1	2.08	0.49
30	28.2	8.0		

[a]Live steam at 100°C.
[b]NSI = nitrogen solubility index
[c]TI = trypsin inhibitor; TIU = trypsin inhibitor units
[d]Protein efficiency ratio, corrected on a basis of PER = 2.5 for casein
Source: Fulmer, R.W., in *Proceedings of the World Conference on Vegetable Protein Utilization in Human Foods and Animal Feedstuffs*, edited by T.H. Applewhite, American Oil Chemists' Society, Champaign, IL, 1989, pp. 55–65.

TABLE 8.4 Applications of Defatted Soy Products in Foods

PDIa	Application
90+	White breat–bleaching agent
	Fermentation
	Soy protein isolates, fibers
60–75	Controlled fat and water absorption
	Doughnut mixes
	Bakery mixes
	Pastas
	Baby foods
	Meat products
	Breakfast cereals
	Soy protein concentrates
30–45	Meat products
	Bakery mixes
	Nutrition, fat and water absorption, emulsification
10–25	Baby foods
	Protein beverages
	Comminuted meat products
	Soups, sauces and gravies
	Hydrolyzed vegetable proteins
Soy grits	Nutrition, meat extender
	Patties, meatballs and loaves, chili, sloppy joes
	Soups, sauces and gravies

[a]Protein Dispersibility Index is a standard AOCS method (Ba 10-65) for measuring the amount of heat treatment used in the processing of soybean meal products.
Source: Fulmer, R.W., in *Proceedings of the World Conference on Vegetable Protein Utilization in Human Foods and Animal Feedstuffs*, edited by T.H. Applewhite, American Oil Chemists' Society, Champaign, IL, 1989, pp. 55–65 and 424–429.

(enzyme-active), 70, 65, and 20, with granulations of U.S. 100 and 200 mesh. Grits are offered in coarse, medium, and fine granulations. The bulk density of heat-treated defatted soy flour is approximately 0.6 to 0.7/g/cm^3 (37 to 44 lb/ft^3).

Refatted or Relecithinated Soy Flours

Soy flours are refatted at levels of 1 to 15% added fat to reduce dustiness and provide fat for a product formula. Refatting an extracted flour allows the use of refined bland oil. Relecithinated soy flours are offered domestically with 3, 6, and 15% added lecithin. Lecithin improves dispersion of the flour and other admixed ingredients in confection and cold beverage products. Generally, oil or lecithin is added to highly toasted flours. The compositions of refatted and relecithinated soy flours are primarily those of the carrier flour, diluted by the amount of oil or lecithin added.

Soy Protein Concentrates

Soy protein concentrates contain at least 65% protein but less than 90% mfb. The older definition of 70% minimum protein was replaced by 65% protein (N × 6.25) in a definition promulgated by the U.S. Department of Agriculture's Food and Nutrition Service (USDA-FNS) in January 1983 (17). Sucrose and total nondigestible oligosaccharides each account for about 8% weight of defatted soy flakes. Protein concentrate yield in processing is about 75% of defatted flake weight.

In making soy protein concentrates, the objective is to immobilize the protein while leaching away the solubles. Products are produced by three basic procedures:

- Extraction of WFs with aqueous 20 to 80% ethyl alcohol
- Acid leaching of WFs or flour
- Denaturing the protein with moist heat and extraction with water (17,53)

By the use of defatted flakes, the concentrate can be produced with a countercurrent continuous chain or belt extractor such as those made by Crown Iron Works, Minneapolis, MN, or DeSmet, Edegem, Belgium, dried by a flash or vacuum drying system, and ground. When WF flour is used, it is ground, suspended at a 1:10 or higher ratio in the solvent, concentrated by decanters, optionally extracted a second time, and spray-dried. Domestic processors prefer nozzle sprayers, whereas rotary atomizers are more common in Europe.

Aqueous Alcohol Process. In 1962, Mustakas et al. (54) reported a process for flash desolventizing soybean meals extracted with 50 to 70% alcohol. According to Campbell et al. (17), the preferred concentration of ethanol is 60% by weight, since soy protein solubility increases on either side of that concentration. The NSIs of pro-

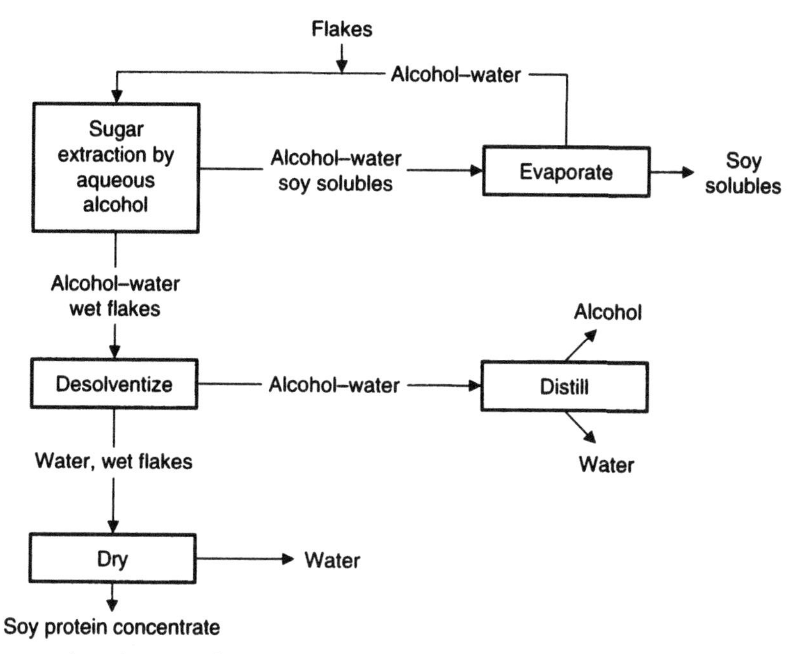

Fig. 8.10. Flow diagram of soy protein concentrate production by aqueous alcohol extraction. *Source:* Campbell, M.F., et al., in *New Protein Foods,* vol. 5, edited by A.M. Altschul and H.L. Wilcke, Academic Press, New York, 1985, pp. 301–337.

tein concentrates made by aqueous alcohol extraction are low, sometimes in the range of 5 to 10, but are not necessarily related to functionality because the mechanism of denaturation is different. Commercial versions can hold about 2.6 times their own weight in water of low-fat meat juice. They are used in meat patties, pizza toppings, and meat sauces and in conditions that stress the product, such as freeze–thaw cycles and extended holding times of precooked or cooked products. Alcohol-extracted concentrates are sometimes referred to as made by the "traditional process." A flow diagram for an alcohol extraction process is shown in Fig. 8.10 (55).

Howard et al. (56), of the A.E. Staley Company, Decatur, IL, patented a method in 1980 to regenerate high NSIs in ethanol-extracted soy protein concentrates. Extracted concentrates are subjected to successive pressure and cavitation, as by centrifugal homogenization, at elevated temperatures and slightly alkaline conditions. The products, along with acid-leached concentrates, are sometimes called "functional soy protein concentrates" and are reported to have very bland flavor.

Acid Leaching Process. The majority of soy proteins are globulins that are insoluble in water at their isoelectric poin of pH 4.5. In the Sair process (19), defatted soy flakes are leached with water at pH 4.5 to remove the soluble sugars, adjusted to neutrality, and spray-dried. Some loss of soluble proteins occurs, but the resulting soy protein concentrate has a relatively high NSI of about 65 to 75 (17).

A typical acid leaching process uses a ratio of about 10 to 20 : 1 water to white flakes or flour, hydrochloric acid for adjusting the pH to 4.5, and 30 to 45 minute

TABLE 8.5 Approximate Composition of Soy Protein Concentrates Made by Three Extraction Processes[a]

Component	Alcohol process[b]	Acid process[c]	Hot-water process[b]
Protein (N x 6.25)[d]	71	70	72
Protein	67	66	68
Moisture	6.0	6.0	5.0
Fat	0.3	0.3	0.1
Crude fiber	3.5	3.4	3.8
Ash	5.6	4.8	3.0
Carbohydrate[e]	17.6	19.5	20.1

[a]Data expressed as percentages.
[b]A.E. Staley Mfg. Co., Decatur, IL.
[c]Griffith Laboratories (Chicago, IL) Technical Data Sheet.
[d]Dry solid basis; all other data expressed on an "as-is" basis.
[e]Percentage by difference.
Source: Campbell, M.F., et al., in *New Protein Foods*, vol. 5, edited by A.M. Altschul and H.L. Wilcke, Academic Press, New York, 1985, pp. 301–337.

extraction at 40°C (104°F). A decanter or centrifuge is used to concentrate the solids to about 20%. A second leach and centrifugation may be employed. The slurry may be dried in acidic form, but it is usually neutralized to pH 6.8 with sodium or calci-

TABLE 8.6 Amino Acid Composition of Soy Protein Concentrates, Soy Solubles, and Soy Flours[a]

Amino acid	Soy flour[b]	Soy protein concentrate: Alcohol[c]	Soy protein concentrate: Acid wash[d]	Soy solubles from alcohol extraction[b]
Alanine	4.0	4.86	4.03	3.94
Arginine	6.95	7.98	6.46	7.36
Aspartic acid	11.26	12.84	11.28	15.0
Half-cystine	1.45	1.4	1.36	4.14
Glutamic acid	17.18	20.2	18.52	20.7
Glycine	3.99	4.6	4.60	3.47
Histidine	2.6	2.64	2.59	2.5
Isoleucine	4.8	4.8	5.26	2.11
Leucine	6.5	7.9	8.13	3.17
Lysine	5.7	6.4	6.67	3.53
Methionine	1.34	1.4	1.40	3.6
Phenylalanine	4.72	5.2	5.61	5.65
Proline	4.72	6.0	5.32	3.48
Serine	5.0	5.7	5.97	3.38
Threonine	4.27	4.46	3.93	3.36
Tryptophan	1.8	1.6	1.35	7.0
Tyrosine	3.4	3.7	4.37	5.47
Valine	4.6	5.0	5.57	2.12

[a]Data expressed as grams amino acid per 16 g nitrogen.
[b]A.E. Staley Mfg. Co., Decatur, IL.
[c]Procon, A.E. Staley Mfg. Co.
[d]Griffith Laboratories (Chicago, IL) Technical Data Sheet.

TABLE 8.7 Vitamin and Mineral Fortification Requirements for USDS-FNS Child Feeding Programs

Vitamins and Minerals	Min./Gram Protein
Vitamin A, I.U.	13.00
Thiamine, mg	0.02
Riboflavin, mg	0.01
Niacin, mg	0.30
Pantothenic acid, mg	0.04
Vitamin B_6, mg	0.02
Vitamin B_{12}, mcg	0.10
Iron, mg	0.15
Magnesium, mg	1.15
Zinc, mg	0.50
Copper, mcg	24.00
Potassium, mg	17.00

um hydroxide before spray drying at 157°C (315°F) inlet air temperature and 86°C (187°F) outlet temperature.

Hot-Water Leaching Process. A patent was issued to McAnelly (57) in 1964 for a process in which a doughlike mass of soy flour and water is first developed, subjected to heat and pressure to denature the protein, and is then extruded to impart a porous structure that is leached with hot water (17).

Product Characteristics. Approximate compositions of soy protein concentrates made by the three processes are shown in Table 8.5. The most obvious difference is that ash contents are lower in concentrates prepared by acidic or hot-water extraction, indicating more thorough removal of minerals. About 5 to 10% of the carbohydrates remaining in soy protein concentrates after leaching are soluble sugars, with the balance being insoluble polysaccharides.

The amino acid compositions of soy flour, protein concentrates made by the ethanol or acid extraction, and soy solubles from alcohol extraction are shown in Table 8.6. Of the essential amino acids, phenylalanine, tryptophan, methionine, and cystine concentrate in the soy solubles fraction during alcohol extraction.

Soy protein concentrates are offered in powder (95% through U.S. No. 100) or granular (90% retention on U.S. No. 60) forms and are refatted or lecithinated for minimum dustiness or rapid dispersibility. Typical bulk densities are

Powder form: 0.40 to 0.45 g/cm^3 (25 to 28 lb/ft^3)

Granular form: 0.54 to 0.61 g/cm^3 (34 to 38 lb/ft^3)

(9%) lecithinated form: 0.43 to 0.48 g/cm^3 (27 to 30 lb/ft^3)

Concentrates, isolates, and texturized flours and concentrates used for meeting a portion of the meat or meat alternative requirement in the domestic school lunch or child nutrition programs must be fortified with vitamins and minerals, according to USDA-FNS requirements (Table 8.7). Separate fortification requirements exist for military ground beef applications (PP-B-2120B).

Soy Protein Isolates

Many processing options exist for making soy protein isolates. The preferred approach is to tailor the extraction process to making the isolate optimally compatible with the intended consumer product. This route sometimes is taken if warranted by the potential for a new large protein market. However, a broad variety of commercial isolates and concentrates already exists and can be readily put into use, including products designed for general use in bakery, processed meats, frozen desserts, whipped products, and dry beverage and sauce mix formulations.

pH Extraction–Precipitation. A water extraction curve for proteins from defatted soybean meal, in the pH range 0.5 to 12, is shown in Fig. 8.11 (11). Nondenatured soy protein is most soluble at pH values of 1.5 to 2.5 and 7 to 12 and least soluble at its isoelectric region of pH 4.2 to 4.6.

A traditional isolate production process is depicted in Fig. 8.12 (58). The basic steps include

1. Solubilizing the protein in ground white flakes, at a 1:10 to 20 solids:solvent ratio, in 60°C (140°F) water adjusted to pH 9 to 11 with sodium hydroxide

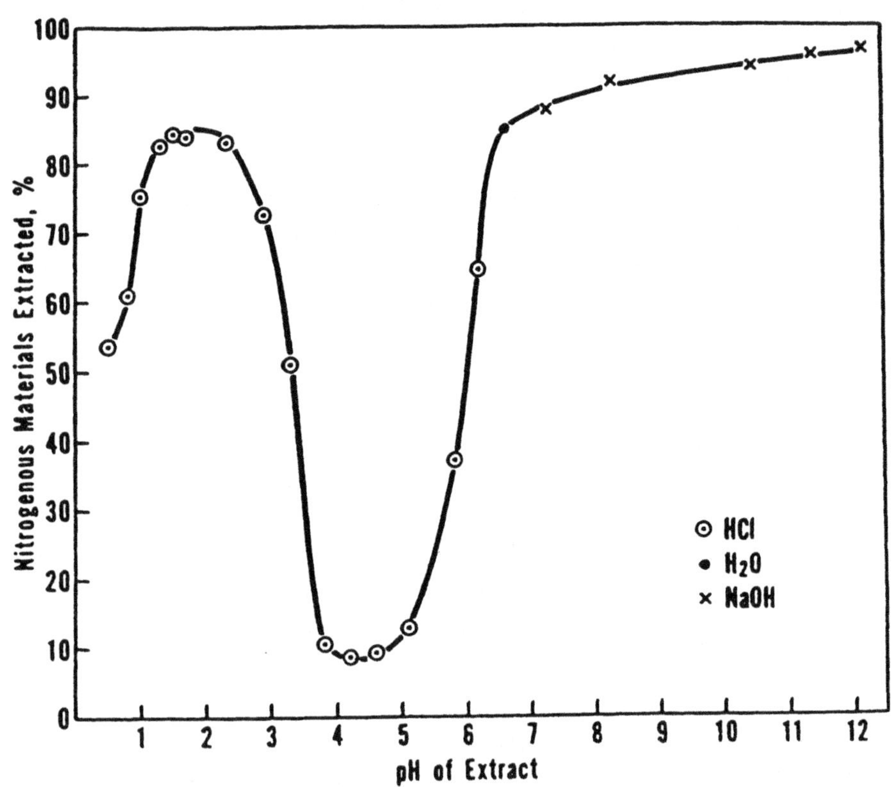

Fig. 8.11. Extractability of proteins in defatted soybean meal as a function of pH. *Source:* Wolf, W.J., and J.C. Cowan, *Soybeans as a Food Source*, rev. edn., CRC Press, Inc., Boca Raton, FL, 1975.

2. Removing the unsolubilized fiber by centrifugation
3. Reconcentration of the protein by acid precipitation (pH 4.2 to 4.5) with hydrochloric acid and mechanical decanting
4. Washing the precipitate (curd) with water and reconcentration by decanting
5. Neutralization to pH 6.8 with sodium or calcium hydroxide
6. Spray-drying at 157°C (315°F) inlet air temperature and 86°C (187°F) outlet

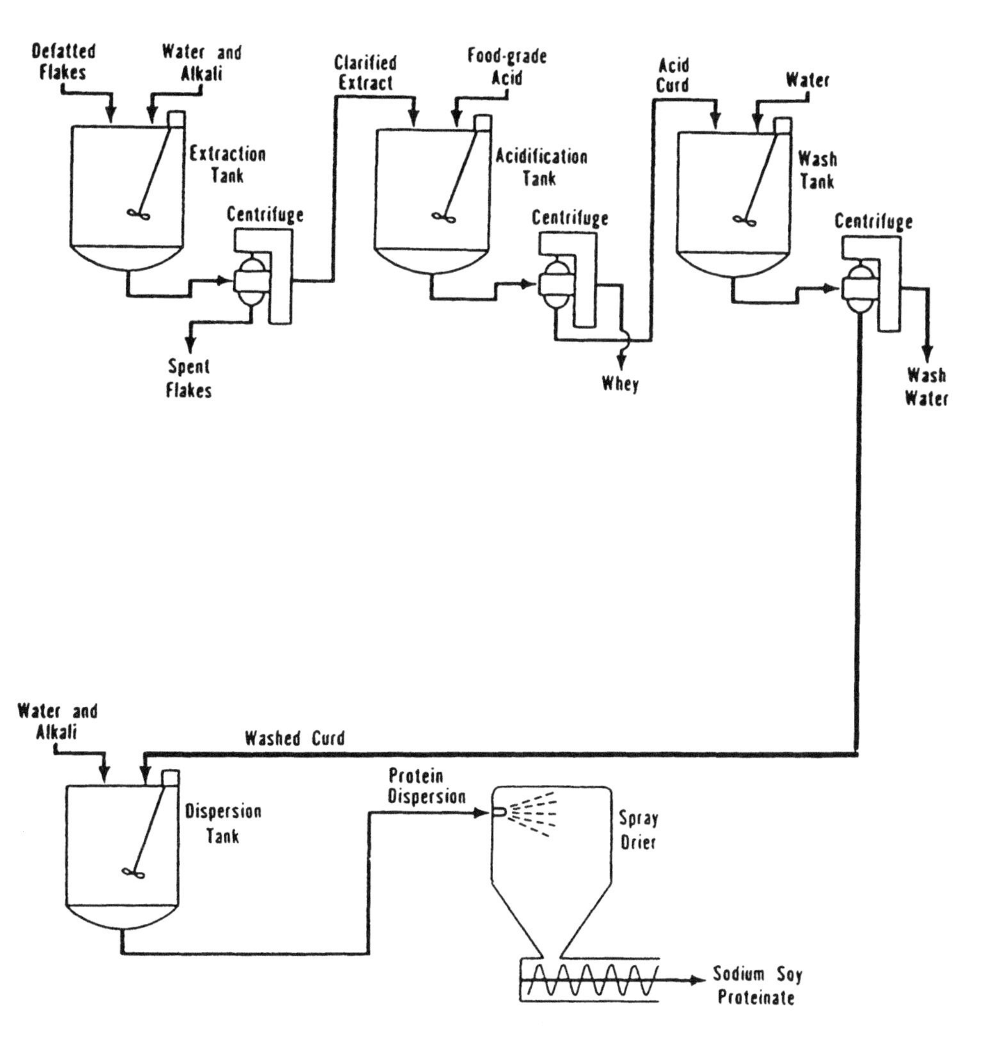

Fig. 8.12. Flow diagram for commercial preparation of soybean protein isolates. *Source:* Wolf, W.J., in *Handbook of Processing and Utilization in Agriculture, Vol II: Part 2,* edited by I.A. Wolf, CRC Press, Inc., Boca Raton, FL, 1983, pp. 23–55.

Some processes wash the fiber a second time to improve protein yield. Deionized process water should be used when producing isolates in naturally "hard" or alkaline waters.

High pH values and temperatures and prolonged processes favor production of lysinoalanine in many proteins. This reaction compound, formed mostly at the expense of lysine and cystine in soy protein (59), causes nephrocytomegaly in rats (60). While its effect on humans is not known (61), conditions favoring production of lysinoalanine (62) should be minimized to those needed for making the intended isolate. Sometimes, the objective may be to obtain an isolate with specific pH solubility characteristics, rather than optimization of total yield. In this case, one product may be harvested at the selected pH, with a second product then taken at the isoelectric pH.

An opportunity also exists to reduce phytate content, whose differential solubility from soy protein occurs at about pH 5.0 (Fig. 8.13). A wash with water at pH 5.0 has the potential for removing about 75% of the phytates in soy protein concentrates. Precipitating alkaline soy protein isolate solution first at pH 5.0 would yield a 11S rich isolate low in phytate content (63).

The majority of isolates sold are prepared by extraction, reprecipitation, and neutralization under pH-controlled conditions and spray drying. They may be further supplemented with calcium, if intended for use in dairy product replacement applications; agglomerated to increase density; and lecithinated to improve dispersion.

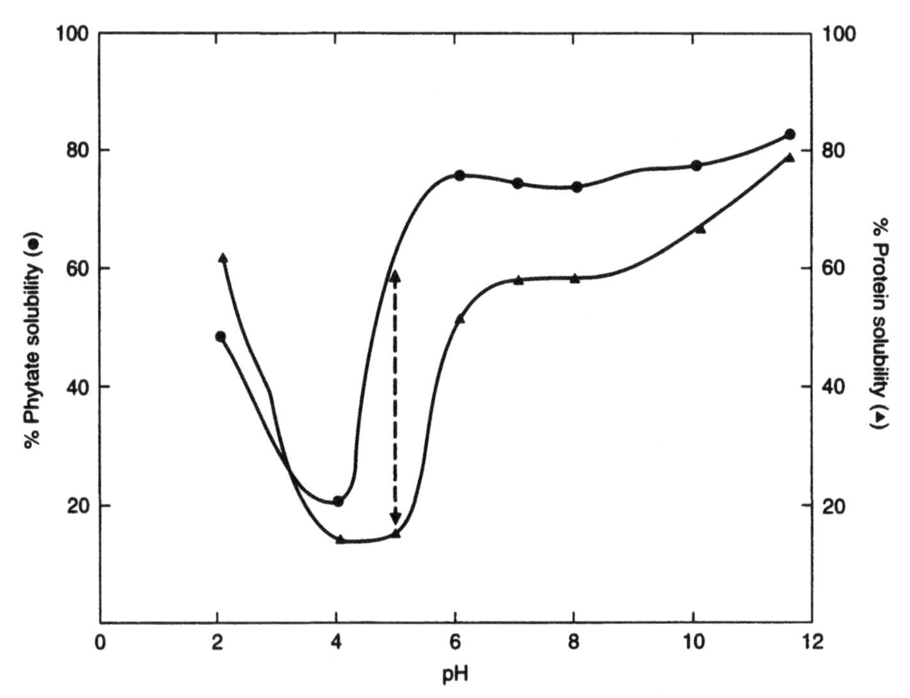

Fig. 8.13. Effects of pH on solubility of protein and phytate in defatted flour. *Source:* Rhee, K.C., and Y.R. Choi in *Annual Progress Report of Food Protein Research and Development Center,* Texas A&M University, College Station, TX, 1981, pp. 203–233.

Separation by Molecular Weight. Wolf et al. (64) reported the following soybean globulin fractions determined by ultracentrifuge sedimentation:

2S type: 22% of total, MW (molecular weight) 8,000 to 21,500
7S type: 37% of total, MW 180,000 to 210,000
11S type: 31% of total, MW 350,000
15S type: 11% of total, MW 600,000.

The 7S (β-conglycinin) and 11S (glycinin) proteins are the major proteins of soybeans, differing in their functional properties. For example, the 11S protein plays an important role in crosslinking with divalent cations to form curds like tofu. The presence of salts slightly raises the isoelectric point of soy protein components. Figure 8.14 (65) shows that each fraction has a slightly different precipitation curve, with the maximum for 7S at about pH 5 and that for 11S at pH 5.8 in a low-ionic-strength solution (0.03 M).

Various processes have been patented to separate 7S and 11S proteins on the basis of differences in their isoelectric points and the tendency of 11S protein to precipitate at low temperatures (66). A process by Davidson et al. (67) includes extracting soy

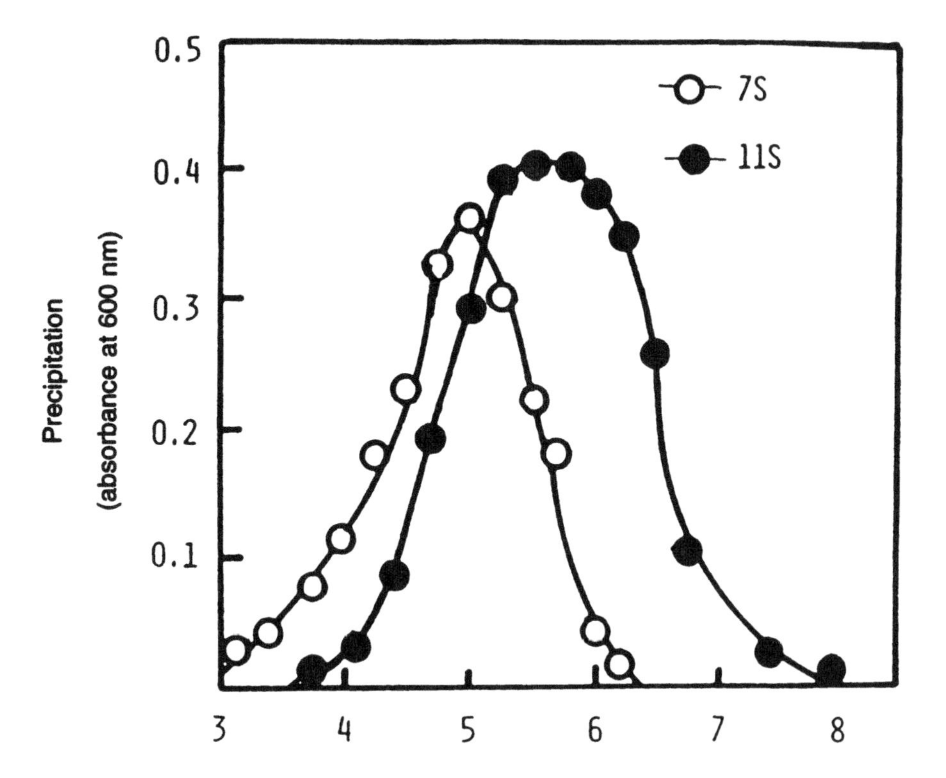

Fig. 8.14. Susceptibility of 7S and 11S soy protein tractions of pH precipitate from solutions at low ionic strength (0.03M). *Source:* Kinsella, J.E., *J. Amer. Oil Chem. Soc. 56:* 242 (1979).

flakes in 5 parts of water at 55 to 70°C (131°C to 158°F) without pH adjustment, cooling the extract slowly to favor precipitation of 11S protein, and collecting the remaining proteins by precipitation at pH 4.5 or by ultrafiltration. The 11S-rich fraction was reported to have thermoplastic properties that may be useful in imitation cheese.

Shermer (68) patented a process for making an isolate rich in 7S globulins by extraction at pH 5.1 to 5.9; under those conditions the 11S protein is only slightly soluble. Howard et al. (69) and Lehnhardt et al. (70) enhanced selective precipitation and selective extraction of 7S and 11S fractions, respectively, by addition of salts and sulfurous ions. Using 0.03 M sodium chloride and 0.77 mM sodium bisulfite solutions in a pH 8 extraction, yields of 95% pure 7S protein and 11S proteins were obtained at 25.3% and 22.8% of the protein in the starting soy grits. The separation of an intermediate 30% 7S:70% 11S fraction was necessary to obtain pure 7S and 11S fractions (68). Similar separations by pH-selective extraction from isoelectric-precipitated soy proteins also was necessary (69). Gibson and Yackel (71) further described the fractionation of 7S and 11S proteins. Essentially pure glycinin fraction soy proteins are available domestically in commercial quantities.

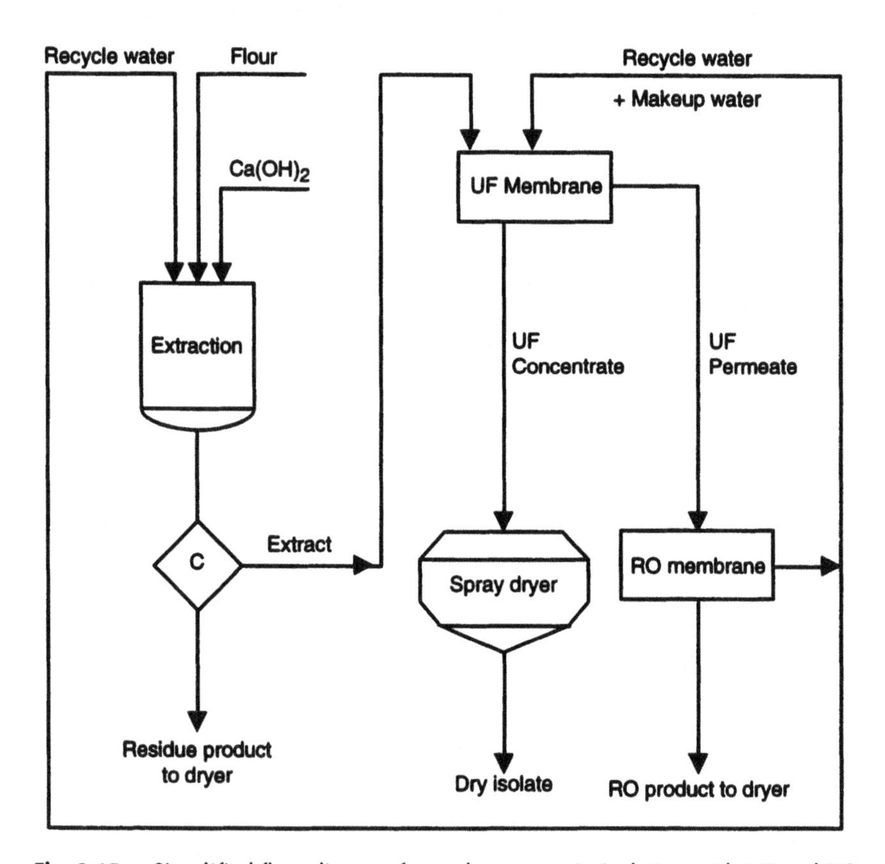

Fig. 8.15. Simplified flow diagram for soybean protein isolation with UF and RO membranes. *Source:* Lawhom, J.T., et al., *J. Amer. Oil Chem. Soc. 58:* 377, (1981).

Membrane Processing. Ultrafiltration (UF) and reverse osmosis (RO) are another form of separation based on molecular size. UF typically is employed to retain or permeate molecules according to the size of membrane pores selected, and RO is used for dewatering and concentration. It has been estimated that energy costs can be reduced by up to 90% through use of RO membranes for dewatering, compared to evaporative processes, and up to 70% when a single-effect evaporator is used. The principles of vegetable protein membrane and adsorptive separations were reviewed by Koseoglu and Lusas (72). It is common practice to incorporate diafiltration (maintaining a constant ratio of water or solvent to solids in the feed stock) to minimize problems of retentate-side concentration and surface fouling during ultrafiltration.

Lawhon et al. (73) described a process for making soy protein isolates from defatted soybean flakes using UF-RO and diafiltration (Fig. 8.15). Ground flour is extracted (a single extraction of 30:1 water : flour ratio; or two extractions, 10:1 followed by 8:1) with water adjusted to pH 8 to 9 by calcium or sodium hydroxide (Calcium hydroxide is preferred as the base for solubilizing protein, because it produces greater isolate yield than sodium hydroxide). The extraction temperature is 43°C (110°F) for high-NSI flour or 55°C (132°F) for toasted flour. After 40 minutes' extraction, the material is centrifuged to remove the fiber and passed through a cross-flow 70,000 mwco (molecular weight cut-off) membrane. The retentate protein fraction is concentrated by RO and spray-dried. The permeate (soluble sugars, minerals, and small protein molecules) may also be concentrated by RO and spray-dried.

Advantages of membrane processing include the following:

1. The ability to recover certain proteins without alkali solubilization–acid precipitation and accompanying protein damage
2. The potential for recovering small (12,000 to 20,000·mw) proteins if a membrane with small enough pore size is selected
3. Opportunities for removing small molecules such as phytates or lysinoalanine
4. Opportunities for greatly reducing plant water consumption and processing discharge streams that contain significant BOD (biological oxygen demand)

Soy protein concentrates have also been prepared by UF/RO processing.

Lawhon (74) was granted a patent for preparing light-colored and bland soy and oilseed protein isolates, using ultrafiltration membranes, of 70,000 to 100,000 mw. In the process, flavor- and color-producing compounds apparently associate preferentially with the smaller (20,000 mw and less) proteins and pass through the membrane, leaving the larger, bland, and light-colored molecular weight fractions behind.

Primary membrane processes for production of soy protein concentrates and isolates have not been publicized as being practiced in the United States but have been reported in use in Japan and Europe (66).

Aqueous Extraction Processing. Lawhon et al. (73) described an aqueous extraction process (AEP) for removing oil and preparing vegetable protein concentrates and isolates from soybeans using water. In soy protein preparation by AEP (Fig. 8.16) cleaned soybeans are dried at 70°C (158°F) to 6% moisture, dehulled by cracking and aspiration, and reduced in particle size (to 99% < 70 mesh U.S. Sieve Series) by a contraplex pin mill. Oil extraction is conducted at a solids-to-water ratio of 1:12 in water at 60°C (140°F), pH 9, containing 0.01% hydrogen peroxide to inactivate lipoxygenase. After 30 min extraction, the slurry is centrifuged to separate an aqueous phase, a solids phase, and an oil/emulsion phase, which later is broken to obtain the oil. The aqueous phase is adjusted to pH 4.5 with hydrochloric acid to precipitate a protein curd, which is separated by centrifugation. Washing the curd before drying increases the protein content to almost 90% mfb. The alkali-neutralized, spray-dried protein isolates contain as high as 8 to 10% residual oil. Residual oil levels in the isolates appear related to the natural phosphatide contents in oil of the specific seed species, with peanuts yielding AEP isolates containing about 2% residual oil.

Optionally, the solids phase from the first alkali extraction may be extracted a second time, as a solids-to-water ratio of 1:5 at pH 9, and recentrifuged, with the resulting fractions combined with those from the initial centrifugation. Although the process is not sufficiently efficient to become the main commercial means for extracting soybean oil, the resulting protein concentrates and isolates are extremely stable to oxidation and have properties that may be functionally useful in selected

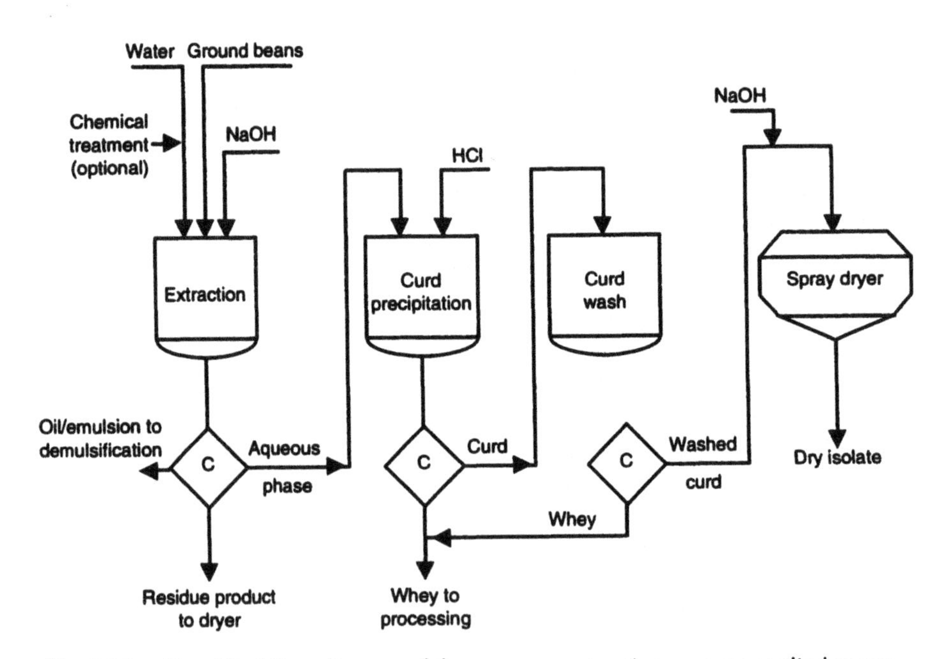

Fig. 8.16. Simplified flow diagram of the aqueous extraction process applied to soybeans. *Source:* Lawhon, J.T., et al., *J. Amer. Oil Chem. Soc. 58:* 377, (1981).

applications. UF and RO membrane techniques have also been tried with AEP.

Salt Extraction. The 7S and 11S proteins are dimers of several subunits and can be dissociated by salt solution. Murray et al. (75) patented a process that employs salt for extracting a soy protein isolate at ionic strengths of 0.3 to 0.6 M, pH 5.0 to 6.8 and 15 to 25°C (60 to 78°F). The extract is then concentrated to one-fourth to one-third its volume and diluted to an ionic strength of less than 0.2 M to form protein micelles that precipitate into an amorphous mass and are dried or further processed.

Separation of Intact Protein Bodies. Storage proteins in soybean cotyledon cells are deposited in protein bodies. Attempts have been made to separate them from other cellular constituents by fine milling and density flotation using glycerin, other polyhydric alcohols, sodium chloride, sucrose, and metal salts of organic acids. A density of 1.2 to 1.5 g/mL is required to float protein bodies, and water activity must be maintained at less than 0.85 to prevent hydration from occurring. The separated bodies have a protein content greater than 80% mfb (66).

Enzyme-Modified Protein Isolates. The period after extracting protein isolates and just before drying provides a good opportunity for enzymatic modification. Although there has been considerable research in succinylated and acetylated derivatives, these are not allowed in food use and are restricted to industrial applications.

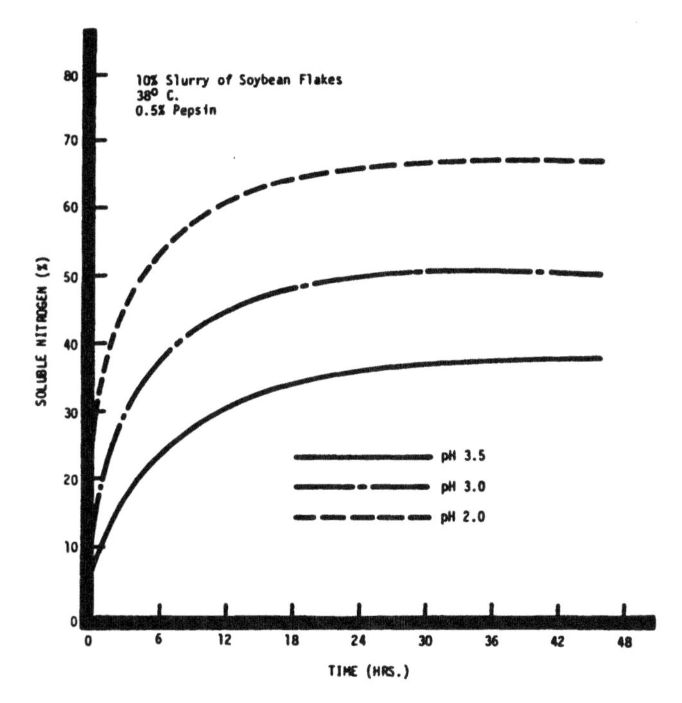

Fig. 8.17. Effects of pH on nitrogen solubility, 0.5% pepsin hydrolysis of 10% soybean flake slurry at 38°C (100°F). *Source:* Gunther, R.C., *J. Amer. Oil Chem. Soc. 56:* 345, 1979.

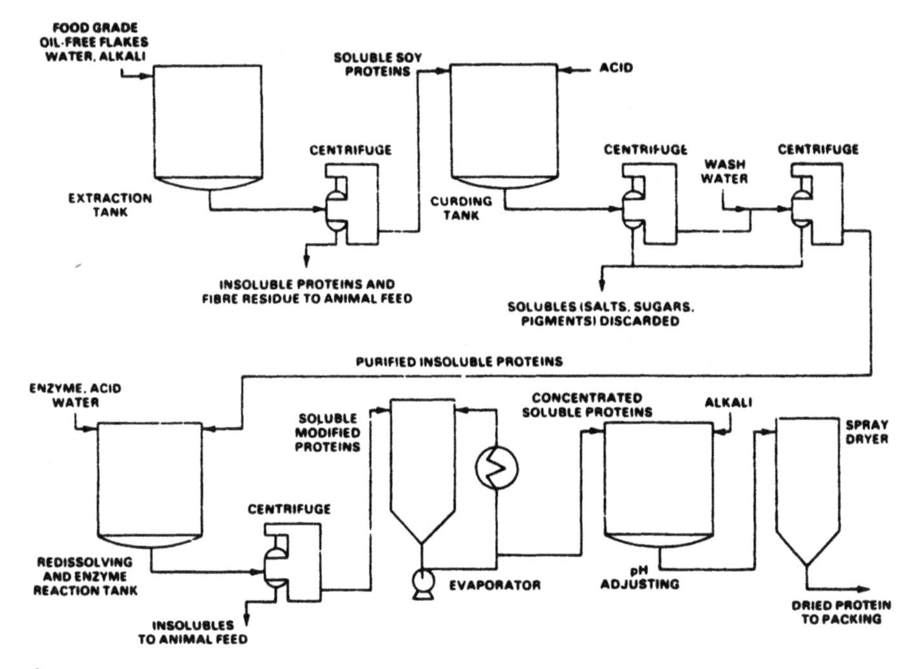

Fig. 8.18. Preparation of enzyme-modified whipping proteins, via soy isolate intermediate process. *Source:* Gunther, R.C., *J. Amer. Oil Chem. Soc. 56:* 345, (1979).

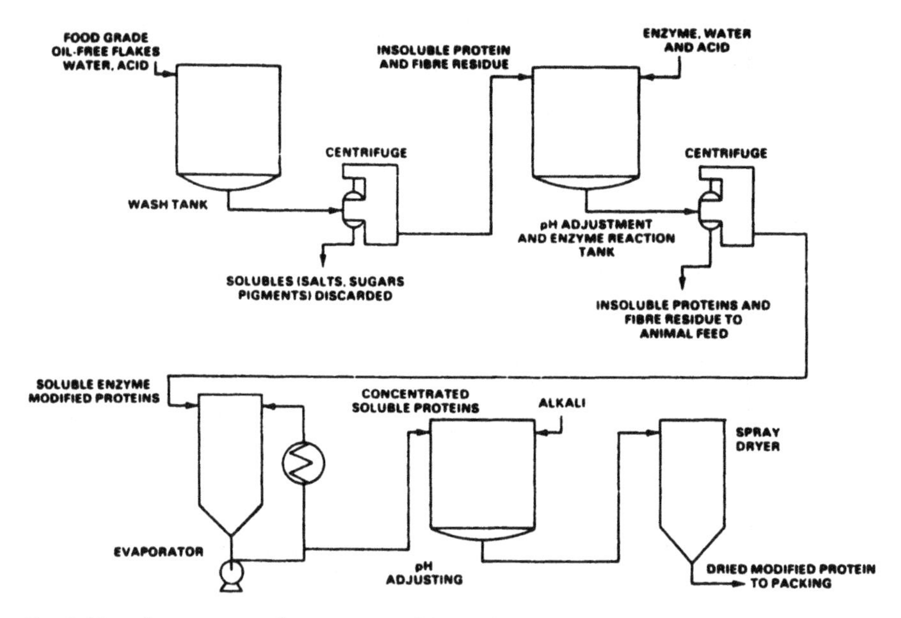

Fig. 8.19. Preparation of enzyme-modified whipping proteins by direct hydrolysis of soy flakes. *Source:* Gunther, R.C., *J. Amer. Chem. Soc. 56:* 345, (1979).

Likewise, plastein synthesis—the reassembling of proteins from peptides—is not practiced in domestic commercially prepared food proteins.

A variety of commercial proteolytic enzymes is available from plant, microbial, and animal intestinal sources. These differ in their affinities for various proteins, the location where they will cleave the peptide bond between different amino acids, and conditions (pH, temperature, inhibitors) affecting their reaction rates.

Whipping Proteins. Soybean enzyme-modified whipping proteins are an example of a special application that has led to a new industry (76). Shortages of egg albumen, caused by World War II, led to a market for three types of whipping (aerating) ingredients: soy albumens, enzyme hydrolyzates made from wet soy isolate, and enzyme hydrolyzates made directly from soy flour. Enzymatic hydrolysis of whipping proteins typically is conducted in the pH 2.0 to 3.5 range, below the isoelectric point of soy protein, and results in molecular weights of less than 14,000. Figure 8.17 shows the change in nitrogen solubility of a 1:10 slurry of soybean flakes over time, during hydrolysis by 0.5% pepsin at pH 2 to 3.5 and 38°C (100°F).

A process for making whipping proteins via an intermediate soy isolate product is shown in Fig. 8.18. The protein in soy flakes is first solubilized by alkali, and the fiber separated by a centrifuge. Protein in the solution is precipitated as an isoelectric curd, which is washed to remove solubles and then solubilized by additional acid. Enzyme hydrolysis is conducted at low pH for 12 to 24 hours under controlled conditions, followed by centrifuging to remove the insoluble residue, concentrating the solubles, neutralization to about pH 5.2, and spray-drying. In an alternative soy flake process (Fig. 8.19) the flakes are first washed to partially remove the solubles and sugars, then directly treated with acid and enzyme to solubilize and hydrolyze the protein. The insolubles are separated by centrifugation, concentrated by evaporation, partially neutralized to about pH 6.6, and spray-dried.

Whipping preparations are used with different background formulas and sometimes contain sucrose, sodium hexametaphosphate, or a polysorbate emulsifier (Tween 60) to improve stability. Many of these preparations can whip to twice the volume of egg whites and are substituted at 25 to 100%. They differ from egg whites in not being heat-setting alone, but they will extend heat-set egg whites at lower levels.

Dietary Fiber Products

Two types of edible fiber coproducts are produced from soybean processing operations.

Soy Cotyledon Fiber

The production of soy protein isolate involves removal of soluble materials from dehulled soybean cotyledons, first by solvent defatting and then by aqueous leaching. In the process, part or all of the protein fraction is solubilized, usually at alkaline pH, and the insoluble portion separated by centrifugation. These solids are analogous to the okara coproduct from making soy milk or tofu, but they differ in that they do not contain hulls and have been defatted and treated with mild alkali.

The cotyledon coproduct is processed, dried, and sold as a dietary fiber in competition with such other sources as α-cellulose, psyllium seed, guar gum, locust bean gum, pectin, and wheat, corn, and oat brans. Manufacturer's specifications for a domestic product include

Dietary fiber: 75% on a moisture-free basis (65% noncellulosic polysaccharides and 10% cellulosic)
Moisture: 12%
Fat: 0.2 (PE)
Ash: 4.5% (as is)

Soy Hulls

Soy hulls are mainly used as animal feeds, but a small quantity is cleaned and sterilized for use as a dietary fiber source in breads. The natural grittiness of the product typically requires fine grinding. Product specifications for a domestic product include

Total dietary fiber: 92.0%
Moisture: 3.5%
Fat: 0.5%
Protein: 1.5%
Ash: 2.5%
Caloric content: 0.1 Kcal/g
Water absorption: 350–400%
pH 6.57–7.5

and sieve analysis:

On U.S. No. 80: 0%
On U.S. No. 100: 0%
On U.S. No. 140: 2%
On U.S. No. 200: 7%
Through U.S. 200: 91%

Texturized Products

Two general types of mechanical texturization are used, although some agglomerated soy proteins may also contribute a texturized-like appearance on hydration.

Spun and Fiber-Type Products

Spun products are intriguing because of the many technical skills that were recruited for their production and early marketing. Peanut, casein, and zein proteins were processed into textile fibers and marketed in the period of 1935 to 1945; soy protein textile fibers were also developed at that time but did not reach the commercial market (77). Boyer, who had worked on the protein spinning process in the mid-1940s, modified and patented processes in 1954 (78,79) to produce a fibrous mass simulating meat in texture and appearance. With later inputs from other scientists, techniques were developed to disperse 14 to 18% soy protein isolate in sodium hydroxide at pH 10 to 11, age at 40 to 50°C (104 to 122°F) until the dispersion becomes a "spinnable dope," which can be forced through a platinum spinneret with 15,000 or more holes 0.20 to 0.25 mm (0.008 to 0.01 in) diameter into an acid coagulating bath. The parallel fibers form a tow, which goes through a second heated bath where the fibers receive additional stretching. Egg albumin, fat, flavor, and coloring materials are also added at this point for eventual forming into meat analog products. Reportedly, the toughness of the fibers is controlled by the pH, salt concentration, and temperature of the bath. Although spun fiber food products are no longer sold in the United States, research on improvement of dopes and spun soybean fibers continued in Japan into the late 1980s (80).

A frozen isolated soy protein filament product, with fibrous-like texture, is sold in the United States for improving textural characteristics of fabricated foods, including structuring mechanically deboned meat and poultry. The manufacturer's specifications for the product include

Protein ($N \times 6.25$, mfb): 93%	Fiber (crude): <0.2%
Moisture (as is): 65.0%	Ash: 0.9%
Fat (PE extract): <0.1%	

Extruder-Texturized Products

The meatlike appearance in spun protein isolates results from strands of parallel fibers, but in extruded soy flours, concentrates, and isolates it is created from multilaminate palisade layers. The meatlike appearance of extruder-texturized proteins is readily acceptable to the public, as demonstrated by their essentially complete replacement of spun products. Extrusion texturization also has the advantages of being a less complicated process and of being able to texturize lower-cost ingredients, including soy flours and soy protein concentrates. A relatively small extruder-texturized soy protein isolate industry exists in the United States and sells its products in frozen form.

The principles of extrusion have been described by Mercier et al. (81) and the processing of proteins by Stanley (82) and Rokey et al. (83). "Texturized Vegetable Protein" and "TVP" are registered trademarks of the Archer Daniels Midland Company, Decatur, IL, and the generic terms "texturized soy protein," "TSP," or "texturized vegetable food protein" are used. Two types of products are made:

- Extrusion-cooked *meat extenders*, which are made from soy flour or flakes or

Fig. 8.20. Single-screw extruder used for making full-fat flours and texturized soy flours and concentrates. Courtesy of Wenger Manufacturing Company, Sebetha, KS.

soy concentrates and are rehydrated to 60 to 65% moisture before blending with meats or meat emulsions at levels of 20 to 30%

- Extrusion-cooked *meat analogs*, which have similar appearances but are intended solely for use in meatless products.

The extrusion process restructures protein-based foodstuffs by applying mechanical and thermal energy, causing the macromolecules to lose their native, organized structure and form a continuous viscoelastic mass (or "melt"). The extruder barrel, screws, and die then work to align the molecules in the direction of flow, exposing bonding sites that cross-link into a reformed, expandable structure that creates a chewy texture in fabricated foods. In addition to restructuring vegetable food proteins, extrusion cooking (77) does the following:

1. Denatures proteins, lowers solubility, improves digestibility, and destroys biologically active enzymes and toxic proteins
2. Inactivates residual heat-labile growth inhibitors native to many vegetable proteins in raw or partially processed states

3. Prevents development of raw or bitter flavors commonly associated with many vegetable food sources
4. Creates a homogeneous, irreversible, bonded dispersion of all microingredients in a protein matrix
5. Shapes and sizes the final product into desirable portions for packaging and sales.

An extrusion system consists of several important subsystems:

1. A feed delivery and proportioning system
2. A preconditioning area, which enables the raw materials to equilibrate in moisture content and heat
3. The cooking extruder itself
4. A laminar-flow area or die that allows aligning of molecules to occur
5. A die and cutter to shape and cut the product into pieces
6. A dryer/cooler to reduce moisture in the final product to a microbiologically stable level.

Barrels and screws have evolved over the years into increasingly efficient designs. The recently introduced twin-screw extruders cost more to acquire per unit of throughput capacity, but they provide nonpulsating discharge and a steady operation (80). A single-screw extruder, as used for making texturized soy protein, is shown in Fig. 8.20, and a flow sheet of a production line using a twin-screw extruder in Fig. 8.21.

The general parameters for raw ingredient specifications for texturized flours and concentrates include

PDI range: 80 to 20

Fat level: 0.5 to 6.5%

Fiber levels: up to 7%

Particle size: up to U.S. No. 8 mesh

Perhaps the most exciting development in extruder operations in recent years has been the ability to induce additional shear and to laminate low-NSI proteins that were once considered untexturizable (80).

A variety of texturized soy food proteins is available from manufacturers, including products made from soy flour or concentrate, colored and sized to different specifications. The volatile constituents are customarily added after extrusion by one of several *enrobing* processes. Specifically fortified products are available for use in school lunch and child feeding programs and in military feeding applications.

Applications of Soy Food Proteins

Functionality

Soy proteins have been accepted in many applications because they provide desirable functionalities in fabricated foods at less cost than animal-source alternatives such as

TABLE 8.8 Functional Properties Supplied by Soy Proteins

Functional Property	Mode of Action	Food System Used	Products[a]
Solubiltiy	Protein solvation, pH-dependent	Beverages	F, C, I, H
Water absorption and binding	Hydrogen-bonding of water, entrapment of water (no drip)	Meats, sausages, breads, cakes	F, C
Viscosity	Thickening, water binding	Soups, gravies	F, C, I
Gelation	Protein matrix formation and setting	Meats, curds, cheeses	C, I
Cohesion–adhesion	Protein acts as an adhesive material	Meats, sausages, baked goods, pasta products	F, C, I
Elasticity	Disulfide links in deformable gels	Meats, bakery items	I
Emulsification	Formation and stabilization of fat emulsions	Sausages, bologna, soups, cakes	F, C, I
Fat absorption	Binding of free fat	Meats, sausages, doughnuts	F, C, I
Flavor-binding	Adsorption, entrapment, release	Simulated meats, bakery items	C, I, H
Foaming	Forms film to entrap gas	Whipped toppings, chiffon desserts, angel cakes	I, W, H
Color control	Bleaching (lipoxygenase)	Breads	F

[a]F, C, I, H, W denote soy flour, concentrate, isolate, hydrolyzate and soy whey, respectively.
Source: Fulmer, R. W., in *Proceedings of the World Conference on Vegetable Protein Utilization in Human Foods and Animal Feedstuffs*, edited by T.H. Applewhite, American Oil Chemists' Society, Champaign, IL, 1989, pp. 55–65.

dried milk solids, casein, egg yolks, egg whites, or gelatin. Reviews on soy protein functionality, modifications, and applications have been prepared by Kinsella (84), Kinsella and coworkers (85,86), Cherry (87), Rhee (88), and Lusas and Rhee (89).

The most sought-after functionalities in compounded foods, their modes of action, and the types of soy proteins used are shown in Table 8.8 (50). The soy ingredient is also expected to provide a concentrated source of protein as well as caloric density appropriate to the traditional or light product. The soy ingredient also should not detract from the product in color or flavor, unless texturized and used with the specific objective of imparting a meatlike appearance. Important functionalities not included in Table 8.8 are thermoplasticity—the ability to solidify and remelt repeatedly with temperature changes, as shown by bovine casein—and the ability to form edible films.

Selection of Soy Protein Preparations

If the food manufacturing or feeding institution is large enough, the setting of nutritional objectives usually is done by nutritionists or registered dieticians. Federally supported

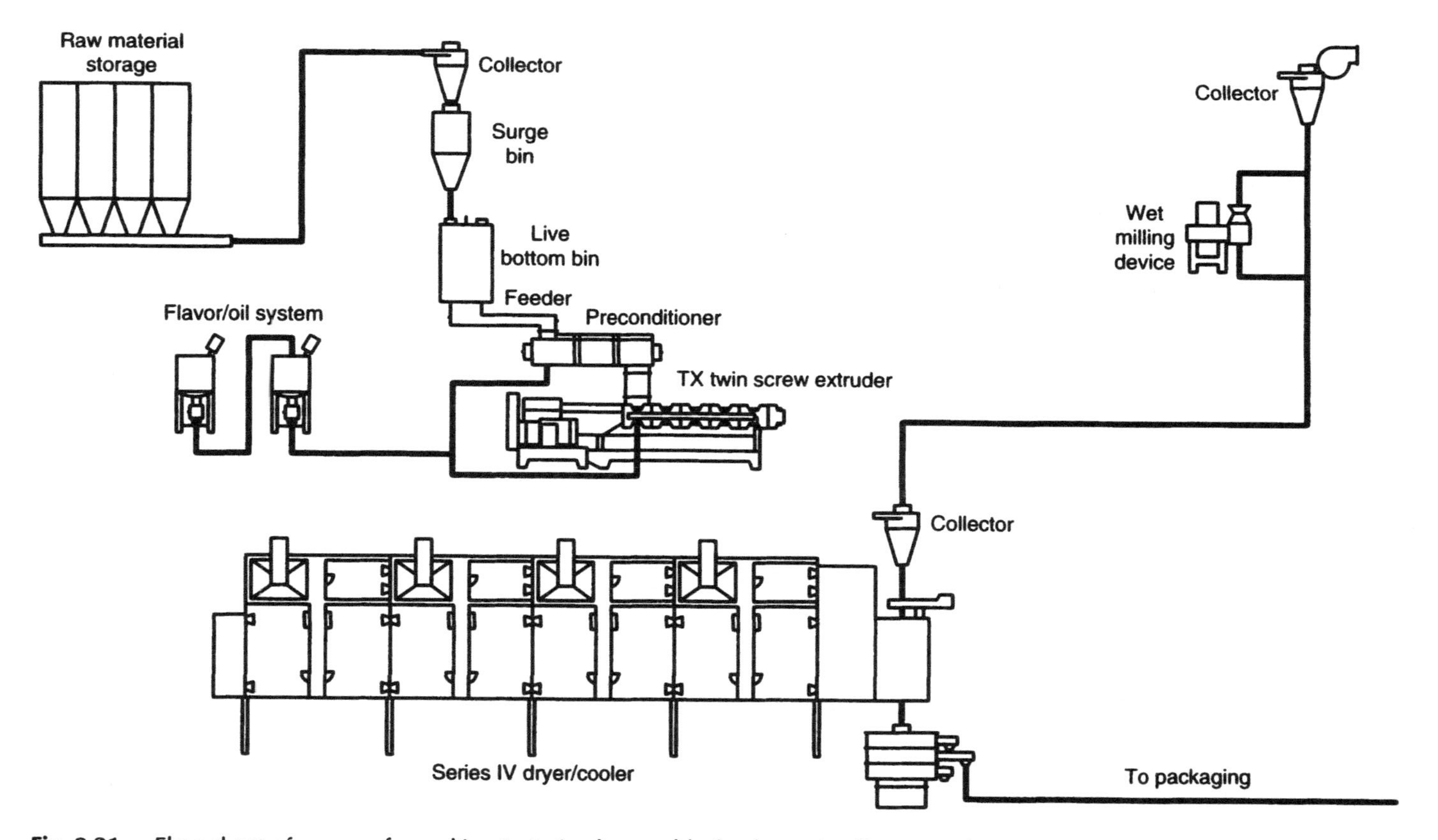

Fig. 8.21 Flow sheet of process for making texturized vegetable food protein. Courtesy of Wenger Manufacturing Company, Sebetha, KS.

feeding programs require professional oversight of menus and approval of the overall diet. However, decisions of which types or forms of ingredients to use are typically left to the formulating technologist as guided by marketing objectives for the product.

Generally, little can be predicted about a soy protein's functional performance by examining chemical compositions in manufacturers' product specification sheets. Matrix tables, showing potential applications of a manufacturer's product line, also teach little. Manufacturer's product specification sheets often tell only whether the ingredient is a flour, concentrate, or isolate and its granulation. However, inquiries to the supplier's technical service department will usually yield additional information about the ingredient's production and its limitations. The formulating technologist is advised to get several opinions of which ingredients to use from competing manufacturers. Processes differ between soy protein producers, sometimes resulting in subtle differences in their performance, and the final selection of any ingredient should be based on its performance in the end product. Users very quickly develop proprietary expertise in soy protein applications exceeding that of the manufacturer.

It is important that formulators keep updated in soy protein developments. Over the years, the flavor of soy flours has improved to where they might be substituted in former concentrate applications, and enzyme modification of concentrates has made them contenders for applications formerly using only protein isolates.

Meat Applications

Meat products are highly prized and attract cost-cutting technologies in all countries. In the United States, soy proteins are used:

1. As processing aids in the manufacture of frankfurters, sausages, and comminuted meat products
2. In marinades and tumbling solutions for restructured meats
3. In injection pumping proteins to increase the weight of intact muscles and cuts
4. In extruder-texturized flours and concentrates that are rehydrated and used at about the 20% level in hamburgers

Processed Meats The U.S. Department of Agriculture permits use of up to 3.5% soy flours or concentrates in standard of identity frankfurters, up to 8% soy flour in scrapple and chili con carne, and up to 2% soy protein isolates in standard of identity frankfurters. Soy flours and concentrates can bind up to three times their weight in water, whereas nonfat dry milk solids bind only an equal weight of water. These ingredients reduce shrinkage due to moisture and fat loss during cooking. The use of soy protein isolates globally, in making skin and fat emulsions for later inclusion in processed meats and other applications, is described in detail by Bonkowski (90). Broad latitudes in formulation for processed meats exist outside of the United States, and also within the United States for non–standard of identity meat products.

Restructured Meats. Principles of restructuring meats are reviewed by Pearson and Dutson (91). Basically, red or poultry meats are flaked or chunked into small pieces, mixed or tumbled with salt and polyphosphates to extract heat-coagulable protein, shaped into loaf pans or other still-forming devices, heat-set at about 68°C (154°F), and cut into desirable shapes and thicknesses. Soy flours, concentrates, and isolates are used at approximately the same levels as in processed meats to improve textural stability and minimize shrinkage.

Pumped Meats. Brines consisting of water, salt, polyphosphates, and soy protein isolates or functional concentrates are prepared and pumped into muscle cuts using stitch pumps. Various domestic federal regulations apply; for example, hams and corned beef can be pumped to achieve cooked yields of 130%, provided a minimum protein content of 17% is maintained. Reviews on meat pumping technology have been prepared by Bonkowski (90) and Rakes (92).

Extruder-Texturized Soy Proteins. Texturized soy flours or concentrates may be rehydrated to 18% protein content (60 to 65% moisture content) and used at levels up to 30% reconstituted soy protein in ground meat blends and hamburgers. However, in domestic practice the reconstituted portion usually has been used at about 20%, because of texture and flavor problems accompanying higher levels of meat substitution. Special vitamin- and mineral-fortified texturized soy protein products are required for school lunch and military feeding. Texturized soy proteins are sometimes included in standard of identity canned meat products above the meat requirement to improve product attractiveness.

Baking Applications

The use of soy proteins in baked foods has been reviewed by Hoover (93), Dubois and Hoover (94), and Fulmer (52). Many applications in this industry have long been served

TABLE 8.9 Bakery Applications of Various Soy Proteins

	White bread and rolls	Specialty breads and rolls	Cakes	Cake doughnuts	Yeast-raised doughnuts	Sweet goods	Cookies
Defatted soy flour	x	x	x	x	x	x	x
Enzyme-active soy flour	x						
Low-fat soy flour			x	x		x	
High-fat soy flour			x	x		x	
Full-fat soy flour			x	x		x	
Lecithinated soy flour			x	x			x
Soy grits		x					
Soy concentrates		x					
Soy isolates		x	x	x	x		
Soy fiber		x					

Source: Fulmer, R. W., in *Proceedings of the World Conference on Vegetable Protein Utilization in Human Foods and Animal Feedstuffs*, edited by T.H. Applewhite, American Oil Chemists' Society, Champaign, IL, 1989, pp. 424–429.

TABLE 8.10 Composition of Dried Low-Fat and Full-Fat Soy Milk and Tofu Sold Domestically

	Dried Soy Milk		Tofu	
Analysis	*Full-Fat*	*Low-Fat*	*Full-Fat*	*Low-Fat*
Protein	38% 2.0	48% 2.0	38% 2.0	48% 2.0
Fat	18% 2.0	9% 2.0	18% 20	9% 2.0
Moisture	10% max.	5% max.	10% max.	5% max.
Ash	7% max.	5% max.	7% max.	5% max.

by soy flours. Bakery applications of various soy proteins are shown in Table 8.9. The increased absorption of lightly toasted (PDI 60 to 80) soy flour requires an additional three-quarters to one pound of water for each pound of flour added. Examples of soy protein uses include:

- *Bread and buns.* 1 to 3% defatted soy flour (on flour basis) increases absorption of water by one pound for each pound of soy flour and improves crumb body, resilience, crust color (from sugars), and toasting characteristics.
- *Cakes.* 3 to 6% defatted soy flour improves batter smoothness and distribution of air cells and gives a more even texture and a softer, more tender crumb.
- *Sweet goods.* 2 to 4% defatted soy flour improves water-holding capacity and sheeting properties. This level should also be used for yeast-raised doughnuts.
- *Cake doughnuts.* 2 to 4% defatted soy flour improves structure, gives an excellent star formation (hole), and reduces fat absorption during frying. The improved moisture retention improves product yield and shelf life.
- *Hard (snap) cookies.* 2 to 5% defatted soy flour improves dough machining and imparts a crisp bite to cookies.

Toasted defatted flours with PDI of about 20 add color to the crumb, and nutty toasted flavor to whole-grain and specialty breads. Up to 15% of toasted defatted flours can be added to leavened quick breads.

Enzyme-active full-fat and defatted soy flours, at 0.5% (flour basis), bleach carotenoid pigments in wheat flours and produce peroxides that strengthen gluten proteins.

Relecithinated soy flours in cakes improve emulsification of fats, ingredient blending, pan release, and machinability, and they partially replace egg yolks.

Dairy and Beverage Applications

Regulatory principles have long required that a new food, intended and promoted as a replacement for a major constituent of the current diet, be at least as nutritious as the product being replaced. The relative nutritional rating of soy protein has improved with abandonment of the PER system for evaluating protein quality. Although soy protein ranks high under the PDCAAS system, it normally contains less calcium than bovine milk and requires supplementation with this mineral.

Soy protein isolates have long been used in coffee creamers, whipped toppings, dairy-type dips, and frozen desserts, including frozen tofus (95,96). Relecithinated soy flours and concentrates are commonly used to assist dispersability of powdered beverages. Another major application has been in formulas for children allergic to milk proteins or showing lactose intolerance as well as in famine relief and nutritional improvement for infants and children at weaning in developing countries.

Other Soy Products

Many proprietary mixtures of soy proteins and dried skim milk, whey, egg yolks, and albumins are offered for the baking industry and others.

Dried Soy Milks and Tofus

Soy milks and tofus are sometimes made specifically for drying. Compositions of dried low-fat and full-fat dried soy milks and tofus sold in the United States are shown in Table 8.10.

Nut-Like Soybean Products

Toasted nut-like full-fat soybeans and soybean butters are also sold in the U.S. Examples of manufacturers' specifications include:

Roasted Soybean Nut-Like Products. Protein, 36.5%; Fat, 25.9%; Fiber, 4.6%; Ash, 3.5%; Moisture, 1.7%; Carbohydrates (by difference), 27.8%; and Calories/ounce (28.57 g), 140.

Dessert Topping. Made by roasting, granulating, and enrobing the granule with confectionery coats. Composition claimed: Protein, 20.1%; Fat, 25.9%; Fiber, 4.5%; Ash, 3.2%; Moisture, 0.9%; Carbohydrate (by difference), 45.4%; and Calories, 142/oz (28.57 g)

Soybean Butter. Protein, 29.9%; Fat, 45.3%; Fiber, 3.9%; Ash, 3.7%; Moisture, 1.6%; Carbohydrates (by difference), 15.6%; and Calories/ounce (28.57 g), 157

References

1. *Soy Protein Products*, Soy Protein Council, Washington, DC, 1987.
2. Circle, S.J., in *Soybeans and Soybean Products, Vol. 1*, edited by K.S. Markley, John Wiley & Sons, New York, 1950, pp. 275–370.
3. Smith, A.K., and S.J. Circle, *Soybeans: Chemistry and Technology, Vol. 1*, Avi Publishing Co., New York, 1972.
4. Hanson, L.P., "Vegetable Protein Processing," *Food Technol. Rev.* No. 16, Noyes Data Corporation, 1974.
5. Martz, M.A., "Protein Food Supplements," *Food Technol. Rev.* No. 54, Noyes Data Corporation, 1981.
6. *J. Amer. Oil Chem. Soc. 51(1):* 6A, (1974).

7. *J. Amer. Oil Chem. Soc. 56(3):* 99, (1979).
8. *J. Amer. Oil Chem. Soc. 58(3):* 121, (1981).
9. Applewhite, T.H., *Proceedings of the World Conference on Vegetable Protein Utilization in Human Foods and Animal Feedstuffs*, American Oil Chemists' Society, Champaign, IL, 1989.
10. Applewhite, T.H., *Proceedings of the World Conference on Oilseed Technology and Utilization*, American Oil Chemists' Society, Champaign, IL, 1993.
11. Wolf, W.J., and J.C. Cowan, *Soybeans as a Food Source*, rev. edn., CRC Press, Inc., Cleveland, OH, 1975.
12. Shurtleff, W., and A. Aoyagi, *Bibliography of Soymilk and Soymilk Products: 2612 References from A.D. 1500 to 1989*, Soyfoods Center, Lafayette, CA, 1989.
13. Shurtleff, W., and A. Aoyagi, *Bibliography of Soy Ice Cream, Yoghurt, and Cheese: 1071 References from 1910 to 1989*, Soyfoods Center, Lafayette, CA, 1989.
14. Shurtleff, W., and A. Aoyagi, *Bibliography of Soy Flour and Cereal–Soy Blends: 308 References from 3rd Century B.C. to 1900.* Extensively Annotated, Soyfoods Center, Lafayette, CA, 1989.
15. Shurtleff, W., and A. Aoyagi, *Bibliography of Tofu*, Soyfoods Center, Lafayette, CA, 1989.
16. Mustakas,G.C., E.L. Griffin, Jr., and V.E. Sohns, *Amer. Chem. Soc. Advan. Chem. Ser. 57:* 101, 1966.
17. Campbell, M.F., C.W. Kraut, W.C. Yackel, and H.S. Yang, in *New Protein Foods, Vol. 5*, edited by A.M. Altschul and H.L. Wilcke, Academic Press, New York, 1985, pp. 301–337.
18. Morse, E.H., U.S. Patent 2,331,619, (1943).
19. Sair, L., U.S. Patent 2,881,076, (1959).
20. *Official Methods and Recommended Practices of the American Oil Chemists' Society*, 4th ed., Champaign, IL, 1993.
21. Central Soya Company, *Soy Flour Product Line Summary*, Central Soya Co., Ft. Wayne, IN, 1988.
22. Rackis, J.J., *J. Amer. Oil Chem. Soc. 58:* 495 (1981).
23. Rackis, J.J., *J. Amer. Oil Chem. Soc. 51*(1)*:* 161A (1974).
24. *Protein Quality Evaluation*. Report of a Joint FAO/WHO Expert Consultation. Food and Nutrition Paper 51. Rome, 1990.
25. Food and Drug Administration. Proposed rules on food labeling, general provisions, nutritional labeling, nutrient content claims, health claims, ingredient labeling, state and local requirements, exemption. (21 CFR 101, 104, 105). *Fed. Reg. 56:* 603092, November 27, 1991.
26. Madi, R.L., *Cereal Foods World 38:* 576 (1993).
27. Liener, I.E., *J. Amer. Oil Chem. Soc. 58:* 406 (1981).
28. Albrecht, W.J., G.C. Mustakas, and J.E. McGhee, *Cereal Chem. 43:* 400 (1966).
29. Wright, K.N., *J. Amer. Oil Chem. Soc. 58:* 294 (1981).
30. Kanzamar, G.J., S.J. Predin, D.A. Oreg, and Z.M. Csehak, in *Proceedings of the World Conference on Oilseed Technology and Utilization*, edited by T.H. Applewhite, American Oil Chemists' Society, Champaign, IL, 1993, pp. 226–240.
31. Johnson, D.W., in *Food Uses of Whole Oil and Protein Seeds*, edited by E.W. Lusas, D.R. Erickson, and W-K Nip, American Oil Chemists' Society, Champaign, IL, 1989, pp. 12–39.
32. Pringle, W., *J. Amer. Oil Chem. Soc. 51*(1)*:* 74A (1974).
33. McDonald, F.M., *Oil Mill Gazetteer 85*(3)*:* 8 (1978).
34. Saio, K., and M. Ariska, *J. Japan. Soc. Food. Sci. Technol. 25:* 451 (1978).
35. Chiba, H., R. Sasaki, M. Yoshikawa, and K. Ikura. *J. Japan. Soc. Food Nutr. 34*(3)*:* 201 (1981).

36. Thomas, R., J.M. deMan, and L. deMan, *J. Amer. Oil Chem. Soc. 66:* 777 (1989).
37. Saio, K., K. Kobayakawa, and M. Kito, *Cereal Chem. 59:* 408 (1982).
38. Saio, K., I. Nikkuni, Y. Ando, M. Ortsura, Y. Terauchi, and M. Kito, *Cereal Chem. 57:* 77 (1980).
39. Smith, A.K. and S.J. Circle, *Soybeans: Chemistry and Technology, Vol. 1, Proteins,* revised 2nd ed., Avi Publishing Co., Westport, CN, 1978.
40. Kouzeh–Kanani, M., D.J. van Zuilichem, J.P. Roozen, and W. Pilnik, *Lebensm, Wiss. Technol., 14:* 242 (1981).
41. Mustakas, G.C., W.J. Albrecht, G.N. Bookwalter, J.E. McGhee, W.F. Kwolek, and E.L. Griffin, Jr., *Food Technol. 24*(11)*:* 1290 (1970).
42. Mustakas, G.C., W.J. Albrecht, G.N. Bookwalter, V.E. Sohns, and E.L. Griffin, Jr., *Food Technol. 25*(5)*:* 534 (1971).
43. Bookwalter, G.N., G.C. Mustakas, W.F. Kwolek, J.E. McGhee, and W.J. Albrecht, *J. Food Sci. 36:* 5 (1971).
44. Jansen, R., and J.M. Harper, *Food and Nutr. 6*(1)*:* 2 (1980).
45. Lorenz, K., G.R. Jansen, and J. Harper, *Cereal Foods World 25*(4)*:* 161, 171 (1980).
46. Harper, J.M., and G.R. Jansen, *Food Reviews International 1*(1)*:* 27 (1985).
47. Molina, M.R., J.E. Braham, and R. Bressani, *J. Food Sci. 48:* 434 (1983).
48. Patil, R.T., D.S. Singh, and R.E. Tribelhorn, *J. Food Sci. Technol. India 27*(6)*:* 376 (1990).
49. Guzman, G.J., P.A. Murphy, and L.A. Johnson, *J. Food Sci. 54:* 1590 (1989).
50. Fulmer, R.W., in *Proceedings of the World Conference on Vegetable Protein Utilization in Human Foods and Animal Feedstuffs*, edited by T.H. Applewhite, American Oil Chemists' Society, Champaign, IL, 1989, pp. 55–65.
51. Vavlitis, A. and E.D. Mulligan in *Proceedings of the World Conference on Oilseed Technology and Utilization*, edited by T.H. Applewhite, American Oil Chemists' Society, Champaign, IL, 1993, pp. 286–289.
52. Fulmer, R.W., in *Proceedings of the World Conference on Vegetable Protein Utilization in Human Foods and Animal Feedstuffs*, edited by T.H. Applewhite, American Oil Chemists' Society, Champaign, IL, 1989, pp. 424–429.
53. Ohren, J.A., *J. Amer. Oil Chem. Soc. 58:* 333 (1981).
54. Mustakas, G.C., L.D. Kirk, and E.L. Griffin, *J. Amer. Oil Chem. Soc. 39:* 222 (1962).
55. Campbell, M.F., R.J. Fiala, J.D. Wideman, and J.F. Rasche, U.S. Patent 4,265,925 (1981).
56. Howard, P.A., M.F. Campbell, and D.T. Zollinger, U.S. Patent 4,234,620 (1980).
57. McAnelly, J.K., U.S. Patent 3,142,571 (1964).
58. Wolf, W.J., in *Handbook of Processing and Utilization in Agriculture, Vol. II: Part 2*, edited by I.A. Wolf, CRC Press, Inc., Boca Raton, FL, 1983, pp. 23–55.
59. Savoie, L., and G. Parent, *J. Food Sci. 48:* 1876 (1983).
60. Karayianis, N.I., J.T. MacGregor, and L.F. Bjelfanes, *Food and Cosmetics Toxicology 17:* 591 (1979).
61. Struthers, B.J., *J. Amer. Oil Chem. Soc. 58*(3)*:* 501 (1981).
62. Freidman, M., *ACS Symposium Series 206:* 231 (1982).
63. Rhee, K.C. and Y.R. Choi, in *Annual Progress Report of Food Protein Research and Development Center*, Texas A&M University, College Station, TX, 1981, pp. 203–233.
64. Wolf, W.J., G.E. Babcock, and A.K. Smith, *Arch. Biochem. Biophys. 99:* 265 (1962).
65. Kinsella, J.E., *J. Amer. Oil Chem. Soc. 56:* 242 (1979).
66. Kolar, C.W., S.H. Richert, C.D. Decker, F.H. Steinke, and R.J. Vander Zanden, in *New Protein Foods, Vol. 5*, edited by A.M. Altschul and H.L. Wilcke, Academic Press, New York, 1985, pp. 259–299.
67. Davidson, R.M., R.E. Sand, and R.E. Johnson, U.S. Patent 4,172,828 (1979).

68. Shermer, M., U.S. Patent 4,188,399 (1980).
69. Howard, P.A., W.F. Lehnhardt, and F.T. Orthoefer, U.S. Patent 4,368,151 (1983).
70. Lehnhardt, W.F., P.W. Gibson, and F.T. Orthoefer, U.S. Patent 4,370,267 (1983).
71. Gibson, P.W., and W.C. Yackel in *Proceedings of the World Conference on Vegetable Protein Utilization in Human Foods and Animal Feedstuffs*, edited by T.H. Applewhite, American Oil Chemists' Society, Champaign, IL, 1989, pp. 507–509.
72. Koseoglu, S.S., and E.W. Lusas, in *Proceedings of the World Conference on Vegetable Protein Utilization in Human Foods and Animal Feedstuffs*, edited by T.H. Applewhite, American Oil Chemists' Society, Champaign, IL, 1989, pp. 528–547.
73. Lawhon, J.T., K.C. Rhee, and E.W. Lusas, *J. Amer. Oil Chem. Soc. 58:* 377 (1981).
74. Lawhon, J.T., U.S. Patent 4,420,425 (1983).
75. Murray, E.D., T.J. Maurice, L.D. Barker, and C.D. Davis, U.S. Patent 4,208,323 (1980).
76. Gunther, R.C., *J. Amer. Oil Chem. Soc. 56:* 345 (1979).
77. Smith, A.K., and S.J. Circle, in *Soybeans: Chemistry and Technology, Vol. 1, Proteins*, rev., edited by A.K. Smith and S.J. Circle, Avi Publishing Company, Inc., Westport, CT, 1978, pp. 339–388.
78. Boyer, R.A., U.S. Patent 2,682,466 (1954).
79. Boyer, R.A., U.S. Patent 2,730,447 (1956).
80. Sogo, Y., S. Dosako, Y. Honda, K. Ahiko, S. Kawamura, T. Izutysu, and S. Tanyea, *Reports of Research Laboratory*, Snow Brand Milk Products Co., Kawagoe, Saitama, Japan *81:* 41 (1985).
81. Mercier, C., P. Linko, J.M. Harper, *Extrusion Cooking*, American Association of Cereal Chemists, Inc., St. Paul, MN, 1989.
82. Stanley, D.W., in *Extrusion Cooking Mercier*, edited by C. Mercier, P. Linko, and J.M. Harper, American Association of Cereal Chemists, Inc., St. Paul, MN, 1989, pp. 321–341.
83. Rokey, G.J., G.R. Huber, and I. Ben–Gera, in *Proceedings of the World Conference on Oilseed Technology and Utilization*, edited by T.H. Applewhite, American Oil Chemists' Society, Champaign, IL, 1993, pp. 290–298.
84. Kinsella, J.E., *J. Amer. Oil Chem. Soc. 56:* 259 (1979).
85. Kinsella, J.E., and W.G. Soucie, *Food Proteins*, American Oil Chemists' Society, Champaign, IL, 1989.
86. Kinsella, J.E., S. Damodaran, and B. German, in *New Protein Foods, Vol. 5, Seed Storage Proteins*, edited by A.M. Altschul and H.L. Wilcke, Academic Press, Inc., New York, 1985, pp. 109–179.
87. Cherry, J.P., *Protein Functionality in Foods, ACS Symposium Series 147* (1981).
88. Rhee, K.C., in *Proceedings of the World Conference on Vegetable Protein Utilization in Human Foods and Animal Feedstuffs*, edited by T.H. Applewhite, American Oil Chemists' Society, Champaign, IL, 1989, pp. 323–333.
89. Lusas, E.W., and K.C. Rhee, *ACS Symposium Series 312:* 32 (1986).
90. Bonkowski, A.T., in *Proceedings of the World Conference on Vegetable Protein Utilization in Human Foods and Animal Feedstuffs*, edited by T.H. Applewhite, American Oil Chemists' Society, Champaign, IL, 1989, pp. 430–438.
91. Pearson, A.M., and T.R. Dutson, *Advances in Meat Research, Vol. 3—Restructured Meat and Poultry Products*. Avi–Van Nostrand Reinhold Co., New York, 1987.
92. Rakes, G.A., in *Proceedings of the World Conference on Oilseed Technology and Utilization*, edited by T.H. Applewhite, American Oil Chemists' Society, Champaign, IL, 1993, pp. 311–319.
93. Hoover, W., *J. Amer. Oil Chem. Soc. 53:* 301 (1979).
94. Dubois, D.K. and W.J. Hoover, *J. Amer. Oil Chem. Soc. 58:* 343 (1981).
95. Wilding, M.D., *J. Amer. Oil Chem. Soc. 56:* 392 (1979).
96. Kolar, C.W., *J. Amer. Oil Chem. Soc. 56:* 389 (1979).

Chapter 9

Handling, Storage, and Transport of Crude and Crude Degummed Soybean Oil

John B. Woerfel

Consultant
Tucson, AZ

Introduction

Crude soybean oil is generally traded in the United States under Trading Rules of the National Oilseed Processors Association (NOPA) (1). Seven types of crude soybeans are recognized, depending on process of recovery (expeller-pressed, hydraulic-pressed or solvent-extracted) and whether or not the oil is degummed or mixed.

An important criterion in handling, storage, and transport of crude soybean oil is *degumming*. In the presence of moisture, the phosphatides in nondegummed oil will hydrate and precipitate as sludge. To avoid this problem oil for export or long-term storage is degummed. Water degumming removes the major part of the phosphatides, which may be recovered as lecithin or returned to the meal (see Chapter 10). This results in oil with less than 200 ppm phosphorus or less than 0.6% phosphatides. In recent years various processes have been introduced that result in lower levels of phosphatides. Such oil is sometimes designated as "super degummed" and may have phosphorus content as low as 20 ppm.

Oilseeds and oils and fats are traded under rules of a number of trade organizations in various countries (2). NOPA Trading Rules specifically cover soybean products. Table 9.1 shows NOPA specifications for crude and crude degummed soybean oils. These trading rules also detail other conditions applying to sales contracts. These may be referred to briefly in the following discussion of technical and operational procedures.

Crude Soybean Oil

Nondegummed crude soybean oil is commonly utilized by U.S. refiners. This is feasible because the oil is usually stored for relatively short periods. Extraction plants operate year-round and refine in adjacent refineries or ship daily to free-standing refineries. Inventories of crude oil at extraction plants may not exceed one week's production. Crude oil inventories at a refinery may be no more than two weeks' requirements. The long-mix caustic refining system used by most U.S. refiners is well adapted to nondegummed oil. In fact, many refiners, when buying crude degummed oil, will mix it with regular crude in storage tanks and refine the mixture.

TABLE 9.1 NOPA Trading Specifications for Soybean Oils

	Prime crude	No. 2 Crude	Crude degummed	AOCS method
Moisture and volatile matter, max %	0.5	0.5	0.3	Ca2d-25
Green color[1]	<B	<B	<B	
Refined and bleached color, max.	3.5	2.8	NA	Cc83-63
Red with discount	6.0	5.0	NA	Cc18b-45
Neutral oil loss, max %	7.5	7.5		Ca9f-57
Unsaponifiables, max %[2]	1.5	1.5	1.5	Ca6a-40
Flash point				Cc9b-55
Max °C	121	121	121	
Max °F	250	250	250	
Free fatty acid,				Ca5A-40
Max % as oleic	NA	NA	0.75	
With discount			1.25	
Phosphorus,				Ca12-55
Max %	NA	NA	0.020	
With discount			0.025	

[1]NOPA tentative method.
[2]Exclusive of moisture and insoluble impurities.
Source: Yearbook and Trading Rules 1992–93, National Oilseed Processors Association, Washington, D.C.

Storage and Handling of Crude Soybean Oil

Deterioration of crude soybean oil is promoted by moisture, impurities such as meal fines, high temperatures, exposure to air and contact with copper or other metals that promote oxidation.

Moisture will cause free fatty acid (FFA) increases, especially in combination with fines and temperatures that promote microbiological and enzymatic activity. Moisture is removed from the oil in the extraction process by the stripper and oil dryer (see Chapter 6). Of particular importance is good vacuum, so that stripping and drying are done without excessive temperature. Final moisture should not exceed 0.2%, and oil should be cooled to 40°C (104°F) for storage. This may be done by heat interchange with partial micella plus water-cooling.

Meal fines are removed from the full micella by a fine mesh-screen or by a *hydroclone*, which separates by centrifugal force. Crude oil from extraction can be maintained below 0.1% fines.

Crude oil to be degummed for lecithin production should be filtered, and this is best done when the oil is fresh. Lecithin that is perfectly clear commands a premium price, but the centrifugal separation of lecithin concentrates fines while removing some fines from the oil.

Degummed oil is dried in a vacuum dryer after degumming and should not exceed 0.1% moisture.

Moisture can be introduced during storage and transport in several ways. Leaking heating coils introduce water that may be contaminated with metals or other deleterious substances. Leakage can occur through improperly sealed manholes or hatches.

Humid air enters partially filled tanks as oil is pumped out or with variations in atmospheric pressure. On cooling, moisture condenses and remains in the tank. Solid impurities, including fines and sludge, settle in tanks and remain on the bottom. In combination with moisture, these can support microbiological and enzymatic activities that cause free fatty acid development.

Oxidation of oil is promoted by high temperatures, and as a rule of thumb, the rate doubles for each 10°C (18°F) increase. Oxidation increases in proportion to the amount of contact of the oil with air. A thin film of oil on the wall of a tank, particularly on an exposed heated coil oxidizes very rapidly. Introduction of air by agitation or during transfer can be very serious. Contact with air can be reduced by bottom-filling tanks; design and operation of agitators to minimize air contact; preventing air leakage on pump suction; and management of tank inventories to provide turnover and keep tanks full where possible.

Storage Tanks

Storage tanks at extraction plants, or terminals, are cylindrical and of mild steel construction. Size and number depends on production capacity and inventory required for orderly marketing and distribution. There should be a sufficient number to allow segregation of oils of different type or quality and to permit tanks to be emptied for cleaning or repairs.

Tanks are usually grouped in a "tank farm" laid out for convenient access and efficient piping. An important consideration is provision to eliminate or minimize commingling of oils from different tanks. Segregated lines, double valving, and blind flanges are helpful in this respect. Clearly marked tank and pipelines are necessary. Pipelines may be cleaned between use for different oils done by draining, blowing with nitrogen or mechanical means.

Environmental regulations require that tanks be surrounded by a dike to retain the total contents of the largest tank in case of a rupture. Controlled drainage of storm water must be provided. Concrete paving within the diked area is desirable for cleanliness and convenience of operators as well as for drainage.

Tanks must be built on foundations suitable for local soil conditions. A 15 m (50 ft) high tank will create a loading of about 15 MT/m^2 (3000 lbs/ft^2). Sedimentary soils in river valleys, or filled land at seaports, may need particular attention.

Agitators are desirable in tanks, to keep contents uniform for oil to be processed or for shipment. Side-entering mixers positioned near the bottom of the tank and pointed off-center are typical. High-efficiency impellers, providing maximum flow with minimum shear or turbulence, are preferred because the degree of mixing is moderate and aeration of the oil is avoided.

Fig. 9.1 shows certain piping features that are desirable for a storage tank (3). Fill pipes should discharge near the bottom of the tank, to minimize splashing and oxidation while filling. A siphon breaker should be installed at the high point in the fill line to avoid accidentally siphoning oil from the filled tank.

The bottom of the tank may be flat or pitched with a sump at the low point to permit the tank to be emptied completely and facilitate cleaning. An alternate connection above the tank bottom avoids picking up sediment that may accumulate.

Heating coils are desirable, because at low temperatures solids may precipitate and the viscosity of the oil will increase. Placing horizontal coils near the bottom, but with sufficient clearance to allow squeegeeing, or cleaning the tank bottom, is good practice. Such coils remain immersed and can be used even when the level is low in the tank. Heating should be with low pressure steam or hot water to avoid overheating. For energy conservation, "flash steam" or hot water from another plant operation may be used.

During hot weather, tanks may heat by exposure to direct sunlight, and painting the exterior with a reflective paint will minimize temperature increase.

Measurement of Crude and Crude Degummed Soybean Oil

Soybean oil is traded on a net weight basis. Where possible, direct scale weights are the preferred method of measurement. Since this is not always possible, volumetric measurement using appropriate techniques can be done with acceptable accuracy.

Tank cars and tank trucks are usually weighed light and heavy when loaded by the shipper and when unloaded by the buyer. In some instances vehicles may be filled from a scale, which allows exact weights to be loaded.

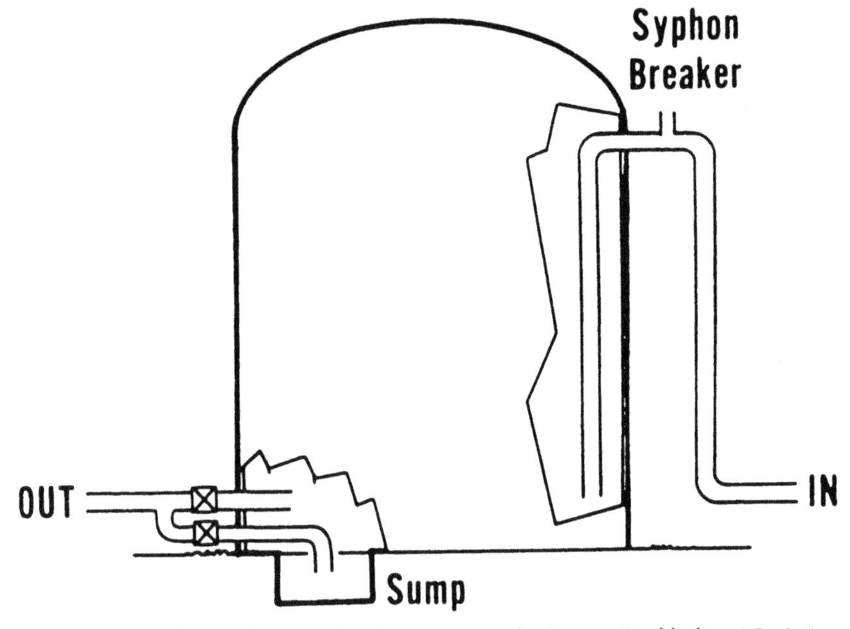

Fig. 9.1. Typical in and out piping for storage tank. *Source:* Burkhalter, J.P., *J. Amer. Oil Chem. Soc. 53:* 1976, pp. 332–333.

Scales may be within the plant, or they may be public scales. In any case, scales should be tested regularly and certified by an appropriate authority, and the weighing itself should be done by a certified weighmaster. Details of certification and official weight requirements are detailed in the trading rules issued by various organizations.

Gauging

Gauging of tanks is frequently done for in-plant inventory control and in ocean shipments. Tank calibration or gauging tables are developed to show the volumetric capacity at different levels from bottom to top at specified temperatures.

The preferred method is to calibrate by weight—that is, by filling the tank with weighed increments of a liquid at a specified temperature and specific gravity and measuring the innage and outage with reference to a permanent fill mark.

Cylindrical storage tanks may be gauged by "strapping". This consists of measuring the circumference of the tank at various levels using a flexible tape and calculating the volume. Tank calibration must be done in accordance with good engineering practices. Considerations include ambient temperature and temperature of contents which affect specific gravity of contents and expansion of tank shells and tape. NOPA Trading Rules reference American Society for Testing and Materials (ASTM) standards for tank calibrations (4).

Overland Transport of Soybean Oil

Tank Cars

Tank cars are of various sizes, construction types, and specifications. The most common sizes used for soybean oil with their normal capacities are:

Standard:
8000 gallons (30 m^3)
60,000 lbs (27 MT)
Jumbo:
20,000 gallons (75 m^3)
150,000 lbs (68 MT)
Super Jumbo:
23,000 gallons (87 m^3)
175,000 lbs (79 MT)

NOPA Trading Rules specify weight tolerances that constitute good delivery in the various sizes.

Fig. 9.2 shows a diagram of a Jumbo tank car suitable for soybean oil. The U.S. Department of Transportation (DOT) classification number indicates that this tank is fusion-welded steel construction with a test pressure of 100 psig (0.7 MPa) (4). While tank cars are generally of similar construction, details and accessories vary. Steam coils may be provided and are needed for soybean oil in cold weather, when the product may partially solidify and become difficult to pump.

20,000 GALLON CAPACITY - NON INSULATED
DOT - 111A100W1
FOR GENERAL SERVICE COMMODITIES
4'' SLOPE TO STRAIGHT CENTER SECTION.

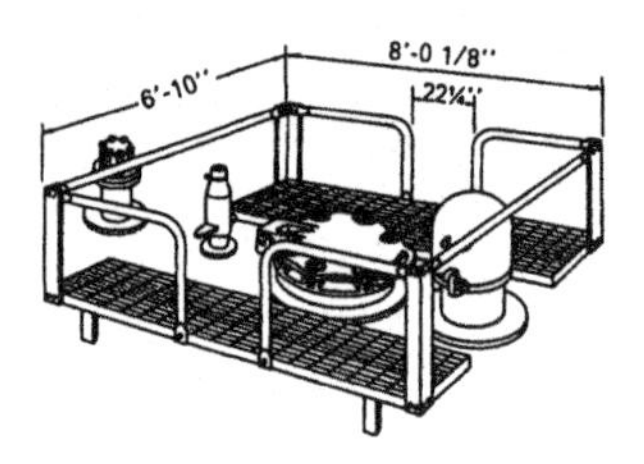

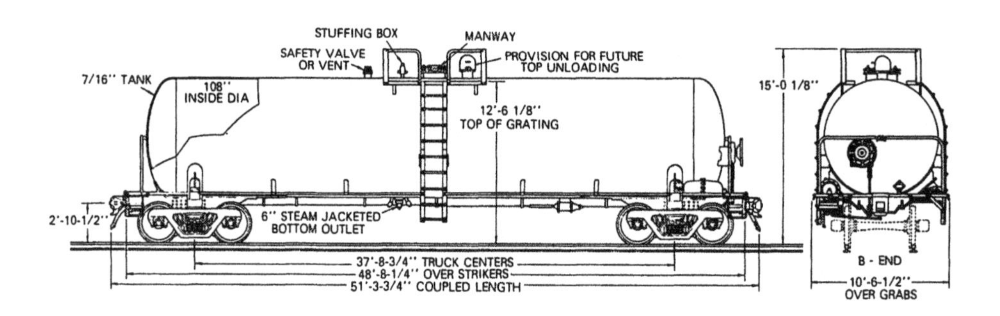

CAPACITY & WEIGHTS

NOMINAL CAPACITY @ 2% OUTAGE - 20,000 GALS.
ESTIMATED LT. WT. (NON COILED) - 57,800 LBS.
RAIL LOAD LIMIT (100 TON TRUCKS) -263,000 LBS.

COMMODITY MAXIMUM DENSITIES

TRUCK CAPY.	WHEEL BASE	NO. OF COILS	COMM. WT./GAL.	NON - COILED COMM. WT./GAL.
70 TON	5'-8''	16	8.02#	8.13#
100 TON	5'-10''	16	9.94#	10.04#

Fig. 9.2. Jumbo tank car. *Source: GATX Tankcar Manual,* General American Transport Corp., Chicago, IL, 1970, p. 87.

The hatch (called a manway) on top is for loading and access for inspection and cleaning. A 4-inch or 6-inch (10 or 15 cm) iron pipe size bottom outlet is closed with an outlet valve and a cap with a test plug. The outlet valve is usually operated from the top by a rod extending through the tank, although some tanks have ball valves operated from the bottom.

Tank Trucks

Tank trucks are usually semitrailers. The manway, with hinged and bolted cover, provides access for loading, inspection and cleaning. A 3-inch (8 cm)outlet valve at the rear provides complete drainage if the tank is parked on a slight incline. Some tanks have the drain in the middle, with the tank bottom pitched for drainage. The valve is easily inspected and is provided with a cover. Provision is provided for applying seals to both manway and valve. Two 3-inch (8 cm) flexible hoses are usually carried in tubes mounted on the trailer.

Heating coils and insulation are not usually required, because trucking transit times are short. Tank trucks designed to carry partial loads have baffles to reduce "sloshing" of contents. These are not necessary on fully loaded tanks and are undesirable for vegetable oil, because they interfere with inspection and cleaning. Procedures for loading tank trucks are similar to procedures for tank cars.

Weighing Tank Cars and Tank Trucks

Weighing is the preferred method for determining the quantity of oil loaded or unloaded. Trading rules require that scales be approved by a recognized authority and that weighing be performed by a certified weighmaster.

When loading, the vehicle is weighed light and heavy; when unloading, the vehicle is weighed heavy and light. In each case the same scale must be used for both weighings.

Tank cars must be weighed free, uncoupled, and centered on the scale. Tank trucks must be weighed on a scale large enough to accommodate both tractor and tank trailer. Double or split weighing is not permitted.

Before weighing, tank cars or tank trucks must be made reasonably free of any snow, ice, or extraneous matter. In the case of trucks, any equipment, as well as the driver, should remain on the truck during both weighings. Any discrepancy between marked tare and light weight should be investigated and resolved.

Volumetric measurements may also be acceptable if made by approved procedures. Gauging tables can be provided by the tank car or truck manufacturer.

Loading and Unloading Cars

Tank cars are positioned at the loading station by a switch engine. Spot adjustments may be made with a power winch and cable. Chocks are placed under the wheels and a sign posted across the track warning not to move the car.

A typical tank car loading procedure used by a major producer of edible soybean oil is reproduced as follows (3).

1. Check car for cleanliness and odor inside and outside. If it is dirty or has an odor, reject it.
2. Close inner valve and remove caps on outlet valve and steam coil.
3. Place five-gallon (20 L) bucket under oil outlet.
4. Have 36″ (90 cm) or 48″ (120 cm) pipe wrench on hand.
5. Secure loading line to car.
6. Start oil into car. Stop pump after a small amount of oil is in car. Check outlet to see whether the inner valve is leaking. If the inner valve is not holding, flush a small amount of oil through the valve into the bucket and then close the valve. This is to flush out any dirt or sand that might be between the valve and the seat. If the inner valve holds, then finish loading. To finish loading, turn on nitrogen sparge line so that the nitrogen will displace all the oxygen that is in the car. Continue to watch the oil outlet to make sure it does not leak.

   During loading of the car, the dome should be covered with a press paper to keep out the dust.
7. After the car is loaded, the lines will have to be blown. To blow all the oil from the loading line, raise the line above the oil level so that the oil will not blow out over the sides. Also, the filter will have to be bypassed, to avoid breaking the paper in the filter and letting dirty oil into the car.
8. Take a trier sample (this must be done to get a representative sample for both the laboratory sample and a keeper sample). First, open the trier and slowly push it down through the oil to the bottom of the car; then close the trier and pull it out of the car, using one hand to wipe as much oil as possible off the outside of the trier. Place the trier into a 5-gallon (20 L) bucket and slowly open the trier. Do this three times to clean the trier to avoid contamination. Avoid oil spills, which would make standing on top of the dome difficult.
9. Check the gaskets on the top and bottom of the car and replace if needed.
10. Secure bottom outlet valve to the car using 36″ (90 cm) or 48″ (120 cm) pipe wrench.

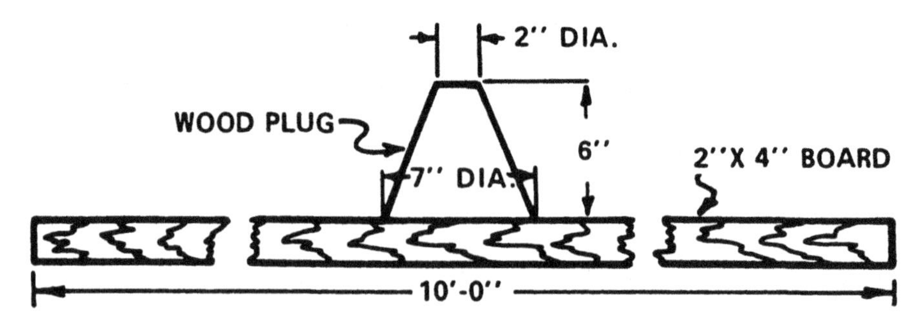

Fig. 9.3. Plug to insert in tankcar outlet to save lading if valve is inoperative. *Source: GATX Tankcar Manual*, General American Transportation Corporation, Chicago, IL, 1979.

11. Close bottom outlet valve; then, go on top of the car and open inner valve. Place 5-gallon (20 L) bucket under bottom valve and drain off 5 gallons (20 L) of oil. (This will carry out any moisture left in the car.)
12. Close all valves on the car and secure steam coil caps.
13. Seal the top and bottom of the car. (Make sure the top and bottom are sealed so that the car cannot be opened without first breaking the seals.)
14. Wash all traces of oil off the car.
15. Be sure proper placards are on the car. Remove all old placards and signs that remain on the car.
16. Record in the proper book the date, kind of oil, seal number top and bottom, condition of car, weather conditions, name of the washer, and name of the loader. If the car is rejected, record reason for rejection.

Tank cars and trucks are unloaded through a flexible hose attached to the car outlet and piped to an unloading pump. An alternative method is to spot the vehicle over a sump and discharge by gravity. The unloading pump is usually permanently installed at the receiving plant, although some tank trucks are equipped with power take-off pumps mounted on the tractor.

Certain precautions are required in unloading for safety and to avoid loss of product. First, the tank should be checked to ensure that there is no internal pressure. The manway is opened, and necessary steps are taken to permit air to enter as the car is unloaded. If heating is required, hook up steam to coils (with soybean oil this will only be required in extremely cold weather). Control is necessary to avoid overheating. Take a sample from car using the official method; although the loading sample is usually designated as Official, the unloading sample may become official under certain conditions. Before unloading, make preliminary tests to verify that the commodity is soybean oil with no obvious defect such as off odor, color, or contamination.

Check the storage tank that is to receive the oil, and all piping and valves leading to it. Check to see that the tank car outlet valve is closed. Then place a container under the outlet and loosen the outlet closure plug. If any leakage occurred during transit, oil will drain out around the closure. If leakage does not stop, try sealing the valve by opening and closing the outlet valve a few times with the closure still in place. If this does not stop the leakage, then the valve or valve mechanism is defective, and other means, such as top unloading, must be used. If the leakage is stopped, the outlet closure can be removed and unloading connections made before the outlet valve is opened. If it is found that the outlet is blocked or damaged for any reason, top unloading is used.

If the outlet and valve are in good order, remove the outlet closure and immediately attach unloading connectors. Then open the valve and proceed with unloading. Check again to see that the oil is going to the proper holding tank. It is recommended that a plug, such as the one shown in Fig. 9.3, be on hand before unloading. If the valve is inoperative and the lading is accidentally released, the plug may be inserted in the outlet nozzle to save the load.

After all the oil is drained and pumped out, an operator may enter the tank and squeegee the remaining oil from the car. Any solid residue that cannot be drained, pumped, or squeegeed from the car is considered *settlings* or *sludge*, which is subject to claims for adjustment.

Water Transportation of Soybean Oil

Barges

Crude and crude degummed oil is transported on major rivers by barge. Shipments originate at extraction plants and move by barge to refineries or large bulk terminals, which store, handle, and ship oil in international trade. A typical barge consists of three compartments with a total capacity of 1,400 short tons (2,800,000 lbs) (1270 MT).

Shipping time is relatively short and seldom exceeds seven to ten days. Barge shipments are not nitrogen-blanketed or sparged.

Tankers

Crude degummed soybean oil is shipped in international trade in seagoing tanker vessels. A single vessel may carry 30,000 tons (27,000 MT) of oils, fats, and other commodities as separate parcels in cargo tanks of various sizes.

Ludwiczak (6) presents a detailed discussion of factors to be considered in booking cargo space and desirable features being incorporated in sophisticated tankers. The following paragraphs give a brief summary.

The ship owner must use due care to protect the cargo from damage and degradation. This includes avoiding shortages, contamination with water or foreign substances, and mishandling such as improper cleaning, comingling of cargo, overheating, or exposure to air. Cargo tanks may be made of stainless steel or mild steel or be coated. Stainless steel tanks or coated tanks with suitable coating are preferred. Cargo tanks must be of proper size for the parcel. A partially filled tank has more potential for exposure to moisture and oxidation and overheating.

The pumping system should be designed to avoid risk of contamination and provide for proper cleaning and maintenance. Some ships have separate pumps built into each cargo tank with individual lines. Tanks should be of improved design with structural supports located outside the cargo tanks. Frames and beams inside the tanks interfere with cargo handling and proper cleaning.

Void spaces (cofferdams) should be used between two bulkheads to separate products requiring high temperatures from edible oils, which are heat-sensitive.

Double bottoms on ships that discharge oils and fats in cold climates eliminate risk of solidification of the product resulting from direct contact of the tank bottom with cold water.

No brass, bronze, or copper should be used in tanks or piping, because these promote oxidation of the fat. Provision for nitrogen blanketing of tanks is desirable, and some tankers have nitrogen generators to provide a continuous supply.

Contamination

A major concern in handling and transporting vegetable oils is to avoid contamination. In particular, substances regulated by pure food laws must be in conformance with established limits, but other substances also need to be controlled. Trading rules include various regulations on cleaning, inspection, and separation practices in transport equipment.

NOPA rules (1) stipulate that "Equipment that has at any time handled materials containing toxic metalo-organic compounds (for example, leaded gasoline or pesticides) or **within the past two years** handled other substances generally recognized as toxic will **not** be used for storage or transfer of soybean oil." NIOP (National Institute of Oilseed Products) rules include detailed lists of acceptable prior cargo and unacceptable prior cargo (2). Backlog (7) discusses contamination and adulteration as well as other concerns about the effects on quality of handling and shipping practices. The best practice is that any tanks used be restricted to use for food products.

Adulteration

Adulteration has been a problem in the edible oil trade for a long time. Rossell (8) discusses the potential for accidental and deliberate adulteration and various tests that have been used to characterize oils.

Sampling and Analysis

Crude oils are sold on specified grade and quality which are a part of the sales contract and are outlined in detail in trading rules.

An Official Sample is taken at the time of loading by either continuous flow or trier methods (9) (AOCS C 1-47). It is divided into three one-quart (0.946 L) samples: one for analysis by the shipper, one forwarded to the consignee, and one retained by the shipper as a referee sample. The emphasis in sampling is on obtaining a truly representative sample. An experienced and qualified sampler is required.

Continuous flow sampling is generally the most satisfactory, but it is applicable only to product that is completely liquid and does not contain any material that might plug the bleeder line.

Trier samples require close attention to detail. A properly designed bomb or core sampler must be used, and the sample taken in a specific pattern depending on the size and shape of the tanks. The size and number of samples must be adequate and composited to provide representation of the entire oil shipment.

Cleaning of Tanks

Trading rules generally require that tanks for transport of soybean oils be suitable for edible products and free of any prior contents. Any person responsible for cleaning and inspection procedures should be fully aware of the detailed trading rules that apply and cleaning techniques.

Storage tanks at plants handling only soybean oil may be in continuous use, with partial filling and emptying determined by production and loading schedules. Such tanks should be fully emptied and cleaned frequently. Excessive delays between cleanings can be counterproductive, because excessive buildup of sludge can make cleaning much more difficult.

Cleaning equipment using high-pressure jets rotating horizontally and vertically is highly effective even in large tanks. High temperature—85°C (185°F)—and an appropriate detergent are desirable. Effective detergents are usually a combination of ingredients and must be suitable for use in food plants. If an appreciable amount of sludge or sediment is present, it may have to be removed manually before washing.

Safety

There are several specialized safety concerns in handling oils. Entering a tank to clean or squeegee is especially hazardous. Precautions include the following:

1. The tank car or truck should not be moved under any circumstances when a worker is on or in it.
2. Steam must be disconnected from coils. The tank must be thoroughly cooled and free of any fumes. Fumes generated by the heating and oxidation of oil can be hazardous.
3. A suitable air supply must be provided.
4. The worker in the tank should wear a harness with safety line attached, and another worker outside the tank should be in attendance at all times to assist the worker inside.
5. Use of suitable protective clothing is essential. Falling or head injuries are possible because of the restricted space, nonlevel and slippery surfaces, and impediments such as coils and braces.
6. Water safety precautions should be taken, including wearing life jackets, by workers on barges or ships.

References

1. *Yearbook and Trading Rules 1992–93*, National Oilseed Processors Association, Washington, DC.
2. Fleming, R., in *World Conference Proceeding, Edible Fats and Oils Processing—Basic Principles and Modern Practices*, edited by David R. Erickson, American Oil Chemists' Society (AOCS), Champaign, IL, 1990, p. 15–24.
3. *Handbook of Soy Oil Processing and Utilization*, edited by David R. Erickson, et al., American Soybean Association (ASA), St. Louis, MO, and American Oil Chemists' Society (AOCS), Champaign, IL, 1980, pp. 309–311.
4. American Society for Testing and Materials. ASTM-D-1220-675; ASTM-D-105-65; ASTM-D-1086-64; ASTM-D-1963-74, American Society for Testing and Materials, Philadelphia, PA.

5. *GATX Tankcar Manual.* General American Transportation Corporation, Chicago, IL, 1979.
6. Ludwiczak, Joseph E., in *Proceedings of the World Conference on Edible Fats and Oils Processing—Basic Principles and Modern Practices*, edited by D.R. Erickson, American Oil Chemists' Society (AOCS), Champaign, IL, 1990, pp. 25–27.
7. Backlog, B.P., in *Proceedings of the World Conference on Edible Fats and Oils Processing—Basic Principles and Modern Practices*, edited by D.R. Erickson, American Oil Chemists' Society (AOCS), Champaign, IL, 1990, pp. 28–30.
8. Rossell, J.R., King, B., and Downes, M.J., *J. Amer. Oil Chem. Soc. 60:* 2, 333–339 (1983).
9. Method C:1-47, *Official and Tentative Methods of the American Oil Chemists' Society*, AOCS, Champaign, IL, 1978.

Chapter 10

Degumming and Lecithin Processing and Utilization

David R. Erickson

Consultant
American Soybean Association
St. Louis, MO

Introduction

Degumming is the process for removal of phosphatides from crude soybean and other vegetable oils. The phosphatides are also called gums and *lecithin*. The latter term is also the common name for phosphatidyl choline, but common usage refers to the array of phosphatides present in all crude vegetable oils. Although all crude vegetable oils contain gums, soybean oil is currently the major source of commercial lecithin, because it contains the largest amount of gums and is also the world's leading vegetable oil (see Chapter 1).

The relative composition of commercial soybean lecithin from conventionally solvent-extracted crude soybean oil is shown in Table 10.1. Changes from this composition, as affected by different extraction practices will be discussed in the following section, followed by a discussion of lecithin production and utilization.

Degumming Processes

Simply stated, degumming is the process used to remove phosphatides, or gums, from crude soybean oil. As shown in Table 10.1, the gums obtained from degumming consist of a mixture of soybean oil and phosphatidyl compounds (1). The chemical structures of the three major phosphatides are shown in Fig. 10.1.

In conventional solvent extraction, only half of the phosphatides present in soybeans are extracted with the composition shown in Table 10.1. Use of preparation processes such as the Alcon process or expanders will change the array of phospha-

TABLE 10.1 Approximate Composition of Natural Commercial Soybean Lecithin (1)

	%
Soybean oil	35
Phosphatidyl choline	16
Phosphatidyl ethanolamine	14
Phosphatidyl inositol	10
Phytoglycolipids, minor phosphatides	17
Carbohydrates	7
Moisture	1

tides in the crude oil by increasing the phosphatidyl choline content by about 30 to 40% and will increase the total extracted phosphatides (2). In the Alcon process double the normal amount of gums is extracted, and use of the expander may approach that level depending on the expander conditions of temperature, time, and moisture addition (for details of these processes, see Chapter 6).

Degumming of soybean oil is done for one of the following reasons:

1. To produce lecithin
2. To prepare degummed oil for long-term storage or transport
3. To prepare a degummed oil for caustic or physical refining

The first reason is obvious and the second is necessary to prevent development of troublesome sludges in storage or transport. Such sludges form because the phosphatides are hygroscopic and become hydrated by moisture from the air. Phosphatides are soluble in dry crude oil, but when hydrated, they become more dense than the triglycerides and precipitate, or settle out, from the crude oil, causing the troublesome sludges. Although this phenomenon is unwanted in storage or transport of crude oils, it is the basis for the process of degumming. Water is added to the crude oil to hydrate the phosphatides and thus prepare them for removal by gravitational forces or centrifugation. The latter is the modern process.

The process of degumming is simple, but the quality of the crude soybean oil has an influence on the efficacy of degumming. The phosphatides in crude soybean oil exist in either the hydratable or nonhydratable forms. The hydratable phosphatides (HP) are readily removed by the addition of water, whereas the nonhydrat-

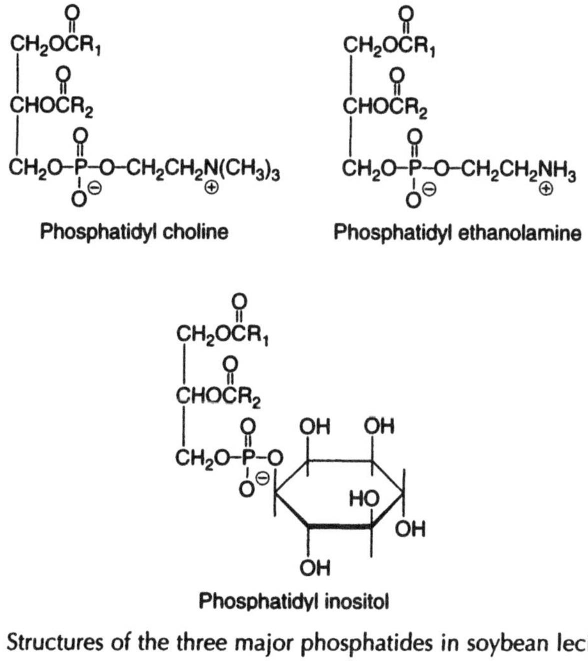

Phosphatidyl choline

Phosphatidyl ethanolamine

Phosphatidyl inositol

Fig. 10.1. Structures of the three major phosphatides in soybean lecithin.

able phosphatides (NHP) are relatively unaffected by water and tend to be more oil-soluble (i.e., remain in the oil phase). The NHPs are generally considered to be the calcium and magnesium salts of phosphatidic acids that arise from the enzymatic action of phospholipases released by damage to the cellular structure of the soybean (3). Such damage may occur with handling, processing practices, or both. The formation of NHPs is shown in Fig. 11.3.

Poor-quality soybean oils are defined by a higher FFA content (>1.0%) indicating a higher than normal NHP content. Other indicators of poor-quality crude soybean oils are a lower phosphatide content (<1.0%) in the extracted oil and a high content (>0.79%) of phosphatides in degummed oils (4). Normal-quality soybean oil from conventional solvent extraction will have about 90% HP and 10% NHP, and their total phosphatide content will range from 1.1 to 3.2% (5). The FFA of good-quality crude soybean oil will be in the range of 0.5 to 1.0%, which will be reduced by 20 to 40% in the degummed oil (6,7).

Phosphatides also form complexes with metals (iron and copper), and their content in degummed and refined oils is reduced as discussed in Chapter 11.

Recognition of the role of calcium and magnesium has led to the use of demineralized (soft) water for degumming and the use of phosphoric or citric acids as aids in degumming. These latter two acids are food-grade and combine with the calcium and magnesium salts, allowing transfer of the phosphatidic acids from the oil to the aqueous phase, thus removing them from the crude oil. The use of acids in degumming is not recommended for gums intended for use as lecithins, because their presence will cause darkening of the lecithin. The effect of the NHPs on caustic refining is discussed in Chapter 11.

Some plants use steam condensates for degumming rather than deionized water; however, the absence of iron in the condensate should be ensured. Some plants also use excess stripping steam in the last stages of miscella desolventizing as a means for hydration of phosphatides, but this is more difficult to control than simple water addition.

Water Degumming

In degumming of soybean oil for lecithin production, it is standard practice to filter the crude oil first to remove meal fines, which affect the clarity of lecithin and contribute to the hexane-insoluble content (8). If the degumming centrifuge is self-cleaning and the gums are destined for return to the meal stream, it may not be necessary to filter the crude oil; however, some plants filter the crude oil routinely to prevent the possible increase of free fatty acids during storage and to reduce refining losses (7).

In the batch water degumming process, soft water at a level of 1 to 2%, depending on the phosphatide content, is added to warm (70°C, 158°F) oil and mixed thoroughly for 30 to 60 minutes, followed by settling or centrifugation. The amount of water to add is about 75% of the phosphatide content (9), but some plants routinely add a fixed percentage based on experience.

A diagram of the process of continuous water degumming is shown in Fig. 10.2 (10). In this case, the oil is heated to about 70 to 80°C (158 to 176°F), soft water is added by an in-line proportioning system, the oil and water are mixed in-line, and then the mixture flows to a retention vessel, where it is held for 15 to 30 minutes and then sent to a centrifuge.

For best results in both batch and continuous degumming, it is absolutely essential to provide sufficient time for the water and the phosphatides to react. Hydration of phosphatides is not an instantaneous reaction and requires initial intensive mixing, followed by sufficient time to allow hydration and coalescence of the hydrated micelles. An excellent general discussion of the chemistry of degumming has been published by Dijkstra (11).

Experience with degumming and caustic refining systems for soybean oils with little or no retention times ("short-mix") has shown that such systems are not adequate for production of optimum-quality soybean oils (12,13,14).

With good-quality crude soybean oil, simple water degumming will reduce the phosphorus content to less than 50 ppm (0.005%), which is well below the 200 ppm (0.02%) level specified in the National Oilseed Processors Association (NOPA) trading rules for crude degummed soybean oil (15). The relation between phosphorus and phosphatide content is

$$\% \text{ Phosphatides} = 30.0 \times \% \text{ Phosphorus content}$$

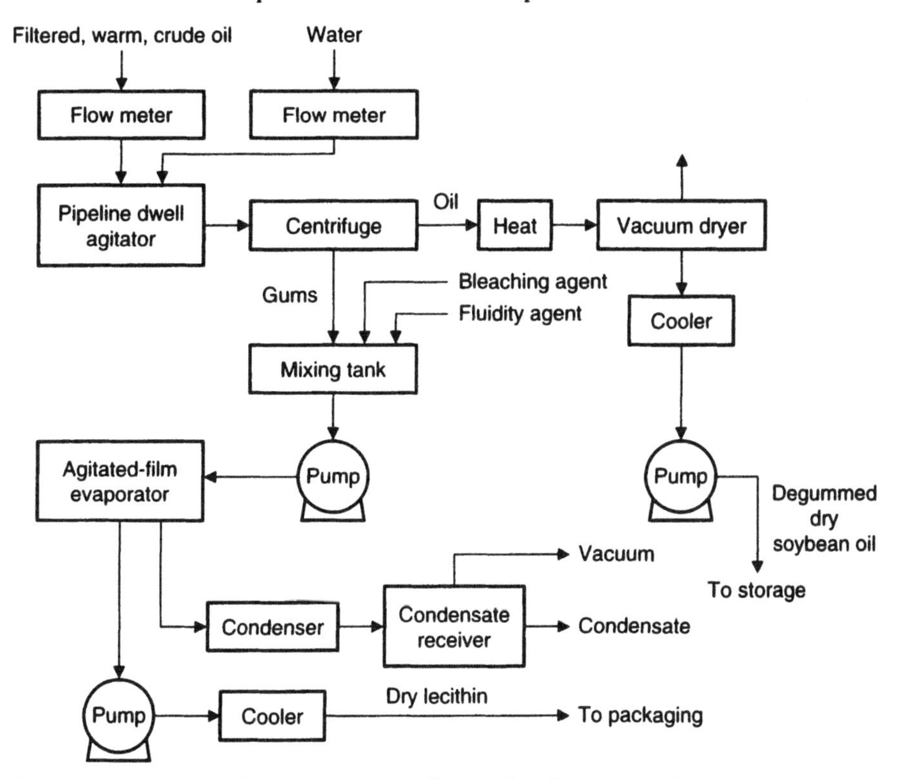

Fig. 10.2. Flowsheet for degumming soybean oil and crude lecithin production. *Source:* Brian, R., *J. Amer. Oil Chem. Soc. 53:* 27, (1976).

Degumming of Nonconventional Solvent-Extracted Oils

As just mentioned, only about 50% of the phosphatides present in soybeans are extracted by conventional extraction processes. However, it has long been recognized that there is an increase in hexane-extractables of soybean meal after the heat and moisture treatment in the desolventizer-toaster. Subsequently, it was discovered and patented by Koch (16) that a heat and moisture treatment of flaked soybeans before extraction essentially doubled the amount of phosphatides in the extracted crude soybean oil. It was also found that there was a different array of phosphatides and that the hydratability was increased, resulting in degummed oils of 0.03 to 0.09% phosphatides (10–30 ppm P). These findings have been commercialized and called the Alcon process (Lurgi Ol. Gas. Chemie Gmbh, Frankfurt, Germany). A description of this process, and plant results, have been published by Penk (17). The lecithin produced by this is of a different composition than lecithin from conventional extraction processes, being enriched 30 to 40% in phosphatidyl choline.

The use of expanders in preparation for extraction gives results approaching those found with the Alcon process, depending on the degree of moist heat treatment occurring in the expander.

Degumming Agents

In addition to water, other additives have been studied and or used to facilitate degumming. The use of phosphoric and citric acid for removal of NHPs has already been mentioned. Other acids studied have been acetic, oxalic, nitric, boric, and tannic (18), but to our knowledge they have never been commercialized. The use of acetic anhydride for degumming of soybean oil has been done on a commercial scale in the United States (19) and is in use in some plants. The advantage is a higher yield of lecithin (i.e., lower gum content of the degummed oil) and ready removal of the volatile acetic acid in drying of the lecithin.

Degumming to Prepare an Oil for Caustic or Physical Refining

Caustic refining of soybean oil may be done on either crude or crude degummed oil. Since the market for soybean lecithins is much less than the potential supply as will be shown in this chapter, it is a practice in the United States simply to caustic refine crude soybean oil, thereby disposing of lecithin into the soapstock, when not making lecithin. The conventional thinking has been that the loss of neutral oil in refining crude oil is less than the combined losses of degumming followed by caustic refining of the degummed oil.

When a refinery is adjacent to an extraction plant, it has the possibility of adding the wet gums to the meal stream and getting at least a meal price for the gums. Such practice will then reduce the refining loss by caustic-refining degummed oil. In addition, there would be less soapstock, and it would give fewer problems in acidulation.

Although it is not necessary to degum and refine soybean oil sequentially, the decision as whether to do so or simply to refine crude oil may be more involved than just considering refining losses, due to the effects just mentioned.

Degumming of soybean oil, as a preparatory step for physical refining, has received considerable attention in the last few years and has recently been well reviewed by Dijkstra (11). The principal impetus for physical refining of soybean oil is to avoid the environmental problems associated with soapstocks from caustic refining. For a discussion of physical refining, see Chapter 14.

In the published results of degumming practices for physical refining, the degummed oils have a P content of 5 to 10 ppm for "good" oils and >30 for "poor" oils (20). To date, even with the use of special degumming techniques, physically refined soybean oils do not meet the flavor and flavor stability requirements of the U.S. market. They have met market requirements in other countries whose consumers are less demanding.

Production of Lecithin

The potential supply of lecithin from soybean oil is about 374,000 metric tons (802 million lbs) on a worldwide basis, but the market for lecithin is estimated at 100,000 to 150,000 metric tons. The other routes for use or disposal of the excess lecithin are to return it to the meal stream at the extraction plant or to caustic-refine crude oil, which disposes of the excess lecithin into the soapstock.

The process for lecithin production is shown in Fig. 10.2, and the important aspects up to and including the centrifugation have been discussed in the foregoing section on degumming. The wet gums coming from the centrifugation will contain about 50% water, with the nonaqueous portion having the composition shown in Table 10.1. Wet gums are susceptible to microbial fermentation and require immediate drying or treatment, for *brief* storage, with a preservative such as a dilute solution of hydrogen peroxide. Required dosage will depend on expected storage time, ambient temperature, and sanitary conditions (microbial types and load). Any storage of wet gums is not recommended, and the brief storage just mentioned is for necessary accumulation for batch drying systems.

In the process flow shown in Fig. 10.2, the wet gums from the centrifuges are transferred to a mixing tank, where bleaching agents, fluidity agents, or both can be added. With or without additives, the wet gums, containing about 50% water, must then be dried down to a maximum of 1% moisture.

The drying of lecithins is a critical step, because the gums tend to darken with heat, and during drying there is a large viscosity increase as the moisture is reduced. This phenomenon is shown in Fig. 10.3, where the increase in viscosity begins at about 20% moisture, peaks at about 8% moisture, and then falls rapidly between 7 and 4% reduction (21).

Batch-type dryers operate under vacuum and are equipped with rotating coils circulating water at about 60 to 70°C (140 to 158°F). More modern continuous dryers utilize agitated-film dryers for moisture removal. A comparison of the conditions used in these two types of dryers is shown in Table 10.2 (21).

From the process flow shown in Fig. 10.2 a variety of lecithins may be produced, and the National Oilseed Processors Association (NOPA) publishes speci-

TABLE 10.2 Average Process Conditions for Drying Lecithin Sludge (Wet Gums) (21)[a]

Process variable	Batch dryer[b]	Continuous agitated-film dryer
Temperature		
°F	140–176	176–203
°C	60– 80	80– 95
Residence time, min	180–240	1– 2
Absolute pressure, mm Hg	20– 60	50–300

[a]Starting product: Wet gums with 50% moisture. End product: Lecithin with less than 1% moisture.
[b]Vacuum dryer with rotating, ball-shaped coils heated with warm water.

fications for six lecithin commercial grades, as shown in Table 2.14. Sullivan and Szuhaj (23) created a useful classification of soybean lecithins as shown in Table 10.3. In their classification, the NOPA products are considered "natural", followed by "refined" lecithins, from custom blending and solvent treatments, and ending with "Chemically modified" lecithins.

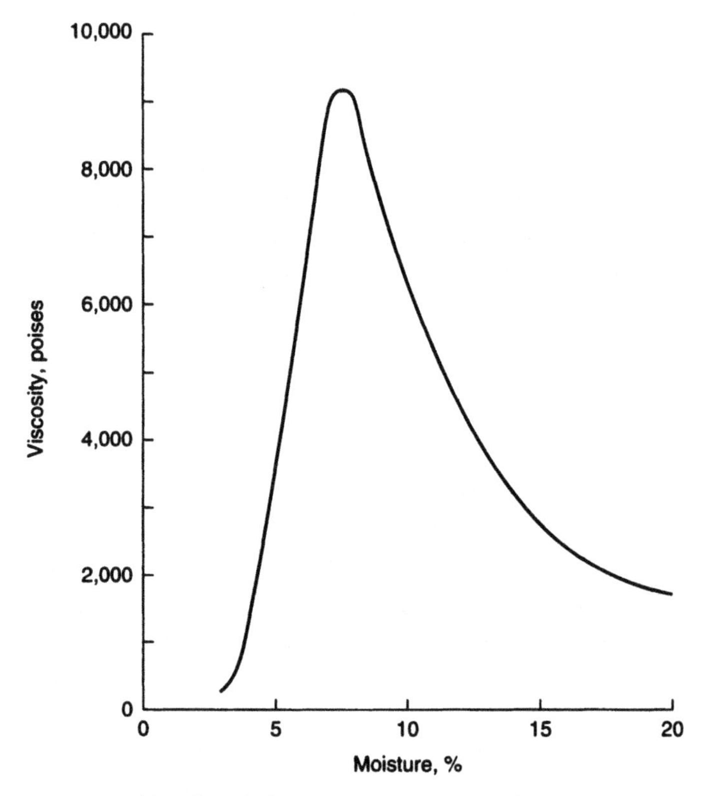

Fig. 10.3. Viscosity of lecithin sludge at 158°F (70°C) in relation to moisture content. *Source:* Van Nieuwenhuyzen, W., *J. Amer. Oil Chem. Soc. 53:* 425, (1976).

TABLE 10.3 Classification of Soybean Lecithins (23)

- I. Natural
  - A. Plastic
    1. Unbleached
    2. Single-bleached
    3. Double-bleached
  - B. Fluid
    1. Unbleached
    2. Single-bleached
    3. Double-bleached
- II. Refined
  - A. Custom blended natural
  - B. Oil-free phosphatides
    1. As is
    2. Custom blended
  - C. Fractionated oil-free phophatides
    1. Alcohol-soluble
       - a. As is
       - b. Custom blended
- III. Chemically modified

Oil-free lecithins are produced by extracting the soybean oil from the natural lecithins with acetone. This is done by both batch and continuous processes and requires high-quality crude lecithin for best results. In turn, alcohol fractionation of an oil-free lecithin can be employed to give an alcohol-soluble fraction high in phosphatidyl choline, and an alcohol-insoluble fraction, enriched in phosphatidyl inositol. The composition of oil-free lecithin and the alcohol fractionation products is shown in Table 10.4. As shown in the table, the lecithin products vary in their emulsifying properties.

Chemically modified lecithins include hydrogenated, hydroxylated, acetylated, sulfonated, and halogenated products. All of the chemical modifications are designed to modify the emulsifying properties of the lecithins and improve their dispersibility in aqueous systems (21).

TABLE 10.4 Approximate Composition of Commercially Refined Lecithin Fractions (23)

Fraction	Oil-free lecithin (%)	Alcohol-soluble lecithin (%)	Alcohol-insoluble lecithin (%)
Phosphatidyl choline	29	60	4
Cephalin	29	30	29
Inositol and other phosphatides, including glycolipids	32	2	55
Soybean oil	3	4	4
Other constituents[a]	7	4	8
Emulsion type favored	Either oil-in-water or water-in-oil	Oil-in-water	Water-in-oil

[a]Includes sucrose, raffinose, stachyose, and about 1% moisture.

Soybean Lecithin Utilization

A comprehensive listing of the uses and functions of phospholipids has been published by Schneider (24) and is shown in Table 10.5.

Lecithin also has unique release properties and as such has been used in pan frying formulations and in pan greases for baking (25). In addition, it is also used industrially as a release agent for ready removal of both wooden and metal concrete casting forms.

A more detailed description of lecithin utilization, including those from other oilseeds can be found in the recent AOCS monograph by Szuhaj (26).

TABLE 10.5 Uses and Functions of Phospholipids (24)

Product	Function
Food	
Instant food	Wetting and dispersing agent; emulsifier
Baked goods	Modification of baking properties; emulsifier; antioxidant
Chocolate	Viscosity reduction; antioxidant
Margarine	Emulsifier; antispattering agent; antioxidant
Dietetics	Nutritional supplement
Feedstuffs	
Calf milk replacers	Emulsifier; wetting and dispersing agent
Industry	
Insecticides	Emulsifier; dispersing agent; active substance
Paints	Dispersing agent; stabilizer
Magnetic tapes	Dispersing agent; emulsifier
Leather	Softening agent; oil penetrant
Textiles	Softening; lubricant
Cosmetics	
Hair care	Foam stabilizer; emollient
Skin care	Emulsifier; emollient, refatting, wetting agent
Pharmaceuticals	
Parental nutrition	Emulsifier
Suppositories	Softening agent; carrier
Cremes, lotions	Emulsifier; penetration improver

References

1. Brekke, O.L., in *Handbook of Soy Oil Processing and Utilization*, edited by D.R. Erickson et al., American Oil Chemists' Society, Champaign, IL, 1978, p. 77.
2. Kock, M., *J. Amer. Oil Chem. Soc. 60:* 210 (1983).
3. Hvolby, A., *J. Amer. Oil Chem. Soc. 48:* 503 (1971).
4. List, G.R., in *Handbook of Soy Oil Processing and Utilization*, edited by D.R. Erickson et al., American Oil Chemists' Society, Champaign, IL, 1978, pp. 355–376.
5. Swern, D., in *Bailey's Industrial Oil and Fat Products*, edited by D. Swern, Vol. I, 4th edn., John Wiley and Sons, New York, 1979, p. 49.
6. Sleeter, R.T., *J. Amer. Oil Chem. Soc. 58:* 239 (1981).
7. Charpentier, R., *INFORM 2:* 208 (1991).

8. *Official Methods and Recommended Practices*, 4th edn., American Oil Chemists' Society, Champaign, IL, 1993. Method Ja 3-87 (1993)
9. Brekke, O.L., in *Handbook of Soy Oil Processing and Utilization*, edited by D.R. Erickson et al., American Oil Chemists' Society, Champaign, IL, 1978, p. 73.
10. Brian, R., *J. Amer. Oil Chem. Soc. 53:* 27 (1976).
11. Dijkstra, A.J., in *Proceedings of the World Conference on Oilseed Technology and Utilization*, edited by T.A. Applewhite, American Oil Chemists' Society, Champaign, IL, 1992, pp. 138–151.
12. Wiedermann, L.H., *J. Amer. Oil Chem. Soc. 58:* 159 (1981).
13. Erickson, D.R., *J. Amer. Oil Chem. Soc. 60:* 351 (1983).
14. Erickson, D.R., and L.H. Wiedermann, *INFORM 2:* 201 (1991).
15. *Yearbook and Trading Rules*, National Oilseed Processors Association, Washington, DC, 1993–1994, p. 86.
16. Kock, M., U.S. Patent 4,255,346 (1981).
17. Penk, G., in *Proceedings: World Conference on Emerging Technologies in the Fats and Oils Industry*, edited by A. R. Baldwin, American Oil Chemists' Society, Champaign, IL, 1986, pp. 38–45.
18. Ohlson, R., and C. Svenson, *J. Amer. Oil. Chem. Soc. 53:* 8 (1976).
19. Meyers, N.W., *J. Amer. Oil Chem. Soc. 34:* 93 (1957).
20. Seger, J.C., and R. van de Sande, in *World Conference Proceedings Edible Fats and Oils Processing*, edited by D.R. Erickson, American Oil Chemists' Society, Champaign, IL, 1990, pp. 88–93.
21. Van Nieuwenhuyzen, W., *J. Amer. Oil. Chem. Soc. 53:* 425 (1976).
22. *Yearbook and Trading Rules*, National Oilseed Processors Association, Washington, DC, 1993–1994, p. 98.
23. Sullivan, D.R., and B.F. Szuhaj, *J. Amer. Oil Chem. Soc. 52:* 152A (1975).
24. Schneider, M., in *Proceedings World Conference on Emerging Technologies in the Fats and Oils Industry*, edited by A.R. Baldwin, American Oil Chemists' Society, Champaign, IL, 1986, pp. 160–164.
25. Dashiell, G., in *World Conference Proceedings Edible Fats and Oil Processing*, edited by D.R. Erickson, American Oil Chemists' Society, Champaign, IL, 1990, pp. 396–401.
26. Szuhaj, B.F., *Lecithins: Sources, Manufacture, and Uses*, American Oil Chemists' Society, Champaign, IL, 1989.

Chapter 11

Neutralization

David R. Erickson

Consultant
American Soybean Association
St. Louis, MO

Introduction

The process of neutralization, *deacidification*, is also called *caustic refining* and *physical* or *steam refining* depending on the process used. In the United States, *caustic refining* means treatment of an oil or fat with an aqueous alkali solution including water washing. This may be confusing, because the term *refining* is also used to describe the entire process involved in making final products from crude fats and oils. In this chapter we will be discussing caustic refining of soybean oils, using both batch and continuous systems, followed by a brief description of physical or steam refining of soybean oil.

The Reaction of Alkalis with Crude Soybean Oil

All crude fats and oils contain variable amounts of undesirable nonglyceride materials. It was discovered that aqueous alkalis could be used to neutralize the free fatty acids (FFA), creating soaps, which in turn could adsorb color and precipitate any gums or mucilaginous substances present in the crude fats and oils. The resulting aqueous phase containing the undesired substances would be heavier than the fat or oil and therefore would settle out with time. This discovery became the basis for the process of caustic refining.

The amount of FFA could be determined by a simple and rapid method and became the standard for determining the amount of alkali to add; hence the term *neutralization.* Experience showed that the best results were obtained by using more than the stoichiometric amount of alkali needed for simple neutralization. This additional amount was called the *excess.*

The alkaline compounds studied for refining include sodium hydroxide (NaOH; caustic soda or lye), potassium hydroxide (KOH; caustic potash), sodium bicarbonate ($NaHCO_3$), and sodium carbonate (Na_2CO_3; soda ash). The only one in current use in almost all soybean oil refining is sodium hydroxide, with some use of the more expensive potassium hydroxide to provide an outlet for the latter soapstock as a fertilizer (1,2). The two methods most used for expressing concentrations of NaOH for use in refining are degrees Baumé (°B) and normality. The corresponding values for °B, normality, and % concentration are shown in Table 11.1.

The selection of the amount and concentraton of NaOH to use for refining is critical. As stated in a preceding paragraph, the calculation of the amount to be used is

TABLE 11.1 Measures of Sodium Hydroxide Concentration

°B at 15°C	Normality	NaOH Content (%)
10	1.64	6.57
12	2.00	8.00
14	2.38	9.50
16	2.76	11.06
18	3.17	12.68
20	3.59	14.36
22	4.02	16.09
24	4.47	17.87
26	4.92	19.70
28	4.40	21.58
30	5.88	23.50

based on the FFA content of the crude or crude degummed oil plus an excess amount based on experience. In the United States the amount (weight) of solution to add is expressed as a percentage of the weight of oil, known as *treat*, calculated as follows:

$$\%\ \text{Treat} = \frac{(\%\ \text{FFA} \times 0.142) + \%\ \text{Excess}}{\%\ \text{NaOH}/100}$$

where the factor 0.142 is the ratio of MW NaOH/MW Oleic acid = 40/282. For example, a soybean crude oil with 0.75% FFA to be treated with 14°B caustic solution with an excess of 0.12% gives the following result:

$$\%\ \text{Treat} = \frac{(0.75 \times 0.142) + 0.12}{9.5\%/100} = \frac{0.2265}{0.095} = 2.38\%$$

Thus, 2.38% of 14° Baumé caustic solution is the proper treatment level for the specified soybean oil.

In the United States the excess is an additional weight of dry NaOH expressed as a percentage of the weight of oil. European practice is to calculate the amount of additional caustic as a percentage of that needed for neutralization and to express this quantity as percent excess. On that basis, the percent excess for the preceding example is calculated as follows:

$$\%\ \text{Excess} = \frac{\text{added} \times 100}{\text{Stoichiometric}} = \frac{(0.12)(100)}{(0.75)(0.142)} = 113$$

The percent treat in the foregoing example can be computed using the percent excess as follows:

$$\begin{aligned}\%\ \text{Treat} &= \frac{(\%\ \text{FFA} \times 0.142)\ (100 + \%\ \text{Excess})/10}{\%\ \text{NaOH}/100} \\ &= \frac{(0.75 \times 0.142)\ (100 + 113\%)/100}{9.5\%/100} = 2.38\%\end{aligned}$$

This difference in calculation and discussion of the meaning of *excess* leads to some confusion and requires definitions of the terms and calculations being discussed.

The use of caustic solutions for removing unwanted nonglyceride substances and FFAs is effective but can cause some hydrolysis, or *saponification*, which will increase refining losses. This is true for both batch and continuous refining, so the refiner is always required to strike a balance, using enough caustic to remove the unwanted substances effectively without causing additional losses. The overall goal in caustic refining is simultaneously to maximize quality and minimize losses. Refining efficiency can be expressed in several ways (3,4) but by definition is simply stated as:

$$\frac{\text{Refined and water washed product output (wt)}}{\text{Crude oil input (wt)}} \times 100 = \%\ \text{efficiency}$$

and

$$100 - \%\ \text{efficiency} = \%\ \text{loss}$$

Batch or Kettle Refining

Batch or *kettle refining* continues in use for some specialty oils, for small production, and in less developed countries. The majority of vegetable oil caustic refining in the world now uses a continuous process utilizing centrifuges. Despite this near demise, it is helpful to review kettle refining, because the principles utilized are applicable to continuous refining.

The basic equipment for kettle refining is both simple and inexpensive. It consists of an open cylindrical tank with a conical bottom, a two-speed agitator, and steam coils or jacket for heating. The tanks will be of various capacities depending on the unit volumes needed by the refiner, such as tank truck size, railroad car size, or batch sizes for sale or further processing. In some cases the kettle is equipped with a closed top and vacuum source, allowing the dual use of the kettle for both refining and bleaching.

The agitator shaft is centered in the vessel and is either suspended from the top or extended to a step bearing on the bottom of the vessel. In the latter case, the refined oil is usually drawn off the top; otherwise the aqueous phase (*soapstock*) may be drained from the conical bottom of the vessel. The agitator shaft is equipped with four or five sweep arms, with paddles canted at an angle that will push the liquid upward during agitation. Suggested agitation rates are 30 to 35 rpm maximum and 8 to 10 rpm minimum.

The agitator shaft and sweeps must also extend down into the conical space to avoid a dead spot in agitation. Another necessary precaution is to avoid any air incorporation during filling or agitation of the vessel contents. Entrained air will cause flotation of soap particles, interfering with the desired settling action.

The process of kettle refining consists of the following steps:

1. Charge (load) the vessel with the fat or oil to be refined. Liquid oils are charged at ambient temperatures 20 to 30°C (68 to 86°F) and fats at temperatures 10 to 15°C (18 to 27°F) above their melting points.
2. Mix the contents thoroughly at slow speed to eliminate any entrained air, and analyze for FFA, refining loss, etc. Calculate the dosage of lye in accordance with the analytical results.
3. Turn agitation to high speed and add the calculated amount of aqueous lye solution, using a spray ring or other device to distribute the lye solution over the surface, and continue agitation until the solution thoroughly mixed (15 to 30 min).
4. Reduce the agitation to slow speed but enough to ensure continued uniformity of the vessel contents. Commence heating to bring the temperature of the charge up to about 60°C (140°F) as rapidly as possible. At or near this temperature, a visible "break" in the emulsion should be observable. The break is the result of de-emulsification and appearance of visible particles in the mixture.
5. After the break has been observed, stop the heating and agitation. The aqueous layer then is allowed to settle to the bottom over a period of several hours. For soybean oil treated with low caustic concentrations (10 to 14°B), the settled aqueous layer (soapstock) should be sufficiently fluid to allow drainage from the bottom of the tank.
6. After removing soapstock, wash the remaining oil by adding about 15% hot soft water, with sufficient agitation for uniform dispersion, and again settling and decanting the aqueous layer. This water wash step may be repeated.

The foregoing is a description of kettle refining of soybean oil and other fats and oils of normal quality. Lower-quality fats and oils may require different concentrations of caustic, may have colors that are difficult to remove, and may form difficult emulsions requiring additives, different conditions, or both to get the required break.

In Europe where fluid, or soft, soapstocks were desired, a so-called "wet" refining method was practiced (4). This method involved first heating of the oil to about 65°C (150°F), mixing the lye solution in, and then washing the precipitated soapstock down with a spray of hot water. The soapstock was then settled out and further water washes and settlings followed.

The approach just described may have led to the subsequent development of the *short-mix* system, which will be discussed subsequently. Alternatively, the foregoing description of the kettle refining technique for soybean oil provides justification for the use of the *long-mix* system for soybean oil.

For more highly colored oils, such as cottonseed, it is often necessary to increase the amount or concentration of caustic to achieve the desired degree of color removal. The consequence of additional caustic may be increased saponification, which increases refining losses. To avoid this, a short contact time between the oil and the caustic mixture was used as just described for "wet" batch refining and was subsequently designed into continuous short-mix refining systems using centrifuges.

Continuous Caustic Refining Systems

In kettle refining, the separation of the oil from the heavier soapstock phase depends on gravity, and separation is slow and sometimes troublesome. In addition, dependence on gravity leads to higher losses and low production rates. A natural progression toward higher efficiency and volumes led to the introduction of continuous refining using centrifuges. This has led to production increases, up to 700 short tons (635 MT) per day in one primary centrifuge.

The first continuous refining system in the United States was described by James (5) in 1934, and the basics of his description remain relevant. Since gravity is replaced by centrifugal forces several thousandfold greater, the separation is much better, and losses are reduced 20 to 30% when compared to kettle refining losses (4).

The basic sequence for continuous refining is addition of the proper dosage to a stream of crude oil, mixing in-line by a mechanical or static mixer, centrifugation to remove soapstock, addition of wash water in-line, mixing, and centrifugation. This basic description is varied with many different configurations as to where heat is applied, concentrations and amounts of caustic solution, retention times, rerefining,

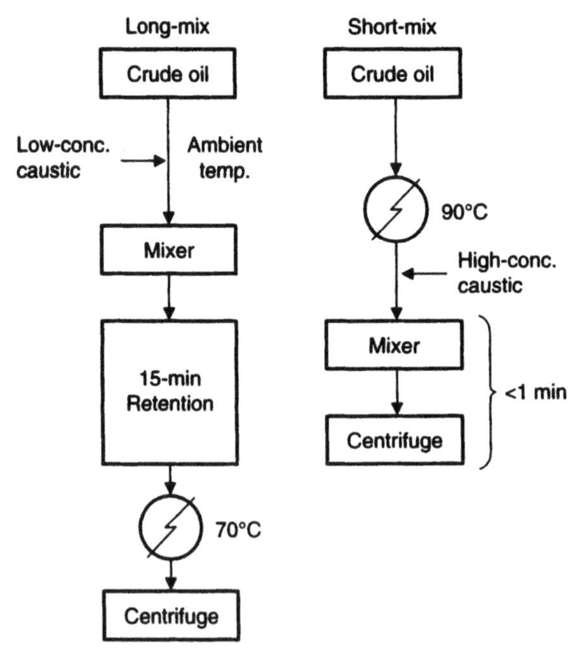

Fig. 11.1. Comparison of long- and short-mix caustic refining systems.

and number of water washes. Basically, the extremes for consideration in soybean oil refining are covered by comparing the long-mix with the short-mix systems as shown in Fig. 11.1.

The process of degumming of soybean oil has been already discussed in Chapter 10; however, some of that bears repeating, because removal of all phosphatides is the key to the production of best-quality soybean oil.

The conventional practices for both degumming and caustic refining of soybean oil were covered in an excellent review paper by Wiedermann (6). Other publications building on that paper are those of Erickson (7), Erickson and Wiedermann (8), and Charpentier (9). A flow diagram of the recommended degumming and caustic refining systems appears as Fig. 11.2. The relationship between degumming and caustic refining practices lies in the type and nature of the phospholipids and the recognition that their hydration is not instantaneous; they require time to react with water.

Soybean Oil Phosphatides and Their Removal

The quality of the starting crude or crude degummed soybean is always variable, and refiners must be prepared to adjust their refining practices to overcome these variations. Such variation in quality is shown in Table 11.2. The refiner has no control of any of the abuses shown in the table except some portion of the crude oil storage.

The nonhydratable phosphatide (NHP) content of soybean oil is of most concern to a refiner. It is generally accepted that the NHPs are the calcium and magnesium salts of phosphatidic acids, which results from the action of the enzyme phospholipase D, which splits off the non–fatty-acid moiety from phospholipids. The phosphatidic acid formed by the calcium and magnesium soaps tend to be more oil-soluble than water-soluble. The formation of NHP is shown in Fig. 11.3.

TABLE 11.2 Crude Soybean Oil Quality (6)

Factors affecting	Increase in*
Weed seed	d,f
Immature beans	f
Field-damaged beans	a,b,c,e
Splits (loading/transport/unloading)	a,b,c
Bean storage (time/temp/humidity)	a,b,c
Conditioning beans for extraction	a,b,d,e
Solvent stripping oil (overheating)	b,d
Oil from stripper (overheating)	b
Crude oil storage (time/temp)	c,d

*a = Total gums/phosphatides
b = Nonhydratable phosphatides
c = Free fatty acids
d = Oxidation products
e = Iron/meal content
f = Pigments

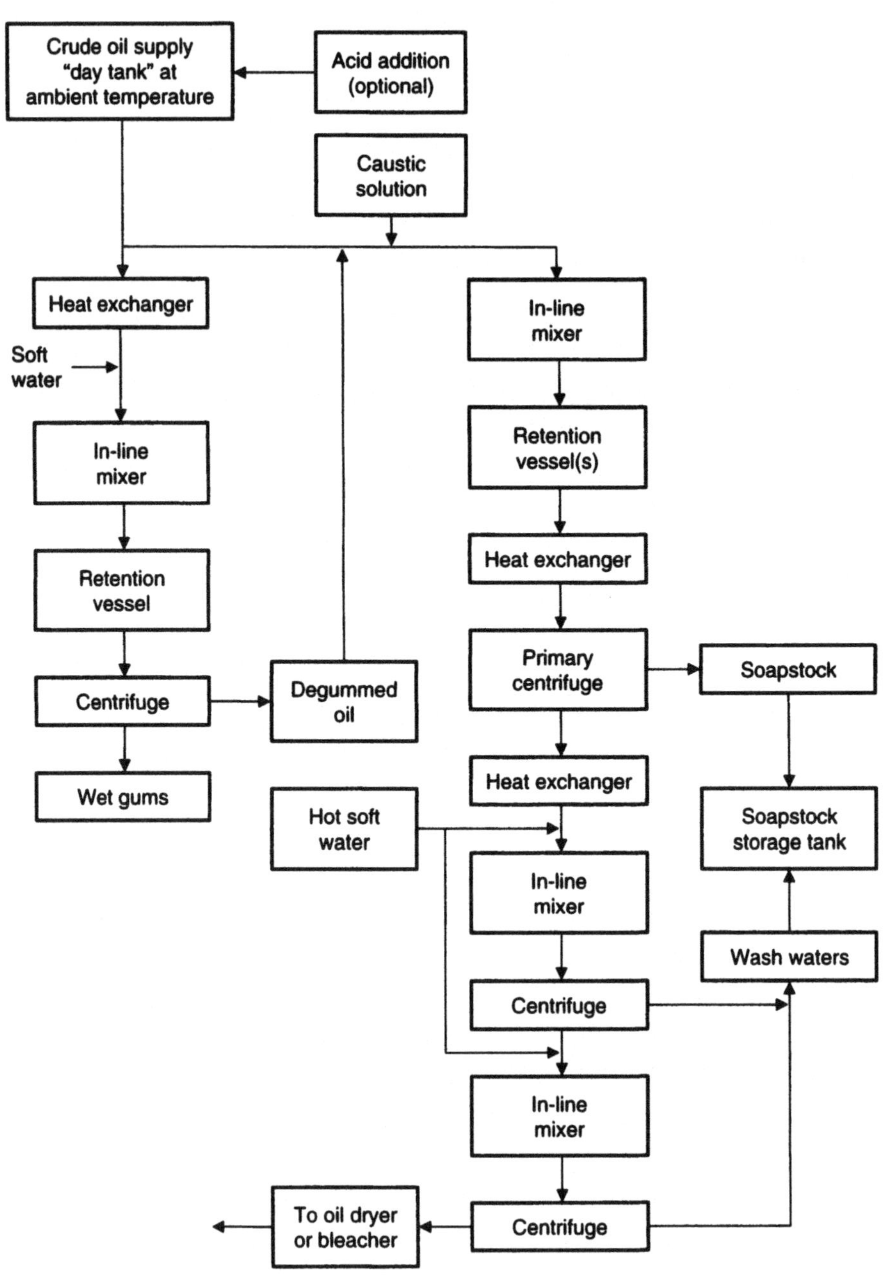

Fig. 11.2. Flow diagram of recommended degumming and caustic refining processes for soybean oil.

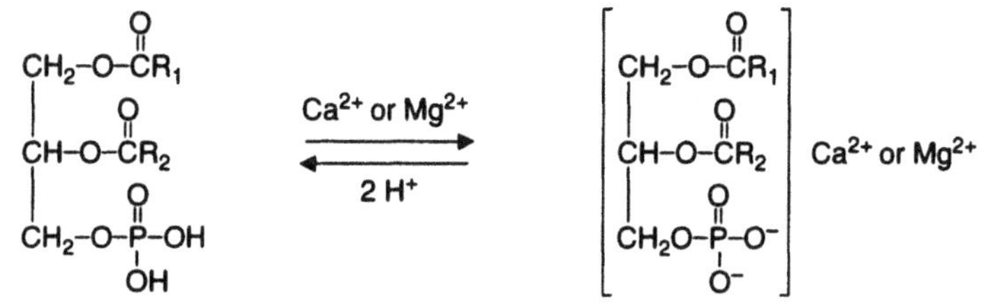

CH_2–O–C(=O)R_1 | CH–O–C(=O)R_2 | CH_2–O–P(=O)(OH)–OH

Ca^{2+} or Mg^{2+} ⇌ $2\,H^+$

[CH_2–O–C(=O)R_1 | CH–O–C(=O)R_2 | CH_2O–P(=O)(O^-)–O^-] Ca^{2+} or Mg^{2+}

Phosphatidic acid

Nonhydratable phosphatide

R_1 and R_2 = fatty acids

Fig. 11.3. Formation of nonhydratable phosphatides.

The reactions involved in degumming and alkali refining are the following (6):

1. Phospholipids (hydratable) + water = hydrated gums
2. Phospholipids (nonhydratable) + acid or alkali = hydrated gums
3. Metal/phospholipid complex + acid = hydrated metal gums
4. Free fatty acids + alkali = soaps

Reaction 1 on this list is that of water degumming to produce lecithin or to produce a degummed soybean oil for long-term storage or transport.

Reaction 2 is used to remove the NHPs, using alkali in a long-mix system and an acid, usually phosphoric, for a short-mix system or as an extra precaution for a long-mix system. Reaction 2 is also operative in preparation of soybean oil for physical refining. Acid treatment before degumming precludes use of the removed phosphatides for lecithin production, because subsequent heat treatment in drying will cause darkening of the acidic lecithin.

Reaction 3 in the foregoing list shows that metal complexes are formed with the phosphatides. The relationship between iron and phosphorus contents in degummed soybean oils is shown in Fig. 11.4 (6). The acid treatment removes iron/copper soaps, which are prooxidants, and calcium and magnesium soaps of fatty acids, which are unwashable like NHPs (6,10). The effect of a phosphoric acid treatment of crude oil on the phosphorus and iron content of soybean oil during processing is shown in Fig. 11.5 (11).

Reaction 4 is simply neutralization of the FFAs.

The foregoing discussion suggests that a phosphoric acid treatment is mandatory for removal of NHPs, but a traditional long-mix system used on crude oil will remove sufficient phosphatides without any acid pretreat. However, many U.S. refiners routinely use an acid pretreat on crude soybean oil to ensure phosphatide removal and to take advantage of a popular theory that the caustic/oil emulsions formed with acid-pretreated oils are less stable and give a better "break". The most convenient method for acid pretreatment of crude oil is addition to the day tank, as shown in Fig. 11.2.

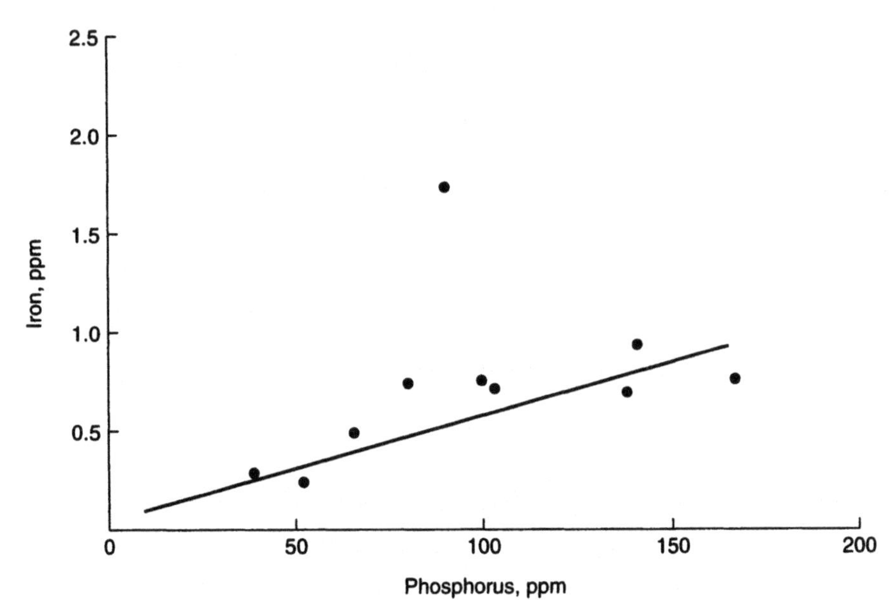

Fig. 11.4. Relation between phosphorus and iron content in degummed oils. *Source:* Wiedermann, L.H., *J. Amer. Oil Chem. Soc. 58:* 159, (1981).

For crude degummed soybean oil a phosphoric acid pretreat is recommended, because the process of water degumming obviously removes the hydratable gums, leaving a much higher proportion of NHPs in the degummed oil. As with crude soybean oil, addition of phosphoric acid to crude degummed soybean oil in the day tank is a convenient method for pretreatment.

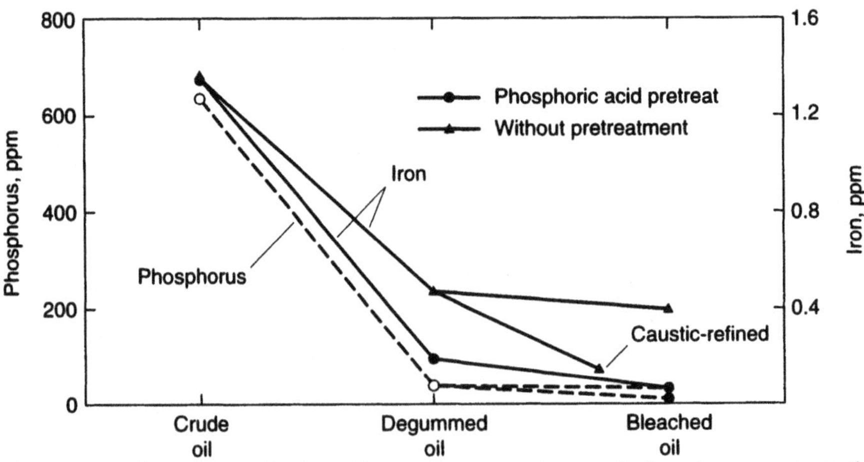

Fig. 11.5. Effect of phosphoric acid pretreatment on iron and phosphorus content of soybean oil during processing (6,11).

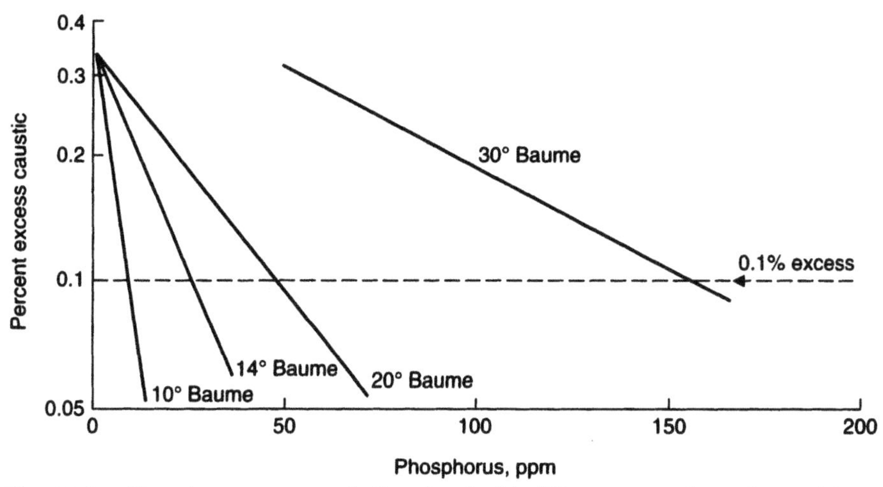

Fig. 11.6. Phosphorus content of oils refined with different strengths and quantities of caustic (6,12).

In addition to phosphoric acid pretreatment, the removal of phosphorus in refining is also a function of the amount of water in the caustic solution as shown in Fig. 11.6 (6,12). There is an operational limit to lowering caustic strength. If too dilute, the emulsion is difficult to separate.

By experience, this lower limit for soybean oil has been found to be about 12 to 14° Baumé. Fig. 11.6 shows why the higher concentrations of caustic typical of short-mix systems are not recommended for soybean oil. This and the lack of time for hydration are two reasons for not using short-mix systems for soybean oil. Addition of a high concentration of caustic (20 to 40° Baumé) to a hot oil (85 to 95°C, 185 to 203°F), as is typical of short-mix systems, probably causes immediate coagulation or coalescence of the soap micelles, effectively taking them prematurely out of the reaction. This may be why it is difficult to get low P values consistently in soybean oils with a short-mix system.

As mentioned previously, kettle refining conditions for soybean oil refining lead toward the optimum configuration for continuous soybean oil refining. This relates to both lower concentrations of caustic solutions and, more importantly, the need for time to hydrate the phosphatides at a lower or ambient temperature. Both the lower caustic concentration and lower temperature decrease the possibility of saponification.

Based on the foregoing discussion, the recommended equipment, procedures, and conditions for continuous caustic refining of soybean oil by the long-mix system are as follows:

1. Install and use an agitated *day tank* or tanks to ensure a supply of a uniform product over a period of time sufficient to stabilize and "fine-tune" the centrifuges, to maximize quality while minimizing losses.
2. Charge the day tank with the oil to be refined at ambient temperature (20 to 30°C, 68 to 90°F), agitate the contents enough to ensure uniformity, and then

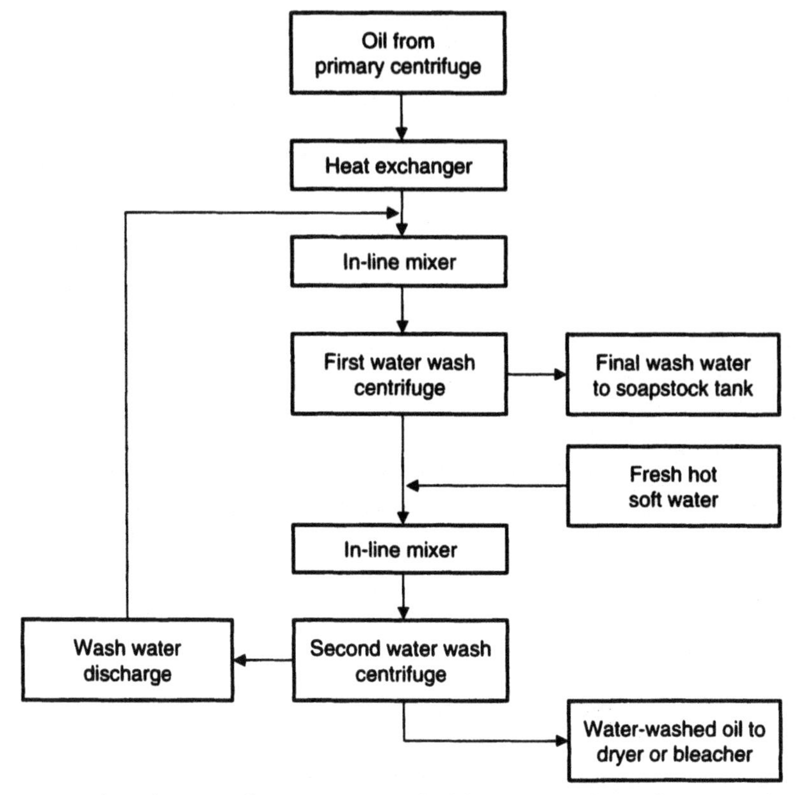

Fig. 11.7. Flow diagram of countercurrent double water washing of caustic-refined soybean oil.

TABLE 11.4 Suggested Working Conditions for Continuous Caustic Refining of Soybean Oil (see Fig. 11.2 and 11.7)

- Charge day tank with crude at ambient temperature [33°C (90°F)], add phosphoric acid (optional), agitate to uniformity, do appropriate analyses, and calculate causticc dosage
- Proportionally add 14 to 18°B caustic to oil plus either 0.12–0.15 excess for crude oil or 0.10–0.12 excess for crude degummed oil
- Mix intensively in line
- Mix in retention vessel(s) for minimum 15 minutes
  heat to about 70°C (160°F) with in-line heat exchanger
- Centrifuge in primary centrifuge
- Heat oil from primary to about 90°C (194°F) and add hot soft water at same temperature in line, mix, and centrifuge
  15% additional for single water wash
  10% each for double water wash (could be less for countercurrent washing)
- Send once-refined oil to dryer or to bleacher.

TABLE 11.3 Primary Centrifuges Used in Vegetable Oil Refining

Model #	Nominal maximum capacity MT/day	10_3 lb/hr	Comments
Alfa Laval[a]			
SRG-214	200	18	Manual cleaning
Westphalia[b]			
Rta-140	240	22	Manual cleaning
Westphalia			
RSA-80	260	24	Self-cleaning
Alfa Laval			Replaced SRG 214
SRP-417	350	32	Self-cleaning
Westphalia			
RSA-150	435	40	Self-cleaning
Alfa Laval			
SRPX-714	490	45	Self-cleaning
Westphalia			Self-cleaning
RSA-200	700	65	Quiet design
Tetra Laval[a]			
PS 60	156	14	New series design
PS 70	288	26	Self-cleaning
PS 80	432	41	Quiet, new paring
PS 90	720	66	Device

[a]Alfa-Laval separation AB, Tumna, Sweden.
[b]Westphalia Separator AC, Oelde, Germany.

analyze the charge for FFA, neutral oil and loss, phosphorus, and so forth to determine starting caustic dosage. If necessary, the oil should be cooled down or heated to the range of temperatures just mentioned.

3. Acid pretreatment may be done by addition of phosphoric acid in the day tank and mixing thoroughly; this may take several hours, depending on the mixing efficiency. The addition would normally be in the range of 0.05 to 0.2% of 75% phosphoric acid, depending on the Ca/Mg content of the oil. Alternatively, experience may justify use of a standard amount. Either way, addition of such small amounts to a large volume of oil usually requires a long mixing period to ensure dispersion. Acid pretreatment is optional for treatment of crude soybean oil and is recommended for crude degummed oil, alone or blended with crude. Acid treatment temperatures should not exceed those given, because it has been reported that treatment at 60°C (140°F) for longer than 20 min will cause the phosphatidic acid to revert to the nonhydratable form (9).
4. On startup, caustic-proportioning system is set to deliver the chosen amount of 14 to 18° Baumé, including that amount necessary to neutralize any added phosphoric acid. The caustic solution/oil mixture then goes to an in-line mixer that is capable of intense and thorough mixing. This mixer can be either mechanical or static; the latter requires a certain flow rate to function properly and may not be effective enough at reduced flow rates. This drawback has led some refiners to prefer mechanical mixers, which are not so affected.

5. Once mixed, caustic solution/oil emulsion flows to one or more retention vessels, which will give a minimum of 15 minutes retention time. The design of the retention vessel(s) provides minimum agitation to ensure uniformity of the emulsion while preventing overemulsification and allowing growth of the soapstock micelles, which contain the unwanted substances. Flow through the retention vessels should be plug or laminar flow (i.e., "first in–first out"). Retention vessels represent an initial capital cost plus some small energy cost of operation, but they have no effect on production rates, because they simply represent a "balloon" in the refining line. With proper design of retention vessels, the actual retention time can be varied for different types of oils.
6. After leaving the retention vessel(s), the emulsion flows through a heat exchanger where its temperature is raised to 70 to 75°C (160 to 170°F), which "breaks" the emulsion just before it is fed into the first, or *primary*, centrifuge.
7. The most modern primary centrifuges are disc-bowl types that are self-cleaning and capable of production rates up to 700 MT per day or more. These are listed in Table 11.3.

   In the United States the majority of primary centrifuges are of the hermetic type, which allows control of the separation zone by adjustment of back pressure during operation. These types are well adapted for handling the viscous soapstocks from the usual long-mix system used on crude soybean oil. Proper operation of this type of centrifuge can give a soapstock containing 20% or less neutral oil on a dry basis.

   The open-type centrifuge differs in being less complex because seals are not required. The light oil phase and heavier soapstock phase are discharged through two centripetal pumps. Small adjustments during operation are accomplished by adjusting back pressure on the light phase. For larger adjustments, the centripetal pump discharging the heavy phase must be replaced with one of different diameter, which requires operations to be stopped for the change-out.

   The two streams from the primary centrifuge are the light oil phase and heavy soapstock phase. For soapstock handling, see Chapter 17.
8. Following discharge from the primary centrifuge, the light oil phase contains levels of 200 to 400 ppm soap and 5 to 10 ppm P, which are further lowered in the water wash centrifuge(s). Typically, the oil is heated to 90 to 95°C (194 to 203°F) by pumping through an in-line heat exchanger; soft water at or near the same temperature is proportioned into the oil stream at a 10 to 15% level; and the water and oil are mixed thoroughly by an in-line mixer and fed to one or more water wash centrifuges.

   With one water wash, about 95% reduction in both soap and P is achievable, resulting in an oil with 10 to 20 ppm soap and <1.0 ppm P. After going through the long mix, there is no need for any retention time after injection of the water, and the results just given are achievable.

   Another option is use of a second water wash, which will reduce the soap and P values even further. In the latter case, two additions of 10% each are used. The second water wash represents more capital investment and operation cost, but it may be offset by less use of bleaching earth, because of less need to scavenge residual soap and phosphatides (see Chapter 12). Wash waters represent additional moisture in the soapstock, which adds to pollution control cost. A newer option is to use two water washes but add the wash water countercurrently, as shown in Fig. 11.7. In this case, treatment with as little as 7% water is possible, reducing wash water loads by 53 to 65%.

An additional option is addition of citric acid (400 to 500 ppm) to the wash water, which practically ensures zero soap and P contents and will sequester any prooxidant metals present (9). A final option is to use silicate addition as a replacement for the water wash. This product will also ensure removal of all soap and phosphatides (13), but its effects on quality are not entirely known (see Chapter 12).

In all of the water wash methods just mentioned, refiners need to evaluate each option in terms of cost, quality, and environmental situation before making their decision.

9. After water washing, the oil will be at 90 to 95°C (194 to 203°F) and will contain 0.2 to 0.4% moisture. If the oil is to be stored, or sold as a once-refined oil, it needs to be dried down to a maximum water and volatile content of 0.1% (14). This is normally done by transferring the oil at its exit temperature to a vacuum dryer operated at about 35 mm Hg absolute pressure where it is dried down to less than 0.1% moisture and volatiles. After drying the oil should be cooled down to a least 60°C (140°F), and preferably lower, to prevent deterioration and sweating in storage tanks. Storage of once-refined oil at a moisture content higher than 0.1% is not recommended, because of the possibility of hydrolysis and other quality concerns.

   An option to drying and its expense is to reduce the temperature of the oil to about 70 to 80°C (158 to 176°F) and go directly to bleaching, where the moisture will help the acidic effect of acid-activated earth and will be removed in the bleaching process. Sending the oil directly to bleaching without cooling runs the risk of flashing off the moisture from the oil and earth prematurely (see Chapter 12).

A tabular summation of the long-mix system for and caustic refining of soybean oil is shown in Table 11.4.

Practical Application of Optimized Refining

A recent publication reports actual plant results where refining practices have been optimized (9). In this plant the crude oil is first filtered, then degummed and caustic-refined using a long-mix system. The filtration was installed in anticipation of lecithin production but was found to be a key factor in reducing FFA buildup in crude oil storage as well as reducing losses in refining and downtime for cleaning the centrifuges (non-self-cleaning).

Contrary to conventional thinking, the study found less refining loss with both degumming and caustic refining than with caustic refining of crude oil. The explanation was the finding that phosphorus could be removed from the degummed oil

TABLE 11.5 Quality of Once-Refined Soybean Oil

FFA	0.035 to 0.060%
Phosphorus	0 to 1.1 ppm
Soap	0 ppm
Peroxide value	0.9 to 3.3 mEq/kg

using 70 to 75% of the calculated caustic dosage based on FFA content. About 500 ppm citric acid was also added to the wash water. The quality of the once-refined oils is shown in Table 11.5.

Long-Mix System Refining Losses

As mentioned earlier, there are several ways to express and determine refining losses (3,4). Typically in the United States, the older cup or Wesson loss methods have been supplanted by the *Neutral Oil and Loss Method* (AOCS Method Ca 9f-57), also known as *chromatographic loss*. This method is now referred to by its initials *NOL* and is loosely called "Neutral Oil Loss," which may be misleading. The method determines the weight of the neutral oil, consisting of triglycerides and the unsaponifiables (nonpolar compounds), in an oil sample minus the more polar compounds retained on the chromatographic column. Such retained compounds are basically the FFAs and phosphatides that are targets for removal in the refining process. The neutral oil then becomes the *theoretical amount*, and the difference between it and the actual amount of oil coming from refining gives a measure of the plant efficiency, as follows:

$$\%\ \text{Efficiency} = \frac{\text{Refined oil yield (wt)}}{\text{Crude oil input (wt)} \times \%\ \text{Neutral oil}} \times 100\%$$

In the United States, an average target loss in soybean oil refining is expressed as follows:

$$\%\ \text{Loss expected} = \text{NOL} + 0.5\%$$

Unconventional Continuous Caustic Refining Systems

There are at least two unconventional continuous caustic refining systems differing from the systems just described. One is called *miscella refining*, and the other is the *Zenith system*. Both of these systems have been described in detail by Cavanagh (15).

Miscella Refining

Refining of vegetable oil combined with solvent is called *miscella refining*. Such refining has been studied for many years and has been commercially practiced on cottonseed oil (15). Studies on other oils, including soybean, have been done on a pilot scale but not on a commercial scale. An advantage for miscella refining is that it can be done on freshly extracted oil, thus avoiding any deleterious effects related to oil storage. Another advantage is that miscella is less viscous, and the differences in density between the heavy phase (soapstock) and light phase are greater, giving better efficiency in centrifugal separation and lower losses. The disadvantage is that the refining space and equipment must be explosion-proof and located adjacent to a solvent extraction plant.

Zenith Refining System

The so-called Zenith refining process was developed in Sweden in 1960 (15). The process was originally designed for rapeseed oil but is now used on practically all oils and fats. The plants range in capacities from 20 to 200 MT of crude oil per day. The Zenith process is shown schematically in Fig. 11.8 (15). All equipment in the system is made with stainless steel.

In this system, soybean oil is treated with 0.1 to 0.4% by weight of an 80% phosphoric acid solution under vacuum in the *P unit*. This unit is divided into three or four trays with deaeration in the first tray and heating to the required reaction temperature in the succeeding trays. This is a semicontinuous system. The trays are agitated by paddles on a common drive shaft, and the tray contents are drained successively by pneumatically operated valves. The precipitated gums are then removed in a specially designed hermetic centrifuge.

The degummed oil is then heated to about 90°C (194°F) and delivered to the bottom of the *neutralizer vessel*, where it is dispersed as small droplets (1.5 to 2.0 mm) and allowed to rise up through a dilute NaOH solution (2.0° Baumé, 0.35 N). The designed flow rate on the bottom is 1 to 1.5 MT/h and the height of the vessel standardized to provide a 3-m lye solution column. This height gives a 30 to 40 second residence time during the rise of the droplets to the top, which is sufficient time to ensure neutralization of the droplets. Increases in capacity are accomplished by increasing the diameter of the neutralizing vessel or by adding more vessels.

The neutralized oil is directed from the top of the neutralizing vessel to the C unit for bleaching. There is said to be no need to water wash the neutralized oil because of the use of the dilute caustic solution. The neutralized oil contains about 1% water and 50 ppm soap, which requires use of citric acid in the C unit and as much as 1.0% bleaching earth.

Physical/Steam Refining of Soybean Oil

Physical or *steam refining* (PR) is not a new process and was originally used on fats and oils with high FFAs and low gum contents, such as palm oil and tallow, and the lauric acid oils, such as palm kernel and coconut oils (4). The process is the same as that used in deodorization (see Chapter 14).

The success with, and knowledge gained in, the use of physical refining of palm oil and the growing concern about the water pollution potential of soapstock acidula-

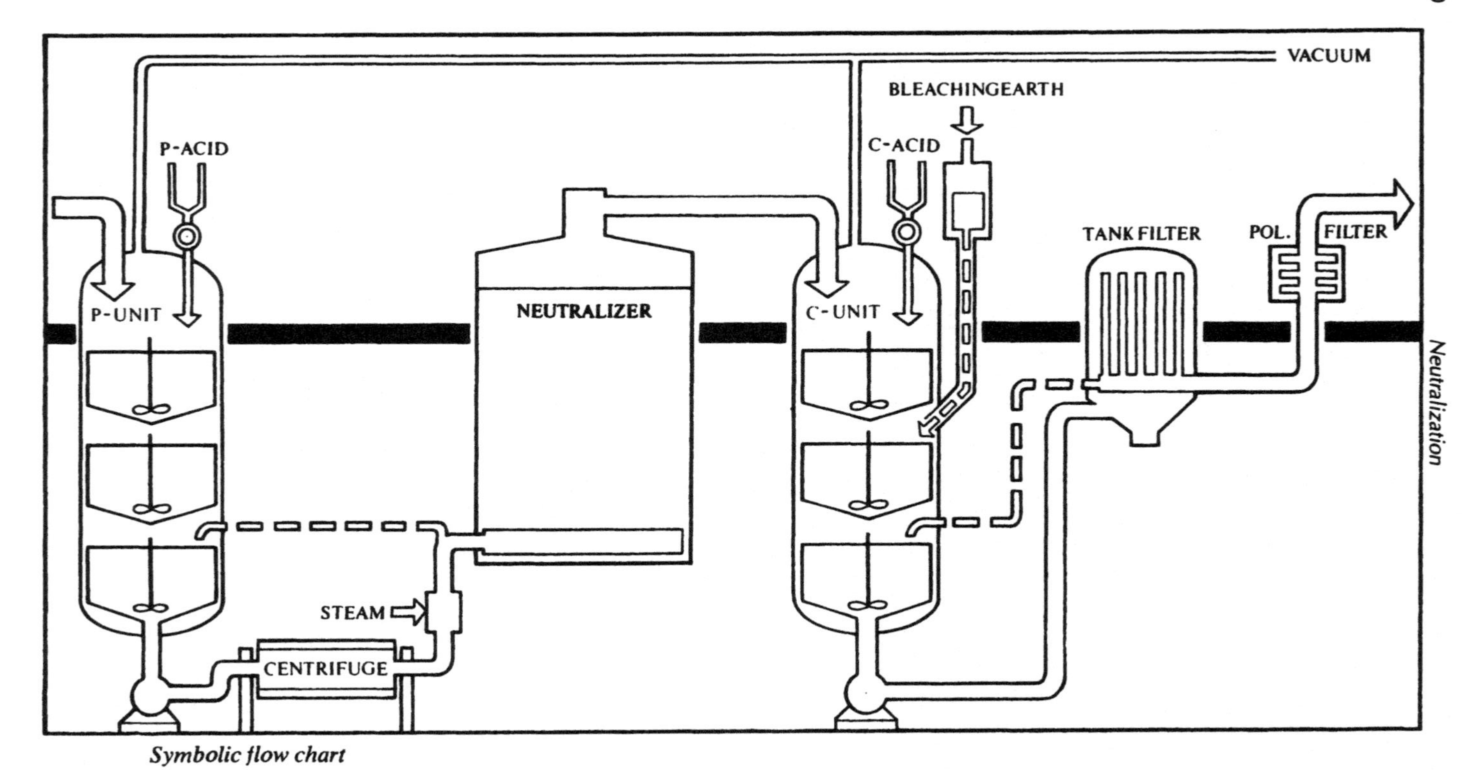

Fig. 11.8. Zenith process for continuous caustic refining. *Source:* Cavanagh, G.C., in *Proceedings World Conf. Edible Fats and Oils Processing,* edited by D.R. Erickson, American Oil Chemists' Society, Champaign, IL, 1990, pp. 101–106.

tion has led to many studies and some use of PR for soybean oil, which, in contrast to the earlier applications (high FFA and low gum contents), has a low FFA and high gum content. Recognition of the gum content problem has led to many studies and proposed processes for the exhaustive degumming of soybean oil (16). Some literature has suggested that reduction to values of 3 to 5 ppm P in soybean oil after bleaching are sufficiently low to produce PR oils that are comparable in quality to caustic-refined soybean oils. More careful examination of such data shows that the comparisons are with less than optimally processed caustic-refined soybean oil (17,18).

Interest in PR has triggered studies by U.S. companies. To date, physically refined soybean oils are not acceptable in the U.S. market because of lack of flavor stability. Well-degummed oils (<5 ppm P) are comparable to caustic-refined oils when freshly deodorized but show considerably less shelf life before going off flavor. In addition, even 1 to 2 ppm P in PR soybean oils reduces the activity of hydrogenation catalysts (19).

The differences between the costs of PR and those of caustic refining are subject to some interpretation but may not be very great (20). The major problem in caustic refining is that associated with soapstock handling and utilization as it relates to potential water pollution problems (see Chapters 17 and 25). For this reason, it is expected that work on physical refining of soybean oil will continue in the United States. Physical refining of soybean oil is practiced in markets that will accept oils with lower quality than that demanded in the competitive U.S. market.

References

1. Daniels, R.S., U.S. Patent 4,836,843 (1989).
2. Anon., *J. Amer. Oil Chem. Soc. 66:* 885 (1989).
3. Sullivan, F.E., *J. Amer. Oil Chem. Soc. 45:* 564A (1968).
4. Norris, F.A., in *Bailey's Industrial Oil and Fat Products*, edited by D. Swern, 4th edn., Vol. II, John Wiley and Sons, New York, 1982, pp. 268–288.
5. James, E.M., *Oil and Soap 8:* 137 (1934).
6. Wiedermann, L.H., *J. Amer. Oil Chem. Soc. 58:* 159 (1981).
7. Erickson, D.R., *J. Amer. Oil Chem. Soc. 60:* 351 (1983).
8. Erickson, D.R., and L.H. Wiedermann, *INFORM 2:* 201 (1991).
9. Charpentier, R., *INFORM 2:* 208 (1991).
10. Braae, B., U. Brimberg, and M. Nyman, *J. Amer. Oil Chem. Soc. 34:* 293 (1957).
11. List, G.R., T.L. Mounts, and A.J. Heakin, *J. Amer. Oil Chem. Soc. 55:* 280 (1978).
12. Beal, R.E., E.B. Lancaster, and O.L. Brekke, *J. Amer. Oil Chem. Soc. 33:* 619 (1956).
13. Welsh, W.A., J.M. Bognador, and G.J. Toeneboehn, in *Proceedings of the World Conference on Edible Fats and Oils Processing*, edited by D.R. Erickson, American Oil Chemists' Society, Champaign, IL, 1990, pp. 189–202.
14. *Yearbook and Trading Rules*, National Oilseed Processors Association, Washington, DC, 1993–1994, p. 87.

15. Cavanagh, G.C., in *Proceedings of the World Conference on Edible Fats and Oils Processing*, edited by D.R. Erickson, American Oil Chemists' Society, Champaign, IL, 1990, pp. 101–106.
16. Seger, J.C., and R.L. K.M. van de Sande, in *Proceeding World Conference on Edible Fats and Oils Processing*, edited by D.R. Erickson, American Oil Chemists' Society, Champaign, IL, 1990, pp. 88–93.
17. Young, F.V.K.,in *Proceedings of the World Conference on Edible Fats and Oils Processing*, edited by D.R. Erickson, American Oil Chemists' Society, Champaign, IL, 1990, pp. 131–132.
18. Penk, G. in *Proceedings: World Conference on Emerging Technologies in the Fats and Oils Industry*, edited by A.R. Baldwin, American Oil Chemists' Society, Champaign, IL, 1982, p. 38.
19. Strecker, L.R., J.M. Hasman, and A. Maza, in *Proceedingss of the World Conference on Emerging Technologies in the Fats and Oils Industry*, edited by A.R. Baldwin, American Oil Chemists' Society, Champaign, IL, 1982, pp. 51–55.
20. Tandy, D.C., and W.J. McPherson, in *Handbook of Soy Oil Processing and Utilization*, edited by D.R. Erickson et al., American Oil Chemists' Society, 1981, pp. 217–228

Chapter 12

Bleaching/Adsorption Treatment

David R. Erickson

Consultant
American Soybean Association
St. Louis, MO

Introduction

The classical approach to the process of bleaching has been to treat it in terms of color reduction—hence the term *bleaching* itself. However, barring the sporadic and infrequent occurrence of high chlorophyll levels, the color of soybean oil is normally not of great concern. Reasonable-quality soybean oils are readily reduced to Lovibond colors less than 20 yellow and 1 red (AOCS Method Cc 13e-92) in fully refined products.

In soybean oil processing, color reduction occurs at each step: degumming, caustic refining, bleaching, hydrogenation, and deodorization. In fact, some refiners have eliminated the bleaching step because they feel that because of its relatively low inherent color, soybean oil does not require bleaching for adequate color reduction. This is a misguided practice, because it was shown many years ago that bleaching improves the *flavor* of refined soybean oil (1). It was also found that acid-activated bleaching earths are more effective in improving flavors than the more neutral earths are (2,3,4).

Modern approaches to bleaching now focus on both color reduction and the importance of an acid treatment of the oil, brought about either by moisture-containing acid-activated earths or by actual addition of moisture and acid in conjunction with weakly acid or neutral earths. Such considerations suggest that for soybean oil, *adsorption treatment* is a more appropriate term for the process than *bleaching*.

Preparation of Oil for Bleaching

The importance of a clean feedstock of neutralized oil going to the bleaching process cannot be overemphasized. Any soap or phosphatides present in the neutralized oil will require adsorption by some portion of the earth, thus requiring more earth than that needed for the basic adsorption treatment. Neutralized oils containing 5 to 10 ppm phosphorus (P) and 10 to 30 ppm soap have been reported as typical, but levels of <1 ppm P and no soap are achievable (see Chapter 11).

The economic benefit in providing clean feedstocks for bleaching is minimal earth usage. This means less earth cost, longer filter cycles, less neutral oil loss, and less spent earth, which requires disposal.

The economic benefit of a decrease of 0.1% earth usage for a 200-short ton/d (TPD) (180 MT/d) plant may be partially estimated by the following calculation:

Production rate = (200 short tons/d)(300 d/yr) = 60,000 short tons/yr (TPY)
Current value of oil = $600.00/short ton
Current cost of bleaching earth = $400.00/short ton

For each 0.1% reduction in bleaching earth, assuming 10% moisture in the purchased earth and 30% loss of neutral oil in the spent earth, the calculated savings is:

(0.001)(60,000 short ton/yr) = 60 short tons (54 MT) earth usage saved/yr
(0.9)(60 short tons)(0.3) = 16.2 short tons (14.7 MT) oil loss saved/yr
(60 short tons bleaching earth) ($400.00/short ton) = $24,000
(16.2 tons oil/yr)($600.00/ton) = $9,720
Total for each 0.1% reduced earth usage = $33,720/yr

The economic benefits of having a clean feedstock leading to reduced earth usage are readily apparent from this example. Other coincidental benefits include the potential for reduction or elimination of phosphatides and soap during the high temperatures of bleaching. The fact that soaps and phosphatides are adsorbed in bleaching does not mean that they cannot exert a deleterious effect in their adsorbed state. This aspect has not been studied but should be. Also, the less earth used, the less chance of hydrolysis or other effects that would increase oil losses or reduce the quality of the finished oil.

Bleaching Functions

Briefly stated, the functions of the bleaching step are to remove or reduce the levels of the following in neutralized oil:

- Pigments (color)
- Oxidation products
- Phosphatides
- Soaps
- Trace metals

As previously stated, bleaching theory has developed around the efficacy of color removal. This remains helpful, because color reduction parallels the process of removing or changing oxidation products, which is at least as important as color reduction. In fact, color removal in soybean oil could be considered incidental to the bleaching process in terms of flavor improvement. The exception to this would be the infrequent occurrence of excess quantities of chlorophyll, which then require special or extra treatment above that for normal soybean oils. This is an example of the adjusted treatment practices mentioned in Chapter 5.

If color removal in soybean oil is incidental to the main goals in bleaching, the previous list of bleaching goals can be changed to the following in order of importance:

Decompose peroxides
Remove or change oxidation product.
Remove trace phosphatides (gums) and soaps
Reduce metal content
Decolorize

It has been recognized for some time that bleaching improves the flavor of deodorized soybean oil and that the more acid-activated bleaching earths are more effective than neutral earths (1,2,3). There have been no reported research results that delineate the actual mechanism of how the activated earths exert their effects on flavor other than in a directional sense. The improvement in flavor of refined, bleached, and deodorized (RBD) soybean oil has been recognized and documented, as already mentioned. What has not been studied or documented is the similar improvement that can be expected in hydrogenated oils. The improvement noted in actual experience has been a significant reduction or elimination of the typical "hydrogenated flavor" in deodorized hydrogenated products, probably by the reduction or elimination of oxygen-containing compounds at the high temperatures of hydrogenation. The flavor improvement is accompanied by reduced accumulation of polymers and other waste products in the hydrogenation vessels, allowing longer cycles between cleanings.

The earliest publication on modern commercial bleaching practice and its results was that of Wiedermann (4), who postulated that oxygen compounds were removed, changed, or both during the bleaching process so that they did not cause oxidation in further processing or premature oxidation of the final deodorized soybean oil or soybean oil products. Wiedermann postulated that, in addition to the physical adsorption associated with color removal, there were also processes involving chemisorption and subsequent chemical reactions occurring on the earth surface. Such a concept was depicted by Wiedermann as in Fig. 12.1 and was called *decomposition and dehydration* or *pseudoneutralization* of peroxides and secondary oxidation products. This possibility has also been discussed by Frankel (5).

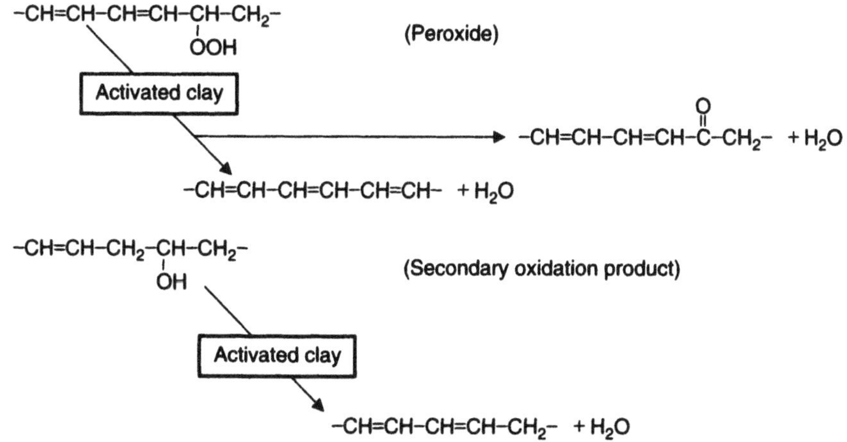

Fig. 12.1. Decomposition and dehydration, or pseudoneutralization, of peroxides and secondary oxidation products in reaction of oils with acid-activated bleaching clays (4). *Source:* After Wiedermann, L.H., *J. Amer. Oil Chem. Soc. 58:* 15 (1981).

Whatever the actual mechanism of bleaching with acid activated earth might be, Wiedermann considered bleaching to be another important step in what he called an "adjusted treatment philosophy," to be applied throughout the refining of soybean oil (4) (see also Chapter 5). In other words, the amount of bleaching earth has to be adjusted to compensate for varying qualities of the starting crude oil.

In actual practice, the "adjusted amount" was found to be that amount necessary to achieve a zero peroxide value in the oil coming out of the bleaching press. By using this criterion, the quality of the finished products would be both high and reasonably uniform, thus overcoming the normal expected variations in crude soybean oil quality. This is a very important concept, because bleaching and deodorization are the last chances refiners have to obtain maximum quality in their finished products, assuming that the succeeding unit processes are also done optimally. Although this approach is somewhat empirical, the efficacy of bleaching to zero peroxide on maximizing the quality of finished soybean oil products has been proved time and again, literally on a worldwide basis (4,6,7).

It is important to note that the desired bleaching end point of "zero peroxide" is a convenient analytical tool that shows that adequate adsorptive treatment has occurred. Other more elaborate tests have been done to assess the efficacy of bleaching on the quality of the finished oil, and only fortuitously was it found that a zero peroxide value in the oil coming out of the bleaching press correlated well with the other tests (4). This has now been confirmed in part by the work of Henderson and coworkers (8).

Wiedermann also noted that the peroxide reduction was to be obtained in the bleaching step and not in the deodorization step (4). The high temperatures of deodorization practically ensure that the peroxide content of the deodorized product will also be zero; however, this is strictly a heat effect and is an entirely different reaction mechanism from that found in bleaching.

In summary, the primary function of bleaching soybean oils is to reduce or change the oxidation products; remove trace amounts of soaps, phosphatides and metals; and decolorize. Such treatment may be considered more as an acid/adsorptive treatment involving moisture, acid, and an adsorbent. When an acid-activated earth is used, this reaction is possible with no added acid or water if there is sufficient acidity and moisture in the earth. Another approach to bleaching without the use of acid-activated earths will be discussed under the section on wet bleaching.

It was mentioned earlier that color removal in soybean oil by bleaching parallels other desired reactions. This is important, because most of the literature on the bleaching reaction is based on adsorption isotherms and other analytical tools developed and used for studying color removal (9). Another aspect of color removal is that the National Oilseed Processors' Association (NOPA) trading rules for soybean oil include color specifications on both once-refined and fully refined oils. These values are rather generous in view of modern practice, being not more than 3.5 red (bleached) for once-refined and maximum 2.0 red (no green) for fully refined (10).

The NOPA specifications, while generous, recognize that increased color and, especially, colors resistant to bleaching are indicators of poorer-quality oils. In the

TABLE 12.1 Color Reduction in Soybean Oil Processing (4)

Processing step	Color removal
Neutralization	7–8 red
Bleaching	2–2.5 red
Hydrogenation	Variable*
Deodorization	0.3–0.6 red

*Reduction of color in hydrogenation depends on degree of hydrogenation.

United States, lighter soybean oils (about 0.3 red) have a marketing advantage over more highly colored oils at the retail level. This has necessitated better control of processing and evaluation of crude oils. A simple test on the bleachability of a crude oil will both assess its quality and predict whether or not it will meet color and quality goals in the final product. Bleaching is only one of the processing steps that remove color in soybean oil. An example of the color removed in each step of processing is shown in Table 12.1.

The removal of chlorophyll from soybean oil is a special case. Normal chlorophyll levels in crude soybean oil are in the range of 100 to 200 ppb. Pritchett et al. (11) evaluated the reduction in chlorophyll in processing as shown in Table 12.2, from which the normal content of chlorophyll in RBD oils would be expected to be in the range of 10 to 20 ppb. Chlorophyll values greater than 50 ppb after bleaching may result in a green/gray tinge in the RBD oil. This occurs because a faint green tinge is masked by yellow and red colors. The final step of deodorization may reduce the yellow and red to levels low enough to allow the green/gray color to become apparent. At that point, it is too late to adjust any treatment practice, and the oil is best rebleached if the green/gray color is a problem. Alternatively, the green/gray color may be overcome by blending with normal oils.

The basic problem with chlorophyll in soybean oil is knowing that it is present at a potentially troublesome level in the crude or crude degummed oil. This is best done by testing crude receipts, usually spectrophotometrically. Once it is known that excess chlorophyll is present, the bleaching dosage can be adjusted upward to ensure its removal.

Products for Bleaching/Adsorptive Treatment

The products used for soybean oil bleaching are neutral earth, acid-activated earths, activated carbon, and silicates. For some filtration systems it is necessary to use a filter aid (*body feed*), which is usually diatomaceous earth or other inert material.

Neutral Earths

Neutral earths are also sometimes called *natural clays* or *earths*, and the original was known as *fuller's earth*. They are basically hydrated aluminum silicates and naturally vary in their ability to adsorb pigments.

TABLE 12.2 Chlorophyll Reduction in Soybean Oil Processing

Processing Step	Effect on chlorophyll content
Crude oil (no processing)	150 ppb (μg/L)
Caustic refining	25% reduction
Deodorization	Little, if any, effect
Hydrogenation	Variable, depending on degree
Bleaching (0.25% earth)	74% reduction
Bleaching (0.75% earth)	91–92% reduction

Acid-Activated Earths

Acid-activated clays are bentonites or montmorillonites that have been treated with hydrochloric or sulfuric acid to improve their ability to adsorb color and the other undesirable components. The degree and method of acid activation is proprietary with the earth manufacturers. The important characteristics of an acid-activated earth are as follows:

Total acidity (titratable acidity, usually expressed as mg KOH/g of earth under standardized methodology)

pH (aqueous slurry)

Moisture

Bulk density

Effective surface area (particle size)

Oil retention

Typical values for an acid-activated earth widely used in the United States are shown in Table 12.3.

The chemical and physical changes on earths caused by the acid activation step have been explained in several publications (2,3,12). The efficiency of an acid-activated earth in relation to its acidity was shown by Wiedermann (4) in a plant test as shown in Table 12.4. The effectiveness of using more acid earth is very evident in terms of both color removal and peroxide reduction.

The role of moisture in a bleaching earth was recognized many years ago, when dried earths were found to be much less effective in removing color (3,9). In addition, a lack of moisture would probably adversely affect the acid/adsorbent reaction necessary for reducing or changing oxygen-containing compounds. The latter has not been studied but seems logical in view of the "wet bleaching," which will be discussed later. The importance of moisture in bleaching earth leads to some important considerations in the process itself, which will be discussed under that section.

Activated Carbon

Activated carbons have a long history of use in edible oils, but they are more expensive, retain more oil, and are difficult to handle. One property of recent interest is their ability to adsorb polyaromatic hydrocarbons in coconut and fish oils that are not

TABLE 12.3 Acid-Activated Earth Properties

Property	Typical range
Free moisture (wt%)	10–15
Titratable acidity (mg KOH/g earth)	4–5
pH (2% slurry)	3–4.5
Surface area (m^2/g)	200–350
(ft^2/oz)	61,000–92,000
Aparent bulk density (kg/m^3)	550–750
(lbs/ft^3)	34.3–46.8
Particle Size (wt%)	
200 mesh	70–90
325 mesh	60–75
Average Particle Diameter (μm)	15–30
Oil Retention %	30–40

removed by earths. Activated carbons are also effective in removing soaps and pigments, especially chlorophyll. Their effect on peroxide and other oxygen-containing compounds have apparently not been studied. When used, activated carbons are added in conjunction with other earths at a 5 to 10% level of the total earth used.

Silicates

Several synthetic silicates now available are particularly effective in removing soaps, phosphatides, and trace metals. They are relatively ineffective in removal of pigments, especially chlorophyll (13).

For oils other than soybean and canola, the silicates are promoted for removal of soaps and phosphatides in conjunction with heat bleaching for color removal. For soybean and canola oils, addition of some activated earth is necessary to accomplish chlorophyll removal; this can be done by continuous addition or by establishing a packed bed of activated earth in the bleaching press prior to bulk filtration of the oil/silicate mixture (13).

TABLE 12.4 Plant Bleaching Test Comparisons of Two Acid-Activated Bleaching Earths[a] (4)

Refined oil to bleacher	Batch#	PV 2.2	Color[b] 7.9
Clay "A"	1	1.7	3.5
(1.7 titratable acidity)	2	1.3	3.2
(3.5 pH)	4	1.5	3.3
	6	0.5	3.1
Clay "B"			
(4.8 titratable acidity)	1	0.4	2.0
(3.0 pH)	2	0.0	2.0
	4	0.0	1.8

[a]0.5% earth added at 82°C (180°F), temperature raised to 104°C (220°F), held 20 minutes, and filtered; atmospheric conditions; 6-batch-capacity filter press.
[b]Lovibond red index (AOCS method Cc 13e-92).

It is not known whether the silicates have the same effect as activated earths on the peroxides and oxygen-containing compounds. From the data available, it is not possible to determine the relative efficacy of silicates vs. acid-activated earths in improving the flavor of finished soybean oil or soybean oil products when optimum refining practices are applied. Some data is available on less well-refined soybean oils that indicates comparable flavors and (AOM) values when comparing silicates and acid-activated earths (13).

Practical Aspects in Choosing a Bleaching Clay

Price, availability, and evolution of newer types of bleaching clays means that a refiner needs to be prepared to evaluate the performance of new or different products periodically. Price and a "one-shot" plant bleaching test are not good bases for choosing a new clay. The following is a listing of the items for consideration when evaluating a new clay (Davidson, R., private communication, KFI, Memphis, TN):

Flowability. Does the clay cause any handling problems in the feeder system, such as bridging and channeling ("rat-holing")?

Dusting. This is not as important in a bulk system, but it can be a serious breathing and housekeeping problem in a bag feed system.

Clay quality. A Certificate of Analysis should be received with each load and checked against in-house lab analyses using the same procedures as the clay supplier. Analyses should include pH, titratable acidity, moisture, oil retention, screen analysis, and a lab bleach test. This will assure the processor that delivery is to specification and will prevent plant problems caused by using substandard product.

Auto ignition. All spent clays are prone to autoignition, but some are more so than others. A good blowdown procedure should be developed and followed, including the capability to wet down the spent clay after removal from the press. New clays should exhibit no additional problems in normal operations.

Filterability. An acceptable clay should filter to a full press without having to shut down because of excessive pressure differential. The ideal situation is to have some pressure leeway with a full press. This will allow the pressure to build slowly and not cause premature blinding of the screens. Laboratory filterability tests can sometimes take the place of expensive plant testing if a correlation has been established.

Many processors no longer use filter aid and depend only on the filterability of the clay. This again stresses the importance of clay quality.

One of the main factors in clay choice is the type of filter being used. A clay that runs well on a plate-and-frame press may not on a pressure leaf filter.

Other considerations. In addition to the foregoing, the following are further considerations:

- *Deodorization.* Any lab/plant bleaching test should include deodorization and comparison of the final product against a control for FFA, colors, flavor and flavor stability, and the other characteristics desired.
- *Color break.* Many bleaching operators have learned that if they slug a fresh press with 20 to 25% of the total clay to be used, the color/chlorophyll content of the bleached oil will drop quickly (color break), allowing less recirculation time, thereby getting the press on-line more quickly. These operators will then cut back the clay feed and "ride the press effect" until the end of the run. Some clays show a color break earlier than others, and the sooner the break, the better the press effect (see the later discussion of "press effect"). If the operators are still ahead of the color break when the press is full, they will stop the clay feed and continue press bleaching until the chlorophyll or peroxide value starts to rise.

The Bleaching Process

The bleaching process is relatively straightforward, in that the neutralized oil is mixed with the appropriate dosage of earth, heated to a bleaching temperature, and then filtered.

Although many plants use mild steel construction for bleaching equipment, it is not a recommended practice. Because of the acidic earths and high temperatures, stainless steel is the construction material of choice.

Earth Dosage

Traditionally, the amount of earth to use has been dictated by the amount of color reduction desired. This has now been supplanted, at least for soybean oil, by choosing the amount that will give a zero peroxide value in the oil coming out of the bleaching filter press (see comments on chlorophyll earlier in this chapter). For normal-quality crude oil this will range from 0.3 to 0.6%, depending on the quality of the neutralized oil and on the relative acidity and other characteristics of the earth.

Atmospheric Batch Bleaching

The original bleaching process was simply to mix the chosen dosage of earth with the dry oil in an open vessel, followed by agitating and heating the mixture to the bleaching temperature of 110 to 120°C (230 to 248°F), holding for 20 to 30 min, and then cooling to 70 to 80°C (160 to 180°F) and filtering.

This original process embodies a very important sequence that applies to all bleaching processes. This sequence brings the oil and earth together at temperatures well below the boiling point of water, which means that the earth still contains the moisture necessary for its action. Once the earth and the oil are in contact, the heating drives off the moisture, allowing the earth to do its job. Earth without moisture is ineffective (3,4,6,7), and adding earth to hot oil decreases the earth's effectiveness, because moisture is driven off prematurely.

The precautions using atmospheric batch bleaching are intended to avoid any incorporation of air, either from the bleaching earth or during agitation. The air can be eliminated from the earth by slurry addition and from agitation by proper design of agitators and baffles to avoid vortexes. The time at top temperature need be only that necessary to drive the moisture from the earth (15 to 30 min).

When the holding time has passed, the oil/earth mixture should be cooled to 70 to 80°C (160 to 200°F) and filtered as quickly as possible. The filtered oil should be further cooled and protected under conditions that will prevent oxidation and subsequent peroxide development. Alternatively, the bleached oil could go directly to deodorization or hydrogenation. The advantages of atmospheric batch bleaching are the low cost of equipment and the assurance that the oil is completely and uniformly treated. Batch bleaching also fits into a batch refining and deodorization system. The disadvantages are the risk of exposure of oil to air at high temperatures and the operator skill needed to control temperatures and time.

Vacuum Batch Bleaching

Vacuum batch bleaching is an improvement over atmospheric, because exposure to air is eliminated and there is less dependence on operator skill. The same operating principles apply, except that the top temperatures are now decreased somewhat in proportion to the vacuum applied. It is still necessary to add the earth at a temperature well below the boiling temperature of water for the same reasons given for atmospheric bleaching. Typical vacuum batch bleaching conditions would be a vacuum of 50 mm Hg absolute, top temperature of 100 to 110°C (212 to 230°F), and holding time of 20 to 30 min.

Continuous Vacuum Bleaching

Continuous vacuum bleaching is the modern method for bleaching edible oils. It is more effective than vacuum batch bleaching, because bleaching action starts with a spray of oil into a vacuum rather than applying a vacuum to a large volume of oil in a tank. The continuous system also fits in well with modern continuous neutralization, hydrogenation (when used), and semicontinuous and continuous deodorizers.

A typical continuous vacuum bleaching system is depicted in Fig. 12.2. The system adheres to the principles just noted in that the oil and earth are brought together in slurry form at a low temperature, followed by deaeration and heating to bleaching temperature. The mixture is then sprayed into the agitated holding vessel, cooled, and filtered. Temperatures and times are much the same as just given for the batch systems.

The advantages of continuous vacuum bleaching are that bleaching occurs in the absence of air and that these systems can be operated by automation and instrumentation, practically eliminating the chance of operator error. The only disadvantage is that there is some intermixing in the bleaching vessel, which could result in a small portion of the oil being overbleached and a small portion being underbleached. The extent of

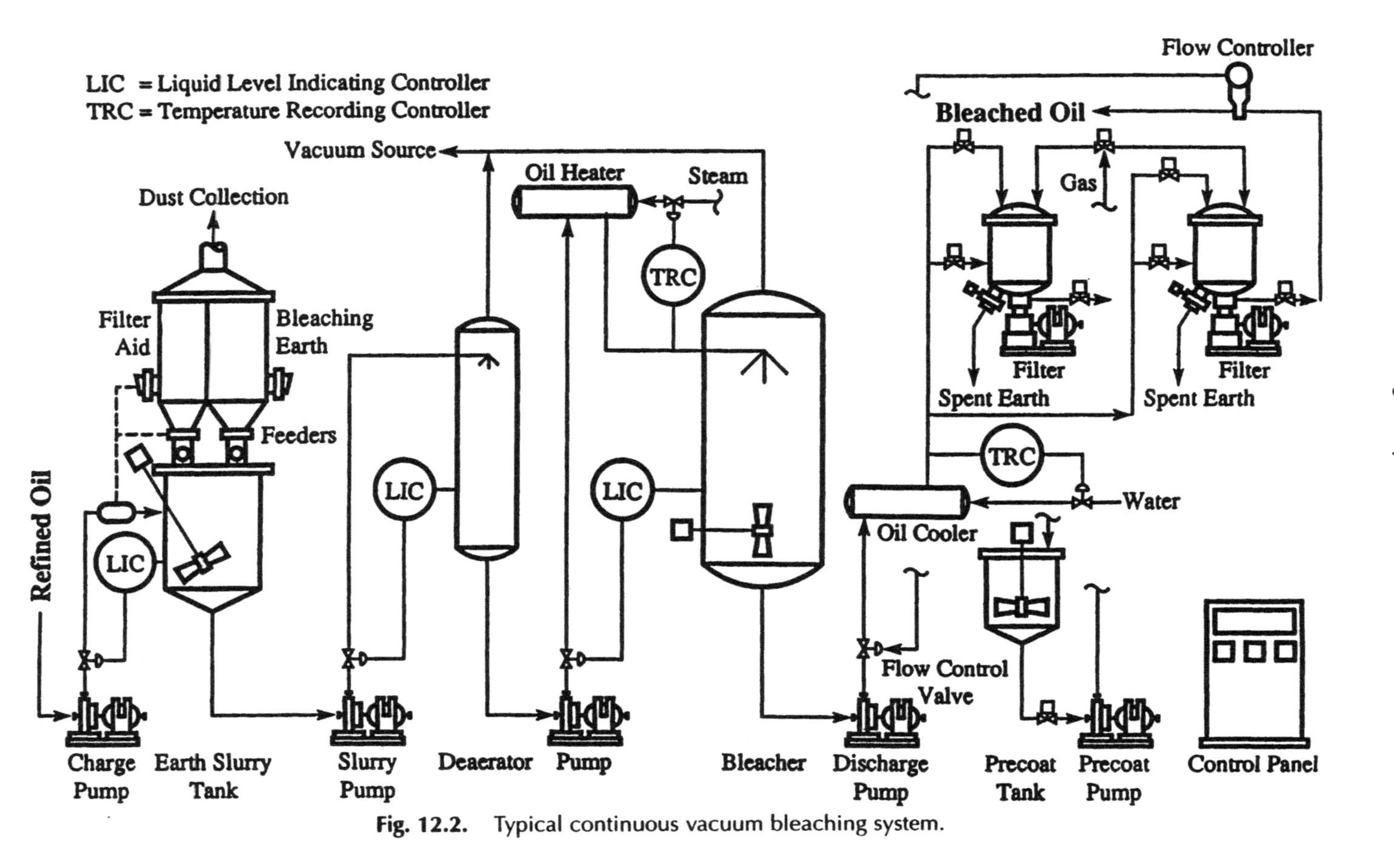

Fig. 12.2. Typical continuous vacuum bleaching system.

this has not been determined and depends on the agitation pattern. In the system shown in Fig. 12.2, a mechanical agitator is shown. Other systems use sparging steam for agitation. In actual practice this "disadvantage" may be of no consequence, but nevertheless, the possibility of intermixing is inherent in the system design.

Filtration

The necessity for good filtration is paramount in ensuring good-quality oil products. Even if bleaching has done its job, poor filtration will reverse any advantage gained by good bleaching practices. In fact, good bleaching practice has to include excellent and foolproof filtration.

The subject of filtration and the equipment used in edible oil processing has been reviewed by Latondress (14) including that used in bleaching. The two basic types of filters usually used are plate-and-frame and pressure leaf filters. Automated, self-cleaning models of both types are now available. The general rule for determining the needed capacity ranges from 250 to 484 kg oil flow/hr/m^2 of filter surface (50 to 100 lbs/hr/ft^2), with the lower flow rate being preferred.

Any spent bleaching earth going forward with the bleached oil will cause problems in hydrogenation and in deodorization. In addition to catalyzing oxidative reactions in both downstream processes, spent earth particles will foul both hydrogenation and deodorization equipment surfaces. *The importance of ensuring exhaustive removal of all the bleaching cannot be overemphasized.* Fig. 12.2 shows two filters operating alternatively, which is a normal and good practice. What is not shown, and should be, is a *safety* or *polish filter*, downstream from the main presses. In normal practice, the oil/earth mixture from the bleacher is recirculated to build up cake in the filter press. The progress of the cake buildup is monitored by observation through a sight glass in the recirculation line. Once the oil appears clear, it is sent forward to the bleached-oil storage tank or to further processing. This practice is subject to observer error and may not show up any particles going forward after the initial observation. Any particles going forward from the press, due to holes in the cloths or screens or to pressure fluctuations, will probably not be observed. The installation of a catch or polish filter will help and should be an automatic addition to a continuous filter system. Perhaps a better long-range solution would be installation of a suitable particle detector. Such a particle detector could also be used for automatic diversion or recirculation by proper instrumentation if so desired.

Press Bleach Effect

It has long been recognized (9) that a significant portion (30 to 40%) of the bleaching process takes place in the filter press. This has been called the "press bleaching effect" and has been studied by Henderson and coworkers (8). The effect is seen in Table 12.4 where successive batches of both the acidic and less acidic bleaching earth show a lessening of both color and peroxide value.

This press bleaching effect is a significant contribution to the bleaching process and can only really be assessed on plant scale. This is another case where laboratory results do not correlate well with what can be accomplished in the plant. Because of this, laboratory data on bleaching media have to be examined carefully.

For press operation and spent earth disposal and utilization, see Chapter 17.

Newer Bleaching Practices

The recognition of the importance of moisture in acid-activated earth treatment led to the energy-saving step of bypassing the drying of neutralized oil. The soybean oil coming from caustic neutralization and water washing normally contains about 0.2 to 0.4% water. This amount of water is easily removed in a continuous vacuum bleaching system, and the extra water may increase the acidity effect of the bleaching earth. The one disadvantage might be a slight increase in free fatty acid (FFA) levels, but this is offset by the energy savings.

This measure will work only if the continuous bleaching system matches, or nearly matches, the capacity of the neutralization system. Otherwise, accumulation and storage of wet neutralized oil would be required, and this is not recommended unless there is assurance that the FFA does not rise during such storage.

Wet Bleaching

A further innovation has been the recognition that citric acid and water added to a relatively neutral earth show the same desirable effects as does the use of an acid-activated earth. This has been studied on a commercial scale (Bazaldua, D., consultant, Mexico City, private communication) and in the laboratory by Henderson (8).

The process as described by Bazaldua is shown in Fig. 12.3 and designated by him as *wet bleaching*. The neutralized and water-washed oil discharges from the last centrifuge into an open kettle, where the bleaching earth (0.7 to 1.7%), citric acid (100 to 300 ppm), and water (0.5 to 2.2%) are added and kept under agitation for a minimum of 20 min at a temperature no higher than 80°C (176°F). The mixture is then pulled into the vacuum bleacher, and the temperature is raised to 100°C (230°F) and held for 30 to 40 min under 50 mm Hg vacuum. The oil is then filtered in the normal manner. The RBD soybean oil from this process was found to have an 18-h AOM after deodorization and, with 50 ppm citric acid added, and packaged in PVC without nitrogen, had a "simulated" shelf life of 16 mo.

Henderson (8) used a neutral clay, to which he added a 20% solution of citric acid with the dosage of citric acid (on a dry basis) at 4% of the clay. He found that the neutral earth usage to be about 30% more than acid-activated earths. The citrated neutral earth was almost as effective as the acid-activated earths on pigment removal, but much less effective in peroxide reduction, according to Henderson.

It is apparent from these results that directionally it may be possible to use an acid/water/neutral earth combination that would be as effective as an acid-activated earth. Whether or not this could be cost-effective is problematical because of such factors as the cost of chemicals and extra neutral oil loss because of increased earth use. Until more information is forthcoming, or personal experience is gained, it is recommended that refiners continue with acid-activated earths for soybean oil bleaching.

Bleached Oil Handling

Ideally, bleached oil would go directly to hydrogenation or deodorization with little or no storage time, because bleached oils are very susceptible to oxidation. If storage is required, all precautions should be taken to prevent such oxidation, and this can be done by cooling, storing under nitrogen protection, or both. It is a mistake to consider storage or shipment of bleached oils unless every precaution is taken to prevent oxidation.

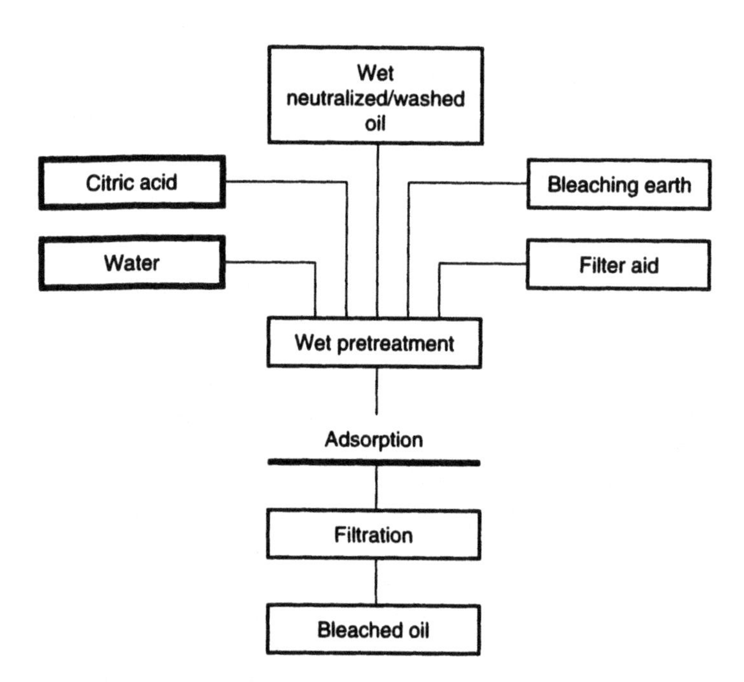

Fig. 12.3. Block diagram of wet bleaching process for soybean oil.

If the peroxide value (PV) of bleached oil is allowed to rise, then the reduction or change in oxygen-containing compounds accomplished in the bleaching step is reversed. If the PV rises as little as 1 to 2 meq/kg, then rebleaching should be considered, because the quality of the finished product has been compromised by reintroduction of oxygen-containing compounds.

References

1. Cowan, J.C., *J. Amer. Oil Chem. Soc. 43:* 300A (1996).
2. Rich, A.D., *J. Amer. Oil Chem. Soc. 47:* 564A (1970).
3. Richardson, L.L., *J. Amer. Oil Chem. Soc. 55:* 777 (1978).
4. Weidermann, L.H., *J. Amer. Oil Chem. Soc. 58:* 15 (1981).
5. Frankel, E.N., in *Handbook of Soy Oil Processing and Utilization*, edited by D.R. Erickson, et al., American Oil Chemists' Society, Champaign, IL, 1981, pp. 232–233.
6. Erickson, D.R., *J. Amer. Oil Chem. Soc. 60:* 351 (1983).
7. Erickson, D.R., and L.H. Wiedermann, *INFORM 2:* 200 (1991).
8. Henderson, J.H., R.F. Ariaanz, D.R. Taylor, and C.B. Ungermann, in *Proceedings International Meeting on Fat and Oil Technology*, edited by D. Barrera, Campinas, Brazil, 1991, pp. 34–47.
9. Norris, F.A., in *Bailey's Industrial Oil and Fat Products*, 4th edn., edited by D. Swern, vol. II, John Wiley and Sons, New York, 1982, pp. 292–314.
10. *1993–1994 Yearbook and Trading Rules*, National Oilseed Processors Association, Washington, DC, p. 82.
11. Pritchett, W.C., W.G. Taylor, D.M. Carroll, *J. Amer. Oil Chem. Soc. 24:* 225 (1947).
12. Mag, T.K., in *Proceedings World Conference on Edible Fats and Oils Processing*, edited by D.R. Erickson, American Oil Chemists' Society, Champaign, IL, 1990, pp. 107–116.
13. Welsh, W.A., J.M. Bogdanor, and G.J. Toeneboehn, in *Proceedings World Conference on Edible Fats and Oils Processing*, edited by D.R. Erickson, American Oil Chemists' Society, Champaign, IL, 1990, pp. 189–202.
14. Latondress, E.G., *J. Amer. Oil Chem. Soc. 60:* 257 (1983).

Chapter 13

Hydrogenation and Base Stock Formulation Procedures

David R. Erickson

Consultant
American Soybean Association
St. Louis, MO

Michael D. Erickson

Sr. Research Scientist
Kraft Food Ingredients Technology Center
8000 Horizon Center Blvd.
Memphis, TN

Introduction

The first hydrogenation patent was issued to W. Normann in 1903. Joseph Crossfield and Sons, a British firm, received title to the patent in 1906, for treating whale oil.

A milestone in the edible oils industry was the acquisition of the American rights to the Crossfield patent by the Procter and Gamble Company, Cincinnati, OH, in 1909. Although no longer cottonseed oil-based, the process gave rise to arguably the most recognized branded edible oil product in the industry: Crisco®. Hydrogenation is now the largest single reaction (other than caustic refining) in the edible oils industry (1).

Basic Hydrogenation Reaction and Its Goals

The process is used to increase the solids (crystalline fat) content of edible fats and oils and improve their resistivity to thermal and atmospheric oxidation. These benefits are achieved by reducing the oil's relative unsaturation as well as by promoting coincidental geometric and positional isomerization. The most common misconception about hydrogenation is that the process merely converts an unsaturated oil into a saturated fat. The basic hydrogenation reaction is typically represented as shown in Fig. 13.1, which suggests that the only result of hydrogenation is saturation.

The primary objective of this chapter is to demonstrate the benefits of optimizing hydrogenation as a unit process, offering a processor the operational flexibility and efficiency required to remain competitive in value-added markets. It will be shown that practical experience in hydrogenation clearly refutes the misconception just mentioned.

Hydrogenation is the process of treating fats and oils with hydrogen gas in the presence of a catalyst. The result is the conversion of liquid oils (usually) to fluid

```
  H     H                              Ni            H  H  H  H
  |     |                                            |  |  |  |
R-C-C=C-C-R      +     H2          ------->        R-C--C--C--C-R
  | | | |                                            |  |  |  |
  H H H H                                            H. H  H  H

    cis          Hydrogen          Nickel
unsaturation        gas           catalyst           Saturation
```

Fig. 13.1. Basic hydrogenation reaction.

(opaque), semisolid, or plastic fats suitable for use in any edible oil application. The reaction proceeds when oil (usually liquid), hydrogen gas, and the catalyst are brought together under appropriate agitation and temperature conditions.

The hydrogenation rate is governed by temperature, the type of oil being hardened, catalyst activity, catalyst selectivity, catalyst concentration, and available hydrogen gas (pressure). The goals in hydrogenation are to achieve the highest possible activity consistent with control of selectivity while using the least possible amount of catalyst.

Such goals are met only by ensuring the following:

1. The cleanest possible feed stock (no soap, phosphatides, sulfur compounds, or oxygen-containing compounds)
2. Hydrogen gas as pure as possible (no sulfur, carbon monoxide, nitrogen compounds, or moisture)
3. Clean, airtight hydrogenation equipment whose operation is understood and repeatable
4. Catalyst control, in terms of both original purchase and reuse when employed
5. Excellent control of the catalyst filtration process
6. Meticulous record-keeping and base stock programs

Adherence to these practices offers several "hidden" economic incentives in addition to ensuring the first priority: consistent, high-quality products. These include improved first-pass yields, minimal catalyst use, and less downtime for vessel cleaning.

Selectivity

Selectivity, in the chemical reactivity sense, refers to the order in which certain classes of fatty acids will be reduced or isomerized. Historically, selectivity was an indication of how fast linoleic acid was converted to a monoene compared to how fast the monoene was converted to stearic acid (2). Modern use of the term generally describes the conditions of hydrogenation. Oils normally contain mixtures of mono-, di-, and triunsaturated fatty acids. Perfect selectivity provides sequential elimination of unsaturated fatty acids as shown in Fig. 13.2.

Fig. 13.2. Selectivity of fatty acids in hydrogenations.

TABLE 13.1 Relative Rates of Oxidation of Fatty Acids (3)

Fatty Acid	Relative Oxidation Rate
Stearic	1
Oleic	10
Linoleic	100
Linolenic	150
α-Eleostearic (*cis*, Δ9; *trans*, Δ11; *trans*, Δ13)	800

Table 13.1 shows the propensity for the different classes of fatty acids to react, supporting an important chemical axiom: that which is most reactive will react first and in the greatest quantity. The relative rates of oxidation of fatty acids are also reflected in their relative hydrogenation reactivity, shown in Table 13.2. Understanding these relative reactivities allows manipulation of the reaction conditions in order to achieve or control the fatty acid profile of the desired end product. Said another way, process conditions, adjusted for the oil being hardened, are used to target certain fatty acids "selectively".

Conversely, the absence of selectivity results in a reduction of double bonds without regard to the types or amounts of those fatty acids present. Therefore, it is the conditions of hydrogenation (including the type of catalyst used) that ultimately determine the extent of hydrogenation selectivity. Although actual commercial hydrogenation of oils approaches a high level of selectivity, it fails to meet the theoretical selectivity (4). This is substantiated in Fig. 13.3.

A detailed investigation of the kinetics of hydrogenation is beyond the scope of this handbook. The reader is strongly encouraged to review the excellent review of the subject by Allen (1).

Hydrogen Gas

Commercial hydrogen is usually measured in terms of standard cubic feet (SCF). SCF is defined as the amount of dry hydrogen in a cubic foot (28.3 L) at 60°F (15.5°C) and 760 mm Hg (14.7 psi). One SCF of hydrogen weighs 2.4 g (0.0053 lb). One cubic foot of hydrogen at standard temperature and pressure (STP) (0°C and 760 mm Hg) is equivalent to 1.057 SCF (1).

Electrolysis

This process shown in Fig. 13.4 separates hydrogen from demineralized water with direct current (DC) electric power, according to the mechanism shown in Fig. 13.5.

TABLE 13.2 Relative Hydrogenation Reactivity of Different Fatty Acid Chains (3)

Fatty acid chain	Relative hydrogenation r\eactivity
Linolenic (Δ9, Δ12, Δ15) and (Δ6, Δ9, Δ12)	40
Linoleic (Δ9, Δ12)	20
Oleic (Δ9)	1

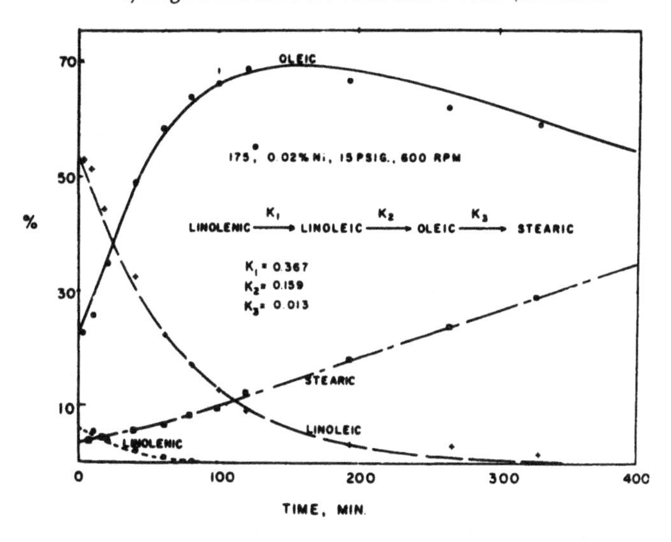

Fig. 13.3. Hydrogenation of soybean oil (5).

The hydrogen is collected at the cathode with essentially 100% purity [the more efficient plants produce ≥99.8% pure (1)], the minor contaminants being moisture and oxygen. Costs associated with production tend to parallel the cost of the electricity used (6). In addition to equipment and facilities for safely bottling the hydrogen, provisions for receiving, compressing, and bottling the oxygen byproduct are also required. In the past, electrolysis was favored because the hydrogen did not require further purification. Modern methods of purification, the significant capital expenditure for electrolysis, and rising energy costs have shifted the preference to other means of production. However, where hydrogen demand is not large or inexpensive power is readily available, this process may be justified.

Hydrocarbon Reforming

Also known as the *steam-hydrocarbon process*, hydrocarbon reforming for hydrogen production begins with either commercial propane or natural gas with a very low nitrogen and sulfur content. Hydrogen is produced by the reaction shown in Fig. 13.6. Figure 13.7 shows a schematic diagram of the apparatus in which the hydrocarbon and water vapor react in the presence of a nickel catalyst at temperatures between 800 and 900°C (1500–1600°F) and pressures of 14.8 to 19.7 Atm (217 to 290 psi) to make hydrogen gas. The cooled, reformed gas is then passed through a *shift converter* to produce additional hydrogen by reacting the carbon monoxide with steam (7). High-purity hydrogen (99.5 to 99.999 + %) is obtained with a *pressure swing adsorption* (PSA) unit. The unit consists of four vessels containing a bed of activated alumina, carbon, and molecular sieves. Hydrogen not recovered through the PSA (20 to 25%) is returned to the reformer burners as fuel. Another option for purification is scrubbing the raw hydrogen steam with monoethanolamine (MEA) to remove most of the carbon dioxide. Residual carbon dioxide and carbon monoxide

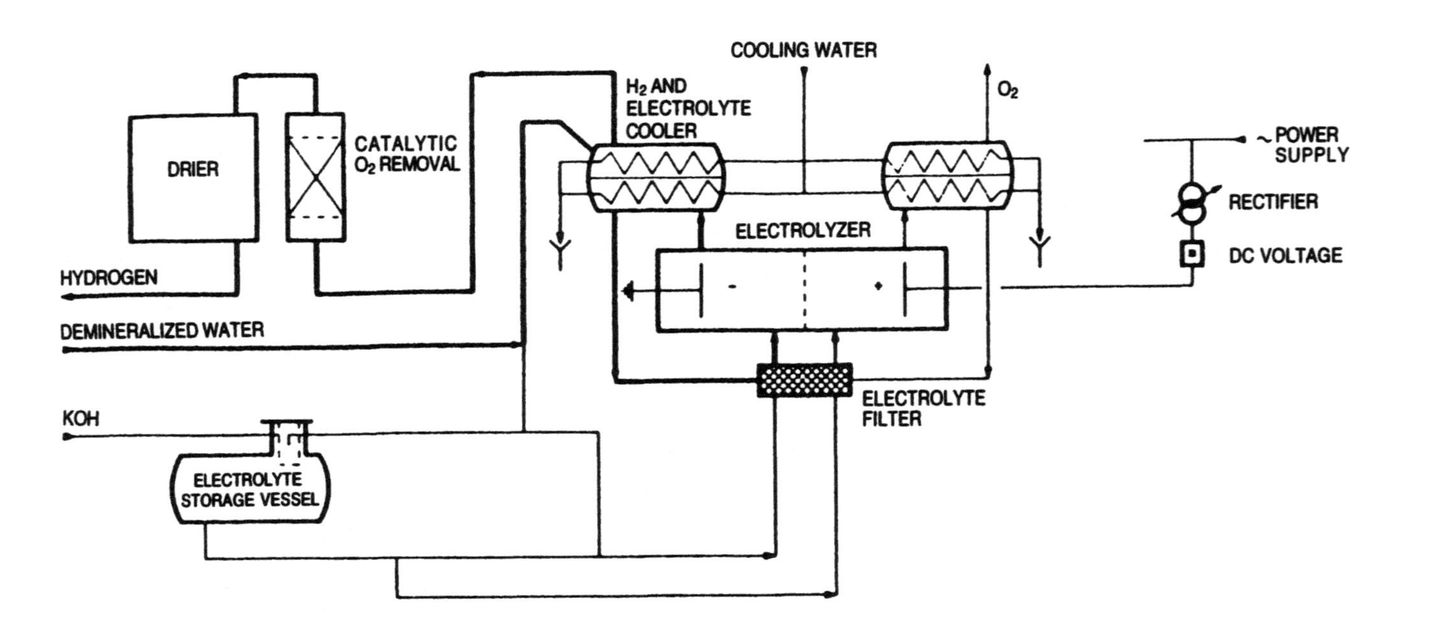

Fig. 13.4. Process flow for electrolytic production of hydrogen. *Source:* Allen, R.R., in *Bailey's Industrial Oil and Fat Products, Vol. 2*, 4th edn., edited by D. Swern, John Wiley and Sons, New York, 1982, p. 90. Reprinted by permission of John Wiley and Sons, Inc.

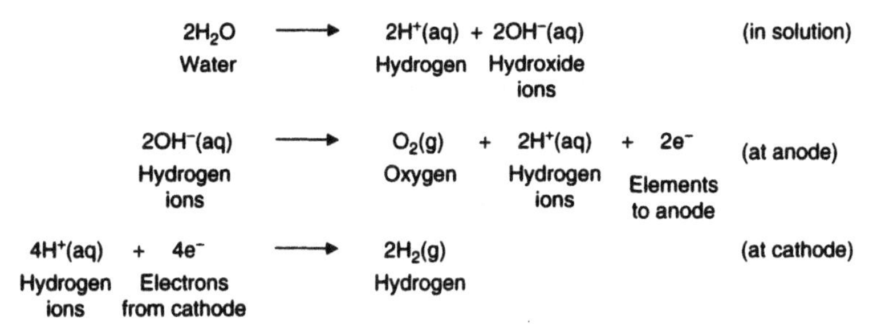

Fig. 13.5. Electrolysis reaction for hydrogen production.

are removed by heating the gas and passing it over a nickel-based methanation catalyst, where the oxides react with hydrogen and form methane (7). Both options are shown in Fig. 13.8.

Purchased Hydrogen

Typical purchased hydrogen purity is 99.999% (7). It is delivered in liquid form and typically stored in one of the ways shown in Fig. 13.9 until used.

Other methods of hydrogen generation include methanol cracking and ammonia dissociation. Both Kuberka et al. (7) and Daum (6) provide excellent discussions of hydrogen supply options, issues, and related costs.

Catalysts

Dry-reduced nickel catalysts are the preferred catalysts for hydrogenation. Prior to their prominence, *wet-reduced* catalysts, from the reduction of nickel formate, suspended in an oil, were used. The oil was then saturated with hydrogen as the nickel was reduced. This resulted in the catalyst being suspended in fully hydrogenated fat. Today, however, most processors purchase active nickel from catalyst suppliers.

Catalyst Activity

Catalyst activity can be viewed as the ability of the catalyst to resist the inhibiting effects of *poisons*. This resistance is related to durability, in one sense, but is not synonymous, because some poisoned catalysts remain very active (8). These poisons and their effects are shown in Table 13.3. Clearly, sulfur poses the biggest common threat to catalyst activity. This is why sulfur-free hydrogen gas is important. However, catalysts are available for special products that are carefully, but intentionally, poisoned with sulfur to increase *trans* isomerization.

$$\underset{\text{Hydrocarbon}}{C_nH_m} + \underset{\text{Water}}{pH_2O} \rightarrow \underset{\text{Hydrogen}}{\left(\frac{m}{2} + n + q\right)H_2} + \underset{\text{Carbon Monoxide}}{(n-q)CO} + \underset{\text{Carbon Dioxide}}{qCO_2} + \underset{\text{Water}}{(p-n-q)H_2O}$$

Fig. 13.6. Hydrocarbon reforming reaction.

TABLE 13.3 Effect of Poisons on Nickel Catalysts (9)

Poison	In the amount of	% Nickel poisoned
Sulfur	1 ppm	0.004
Phosphorus	1 ppm	0.0008
Bromine	1 ppm	0.00125
Nitrogen	1 ppm	0.0014

Evaluation of Hydrogenation Catalyst Activity

AOCS Official Method Tz 1a-78 provides methodology to evaluate catalysts. The method determines the activity of a catalyst relative to a standard catalyst under standard conditions of the test (10).

The rapid acceptance of soybean oil over animal fats in the early 1960s provided an incentive to develop catalysts capable of reducing linolenic acid from 7 or 8% to ≤2% and simultaneously reducing linoleic acid. Further requirements included no appreciable increase in total saturates and minimal *trans* acids (11). Hydrogenation catalysts and their effects on selectivity are reviewed by Gray and Russell (12).

AOCS Official Method Tz 1b-79 determines a catalyst linoleic selectivity, linolenic acid selectivity, or both, either graphically or by computer program capable of solving the kinetic equations of consecutive first-order reactions. It is applicable to all hydrogenation catalysts used to hydrogenate vegetable oils (10).

Other Catalysts

Raney Nickel. Nickel-aluminum alloy (50:50) powder is dissolved in a sodium hydroxide solution to form sodium aluminate and pyrophoric nickel. After washing,

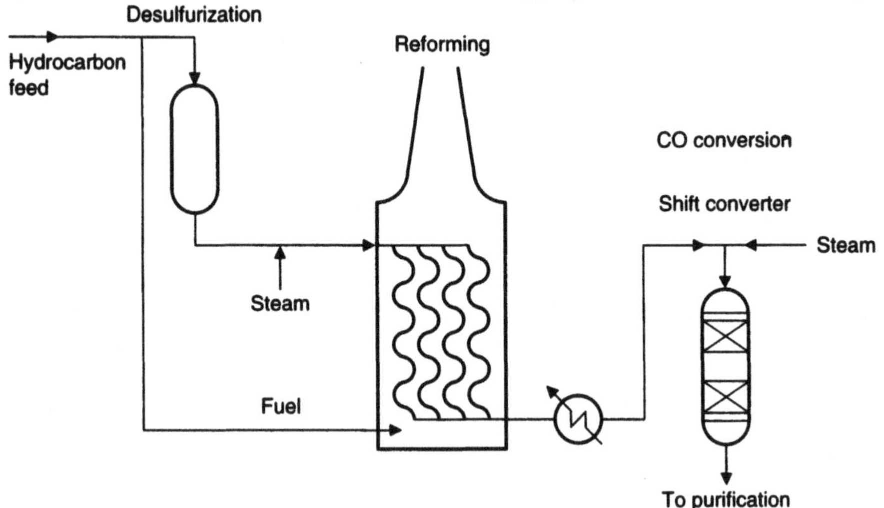

Fig. 13.7. Process flow sheet for hydrocarbon reforming. *Source:* Kuberkz, K.A., et al., *J. Amer. Oil Chem. Soc. 66:* 1289 (1989).

blanketing with an inert liquid, and vacuum drying, the catalyst is ready. Although ease of preparation generated interest in the catalyst, no apparent gain in activity or selectivity was observed (13).

Ziegler Type. Stern et al. (14) patented a hydrogenation process that utilizes a homogeneous "liquid nickel" catalyst. The activities of these catalysts are high, as evidenced by the reduced amounts of catalyst required: 0.005 to 0.009% metal catalyst content. After hydrogenation, water washing removes the catalyst to below maximum acceptable levels (>1 ppm) (15).

Copper-Based Catalysts. Koritala et al. (16) report that selective hydrogenation of soybean oil to reduce linolenic acid is accomplished better with copper than with nickel catalysts. The low activity of copper catalysts at low pressure and the capital requirement for high-pressure batch equipment makes commercial use of these catalysts unattractive.

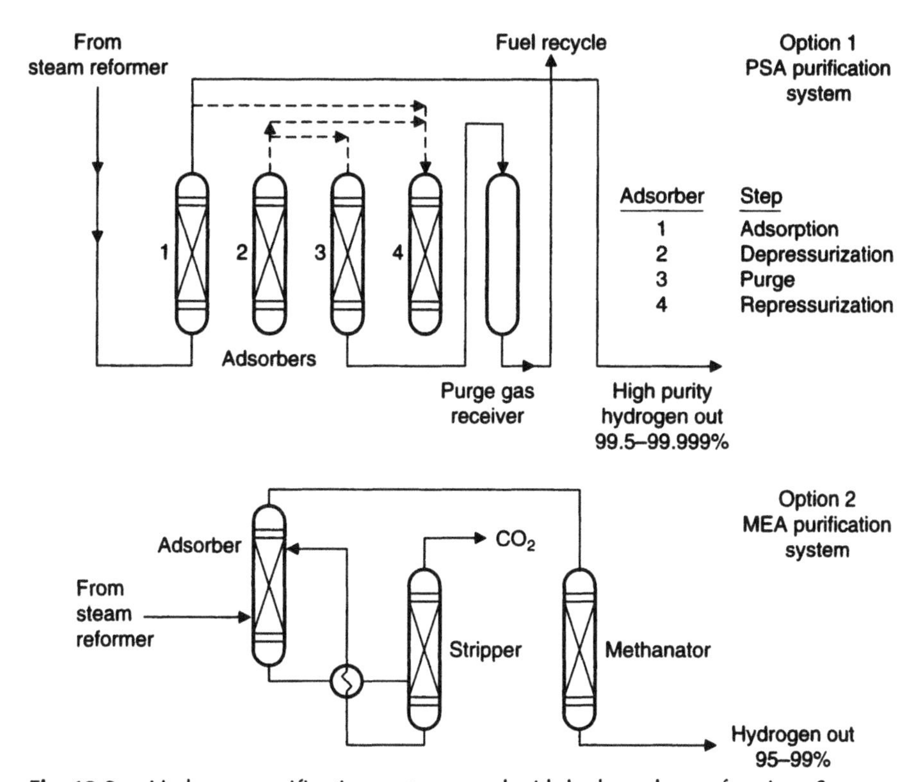

Fig. 13.8. Hydrogen purification systems used with hydrocarbon reforming. *Source:* Kuberkz, K.A., et al., *J. Amer. Oil Chem. Soc. 66:* 1289 (1989).

Hydrogenation Reactors

Batch Production Scale

Due to the nature of the hydrogenation process, by far the most common type of reactor is the batch-type slurry hardener. Primary design considerations for an efficient batch hydrogenation reactor include the following (17):

1. Good contact with the oil and the gas
2. The energy to operate the agitator
3. The hydrogen compressor
4. The recirculation pump
5. Good temperature control

The two major kinds of batch reactors for edible oil hydrogenation are recirculation reactors and "dead-end" reactors.

Recirculation Reactors. In a recirculating-type reactor, fresh hydrogen is introduced at the bottom of the vessel. Heating coils are also located near the bottom. The system captures unreacted hydrogen from the head space, repurifies it, and reintroduces the gas continuously. The system requires a blower or compressor to accom-

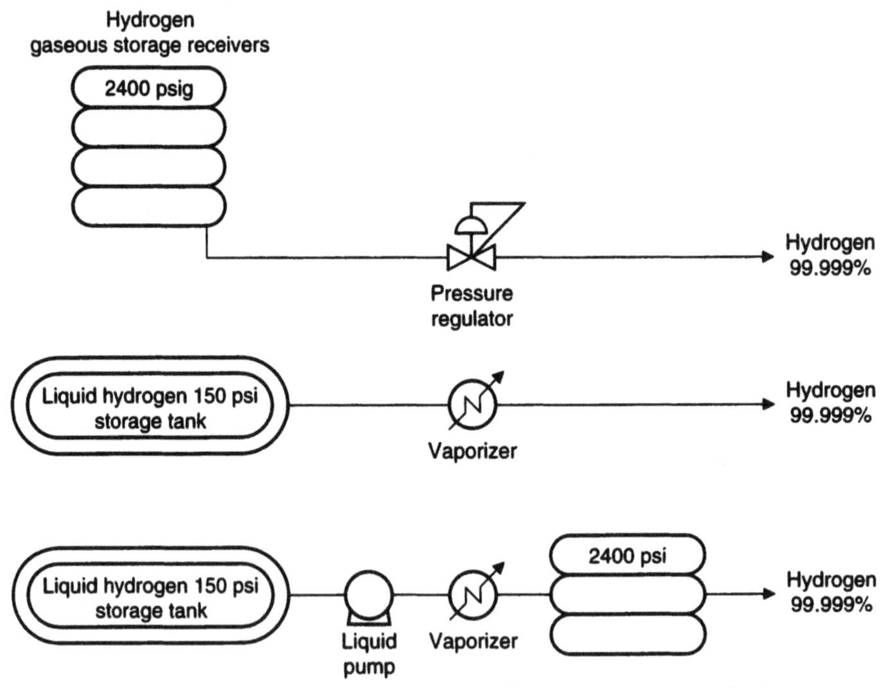

Fig. 13.9. Purchased-hydrogen storage and supply options. *Source:* Kuberkz, K.A., et al., *J. Amer. Oil Chem. Soc. 66:* 1289 (1989).

plish this at a purge rate of 3 to 4% of the hydrogen to avoid accumulation of nitrogen and other impurities. These reactors will process 10,000 to 30,000 kg of oil. A typical batch reactor is shown is shown in Fig. 13.10 (5).

Dead-End Converters. During the initial heating, oils hardened in dead-end converters are first held under vacuum to remove oxygen and moisture. When the oil has achieved the desired temperature, hydrogen is then introduced. Turbine agitators are typically used to draw unreacted hydrogen from the relatively small head space back into (down) the reaction mixture. After reacting, the oil is partially cooled in the reactor prior to removal and filtration.

Dead-end converters have a distinct advantage over recirculation reactors because of the ability to maintain constant reaction temperatures. This allows much tighter process control.

Continuous Production Scale

Continuous reactors tend to lack the utility of the batch process but become practical when very large inventories of single-stock (unblended) products are required.

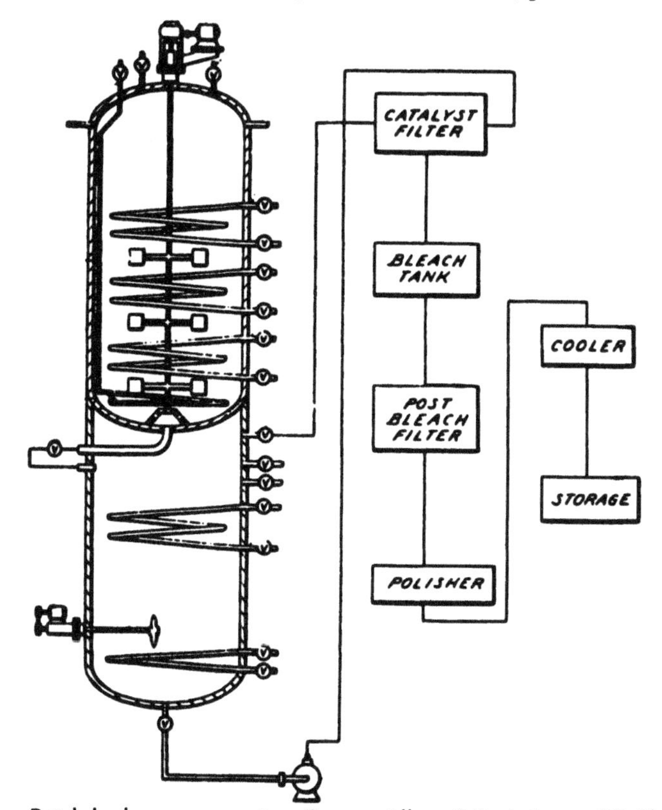

Fig. 13.10. Batch hydrogen apparatus. *Source:* Allen, R.R., *J. Amer. Oil Chem. Soc. 55:* 792 (1978).

For a more detailed treatment of hydrogenation reactors, including discussions on pilot and lab scale (batch and continuous), the reader is strongly encouraged to obtain, the excellent review of the subject by Edvardsson and Irandoust (17).

Aids for Hydrogenation Calculations

Conversion factors for hydrogenation calculations (1) are shown in Table 13.4.

Control of Hydrogenation

Control during hydrogenation is obtained through managing the following variables:

1. Feedstock integrity
2. Catalyst activity and selectivity
3. Catalyst concentration
4. Available hydrogen (pressure)
5. Temperature
6. Agitation

Feedstock Integrity

The importance of clean feedstock to hydrogenation efficiency cannot be stressed enough. Charpentier (18) clearly demonstrates the direct economic advantages when well-refined and bleached soybean oil is used in hydrogenation. The quality of the feedstock going to hydrogenation is shown in Table 13.5; in Table 13.6, Charpentier shows the economic incentive for ensuring that clean feedstock is delivered to the hydrogenation plant; Table 13.7 shows the actual U.S. dollar savings resulting from Charpentier's refining practices.

TABLE 13.4 Aids for Hydrogenation Calculation (1)

One standard cubic foot of H_2	=	0.00532 lb
One liter of H_2	=	0.085 g
To reduce 1000 lb of oil, one IV unit	=	0.0795 lb of H_2
To reduce 1000 lb of oil, one IV unit	=	14.15 ft^3
To reduce 1000 kg of oil, one IV unit	=	0.0795 kg of H_2
To reduce 1000 kg of oil, one IV unit	=	883.3 L of H_2 at STP
One pound of oil reduced by one IV unit produces 1.6–1.7 Btu.		
One kilogram of oil reduced by one IV unit produces 888–943 cal.		

Source: Allen, R.R., in *Bailey's Industrial Oil and Fat Products, Vol. 2,* 4th edn., edited by D. Swern, John Wiley & Sons, New York, Copyright © 1982, pp. 1–95. Reprinted by permission of John Wiley & Sons, Inc.

TABLE 13.5 Hydrogenation Feedstock Quality (18)

Characteristic (Bleached Oil)	Value
FFA	0.05–0.08%
Phosphorus	0 ppm
Soap	0 ppm
Color (Gardner)	4–5
Peroxide value	0–0.2 meq/kg
Iodine value	129–131
Moisture	0%

Catalyst Activity and Selectivity

As mentioned previously, catalyst activity is related to its resistance to poisons. A catalyst with good activity can be used repeatedly. Since the integrity of the catalyst diminishes with each use, the number of times it can be reused is a function of the feedstock integrity (20).

Agitation

Agitation is usually constant. To be sure that agitation is not a factor, the level of oil in the reactor must be kept constant.

Effect of Hydrogenation Conditions

As mentioned earlier in the chapter, the current use of the terms *selective* and *non-selective* describes the conditions of hydrogenation as shown in Table 13.8 (21). Table 13.9 summarizes how increases in different hydrogenation conditions affect reaction selectivity, rate, and *trans* fatty acid formation.

It is the proper management of these parameters that allows the processor maximum flexibility in producing any edible oil product. This is done through an integrated base stock program. Operational efficiency increases when the number of product

TABLE 13.6 Materials Consumption Before and After Implementing Feedstock Quality Controls (18)

	August 1988	August 1990	% Decrease
Hydrogenation:			
Phosphorus in bleached oil	1.2 ppm	0 ppm	100
Soap in bleached oil	15.0 ppm	0 ppm	100
Moisture	0%	0%	—
IV drop	20	20	—
Catalyst (25% Ni)	0.60 kg/MT	0.25 kg/MT	58.3
Filter aid	0.60 kg/MT	0.60 kg/MT	—
Postbleaching:			
Bleaching clay	5 kg/MT	0.60 kg/MT	88
Filter aid	—	0.1 kg/MT	—

TABLE 13.7 Potential Cost Savings Using Clean Hydrogenation Feedstock (19)

Catalyst Savings:		
1988 (7,000 MT/yr) x (0.6 kg/MT)	= 4,200 kg	
1990 (7,000 MT/yr) x (0.25 kg/MT)	= 1,500 kg	
Difference	= 2,700 kgs	
2700 kg x $10	=	$27,000
Postbleach Savings:		
1988 (7,000 MG/yr) x (5 kg/MT)	= 35,000 kg	
1990 (7,000 MT/yr) x (0.6 kg/MT)	= 42,000 kg	
Difference	= 30,800 kg	
(30,800 kgs/1000 kg/MT) x $300/MT	=	$ 9,240
Total Saving from Hydrogenation	=	$36,240

or stock changes is minimized. Therefore, optimum efficiency prohibits formulating hydrogenated products on a per order basis. Individual formulations rarely use exact multiples of hydrogenation batches. This means that a greater number of storage tanks may be required, or the size of hydrogenation batches has to be adjusted.

Troubleshooting

Table 13.10 reflects some of the common problems encountered during hydrogenation (Kuss, G.E., retired, Humko, Memphis, TN, private communication, 1994).

Catalyst Reuse Considerations

In reusing catalysts, the following considerations should be taken into account (Kuss, G.E., retired, Humko, Memphis, TN, private communication, 1994). Catalyst reuse can cause fluctuations in the concentration of available nickel. This can affect control and may cause problems with selectivity (*trans* formation). This may be due to the unavoidable introduction of poisons that are known to promote *trans* formation. Too much reuse (second or third *reuse*) may compromise ingredient statements unless reuse is always done with the same oil. This may be suspected if strange fatty acid profiles suddenly appear for normal base stocks. Spent catalyst can be used to absorb poisons for oils that are consistently difficult to harden. Introducing the spent catalyst into the bleaching clay has been considered.

TABLE 13.8 Selective and Nonselective Hydrogenation Conditions (21)

Processing control	Selective	Nonselective
Temperature	High: 170°C	Low: 120°C
Hydrogen pressure	Low: 1 Atm	High: 3–? ATM
Agitation	Low	High
Catalyst concentration	High: 0.05+ as Ni	Low: 0.02+ as Ni
Trans acids	High	Low
Selective catalyst	Yes	No

TABLE 13.9 Effect of Hydrogenation Conditions

Increase in	Result
Temperature	• Increase in selectivity • Increase in *trans* formation • Increase in hydrogenation rate
Hydrogen pressure	• Decrease in selectivity • Decrease in *trans* formation • Increase in hydrogenation rate
Catalyst concentration	• Increase in selectivity • Increase in *trans* formation • Increase in hydrogenation rate
Agitation	• decrease in selectivity • Decrease in *trans* formation • Increase in hydrogenation rate

TABLE 13.10 Troubleshooting

Problem	Potential cause	Corrective action
Reaction won't start	• Reaction initiation temperature too low ("light-off" temperature)	• Increase oil temperature. • Verify accuracty of process thermometers. • Recalibrate thermometers if necessary.
	• Unusually high acidic material present	• Check free fatty acid. • Dilute with additional feedstock.
Slow reaction rate	• Impurities in H_2 gas	• Increase H_2 purge (bleed rate) to get through the batch[a]. • Depending on H_2 delivery system, check pressure swing absorption (PSA) unit.
	• Impurities in N_2 gas	• Check purity and replace if necessry.
	• Dirty converter	• Remove polymerized oil (usual cause).
	• Moisture	• Check for steam leaks in heating coils.
	• Residual acid (from mineral/organic acid treatments before hydro)	• Check free fatty acid. • Dilute with additional feedstock.
	• Diminished catalyst activity	• Add more fresh catalyst (up to half the amount added initially)[b].
Color[c]	• Nickel soap derivatives	• Rebleach and blend with product that has acceptable color.
	• Oxidation	• Rebleach and blend with product that has acceptable color.
	• N_2 impurities	• NO_x are very detrimental to oil color.
	• Moisture	• Check for steam leaks in heating coils.
	• Unusually high acidic material present	• Check free fatty acid. • Dilute with additional feedstock.
	• Inadequate Ni removal	• Check postbleach.

[a]If increasing the purge does not increase the reaction rate, then add more catalyst as described at bottom of this problem section. If 0.5% addition of unused catalyst also fails, it is improbable that any more will help.
[b]Sulfur-poisoned-type catalysts have inherently slow reaction rates.
[c]Use of neutral bleaching earths plus citric acid can contribute to color problems if citric acid is overused.

Base Stocks and Formulation (21,22)

Optimum operational efficiency is achieved with the development and use of a base stock program. Such a program results in a limited number of hydrogenated products, allowing

1. Full use of hydrogenation capacity
2. A limited number of storage tanks
3. Accommodation of normal variations resulting from day-to-day hydrogenation operations

This helps streamline plant operations, because the scheduling of hydrogenation is dictated by appropriate base stock tank inventories, not by the scheduling based on orders received.

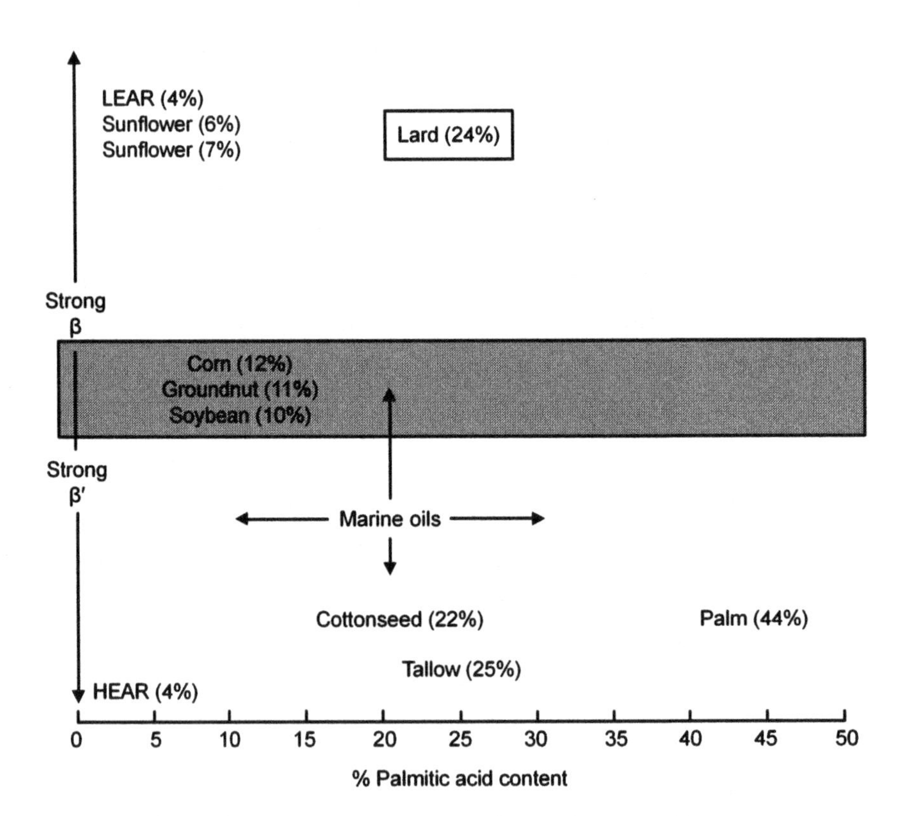

Fig. 13.11. Crystal habit as a function of palmitic acid content. *Source:* Orthoefer, F., *American Oil Chemists' Society Short Course on Science and Technology of Fats and Oils,* American Oil Chemists' Society, Champaign, IL, 1993.

TABLE 13.11 Classification of Fats and Oils According to Crystal Habit (23)

β-type	β′-type
Soybean	Cottonseed
Safflower	Palm
Sunflower	Tallow
Sesame	Herring
Peanut	Menhaden
Corn	Whale
LEAR[a]	HEAR[b]
Olive	
Coconut	Milk fat
Palm kernel	
Lard	Modified lard[c]
Cocoa butter	

[a]Low-erucic-acid rapeseed (canola).
[b]High-erucic-acid rapeseed.
[c]Interesterified lard.

Quality Control in Base Stock Programs

Qualities usually controlled in fat and oil products include consistency, color, color stability, flavor, free fatty acid, peroxide value, (AOM) stability, and, occasionally, a functionality requirement. For the purpose of this chapter it is assumed that the normal requirements will be met; the major remaining concern with base stocks is control of their liquid/solid ratios—solid fat index (SFI), (SCI) or nuclear magnetic resonance (NMR)—over a range of temperatures, or *solids profile*. This amounts to controlling the following physical properties (see Chapter 15):

Solids/liquid ratios (SFI, SFC, NMR)
Melting points
Crystal habits
Consistency
Plastic range

Although melting point and similar-type measurements are helpful, they do not give a complete picture. *Consistency* is a measure of the hardness of the product at a temperature, and the change in this consistency over a temperature range is the *plastic range*. *Crystal habits* are inherent in the fat and oil as shown in Table 13.11. This is somewhat misleading, because there are strong β forms such as LEAR, and strong β′ forms, such as HEAR. For most fats and oils the crystal tendency is as shown in Fig. 13.11 (23).

As a fundamental concept, the more a triglyceride approaches a pure chemical compound, the more it will tend to crystallize in the β form. Therefore, although soybean oil is roughly 10% C_{16} and 90% C_{18} when it is hydrogenated, the formation of *trans* fatty acids of the C_{18} class results in a greater mixture of different fatty acids and thus more divergence from a pure compound. Such hydrogenated soybean oil stocks, when blended together, tend away from strong β formation toward β′ formation.

The key quality control tool for base stock programs is determination of liquid/solid ratios. If dilatometry is used, the value obtained is SFI or SCI. If NMR is used, the result is a direct measure of liquid/solid fats present. Both of these give solid/liquid ratios over a temperature range. Melting point, plastic range, and consistency are all related to SFI. SFI and NMR, however, are too time-consuming for routine control of hydrogenation. Other, more rapid, methods are required that, in turn, can be related to SFI under the particular hydrogenation reaction parameters being employed. Refractive index and congeal points are used most. Measurement of the amount of hydrogen uptake may also be used.

Some variation in degree of hydrogenation may be adjusted through judicious blending of the various base stocks.

Characterization of the Hydrogenation Reactor

It is axiomatic that hydrogenation curves for various oils be developed for any particular type hydrogenation apparatus. Such curves should be developed based on the type of catalyst (selective/nonselective and activity), temperature, and hydrogen pressure. The curves developed should be reproducible under similar conditions. Any deviation from then under careful control of reaction conditions will alert the operator to any developing problems.

If a laboratory or pilot plant reactor is used to test catalysts, develop new products, or test new sources of oils, the lab equipment should give results that are transferable to plant hydrogenation.

TABLE 13.12 Typical Base Stock Program for Soybean Oil

	Shortening bases			Margarine bases		Shortening or margarine base
Base stock number[a]	1	2	3	4	5	6
Hydrogenation conditions						
Initial temperature °C	150	150	150	150–163	150–163	140
Hydrogenation temperature °C	165	165	165	218	218	140
Pressure (atm)	1.0	1.0	1.0	0.3	0.6	2.7
Catalyst concentrate (% nickel)[b]	0.02	0.02	0.02	0.05[b]	0.02[c]	0.02
Final iodine value[d]	83–86	80–82	70–72	64–68	73–76	104–106
Final congeal point °C	—	—	25.5–26.0	33.0–33.5	24.0–24.5	—
SFI						
10.0°C	16–18	19–21	40–43	58–61	36–38	4 max.
21.1°C	7–9	11–13	27–29	42–46	19–21	2 max.
33.3°C	—	—	9–11	21 max.	2.0 max.	—

[a]Properly refined and bleached to remove soap, phosphatides, and peroxides.
[b]Based on weight of oil.
[c]Very selective catalyst.
[d]Final IV approximate.

TABLE 13.13 Formulation for Shortenings

	Required SFI for shortening				Formulation	
Shortening type	10.0°C	21.1°C	33.3°C	40°C	Base stock	%
All-purpose #1[a]	22–24	18–20	13–15	10–12	#1	88–99
					Hardfat[b]	11–12
All-purpose #2[a]	24–27	18–20	12–14	6– 8	#2	92–93
					Hardfat[b]	7– 8
Frying (heavy-duty)	41–44	29–30	12–14	2– 5	#2	97
					Hardfat[b]	3
Frying (fluid)	5– 6	3– 4	2– 3	1– 2	#6	98
					Hardfat[c]	2
Specialty (nondairy)	43–47	27–30	6– 9	1– 5	#4	30± 5
					%5	70± 5

[a]Emulsified shortenings made by adding proper type and amount of emulsifier.
[b]β'hardfat (1–8 IV).
[c]βhardfat (1–8 IV).

Typical Base Stock Program for Soybean Oil

To make the largest variety of products while keeping the number of base stocks as low as possible requires only six hydrogenated soybean oil stocks. Suggested hydrogenation conditions and appropriate analyses are shown in Table 13.12.

The most difficult base stock to produce is #4, which requires a very selective catalyst and selective conditions. Base stock #6 is nonselective, because above IV 100, very few saturates are formed; hence *trans* formation can be suppressed. Also, the hydrogen pressure should be the highest attainable in the equipment being used.

The base stocks produced in Table 13.12 are used to formulate shortenings and margarines as shown in Tables 13.13 and 13.14.

The contribution to solids at any given temperature of each base stock can be calculated algebraically from SFIs, as shown in the following example (24). This method is used on products where two or more hydrogenated stocks (not liquid oil or hardfat) are blended.

TABLE 13.14 Formulation for Margarines[a]

	Required SFI for blend				Formulation	
Margarines	10.0°C	21.1°C	33.3°C	40°C	Base stock	%
Soft stick	20–26	13–17	1.4–3.0	0	#4	50±5
					Unhydrogenated soybean oil	50±5
Regular stick	30.0 max.	17.5 min.	3.5 max.	0	#4	38±5
					#5	20±5
					#6	42±5
Tub	12–13	7–8	3 max.	0	#3	20–30
					Unhydrogenated soybean oil	70–80

[a]These formulations are all soybean oil suitable for refrigerated product. For maximum resistance to thermal shock, use 5 to 10% β' fats with SFI profile similar to base stock #4.

The product specifications are given in Table 13.15, and the calculations are as follows:

10.0°C: $50.8x + 30.2\,(1 - x) = 41.0$ (midpoint of range 39 to 43)
$50.8x + 30.2 - 30.2x = 41.0$
$20.6x = 10.8$
$x = 0.52$ Blend 52% IV = 70 with 48% IV = 80

21.1°C: $37.1x + 14.8(1 - x) = 24.0$ (midpoint of range 22 to 26)
$37.1x + 14.8 - 14.8x = 24.0$
$22.3x = 9.2$ Blend 41% IV = 70 with 50% IV = 80
$x = 0.41$

33.3°C: $13.2x + 0.2(1 - x) = 5.2$ (midpoint of range 3.5 to 7.0)
$13.2x + 0.2 - 0.2x = 5.2$
$13x = 5.0$
$x = 0.38$ Blend 38% IV= 70 with 62% IV = 80

40.0°C: Unnecessary, because of 2 maximum specification.

Average the percent composition for all temperatures:

IV = 70: $(52 + 41 + 38)/3 = 43.7 \approx 44\%$ IV = 80: $100 - 44 = 56\%$

Calculate the SFI from a 44/56 blend and check against desired finished product specification:

10.0°C: $0.44(50.8) + 0.56(30.2) = 39.3$ (39–43)
21.1°C: $0.44(37.1) + 0.56(14.8) = 24.6$ (22–26)
33.3°C: $0.44(13.2) + 0.56(0.2) = 5.9$ (3.5–7.0)
40.0°C: $0.44(0.9) + 0.56(0) = 0.4$ (2 max.)

In this case, it is advisable to address the close tolerance at 10.0°C by increasing the IV = 70 stock and recalculate the SFI.

Specialty Products

In addition to the array of possible products shown in Tables 13.13 and 13.14, there will be the occasional request for a product not made routinely. In some cases, it may be necessary to require special hydrogenation and blending outside of those already discussed, but quite often the new request can be made by blending existing base stocks.

Formulation of Specific Shortening Products

The basic principle of base stock formulations has been presented. A further consideration for some types of shortening is the addition of emulsifiers. Also, control of

TABLE 13.15 Specifications for Base Stock Solids Example

°C	IV = 80 SFI	IV = 70 SFI	Desired finished product specification
10.0	30.2	50.8	39–43
21.1	14.8	37.1	22–26
33.3	0.2	13.2	3.5–7.0
40.0	0.0	0.9	2 max.

crystallization and tempering procedures is sometimes as important as the basic formulation (see Chapters 15 and 20).

All-Purpose Shortening, Nonemulsified. Nonemulsified all-purpose shortening may be compounded as shown in Table 13.13. The suggested crystallization and tempering procedure would be as follows:

"A" unit exit temperature: 18 to 21°C (65 to 70°F)
"B" unit exit temperature: 27 to 29°C (80 to 85°F)
Tempering conditions: 27°C (80°F) for 24 to 28 h

This shortening is suitable for both baking and frying.

All-Purpose Shortening, Emulsified. Compounding, crystallization and tempering conditions would be the same as those given for all-purpose shortening, nonemulsified.

Cake/Bread Shortening. A hard monoglyceride is added at a level of 2 to 3% (calculated as α-monoglyceride).

Icing Shortening. A soft monoglyceride is added at a level of 2 to 3% (calculated as α-monoglyceride). This product would be good for both cream icing and enrobing icing formulation.

All Purpose (Cake/Icing) Shortening. This product would represent a compromise between cake/bread shortening and icing shortening and would use an intermediate-hardness monoglyceride.

Specialty Cake Shortening, Extra Moist. For specialty cakes that are to be moist, liquid soybean may be added to the base formulation at levels of 20 to 30% along with 4 to 5% propylene glycol monoester and 2 to 3% monoglyceride of intermediate hardness. This formulation would also be very suitable for pound cakes, where dryness or moistness may be controlled by adjusting the level added soybean oil.

Frying Fats

The nonemulsified all-purpose shortenings described above are also suitable for most frying applications.

Fluid/Pourable Frying Fat. This type of product is popular in the United States because of convenience in handling and suitability for heavy-duty frying. It is made by combining base stock #6 with about 2% fully hydrogenated soybean oil. The product requires a β hardfat to maintain fluidity. The product will be pourable down to about 15°C (60°F), and if it should go to lower temperatures it will thaw back to its original physical state (pourable) at room temperature or above.

Extra Heavy-duty Frying Fat. This product is described in Table 13.13. Crystallization and tempering conditions are not critical, and only those conditions that give a good appearance in the package are needed. In both of the above frying shortenings it is common to add an antifoam (2 to 5 ppm dimethylpolysiloxane, for example) and antioxidants.

Note: Antifoams should *not* be added to all-purpose shortenings. If an all-purpose shortening with an antifoam is used for baking, the batter may collapse.

Formulation of Margarine

Table margarine formulations are shown in Table 13.14. Margarines for baking may be made following shortening formulations using table margarine formulations.

References

1. Allen, R.R., in *Bailey's Industrial Oil and Fat Products, Vol. 2*, 4th edn., edited by D. Swern, John Wiley & Sons, New York, 1982, pp. 1–95.
2. Richardson, A.S., C.A. Knuth, and C.H. Milligan, *Ind. Eng. Chem. 16:* 519–522 (1924).
3. Beckmann, H.J., *J. Amer. Oil Chem. Soc. 60:* 234A (1983).
4. Weiss, T.J., *Food Oils and Their Uses*, 2nd edn., Avi Publishing Co., Westport, CN, 1983, p. 77.
5. Allen, R.R., *J. Amer. Oil Chem. Soc. 55:* 792 (1978).
6. Daum, P., *INFORM 4:* 1394 (1993).
7. Kuberka, K.A., M.K. Weise, and C.A. Messina, *J. Amer. Oil Chem. Soc. 66:* 1289 (1989).
8. Patterson, H.B.W., in *Hydrogenation of Fats and Oils*, Applied Science Publishers, London and New York, 1983, pp. 114–115.
9. Beckmann, H.J., *Harshaw Technical Bulletin* (1983).
10. *Official and Tentative Methods of the American Oil Chemists' Society*, American Oil Chemists' Society, Champaign, IL, 1989.
11. Patterson, H.B.W., in *Hydrogenation of Fats and Oils*, Applied Science Publishers, London and New York, 1983, p. 119.
12. Gray, J.I., and L.F. Russell, *J. Amer. Oil Chem. Soc. 565:* 34 (1979).
13. Patterson, H.B.W., in *Hydrogenation of Fats and Oils*, Applied Science Publishers, London and New York, 1983, p. 120.
14. Stern, R., et al., U.S. Patent 4,038,295 (1977).
15. Mounts, T.L., in *Handbook of Soy Oil Processing and Utilization*, edited by D.R. Erickson, et al., American Oil Chemists' Society, Champaign, IL, 1980, p. 142.
16. Koritala, S., K.J. Moulton Sr., J.P. Friedrich, E.N. Frankel, and W.F. Kwolek, *J. Amer. Oil Chem. Soc. 61:* 909 (1984).
17. Edvardson, J., and S. Irandoust, *J. Amer. Oil Chem. Soc. 71:* 235 (1994).
18. Charpentier, R., *INFORM 2:* 208 (1991).
19. Erickson, D.R., and M.D Erickson, in *Proceedings from the 4th Latin American Congress on Fats and Oils Processing*, Associación Argentina de Grasas y Aceites (ASAGA), Buenos Aires, Argentina, November 1992, pp. 33–45.
20. Mounts, T.L., in *Handbook of Soy Oil Processing and Utilization*, edited by D.R. Erickson et al., American Oil Chemists' Society, Champaign, IL, 1980, p. 139.
21. Erickson, D.R., in *American Oil Chemists' Society Fundamental Short Course on Frying Fats*, American Oil Chemists' Society, Champaign, IL, 1993.
22. Latondress, E.G., *J. Amer. Oil Chem. Soc. 58:* 185 (1981).
23. Wiedermann, L.H., *J. Amer. Oil Chem. Soc. 55:* 823 (1978).
24. Orthoefer, F., in *American Oil Chemists' Society Short Course on Science and Technology of Fats and Oils*, American Oil Chemists' Society, Champaign, IL, 1993.

Chapter 14

Deodorization

Calvin T. Zehnder

5502 Hidden Road
Louisville, KY 40291

Introduction

The role of deodorization in processing and refining of edible fats and oils has long been accepted as the last step in preparing the oil for use as an ingredient in margarine, shortening, salad oil, cooking oil, hard butters for the confectionery industry, and many other products in the food industry. The finished, deodorized oil can be classified as "acceptable" over a wide range of specifications that depend on the market likes and dislikes of the people within the particular country or region in which the products will be consumed. Some areas will prefer the oil to have a more pronounced flavor and odor; others will want a more characteristic oil color. Judicious advertising can alter the preferences of the individual; the majority of the world, however, desires that the oil be essentially bland-tasting, colorless or as water-white as possible, odorless, and with a free fatty acid (FFA) content of less than 0.03%. Further, this oil would carry a Flavor Score of 7.5 or above when taste-tested in accordance with the American Oil Chemists' Society (AOCS) accepted panel of 10 qualified oil flavor and odor testers. Unfortunately, there are presently no scientific methods for testing the oil organoleptically; therefore, this most critical evaluation remains a human subjective test.

It is important to state that all processing prior to deodorization must be performed with extreme care to ensure quality product. Deodorization, as the final refining process, cannot correct improperly caustic-refined oil, bleached oil with high peroxide value or residual bleaching earth, or hydrogenated oil containing residual colloidal nickel. A quality product must receive quality care in all stages of its processing (see Chapter 5).

Deodorization Theory

Deodorization is a steam stripping process wherein a good-quality steam, generated from deaerated and properly treated feedwater, is injected into soybean oil under low absolute pressure and sufficiently high temperature to vaporize the FFA and odoriferous compounds and carry these volatiles away from the feedstock. The detailed theory and mathematical equations are not included in this chapter but can be reviewed by referring to Bailey (1), Bates (2), Gavin (3) and Zehnder and McMichael (4).

Deodorization Operating Conditions

Table 14.1 presents the current ranges of operating conditions used by refiners in producing high-quality soybean oils. Each of the operating variables will be reviewed as to its effect on the deodorization process.

Absolute Pressure

Years of experience have proven that pressure has no measurable affect on quality when the deodorizer is operated at absolute pressures within the range of 1 to 6 mm Hg. It has been proven that consistent operation above 6 mm Hg, even in the range of 7 to 9 mm Hg, will result in such quality problems as off-flavors and odors. At the usual deodorization temperature, the vapor pressures of the volatile ingredients (aldehydes, ketones, alcohols, hydrocarbons, and other organic compounds) are sufficient to be unaffected by pressure below 6 mm Hg.

The majority of refiners have chosen to operate deodorizers at 3 mm Hg. This choice primarily relates to purchasing an insurance policy. Most refineries, unfortunately, are faced with fluctuating steam pressures and steam jet ejector vacuum equipment that is very intolerant of steam pressure below design pressure; fluctuations above 3 mm Hg are not as serious as fluctuations above 6 mm Hg. Also, the capital cost and operating cost for a 3 mm Hg system are not significantly more than for a 6 mm Hg ejector. Once an operating pressure is chosen and the ejector equipment selected for a refinery, this operating condition is no longer a variable for a specific deodorizer load of stripping steam and distillate (volatiles and neutral oil).

Temperature

Deodorizing temperature is normally the operating condition that can truly be classified as a variable. It directly affects the vapor pressure of the volatile constituents to be removed; therefore, increasing or decreasing the temperature produces a corresponding higher or lower rate of removal for the odoriferous compounds. In addition to establishing the desirable vapor pressure, temperature is the prime factor in the thermal decomposition of the carotenoid pigments, resulting in what is commonly referred to as "heat bleaching" of soybean oil.

Even though temperature is the primary processing variable, its relationship to "holding" or deodorizing time must not be overlooked. When contemplating fairly wide ranges in temperature, it is important to remember that the higher the deodorizing temperature, the shorter the holding time, and vice versa. During the deodorization process, many desirable reactions are taking place to achieve a quality end

TABLE 14.1 Current Deodorization Conditions for Soybean Oil

Absolute pressure	1–6 mm Hg
Temperature	(485–510°F) 252–266°C
Holding time at temperature	15–60 min
Stripping steam as wt. % of oil	1–3%

product, but some undesirable reactions, such as fat splitting and polymerization, are also occurring. Therefore, it is strongly recommended that the deodorization temperature be no higher than necessary to obtain the required oil quality specifications.

The following outline is presented to assist in evaluating the general effects of temperature on the deodorization process for soybean oil. Increasing temperature:

- More rapidly removes flavor and odoriferous compounds.
- Produces a lighter-colored oil due to greater decomposition of carotenoid pigments.
- Removes more of the sterols and tocopherols (operation at 260°C (500°F) will result in removal of 40 to 50% of these compounds).
- Permits some increase in deodorizer throughput capacity by allowing shorter holding time.
- Increases neutral oil losses due to greater fat splitting and fatty acid generation.
- Increases energy consumption (wasteful if temperature is higher than required for acceptable product quality).

Deodorizing ("Holding") Time

Deodorization holding time is defined as the time during which the feedstock is at deodorizing temperature and subjected to the designed stripping steam flow rate. This does not normally include the time of heating and cooling the oil. This holding time can vary with various equipment manufacturers and reflects the efficiency of the stripping mechanisms used in these deodorizers. Stripping efficiency relates to the ability of the mechanism to mix the steam and the oil intimately. Current available equipment requires holding times in the range of 15 to 60 min. These times are required for deodorizing feedstocks that have been chemically (caustic) refined and contain FFA of less than 1%. If the feedstocks contain higher levels of FFA, such as 1.5 to 3% or higher (when physical refining in the deodorizer), the holding times will increase to as much as double that required for the chemically refined stocks. To better understand the relationship between stripping steam apparatus and holding time, the following schematics and figures are presented:

- Fig. 14.1 depicts the most basic sparger design, which consists of pipes containing drilled holes for admitting the steam into the oil. The pipes are located in the lower part of the deodorizing compartment and evenly spaced across it. Sufficient holes are drilled to distribute the steam effectively throughout the oil and direct it toward the bottom to roll the oil upward, thus eliminating any dead (undisturbed) pockets of oil. Deodorizers with these or similar spargers, including bubble cap spargers, will usually require 45 to 60 min holding time.
- Fig. 14.2 depicts a sparger design developed by Votator (5) (S.A. Extraction DeSmet, Atlanta, GA) and known as a mammoth pump. This sparger achieves a very high degree of steam-oil intermixing and thus is an efficient stripper. This invention, coupled with about a 5% increase in deodorization temperature, resulted in 15 min or less holding time. Steam exits the steam chest through the

sparger holes (all located below the quiescent oil level) and is intimately mixed with the oil contained in the annulus (ring) area around the steam chest. The hydraulics are such that the reduced-density mixture of steam and oil rapidly rises (explodes) upward, strikes the contoured cover, and is deflected downward in a spray pattern onto the pool of oil in the compartment. The abrupt changes in direction and speed further add to the intermixing and permit the lighter vapors to separate from the stream and exit the deodorizing area. Oil from the outside compartment, in turn, enters the steam-oil mixing chamber by flowing under the annulus area wall. Thus a violent pumping action is created. Instead of the single mammoth pump, some deodorizers are equipped with similar multiple small steam-oil mixers in the deodorizing compartments.

- Fig. 14.3 depicts a packed column deodorizer, which utilizes a column filled with structured stainless steel packing. Oil enters at the top and flows by gravity through the packing; stripping steam injected into the bottom passes upward and intimately mixes with the oil. This has been claimed to be the most efficient strip-

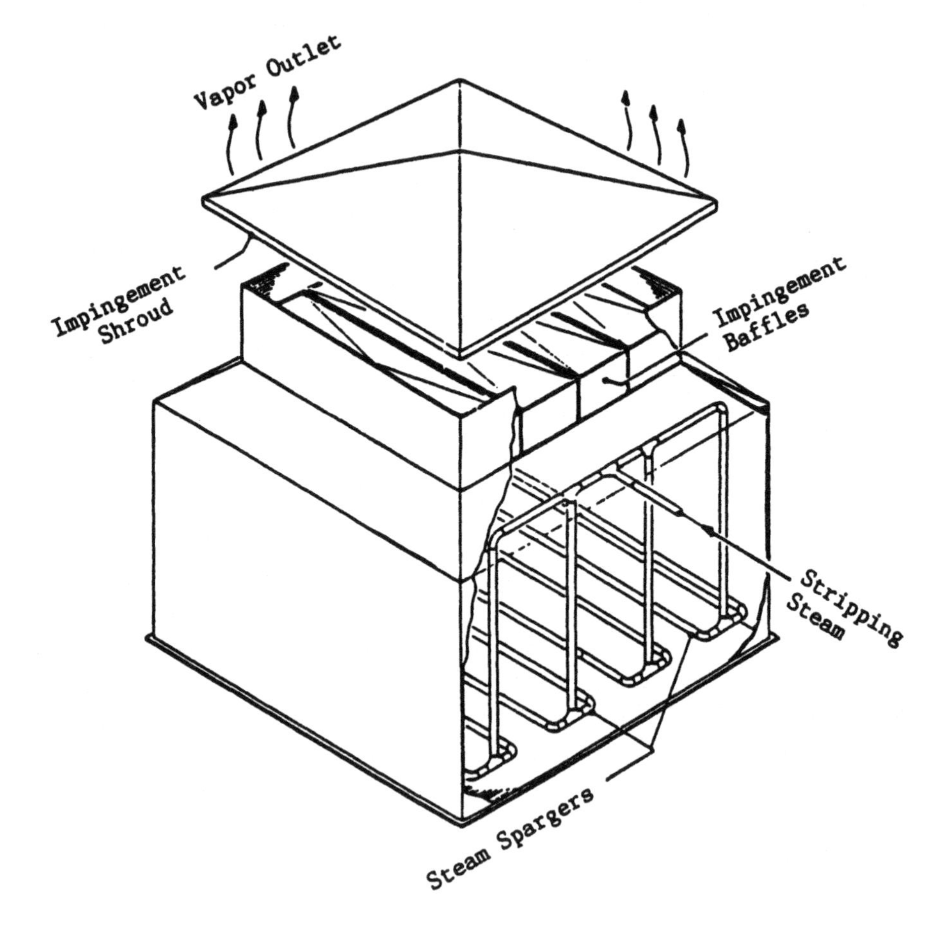

Fig. 14.1. Basic vapor-liquid contacting spargers.

ping steam apparatus on the current market. This equipment has been most successful in deacidifying plants, such as physical refining of palm oil, and usually requires the addition of holding chambers (especially to achieve the heat bleaching of soybean oil), because the residence time in the packing chamber is relatively short (unless the deodorizer is extremely tall). Holding time is comparable to that of the mammoth pump but offers some savings in stripping steam quantity.

Stripping Steam

Stripping steam is the primary motive force in the deodorization process in that it is the carrier medium for moving the vaporized FFA, ketones, aldehydes, and other volatiles from the feedstock to the vacuum ejector and distillate recovery system. Raoult's law shows that steam volume rather than weight is important for the steam stripping process. Therefore, in discussing the quantity or percentage of stripping steam required for a particular deodorizer, it is necessary to relate this to the system absolute pressure. For many years, deodorizers were designed for 3% (by weight) at a system pressure of 6 mm Hg absolute. As stated in the discussion of absolute pressure, the industry change to lower operating pressures results in reduced percentages of steam, varying in direct ratio to the absolute pressure. Thus, 3% steam at 6 mm Hg has the same volume as 1.5% steam at 3 mm Hg. Again, the particular deodorizer design and efficiency can affect the stripping steam quantity required. Current systems being offered will require stripping steam in the range of 0.6 to 1.2% of the oil throughput at an operating pressure of 3 mm Hg absolute.

In addition to quantity, two additional characteristics of stripping steam are purity and quality.

Purity relates to the water used for generating the steam, which in all cases should be treated and deaerated in accordance with good steam boiler/generator prac-

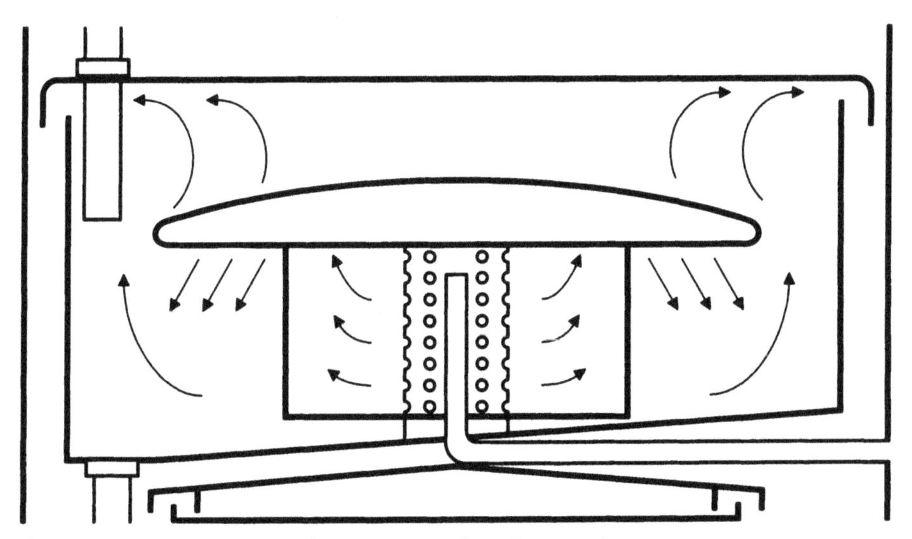

Fig. 14.2. Votator mammoth pump vapor–liquid contacting sparger.

tices. Any dissolved air in the steam will seriously degrade the oil quality at the elevated deodorization temperatures.

Quality relates to boiler industry standards for dry steam (that is, free of unvaporized water or condensate). Stripping steam entering the deodorizer should be dry with a quality of 98% or better. Wet steam can be a major cause of higher than normal deodorization (losses of neutral oil from the product) due to the increased volume of water versus vapor expanding to the reduced absolute pressure.

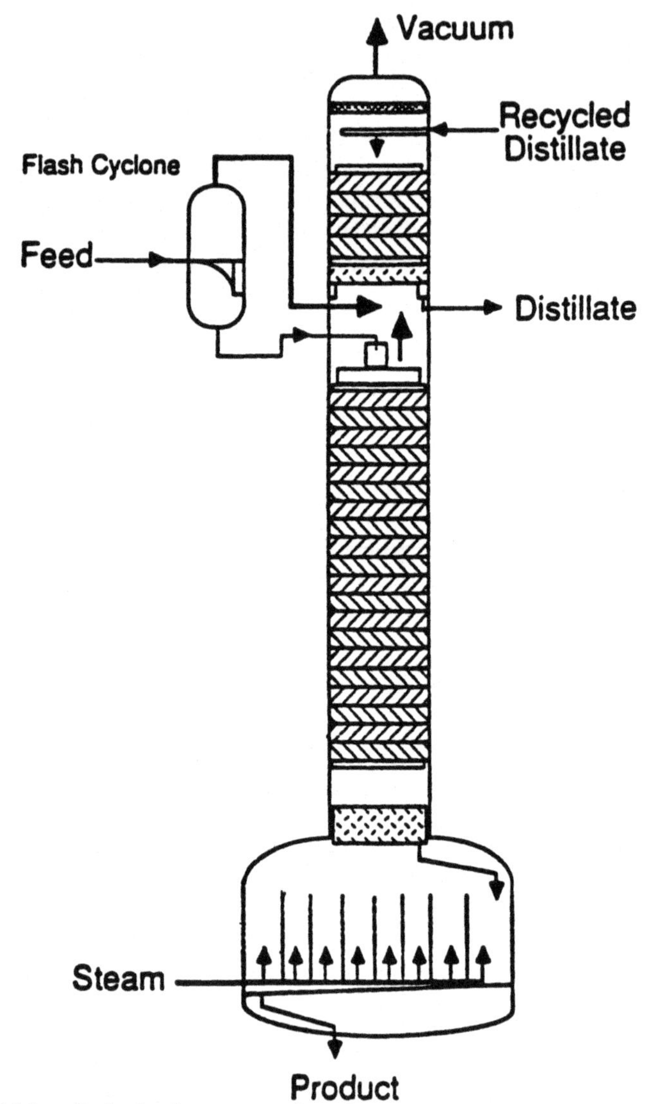

Fig. 14.3. Packed-column vapor-liquid contacting stripper. *Source:* Johnson-Loft Division, Tetra-Laval Foods, Novato, CA.

Feedstock Quality

To help ensure that the deodorized product will be of the highest quality, it is imperative that the feedstock likewise be of the best quality. All steps of prior handling and processing, from harvesting to deodorizer feedstock, should be properly executed. Mishandling or shortcut processing at any step can result in quality problems and even permanently damaged feedstock (see Chapter 5).

Good quality soybean feedstock can be defined as well-refined and bleached oil; free of soap, bleaching clay, or residual catalyst (if hydrogenated); FFA of less than 0.1%; phosphorus of less than 1 ppm; and iron less than 0.1 ppm.

To achieve these specifications, chemical (caustic) refining provides the most positive and consistent method of refining. The deodorization system can be utilized as a part of the physical refining process, but it is very critical that the pretreatment be thoroughly executed to attempt to achieve less than 1 ppm phosphorus and 0.1 ppm iron. Physically refining deodorizers are required to handle considerably higher levels of FFA. This usually requires longer holding times, with a resultant reduction in capacity and the need for considering higher-grade stainless steel to withstand the more corrosive vapors. Physical refining of soybean oil is further discussed in Chapter 11 but here it will be emphasized that the fats and oils industry, particularly in the United States, has chosen not to physically refine soybean oil. Chemical (caustic) refining is much more forgiving in handling the variations in feedstocks that result from soil conditions, weather extremes, harvesting and storage conditions, crushing-expeller-extraction variations, and so forth.

Deodorization Losses

Oil losses during deodorization can be classified into two categories: chemical and mechanical. *Chemical losses* consist of the removal of the undesirable components: FFA, aldehydes, ketones, peroxides, polymers, and other volatiles. Since all but FFA are trace amounts, the quantitative losses can be stated as the reduction in FFA; for example, 0.1% in feed and 0.03% in product results in 0.07% chemical loss. *Mechanical losses* are those associated with carryover in the stripping steam (entrainment). The entrainment quantity for a specific deodorizer is a factor of the amount of stripping steam required and the efficiency of the entrainment separators, such as baffles and demisters. Since no separator is 100% efficient, the lower the stripping steam rate, the lower the loss, as each quantum of steam will entrain some minimum amount of neutral oil. All present deodorizers encompass state-of-the-art separators/demisters, and as a result the losses are expected to be in the range of 0.15 to 0.25%. It should be pointed out that determination of the losses for any unit process is very difficult in refineries, since facilities are not available for weighing in and out. Some refiners use flow meters, but it should be remembered that the accuracy of many liquid flow meters is in the range of 1 to 3%, which could be outside the total deodorization loss.

Deodorization Equipment Components

Materials of Construction

The material of choice for commercial deodorizers operating above 150°C (300°F) is stainless steel in contact with the product. Type 304 stainless has proven satisfactory for caustic refined oils; however, if physical refining is anticipated, type 316 stainless is recommended because of its greater acid resistance. At temperatures below 120°C (250°F), all piping and ancillary equipment may be of carbon steel. It must be remembered that copper or copper-bearing alloys must not be used in contact with edible oils.

High Temperature Heating Media

For about 50 years, deodorization has required a high-temperature heating medium in excess of 270°C (520°F). Standard steam boilers are unable to provide these temperatures with saturated steam; therefore, the edible oils industry standardized on Dowtherm "A" (a diphenol compound) (Dow Chemical Co.,) or mineral oil systems. Dowtherm "A" vapor systems are capable of providing up to 315°C (600°F) at a pressure of about 3.2 bar (46 psig), and in the event of a leak from the coil into the product, the Dowtherm will strip out of the oil. The use of liquid mineral oil is not as popular, because it will not strip out. In the early 1970s the industry became concerned about product contamination with organic chemicals in food processing. Even though Dowtherm "A" strips out of the oil at deodorization temperatures, marketing pressures and competition have the industry shifting to high-pressure steam generators, which require about 80 bar (1150 psig) to provide a temperature of 295°C (560°F). High-pressure steam affords savings in heat transfer due to improved heat transfer coefficient over Dowtherm, but the equipment cost is about twice that for Dowtherm. Fig. 14.4 depicts one such high-pressure natural circulation steam generator.

Deodorizer Vacuum Equipment

Steam jet ejector vacuum systems are used to maintain the absolute pressure for deodorization. For operation at 3 mm Hg, the system usually consists of four steam ejectors and two interstage condensers. The ejector system will handle high-molecular-weight materials such as fatty acids (approximately 280 MW) more readily than steam/water (18 MW). The sizing of ejector systems is determined by the motive steam pressure and condensing water temperature available, stripping steam load, fat and oil load, and air (noncondensable) load. In general, the system will require motive (operating) steam at a rate of 5 to 7 times the stripping steam rate. Systems with barometric condensers (direct-contact water condensing) are the most economical, but they greatly contribute to both air and water environmental pollution problems. The fatty material is condensed in the water, which is circulated through a cooling tower or discharged into sewers or streams (rivers). In the cooling tower the fat becomes an air (odor) pollutant, whereas in the sewer or stream it is a water pollutant. A solution to the problem is to use indirect condensing (cooling), in either a surface condenser or a plate heat exchanger. The surface condenser is supplied with clean cooling water or, in some systems, refrigerated water. To assist in

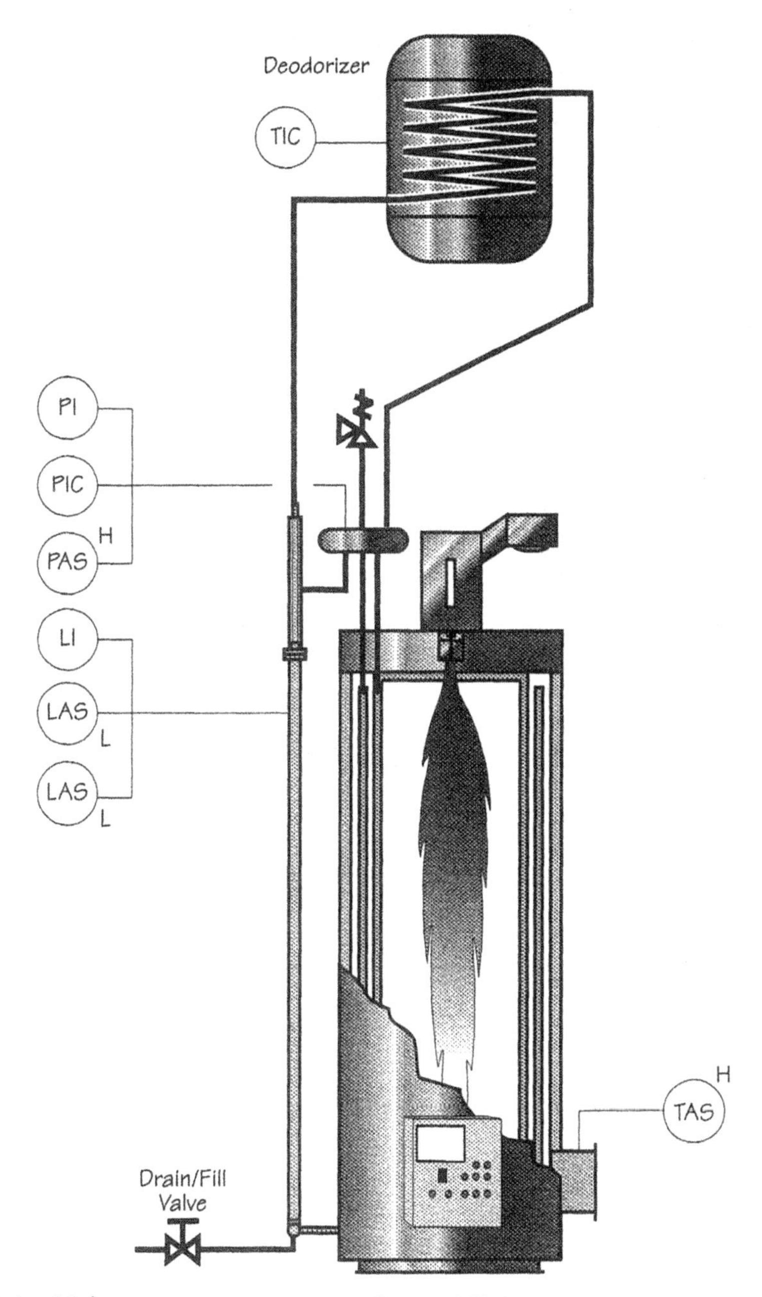

Fig. 14.4. High-pressure steam generator. *Source:* GTS Energy, Inc.

keeping the condensing surfaces clean, caustic is injected into the greasy water stream. One such system is depicted in Fig. 14.5. These systems are considerably more costly, because the condenser must be constructed of stainless steel, since it will be handling the more concentrated FFA (not diluted by condenser water). These costs can be acceptable in order to satisfy environmental inspectors and, in extreme cases, keep the refinery from being shut down.

In order to minimize steam consumption, the air or noncondensable jets can be replaced with a liquid ring mechanical vacuum pump. This will substitute electrical

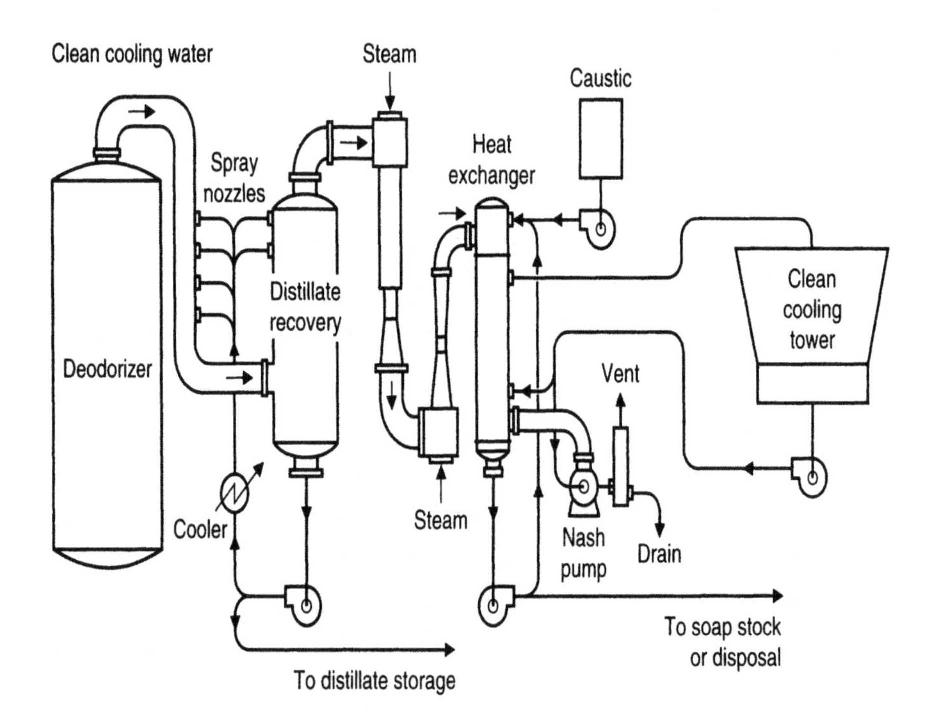

Fig. 14.5. Deodorizer vacuum system with surface condenser.

energy for steam and save about 10 to 15% of the ejector motive steam requirements. This should be individually evaluated as to utility costs and the additional maintenance costs for a mechanical pump.

Deodorizer Distillate Recovery

The concerted effort to recover deodorizer distillate began in earnest in the late 1960s for two reasons: (1) Increasing demand for sterols and tocopherols (natural Vitamin E) by the pharmaceutical companies and (2) the refiners' interest in cleaning up their messy soapstock skimming operations. Soybean oil contains about 0.1% tocopherol, and deodorizing temperatures of 260 to 265°C (500 to 510°F) will remove about 40 to 50%. Unfortunately, the value of distillate seems to follow the general economy closely and, as a result, has fluctuated quite violently. The continued pressure of governmental regulations mandates the use of distillate recovery systems, regardless of byproduct economics.

Distillate recovery systems are normally included with most deodorizer offerings. They pass the deodorizer vapor through a recirculating spray of cooled distillate, in either a vacuum vessel or a packed tower. The fatty material is condensed, while the water vapor (stripping steam) passes on to the ejector system. The efficiency of these systems should be in the range of 85 to 95% when handling soybean oil, with the packed column type being the most efficient.

When the tocopherol and sterol market is strong, disposal of distillate is profitable; however, at other times it must be sold as a soap stock or can be a disposal problem. Some countries permit the use of distillates as an animal feed extender, but in others it is prohibited because of the distillation and concentration of pesticides in the deodorization process.

Deodorization Systems

There are over 12 deodorization equipment manufacturers in the world, and no attempt will be made to review them. The general types of systems will be presented, and just about all of the manufacturers' equipment will come under these classifications. In many cases, the only difference will be the company's approach to stripping steam mechanism, heat recovery (energy conservation), or both. It is for the refiner to evaluate the systems, based on the capabilities required and in-depth discussions with the manufacturer and select accordingly. It can be helpful to solicit an independent, knowledgeable consultant (experienced individual) in evaluation of the refiner's needs and equipment selection. One thought should always be kept in mind: "The lowest price is *NOT* always the best buy."

Batch Deodorization

Batch deodorization is the original method utilized for edible oils, and some few refiners still use it for small odd-lot quantities of specialty oils. Traditionally, these systems consist of a vacuum vessel containing internal heating and cooling coils and stripping steam injection apparatus, steam jet ejector vacuum equipment, high-temperature heating media source, and polish filtration. These systems can also be fitted for heat recovery.

Disadvantages of these systems are high stripping steam and, likewise, high ejector motive steam consumption; difficulty in achieving consistent good quality, due to reflux of stripped components recondensed on the deodorizer head; and long cycle or turnaround times, usually 8 to 10 hours.

Continuous Deodorization

The continuous system generally provides the lower-capital-cost option and the greatest energy efficiency, thus the lowest operating costs. The continuous flow allows uniform temperature gradients during heating and cooling, which permits smaller ancillary equipment, namely the high-temperature heating system. It also offers the greatest potential for energy conservation, with many systems achieving up to 80% heat recovery. But, by definition, a continuous system is to be run continuously and therefore is primarily suited for the refiner processing a minimum variety of feedstocks. If the marketing plan calls for as many as 3 to 4 stock changes per 24-hour day, then the use of a continuous system is highly questionable. The problems encountered will be loss of production, since 20 to 60 minutes or more lost time will be required to ensure against intermixing (commingling) and the quality concerns of incorrect product reaching the customer.

Equipment suppliers have chosen two approaches for continuous deodorizers: either to carry out heating, cooling, and heat recovery in exchangers external to the deodorizer or to do all heating, cooling, and heat recovery within the deodorizer. The internal approach, as depicted in Fig. 14.6, while somewhat less efficient for heat recovery, affords a simpler and more foolproof method for stock changes than does external heat exchange, depicted in Fig. 14.7. The complete draining and blowing of heat exchangers, pipe lines, control valves, pumps, and other components is very time-consuming and critical for effectively minimizing stock contamination. Especially with vertically stacked deodorizers and internal heat exchange, the use of nature (gravity) greatly enhances the operation. But, lest we forget, the purpose of a continuous deodorizer is to operate continuously!

Continuous deodorizer systems that use external exchangers for heat recovery, heating, and cooling must to include two steps in the system for guaranteeing optimum-quality oil. First, the feedstock must be deaerated prior to heating above 120°C (250°F). The oil is sprayed into a small vacuum vessel and, for best results, should be preheated to about 80°C (175°F). With internally heated deodorizers, the oil will have ample time for deaeration as it is heating up to deodorizing temperature. Second, the hot deodorized oil should be cooled under vacuum and in contact with stripping steam prior to reaching about 145°C (295°F) to avoid flavor problems. A sparged compartment or vacuum vessel must be added for packed column units, whereas with a conventional shell-type unit the oil can be recycled to a bottom chamber for final stripping. This additional step reduces the potential heat recovery of the system by limiting the amount of interchange with the incoming feedstock.

Semi-Continuous Deodorizers

The semi-continuous deodorizer was originally developed in 1946 by A.E. Bailey (*Bailey's Industrial Fats and Oils*), an employee of Girdler Corp. (later known

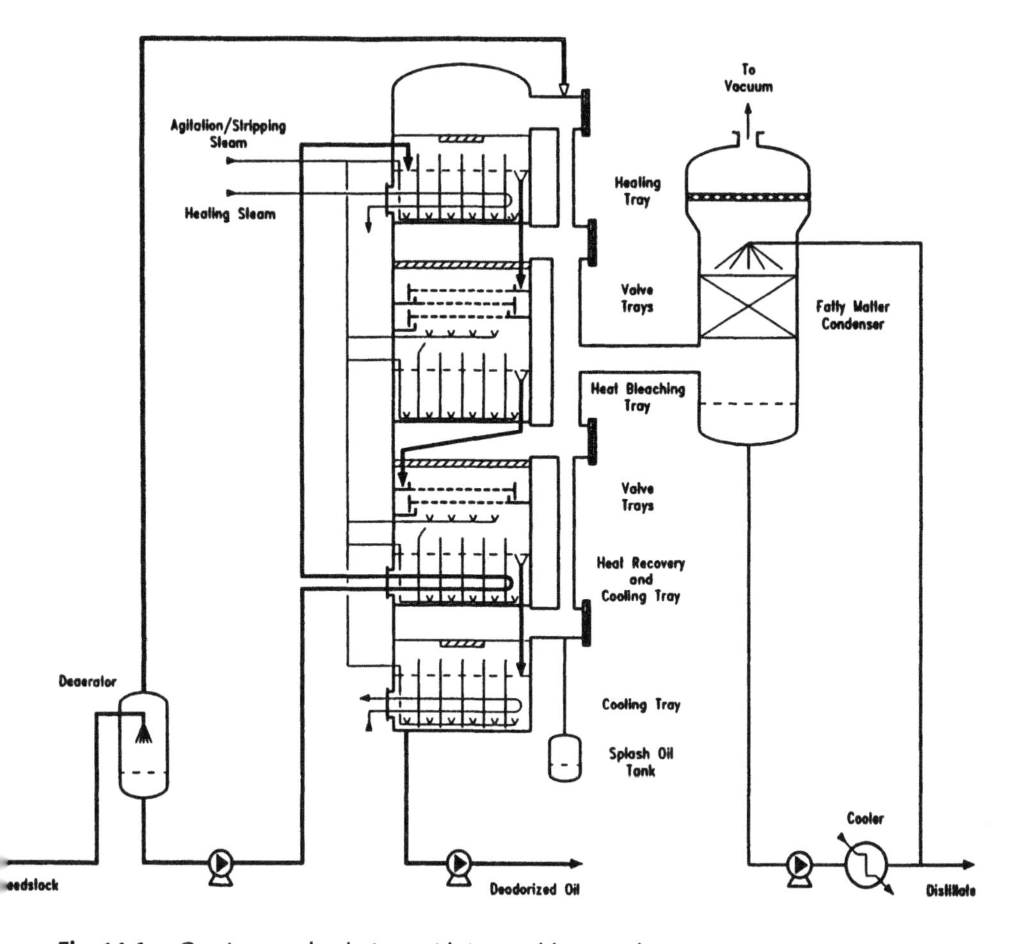

Fig. 14.6. Continuous deodorizer with internal heat exchange. *Source:* EMI Corporation, Des Plaines, IL.

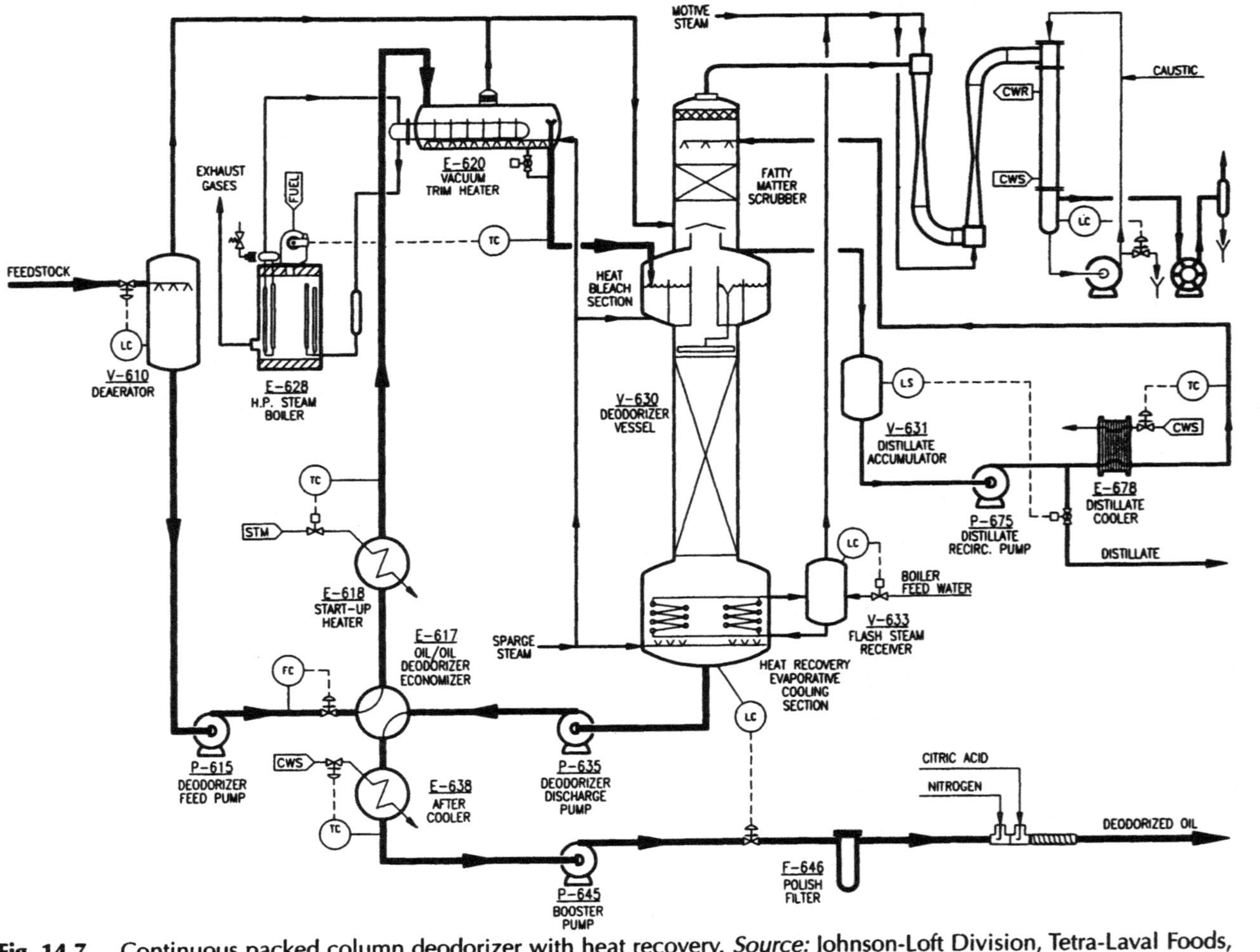

Fig. 14.7. Continuous packed column deodorizer with heat recovery. *Source:* Johnson-Loft Division, Tetra-Laval Foods, Novato, CA.

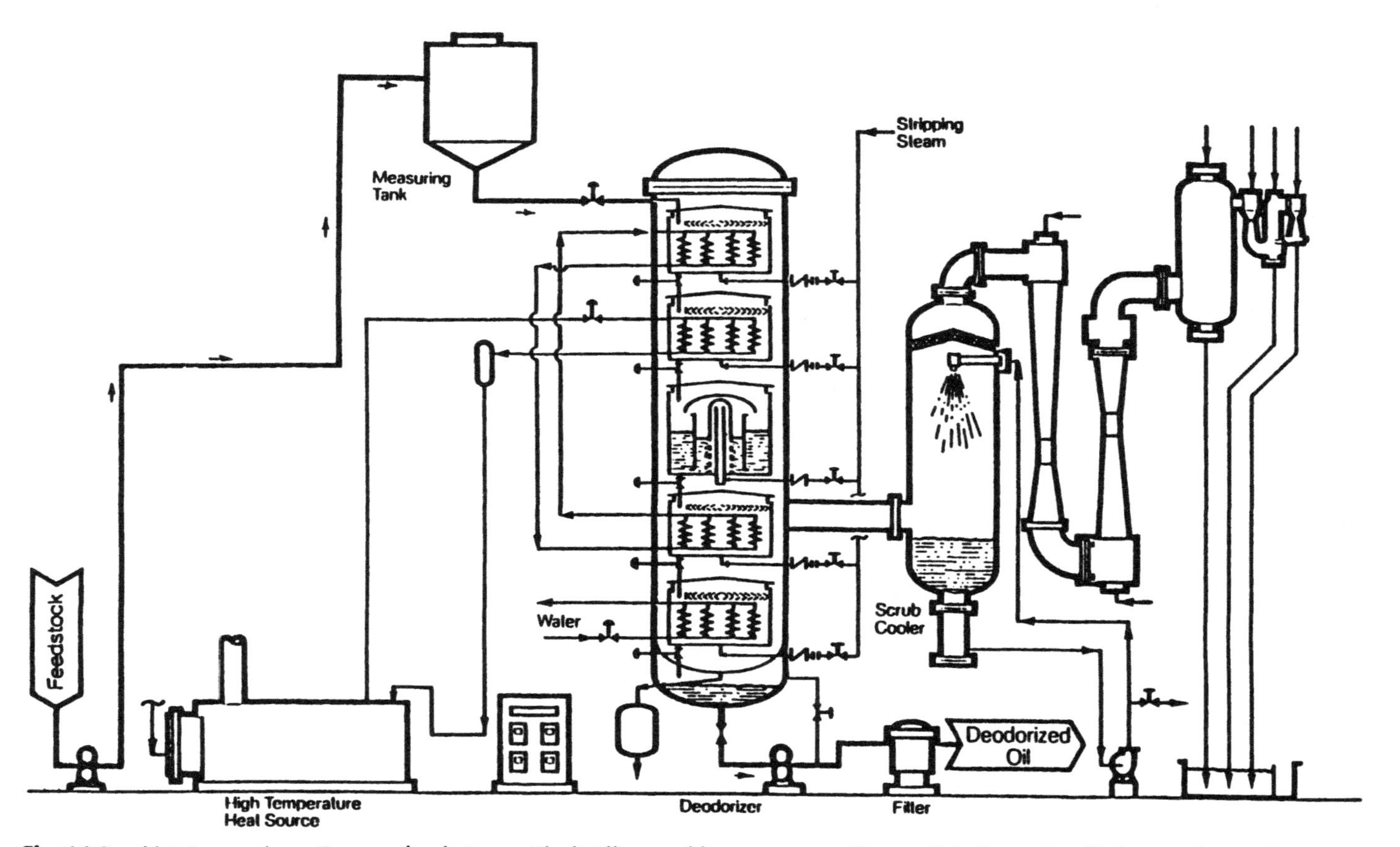

Fig. 14.8. Votator semi-continuous deodorizer with distillate and heat recovery. *Source:* S.A. Extraction DeSmet, Atlanta, GA.

worldwide as Chemetron-Votator). This equipment, depicted in Fig. 14.8, is now under the ownership of S.A. Extraction DeSmet, Belgium.

The concept of successively processing individual batches of fats and oils was to ensure maximum product quality. Each quantum of product would be subject to the identical processing conditions of time, temperature, pressure, and stripping steam—no chance for bypassing, short-circuiting, or dilution as in a continuous system. As refiners' marketing became more complex, especially with the variety of hydrogenated products and multiple oils, the main advantage soon became its ability to handle many varieties of feedstocks with as little as zero lost production and minimum contamination during stock changes. Zero lost production can be achieved by carefully scheduling stocks, including stock quantity, such that a stock can tolerate a small amount of contamination from the preceding stock. Do not schedule a small stock of liquid oil behind a shortening or other hardened stock!

Semi-continuous deodorizers have been known to handle 15 or more stock changes per 24-hour day successfully, and the author has personally witnessed three separate stocks within a deodorizer at the same time.

Semi-continuous operation necessitates larger high-temperature heating media equipment due to peak loads occasioned with batch heating, and the cyclic heating and cooling results in direct heat recovery capabilities of about 50%, as compared with up to 80% for continuous systems.

It should be emphasized that even with the capital cost and energy consumption disadvantages of a semi-continuous system, it should be the deodorizer of choice for the refiner processing a variety of feedstocks. The refiners' only reason for being in business is to make a profit, which requires being able to supply the products desired by the customers—thus, the refiner must have the equipment capable of doing so!

Many of the original semi-continuous units were designed as double-shell; that is, an outer vessel of carbon steel contained internal trays or compartments of stainless steel that were in contact with the product. The double-shell concept was also quality-oriented in that it guaranteed no air (oxygen) contact with the hot oil should there be any leakage into the vacuum vessel. In order to reduce capital costs, the single-shell concept has proven very acceptable when careful attention is paid to design features in handling the effluent vapors to overcome the air degradation potential. A single-shell semi-continuous unit is depicted in Fig. 14.9.

Deodorizer Manufacturers

The following is a list of international companies known to market deodorization systems throughout the world. Some may have been unintentionally omitted.

- Andreotti Impianti S.p.A., Firenze, Italy
- Campro Agra Limited, Mississauga, Canada
- Costruzioni Meccaniche Bernadini, Rome, Italy
- EMI Corporation, Des Plaines, Illinois, U.S.A.
- Fractionnement Tirtiaux S.A., Fleurus, Belgium
- Frans Kirchfeld GmbH & Co. KG, Dusseldorf, Germany

- Fratelli Gianazza S.p.A., Milan, Italy
- H.L.S. Ltd., Petach Tikva, Israel
- Krupp Maschinentechnik GmbH, Hamburg, Germany
- Lurgi GmbH, Frankfurt, Germany
- Masiero Industrial S.A., Saõ Paulo, Brasil
- S.A. Extraction DeSmet N.V., Edegem Antwerp, Belgium

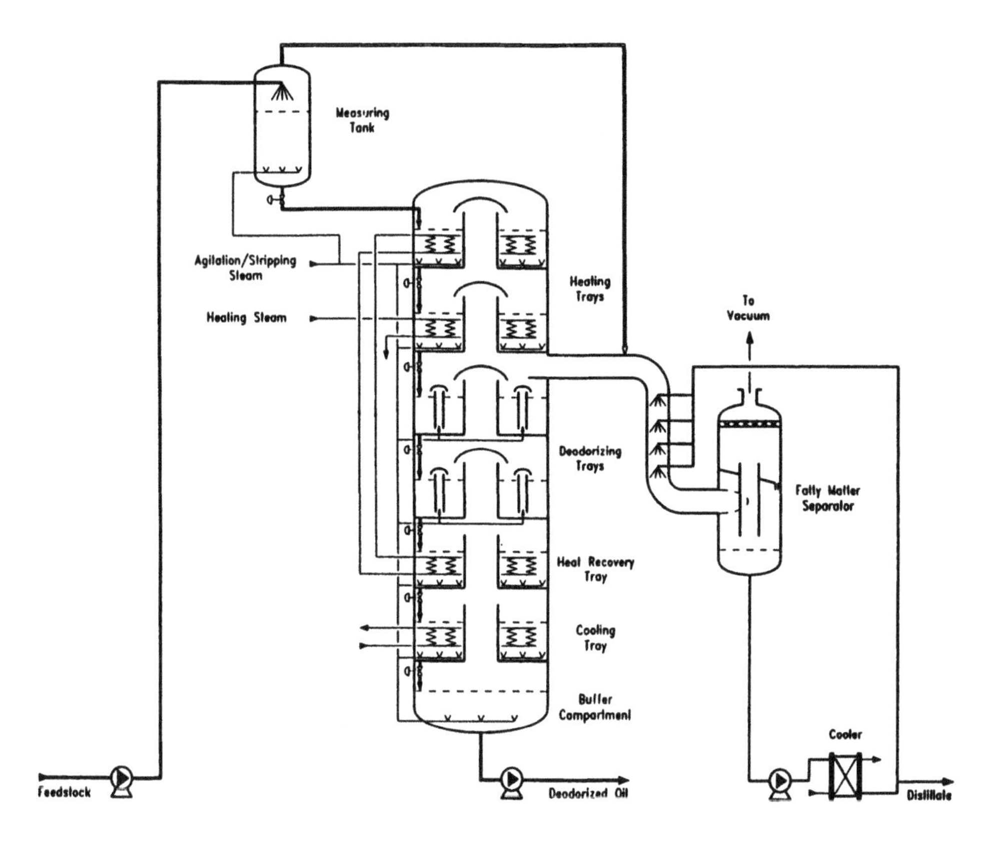

Fig. 14.9. Semi-continuous deodorizer with single shell and internal heat exchange. *Source:* S.A. Extraction DeSmet, Atlanta, GA.

- Tetra-Laval, Tumba, Sweden, and Johnson-Loft Engr. Div., Novato, California, U.S.A.
- Wurster-Sanger Div., Crown Iron Works Co., Minneapolis, Minnesota, U.S.A.

The companies listed may be able to supply whatever type of system the refiner needs. Therefore, the refiner should establish the necessary marketing-processing specifications and then evaluate the offerings solicited from the equipment suppliers.

Maintenance and Operational Considerations

The operation and maintenance instructions received with the purchased system should be the primary guide for plant personnel. The adage "If it ain't broke, don't fix it" should *not* be the rule to protect $1 to 2 million investments.

The author's experience (over 40 years) has shown that the following specific items can be troublesome in the operation and maintenance of deodorizers. These will apply to any type of system.

- Improper control of stripping steam flow rate and steam quality can be prime contributors to increased entrainment (neutral oil) losses. The stripping steam flow rate should be accurately controlled with a reliable steam flow meter, and when once set, *do not disturb*, unless checking or resetting. Adjusting the stripping steam should not be the responsibility, or option, of the deodorizer operator.

  Where the deodorizer is located remote from the steam boiler, (in other words, there is a long steam header), steam quality can be ensured by attention to the piping arrangement. For multiple stripping steam lines, establish a header downstream from the reducing valve, install steam traps on each end of the header, and pipe all lines from the top of the header; that is, the lines should rise from the header, even though the final connection may be lower than the header.
- Inability to maintain good oil quality, assuming time, temperature, and vacuum are normal, can indicate reduced stripping steam flow, plugged sparger holes, or both. If the holes still seem restricted after boil-out with caustic or sodium metasilicate, then it may be necessary to attempt to redrill the holes. Redrilling should be done with a combination drill/countersink bit, which results in a self-relieving hole; that is, the inner part of the hole becomes the orifice, and the outer part is relieved at an angle, a larger diameter than the hole size.
- Difficulty in maintaining the design absolute pressure without increasing the motive steam pressure is a good indication of worn steam nozzles or diffusers. The design bore diameters can be obtained from the ejector manufacturer and used to check the equipment bores. Wet steam can be a major cause of nozzle/diffuser wear; it is a bit like sand-blasting the nozzles. If wet steam is indicated, steps should be taken to add steam traps, separator, or both in the piping adjacent to the ejector system.

- The maintenance schedule should include an annual boil-out and will vary with respect to continuous and semi-continuous units. Again, the manufacturer's recommendations should be the guide. The boil-out can be accomplished with 10° Baumé caustic—sodium metasilicate—or suitable product, boiled for about an hour with stripping steam flowing, then rinsed with water and oil-flushed. With double-shell units (trays within vessel), the shell should not be filled with cleaning agent, only the trays to overflowing. Filling the shell will result in excessive loading (weight) of the vessel and foundations, increased cost due to excessive use of cleaning agent, and having heavy polymerized fat floating into the trays which is very difficult to remove.

  After cleaning, it is recommended that a pressure vacuum pull-down test be performed to assure system tightness.
- The distillate recovery system should be inspected annually to make certain the demister is clean and securely in place. Displaced or partially plugged demisters will lead to higher carry over of fat into the ejector system and loss of valuable distillate. In necessary, the demisters should be removed and boiled out.

References

1. Bailey, A.E., *Eng. Chem. 33:* 404–408 (1941).
2. Bates, R.W., *J. Amer. Oil Chem. Soc. 26:* 601–606 (1949).
3. Gavin, A.M., *J. Amer. Oil Chem. Soc. 55:* 783–791 (1978).
4. Zehnder, C.T., and C.E. McMichael, *J. Amer. Oil Chem. Soc. 44:* 478A–512A (1967).
5. Lineberry, D.D., and F.A. Dudrow, U.S. Patent 3,693,322 (1972).

Chapter 15

Soybean Oil Crystallization and Fractionation

R.D. O'Brien

Introduction

The functionality of solidified edible oil products is influenced by three basic processes:

1. Formulation, which includes the choice of source oils and the hydrogenation techniques used for the base stocks
2. Chilling, which initiates the crystallization process
3. Tempering, where the desirable crystal nuclei are developed

Formulation is the first requirement for consistency control and is reviewed separately in Chapters 13, 18, 19, and 20. Chilling and tempering develop and stabilize the desired crystal structure provided by the product composition.

Chilling and tempering processes are employed when shortening and margarines are packaged. The physical form of edible oil products is important for handling and performance in food products. Many applications depend upon the softness, firmness, oiliness, creaming properties, melting behavior, surface activity, workability, solubility, aeration potential, pourability, and other physical properties peculiar to each plasticized shortening or margarine product.

The crystallization processes for shortening and margarine are somewhat similar. Both involve premixing of ingredients followed by rapid chilling. After chilling, the further processing of shortening and margarine products may be different, but for both, the processes are aimed at controlling consistency, which is extremely important. The importance of consistency to the performance of shortening, margarine, and other fat-related products cannot be overemphasized. In the case of plastic shortenings, consistency is important from the standpoint of usage and performance in bakery products. The importance of the consistency of a fluid shortening is obvious; it must be maintained as a uniform, homogeneous suspension (1).

Fat Plasticity

Edible fat products appear to be soft, homogeneous solids; however, microscopic examination shows a mass of very small crystals in which a liquid oil is enmeshed. The crystals are separate discrete particles capable of moving independently of each other. Therefore, shortening and margarine products have the three conditions essential for plasticity in a material:

1. A solid and a liquid phase
2. Fine enough dispersion of the solid phase to hold the mass together by internal cohesive forces
3. Proper proportions of the two phases.

Plasticity, or consistency, of an edible oil product depends on the amount of solid material; the size, shape, and distribution of the solid material; and the development of crystal nuclei capable of surviving high-temperature abuse and serving as starting points for new, desirable crystal growth. The factor most directly and obviously influencing the consistency of a plastic shortening or margarine product is the proportion of the solid phase. As the solids content increases, an edible fat product becomes firmer. The proportion of the solid phase is influenced by the extent of hydrogenation and attendant isomerization. Shortening and margarine products are also firmer, with smaller crystal sizes, because of the increased opportunity for the solid particles to touch and resist flow. Stiffening is also increased more with the interlacing of long needle-like crystals than with more compact crystals of the same size. The crystal nuclei ("memory") are developed by the further treatment of a newly solidified product. Exposure immediately after chilling to 29°C (85°F) for 24 h or more before cooling to 21°C (70°F) provides a softer consistency, with the ability to withstand widely fluctuating temperatures and still revert back to the original consistency at room temperature (2).

Crystallization

Fat crystals represent lower energy states of molecular configurations. Fats at elevated temperatures retain enough molecular motion to preclude organization into stable crystal structures. Edible oils go through a series of increasingly organized crystal phases with cooling until a final stable crystal form is achieved. This process can occur in fractions of a second or months. The crystal types formed define the textural and functional properties of most fat-based products.

The crystal structure of a shortening, margarine, or other fat-based product is determined by its

1. Source oil composition
2. Processing
3. Tempering or maturing

Crystallization is induced when melted fat is cooled rapidly to initiate the formation of crystal *nuclei* or *seeds*. The seeds form templates on which crystals grow. Formulation, cooling rate, heat of crystallization, and agitation levels affect the number and type of crystals formed.

Each source of oil exhibits inherent crystallization tendencies. A fat may pass through one or more unstable crystalline stages before assuming either β or β′ crystal forms. The differences among the three crystal forms are as follows (3):

TABLE 15.1 Crystal Structure of Hydrogenated Fats and Oils (4)

β′	β
Cottonseed oil	Soybean oil
Palm oil	Safflower oil
Rapeseed (HEAR, high-erucic-acid rapeseed) oil	Sunflower oil
Tallow	Sesame oil
Fish oils	Peanut oil
Modified lard	Corn oil
Milk fat	Canola (LEAR, low-erucic acid rapeseed) oil
	Olive oil
	Lard
	Cocoa butter

- α crystal forms are unstable and will convert into the more stable β or β′ crystal forms.
- β crystals are large, coarse, and self-occluding.
- β′ crystals are small and needle-shaped and can pack together into dense, fine, grained structures.

Each common fat or oil has a definite *crystal habit* depending on four factors:

1. Palmitic fatty acid content
2. Distribution and position of palmitic and stearic fatty acids on the triglyceride molecule
3. Degree of hydrogenation
4. Degree of fatty acid randomization in the triglyceride molecule

Table 15.1 identifies the crystal habits of hydrogenated edible oils (4).

Many edible oil products contain various combinations of β and β′-tending components. The ratio of β to β′ crystals helps to determine the dominant crystal habit, but the higher-melting triglyceride portions of a solidified fat product usually force the fat to assume their crystal form. The crystal form of the solidified fat product has a major influence on the textural properties. Fats exhibiting a stable β′ form appear smooth, provide good aeration, and have excellent creaming properties for the production of cakes, icings, and other bakery-type products. Conversely, the β polymorphic form tends to produce large granular crystals for products that are waxy and grainy and provide poor aeration. These β formers perform well in applications such as pie crusts, where a grainy texture is desirable (5), and opaque fluid shortenings, where the large granular crystals are preferred for stability and maintenance of fluidity.

Shortening Processing

One of the earliest methods of solidifying shortenings involved the use of a *chill roll*, which was internally refrigerated by circulating cold brine. The melted product was fed into a trough from which a thin film of the liquid was picked up as the roll revolved and crystallized into a semi-solid state during a revolution. A blade removed the solidified product, which fell into a trough equipped with a rotating shaft with metal fingers and called a *picker box*, which worked the product to make it homogeneous and to incorporate air. Pumps then picked up the shortening for transfer to filling machines. This process has now become obsolete in the United States, except for some very special products. The reason for obsolescence was the difficulty in controlling the variables to produce a uniform product (6), poor thermal efficiency, and moisture condensation on the chilled roll surface.

Most shortenings are now quick-chilled in closed thin-film scraped-wall heat exchangers. The principle of operation for most of these units is chilling the fat in a very thin film that is continually removed by scraper blades with precise temperature control. The residence time within the heat exchanger tubes is very short, almost always less than 20 s. Since all triglycerides exhibit a high propensity for supercooling, the product exits from the heat exchanger cooled well below the equilibrium crystallization temperature.

If the product is allowed to solidify without agitation at this point, it will form an extremely strong crystal lattice and exhibit a narrow plastic range. This consistency may be desirable for stick margarines, but it is detrimental for products requiring a plastic-like consistency. Therefore, processing after the initial quick-chilling cycle has been adapted to the product consistency or form desired (7,8). The various forms in which shortening and margarine products are currently available are listed in Table 15.2.

Mechanical Agitation

Enclosed worker units with speed controls have replaced the picker box that followed the chill roll in the early plasticization process. The heat of crystallization is dissipated rapidly in these units as they work the product to provide fine crystals. Extrusion valves are employed in most units to deliver a homogeneous smooth product to the filler at 17 to 27 atm (250 to 400 psig) (6).

Gas Incorporation

Air incorporation during agitation in the open picker box has been replaced with the injection of nitrogen into the inlet side of the chiller in precisely controlled quanti-

TABLE 15.2 Shortening and Margarine Product Forms

Shortening	Margarine
Melted	Solid
Plasticized solid	Whipped solid
Opaque fluid pumpable	Plasticized solid
Clear liquid	Squeezable liquid
Flake	Soft spreadable
Bead	Roll-in or pastry

ties, normally 13 ± 1% by volume for standard plasticized shortenings. This provides a white, creamy appearance and increases the product's workability.

The correct gas content is important for appearance and stability. Shortenings containing the proper levels are white and creamy in appearance, with a bright surface sheen. Too low a gas content gives a yellowish, greasy appearance, and shortenings packed with no gas have a petroleum-jelly–like appearance. Too high a gas level causes a dead-white, chalky color with a lifeless surface appearance and often large air pockets within the product. Nonuniform dispersion of the gas gives an unattractive streaked appearance.

Gas levels range from 0 to 30% or more, depending on the product requirements. Table 15.3 compares the typical gas requirements for shortening and margarine products.

Tempering

Immediately after packaging, shortening and bulk margarines requiring a plastic consistency are stored 24 or more hours undisturbed at a temperature slightly above the packaging temperature. In practice, holding at 29°C (85°F) until a stable form is reached is an acceptable compromise (5), usually 24 to 72 h. This conditioning period is referred to as *tempering*. The primary purpose of tempering is to condition the plasticized shortening so that it will withstand wide temperature variations in subsequent storage and still have a uniform consistency when brought back to 21 to 24°C (70 to 75°F), which is the use temperature for a majority of the plasticized shortenings and baker's or roll-in margarines (7).

Most technologists agree that a shortening is tempered when the crystal structure of the hard fraction reaches equilibrium by forming a stable crystal matrix. The crystal structure entraps the liquid portion of the shortening. The mixture of low-melting and high-melting components of the solids undergoes a transformation in which the low-melting fractions remelt and then recrystallize into a higher-melting, more stable form. This process can take from 1 to 10 days, depending on the shortening formulation and package size. After a shortening takes an initial set, some α crystals are still present. These crystals remelt and slowly recrystallize in the β′ form during tempering. The β′ crystals are preferred for most plastic shortenings and margarines, especially those designed for creaming or laminating (8). Therefore, soybean oil–based shortenings requiring a plastic range are formulated with 5 to 20% of a β′-tending hardfat. The β′ hardfat must have a higher melting point than the soybean oil base in order for the entire shortening to crystallize in the stable β′ form.

TABLE 15.3 Typical Gas Content of Edible Oil Products

Shortening product	Amount (% vol.)	Margarine product	Amount (% vol.)
Regular plastic	13 ± 1	Stick	None
Pre-creamed	18–25	Soft tub	4–8
Opaque fluid	None	Whipped tub	30–35
Clear liquid	None	Baker's	13 ± 1
Puff pastry	None	Roll-in	None

The effect of tempering on a plastic shortening or margarine can be demonstrated only by performance testing. In some cases, penetration values undergo changes during tempering, showing a softening of the conditioned product as compared to a nontempered product. The effect of tempering can be identified with the feel or workability of the plastic fat products; the tempered product is smoother, with good plasticity, whereas the nontempered product will be more brittle with less plasticity. Products forming β′ crystals, when transferred to a cool temperature immediately after filling, become permanently hard and brittle; attempts to recondition these products by warming to tempering conditions have not been successful (9).

Quick Tempering

The expense and logistical problems associated with constant-temperature rooms for tempering have led several equipment manufacturers to develop mechanical systems in attempts to eliminate tempering. Most of these systems do not claim complete elimination, but rather a 50% or more reduction, in tempering time.

Most of the so-called "quick tempering" processes add a postcooling and kneading, or working, unit to the conventional-type chilling and working systems that are used with tempering. The theory postulated for these systems is that when liquid fat is forced to crystallize rapidly, it creates many smaller, more stable individual crystals (6,7) instead of accreting onto existing crystals and increasing their size, as happens in normal tempering.

Performance characteristics equivalent to well-tempered shortening have been claimed for the quick-tempered products with only 24 hours of conditioning (9). These systems have received acceptance from some edible oil processors, but the standard tempering procedures are still practiced by other processors for plastic shortenings and margarines.

Shortening Consistency

The final consistency of a shortening is the culmination of all the factors influencing crystallization and plasticization discussed individually: chilling, working, tempering, pressure, and gas incorporation. The effect of each factor on the final shortening is summarized on Table 15.4. Each condition is individually and collectively important; the shortening's final performance can be adversely affected if any of the consistency factors do not conform to the established standards for each individual shortening product.

TABLE 15.4 Factors Influencing Shortening Consistency

	Soft	Firm
Chilling	Cold	Warm
Working	More	Less
Tempering	Warm	Cold
Pressure	High	Low
Creaming gas	High	None

Margarines

Bulk Margarine Processing

Bulk margarines are chilled in the same manner as shortenings. The only difference between shortenings and baker's or roll-in margarine plasticization is that the gas usually incorporated into a shortening (Aeration which gives shortening whiteness and opacity) is eliminated. Margarines contain water or milk in emulsion form and are not normally aerated. The aqueous phase of margarine emulsions has the same effect as gas incorporation on the reflection of light.

Stick Margarine Processing

Until newer, totally closed machines were introduced, stick table-grade margarines utilized an aging tube (or quiescent resting tube) between the scraped-wall heat exchanger and the stick former. The resting tube allowed time for the margarine emulsion to crystallize without working. This unit, usually split lengthwise into two sections, had a perforated end plate for back pressure. A reciprocating valve, to open and close the sections alternately, provided the margarine time to rest and solidify. The margarine was pushed through the perforated end plate as noodles, which were then compressed into prints simulating the open texture of butter. The newer print-forming machines do not use the open noodle former, because smooth-textured margarine is now more accepted. The chilled margarine emulsion is forced directly into the print-forming chamber for discharge to the wrapping machine.

Soft Tub Margarine Processing

Soft tub margarine crystallization more closely resembles shortening chilling than the crystallization process of stick table-grade margarine or even of the bulk margarines, except for a chilling unit temperature like that used with the stick product (10°C or 50°F). Gas is incorporated in these margarines at 4 to 8% for the regular product and 30 to 35% for the whipped tub margarines. These products are also filled at high pressures, as is shortening, to help achieve a smooth consistency. Tempering of both the stick and tub margarines is at refrigerator temperatures, 4°C (40°F), rather than the higher 29°C (85°F) utilized for shortening and bulk margarines (see Chapter 19).

Fluid Opaque Shortenings

Fluid opaque shortenings are distinguished from liquid oils by composition and appearance. Both are pourable, but liquid oils are clear, whereas fluid shortenings are opaque because of their suspended solids. Fluid shortenings are flowable suspensions in liquid oil of hardfats, emulsifiers, or a combination of the two, depending upon the intended use for frying or for production of breads, cakes, or nondairy products.

The liquid oil may be unhydrogenated, lightly hydrogenated, or lightly hydrogenated and winterized, depending on the oxidative stability requirements of the intended use. A β hardfat, such as soybean oil at 5–10 iodine value, serves as quick-forming nucleus that causes solids in the base oil to precipitate in small enough crystals to ensure pourability and prevent separation. Many different crystallization procedures for opaque fluid shortenings have been patented, some of which are briefly summarized here (9,10,11,12):

- Gradual cooling of melted product to form large β crystals with gentle agitation (Requires 3 to 4 d)
- Gradual cooling followed by comminution with a homogenizer or colloid mill
- Rapid chilling with a swept-surface heat exchanger and holding for at least 16 h with gentle agitation to fluidize prior to packaging
- Suspension of finely ground hardfat or emulsifier in a cool liquid oil, with subsequent comminution using a homogenizer, colloid mill, or shear pump
- Quick chilling of a concentrated mixture of hardfat in liquid oil, followed by a holding period in an agitated tank to allow β crystal growth, after which the stabilized concentrate is blended with room-temperature liquid oil seeded with β-forming hardfat
- Quick chilling of a solution of β crystal–forming hardfat in liquid oil to 38°C (100°F) and allowing the heat of crystallization to carry the temperature of the cooled oil to not over 54°C (130°F); complete crystallization reportedly requires 20 to 60 min
- Recirculation of a hot oil solution of hardfat in an agitated holding tank through a scraped-surface heat exchanger and back into the holding tank; the oil in the tank cools slowly, the crystals formed in the early stages melt in the hot oil, and eventually the mass cools to a point at which the crystals do not melt but form α, then β′, and finally stable β crystals; further chilling is stopped and the resulting opaque pourable shortening is packaged
- Double chilling/tempering system of first cooling from 65 to 43°C (150 to 110°F) followed by a 2-h crystallization period with gentle agitation and subsequent supercooling to 21 to 24°C (70 to 75°F), followed by a second crystallization period of 1 h; the opaque shortening is packaged after the final tempering with an expected heat rise of about 9°C (15°F)

The summarized processing procedures for opaque fluid shortenings agree on several points:

1. All are dispersions of solid and liquid oil fractions.
2. β-forming hardfat is the preferred solid fraction.
3. Heat of crystallization must be dissipated before a stable product is achieved.

Another agreement is that aeration of the opaque fluid shortenings must be avoided at all stages of processing before, during, and after crystallization. Air incorporation makes the product more viscous or less pourable and promotes product separation. Storage studies indicate an air content must be less than 1.0% for a stable suspension.

Opaque pumpable shortenings do not require any further tempering after packaging, but the storage temperatures are important. Storage below 18°C (65°F) will cause the pumpable shortenings to solidify with a loss of fluidity. Storage above 35°C (95°F) will result in partial or complete melting of the suspended solids. The solidification due to cool temperatures can be reversed by controlled heating not to exceed the melting point, but the damage caused by high temperature cannot be remedied except by complete melting and reprocessing.

Shortening Flakes

The term *shortening flake* describes the higher-melting edible oil products solidified into a thin flake form for ease in handling, for quicker remelting, or for a specific function in a food product. Flaking rolls, utilized for the chilling of shortening and margarine products before scraped-wall heat exchangers were introduced, are still used for the production of shortening flakes. Chill rolls have been adapted to produce several different flaked products used to provide distinctive performance characteristics in specialty formulated foods. Consumer demands have created the need for such specialty fat products. Specialty high-melting fat flakes have been developed for specific applications with varied melting points, as shown in Table 15.5. Instead of becoming obsolete, chill rolls have fulfilled an equipment requirement to produce specialty fat-based products.

Chill Rolls

Chill rolls are available in different sizes, configurations, surface treatments, feeding mechanisms, and so forth, but most consist of large, hollow metal cylinder with a surface machined and ground smooth to true cylindrical form. Rolls, internally refrigerated with either flooded or spray systems, turn slowly on longitudinal and horizonal axes. Several options exist for feeding the melted oil product to the chill roll:

1. A trough arrangement positioned midway between the bottom and top of the roll
2. A dip pan at the bottom of roll

TABLE 15.5 Typical Flaked Shortenings

Products	Melting points	
	°F	°C
Low-iodine value hardfat	140–150	60–66
Icing stabilizers	110–130	43–54
Shortening chips	110–115	43–46
Confectioners' fats	100–110	38–43

3. Overhead feeding between the chill roll and a smaller applicator roll
4. Double- or twin-drums operating together with a very narrow space between them, where the fat product is sprayed for application to both rolls

The coating of fat is carried over the roll to solidify and removed by a doctor blade, positioned ahead of the feed mechanism utilized.

Flaked Product Crystallization

In the crystallization of hydrogenated edible oil products, the sensible heat of the liquid is removed until the temperature of the product is equal to the melting point. At the melting point, more heat must be removed to allow the crystallization of the product. The quantity of heat associated with this phenomenon is called *heat of crystallization*. Sensible heat (specific heat) of most common hardfat products is approximately 0.27 cal/g·°C (0.5 Btu/lb·°F), and the heat of crystallization is equal to 27.8 cal/g (50 Btu/lb). Thus, the amount of heat that must be removed to crystallize hardened oil is 100 times the amount of heat that must be removed to lower the product temperature by 1°F (13).

Flaking Conditions

The desired shortening flake product dictates the chill roll operating conditions and additional treatment necessary before and after packaging. However, some generalizations relative to chill roll operations and product quality can be made:

Crystal structure. Each flaked product has crystallization requirements that depend on source oils, melting points or degree of saturation, and the physical characteristics desired.

Flake thickness. This depends primarily on four controllable variables:

1. Oil temperature to the roll
2. Chill roll temperature
3. Speed of the roll
4. Feed mechanism

Wet flakes. In this physical condition, the flakes contain liquid oil that has not been completely solidified. Wet flakes can be caused by too low a chill roll temperature, causing "shock" chilling; that is, the oil film against the roll solidifies rapidly and pulls away, insulating itself. The outside surface of the flake remains liquid, creating a wet flake that causes lumping after solidification.

Package temperature rise. Heat of crystallization will cause a product temperature rise after packaging if not dissipated previously. The packaged product temperature can increase enough to melt some of the product partially and cause lumping.

Winterization

Winterization is an old practice that evolved from the observation that refined cottonseed oil stored in outside tanks during the winter, physically separated into a hard and clear fraction. The clear liquid oil on top was decanted for bottling and became known as winter salad oil (14). The bottom hard fraction has become known as stearine.

Winterization is a form of the overall process called fractionation. It has continued to be a major process for producing liquid oils, either those qualifying as a salad oil or a high-stability liquid oil product. *Salad oil* has been defined as an edible fat that will not solidify or cloud at temperatures of 4 to 10°C (40 to 50°F). *High-stability oils* do not have the low-temperature requirement but must be clear liquids at room temperatures.

Winterized Soybean Oil Products

In very few cases do natural fats and oils have completely satisfactory compositions. For example, soybean oil is a natural "winter salad oil," but it has a relatively short flavor stability because of its high degree of polyunsaturation. Studies with trained flavor panels have shown that partial hydrogenation significantly reduces the incidence of offensive odor and flavor development. The improvement in organoleptic ratings was attributed to the reduction of linolenic (C18:3) and linoleic (C18:2) fatty acids. However, hydrogenation to reduce the polyunsaturated fatty acids increases the higher-melting disaturated glycerides and *trans*isomers, which causes clouding (14). Therefore, production of a clear soybean oil product requires winterization to remove the higher-melting fractions and meet the requirements of a salad oil or a high-stability clear liquid oil.

Winterization Process

Basically, all winterization processes involve chilling the oil at a prescribed rate, allowing the solid portions (stearine) to crystallize, and finally separating the solid and liquid phases. The basic and most difficult problem is the development of the crystals into a form easily separated from the liquid fraction.

Fat crystallization occurs in two steps. The first is the crystal formation process, called *nucleation*. The second is crystal growth. The driving potential for both steps is supersaturation, and neither nucleation nor growth will occur in a saturated or unsaturated solution. A supersaturated solution is one that contains more solute than would dissolve under equilibrium conditions. For fats and oils, a supersaturated solution can be obtained only by reducing the temperature of the melt. Simply stated, the rate of nucleation depends on the triglyceride composition of the oil, the rate of cooling, the temperature of the nucleation, and the mechanical power input. The crystal growth rate depends on the temperature of crystallization, the time of crystallization, and the mechanical power input.

Crystal Inhibitors

A crystal inhibitor is a material that retards crystal formation, which can be measured by an increase in cold test results. Cold test measures the length of time required for the first appearance of fat crystals in an ice bath (15). A crystal inhibitor's action is to change the shape of the stearine crystals so that the crystalline mass will retain less oil and permit the oil to be filtered at a faster rate.

Chemically, a crystal inhibitor is a fat-soluble product similar to a triglyceride in general structure but differing from it in some specific way. Since crystals grow by the orderly deposition of similar molecules, the inclusion of a dissimilar molecule can stop crystal growth as long as the different molecule remains in place in the crystal lattice. The dissimilar molecule or crystal inhibitor must have some similarity to the other molecules present, or it will not deposit on the crystal surface in the first place (9).

Lecithin was the first salad oil crystal inhibitor. It is a triglyceride with one fatty acid replaced by a phosphoric acid ester of choline or ethanolamine. Many other chemical compounds have been developed for controlling crystal growth, but only two have been approved by the U.S. Food and Drug Administration (FDA).

Oxystearin. This product is prepared by heating fully hydrogenated soybean or cottonseed oil to a high temperature, with air blowing through the hot oil. The exact chemical composition is not known, but it contains many polymers and breakdown products of triglyceride molecules. The typical usage level is 0.05%, but up to 0.125% is permitted by FDA regulations.

Polyglycerol esters. This emulsifier is formed by reacting fats or fatty acids with polymerized glycerol. U.S. federal regulations forbid the food use of polyglycerol esters that contain more than ten (10) polymeric glycerol units. The food standards allow the necessary level to obtain the desired result, but 0.02 to 0.04% is the typical usage level.

Crystal inhibitor effectiveness depends on the winterized oil quality. A poorly processed oil with a low cold test will be improved only slightly, whereas a well-winterized oil will experience a significant increase in cloud resistance.

Factors Affecting Winterization

A careful selection of process variables is very important. The primary objective is to produce a small number of nuclei, around which the crystals grow in size as cooling continues. If a large number of nuclei are formed, then an equally large mass of small crystals is developed, which is difficult to filter. Or, if the crystals group together in clumps, the liquid phase will be occluded, causing poor separation and low yields.

The effects of the major processing variables on the performance of a winterization process (16) are as follows:

Glyceride composition. Hydrogenation conditions for the soybean oil feedstock have a definite effect on yield. The conditions should be chosen that make the least possible saturates and *trans*isomers but still produce the desired winterized oil characteristics.

Cooling rate. Control is required to produce stable crystal forms while maintaining a low viscosity to permit nuclei movement.

Crystallization temperature. High viscosities, caused by crystallization temperatures that are too low, reduce crystal growth. After crystal growth begins, temperature increases should be controlled for orderly transformation from α to β' to β crystals.

Agitation rate. Gentle agitation should be maintained to enhance crystal growth. High agitation rates provide shear, which may break up the crystals.

Crystallization time. The design of the winterizer system dictates the proper time required for crystallization.

The effect and the order of significance of the processing variables were determined for partially hydrogenated soybean oil (H-SBO). The results are shown in Table 15.6. It is evident that all of the variables need to be carefully evaluated to obtain quality salad oil, with optimum filtration rates and maximum yields.

Soybean Salad Oil Winterization

Soybean oil that has been hydrogenated to improve flavor stability must be winterized to regain clarity at cool temperatures. Most U.S. producers have offered a stabilized soybean salad oil of this type with a 111 ± 3 iodine value. The winterization process employed differs for each producer and sometimes each individual plant, but all include the same basic principles of cooling for crystallization followed by filtration to separate the liquid and solid fractions:

Crystallization. This step can be accomplished with a system of tall, narrow tanks in a chilled room or with programmed chilling in specially designed tanks or with some other type of heat exchanger.

TABLE 15.6 Effect of Processing Variables on the Winterization Process for Hydrogenated Soybean Oil

	Order of significance			
	First	Second	Third	Fourth
Cold test	Holding temperature	Cooling rate	Agitation rate	Holding time
Yield	Holding time	Cooling rate	Agitation rate	
Filtration rate	Holding time	Agitation rate	Cooling rate	Holding temperature

Filtration. This step is performed with filters of many designs, such as plate and frame, tank type or rotary vacuum filters. Separation is also being accomplished with centrifuges of various design.

The equipment used will dictate the process cycle times to produce acceptable winterized oil. A winterization cycle time and temperature sequence used with a programmed chiller is outlined in Table 15.7.

Fractionation

Fractionation and winterization operations for processing of edible oils basically consist of the separation of oils into two or more fractions with different melting points. In the winterization process the oils are cooled in a simple way and kept at low temperature for some time. The liquid and the solid fractions are generally separated by filtration. In fractionation processes the cooling of the oil and the separation of the fractions are done in a more sophisticated manner, under more controlled conditions.

Fractionation processes have a broad application in edible oil technology. High-stability liquid oils, with AOM stability results of 350 h minimum without the benefit of added antioxidants, is one class of commercially available fractionation products other than winterized oils. Another family of fractionated products has been developed for the confection industry. Domestic oil hard butters, for application in confectioners' cocoa powder coatings with no tempering requirement and with a compatibility for cocoa butter, have been produced utilizing solvent fractionation techniques.

Principles of Fractionation

Fractional crystallization is a thermomechanical separation process wherein component triglycerides of fats and oils are separated, usually as a mixture, by partial crystallization in a liquid phase. In this process three successive stages are recognized (17):

1. Cooling of the liquid oil to supersaturation, resulting in the formation of nuclei for crystallization
2. Progressive growth of the crystals by gradual cooling
3. Separation, isolation, and purification of the resultant crystalline and liquid phases

TABLE 15.7 H-SBO Winterization Cycle

Winterization function	Time (h)	Temperature °C	Temperature °F
Cooling	7	12.8	55
Initiate crystal formation	6	12.8	55
Crystal growth	4	3.3	38
Crystal stabilization	2	3.3	38
Separation	6		

The efficiency of separating the liquid and solid fractions depends particularly on the method of cooling, which determines the form and size of the crystals. Rapid cooling causes heavy supersaturation and gives a great number of small crystals, resulting in the formation of an amorphous, microcrystalline, softish precipitate with poor filtration properties. This form will be slowly transformed into the metastable α form, with characteristics of microcrystallinity and a tendency to develop mixed crystals. Gradual cooling of the supersaturated oil results in stable β and β' macrocrystals, which can be separated easily from the liquid phase by filtration.

The separated liquid and solid fractions show a significant difference in physical and chemical properties. However, the distribution of the fatty acids in the separated fractions is less pronounced than might be expected; shifting of fatty acids from one fraction to another is relatively minor, even though the differences in iodine values and melting points between fractions are considerable. These observations are shown on Table 15.8 with the fractionation of hydrogenated soybean oil.

Separating an oil into fractions does not lead to effective separation of the different types of fatty acids. Therefore, the objective of fractionation is, in general, to modify the texture, crystallization, and melting behavior, which are defined by the composition of the triglycerides (18).

Fractionation Processes

Three basic edible oil fractionation processes have found industrial application (18):

1. *Dry Fractionation.* The principle of this fractionation process is based on the cooling of oil under controlled conditions without the addition of chemicals. The liquid and solid phases are separated by filtration. Fig. 15.1 shows the flow diagram of the Tirtiaux system to illustrate a dry fractionation process.

TABLE 15.8 Fractionated Hydrogenated Soybean Oil

	H-SBO	Fractions	
	Base	Liquid	Solid
Iodine value	88.6	91.2	84.6
Mettler dropping point, °C	26.0	20.0	32.7
Solid fat index, %			
At 10°C (50°F)	15.6	9.4	22.5
21.1°C (70°F)	5.6	0.4	11.8
26.7°C (80°F)	2.1	0	7.0
33.3°C (92°F)	0.2	0	1.0
40.0°C (104°F)	0	0	0
Fatty acid composition, %			
C16:0 (Palmitic)	10.7	9.8	12.0
C18:0 (Stearic)	4.8	4.2	6.1
C18:1 (Oleic)	65.4	65.5	64.4
C18:2 (Linoleic)	17.5	18.8	15.8
C18:3 (Linolenic)	0.8	0.9	0.7
C20:0 (Arachidic)	0.4	0.4	0.5
C22:0 (Behenic)	0.4	0.4	0.5

2. *Dry Fractionation with Additive*. The principle of this type of fractionation is similar to that of dry fractionation, based on the cooling of oil under controlled conditions without the addition of a solvent. The liquid and solid phases are separated by centrifugation after an aqueous detergent solution has been added. The surface-active agent replaces the oil phase on the surface of the crystals. The crystals and the aqueous solution form a suspension, which can be separated from the liquid oil phase by centrifugation. The Alfa-Laval Lipofac system flow diagram in Fig. 15.2 is an example of a dry fractionation with additive process.
3. *Solvent Fractionation*. Cooling of a fat diluted with a solvent generally results in the formation of crystals of the stable β and β′ type and reduces the tendency to form mixed crystals. The incoming oil is mixed in a certain ratio with an organic solvent. The resulting solution is subjected to preliminary cooling and is pumped to the crystallization vessels, in which part of the glycerides precipitate due to crystallization. The mixture is separated into a liquid phase of oil and solvent and into a solid phase of glyceride crystals and solvent. Both fractions are separated from the solvent by distillation. Acetone, hexane, and 2-nitropropane are solvents that have been utilized with this process. Fig. 15.3 shows the flow of a commercial solvent fractionation plant.

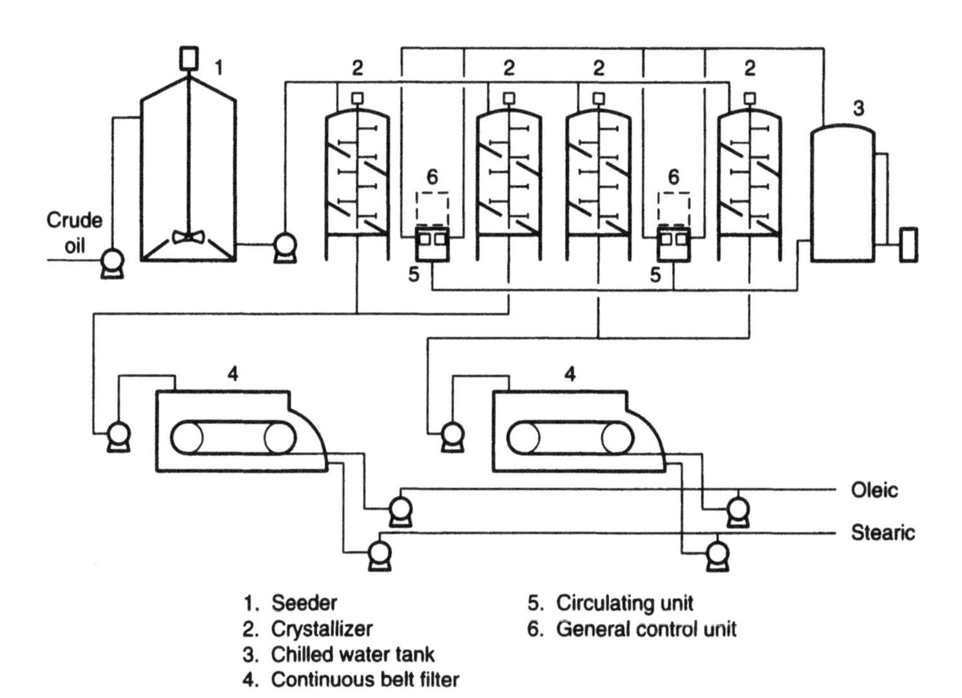

Fig. 15.1. Flow diagram of Tirtiaux fractionation plant. *Source:* Kreulen, H.P., *J. Amer. Oil Chem. Soc. 53:* 394 (1976).

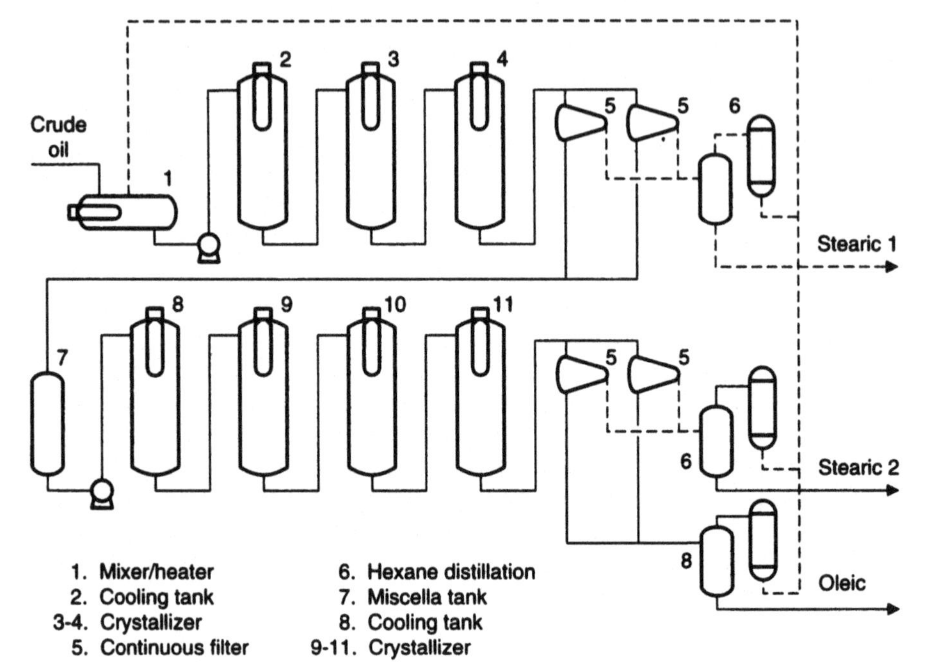

Fig. 15.3. Flow diagram of Bernardini fractionation plant. *Source:* Kreulen, H.P., *J. Amer. Chem. Soc. 53*: 394 (1976).

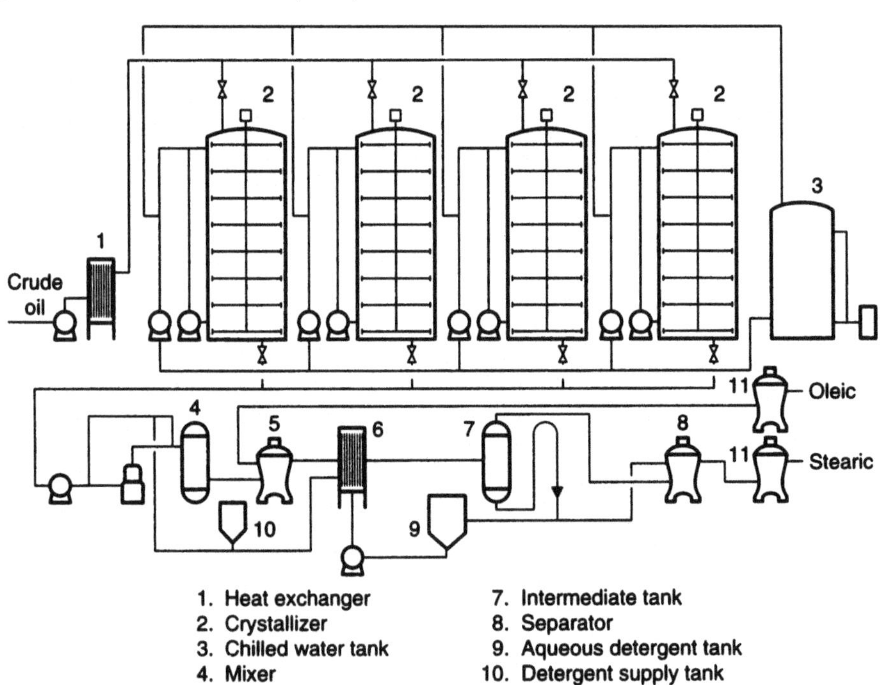

Fig. 15.2. Flow diagram of Alfa–Laval fractionation plant. *Source:* Kreulen, H.P., *J. Amer. Oil Chem. Soc. 53*: 394 (1976).

Fractionation Products

The ultimate objective in the application of the fractionation technology to fats and oils is the commercial production of substances with unique properties. Two product categories have utilized the fractionation process to produce improved products with soybean oil as a portion of the feedstock:

Hard Butters. Hydrogenation and fractionation technology have been combined to produce a cocoa butter substitute that is compatible with cocoa butter and does not require tempering. Table 15.9 shows the analytical characteristics of domestic oil hard butters with three different melting points.

High Stability Liquid Oils. Modification of oils by utilizing hydrogenation with fractionation has permitted the development of liquid oils with high resistance to oxidative degradation. Liquid oils with AOM stabilities of 350 h are available commercially.

TABLE 15.9 Fractionated, Hydrogenated Cottonseed and Soybean Oil Hard Butters (18)

	98	101	103
Wiley melting point, °C	36.7	38.3	39.4
Iodine value	55.8	57.8	60.0
Solid fat index, %			
At 10°C (50°F)	77.0	73.4	69.5
21.1°C (70°F)	70.3	63.6	58.6
26.7°C (80°F)	63.2	56.0	50.0
33.3°C (92°F)	27.3	26.2	21.3
37.8°C (100°F)	0.0	4.5	4.3
43.3°C (110°F)	0.0	0.0	0.0
Fatty acid composition, %			
C12:0 (Lauric)	0.2	0.2	0.6
C14:0 (Myristic)	1.0	0.6	0.7
C16:0 (Palmitic)	23.4	17.5	18.7
C16:1 (Palmitoleic)	0.4	0.4	0.4
C17:0 (Margaric)	—	—	0.2
C18:0 (Stearic)	11.7	14.4	13.3
C18:1 (Oleic)	62.0	66.1	62.1
C18:2 (Linoleic)	1.1	0.4	3.4
C18:3 (Linolenic)	—	—	0.1
C20:0 (Arachidic)	—	0.4	0.5

References

1. Bell, R.J., in *Introduction to Fats and Oils Technology*, edited by Peter J. Wan, American Oil Chemists' Society, Champaign, IL, pp. 197–204.
2. Mattil, K.F., *Bailey's Industrial Oil and Fat Products*, 3rd edn., edited by D. Swern, Interscience Publishers, New York, 1964, pp. 272–281.
3. Best, D., *Prepared Foods*, 168 (May 1988).
4. Wiedermann, L.H., *J. Amer. Oil Chem. Soc. 55:* 825 (1978).
5. Thomas, A.E., *J. Amer. Oil Chem. Soc. 55:* 830 (1978).
6. McMichael, C.E., *J. Amer. Oil Chem. Soc. 33:* 512 (1956).
7. Joyner, N.T., *J. Amer. Oil Chem. Soc. 30:* 526 (1953).
8. Hoerr and Ziemba, *Food Eng.* (May 1965).
9. Weiss, T.J., in *Food Oils and Their Uses*, 2nd edn., AVI Publishing Co., Westport, CN, 1983, pp. 92–97, 113–114, and 126–129.
10. Petricca, *The Baker's Digest*, 39–41, (October 1976).
11. Mattil, K.F., in *Bailey's Industrial Oil and Fat Products*, 3rd edn., edited by D. Swern, Interscience Publishers, New York, 1964, pp. 1064–1068.
12. Cheysam, M.M., in *Bailey's Industrial Oil and Fat Products,* edited by T.H. Applewhite, Wiley-Interscience, New York, 1985, Vol. 3, pp. 100–104.
13. Wan, P.J., and A.H. Chen, *J. Amer. Oil Chem. Soc. 60:* 743 (1983).
14. List, G.R., and T.L. Mounts, in *Handbook of Soy Oil Processing and Utilization*, edited by D.R. Erickson, E.H. Pryde, O.L. Brekke, T.L. Mounts, and R.A. Falb, American Soybean Association and American Oil Chemists' Society, Champaign, IL, 1980, pp. 193–211.
15. *The Official Methods and Recommended Practices of The American Oil Chemists' Society*, 4th edn., American Oil Chemists' Society, 1994, Method Cc 11-53.
16. Sterton, A.J., in *Bailey's Industrial Oil and Fat Products*, edited by D. Swern, Interscience Publishers, New York, 1964, pp. 1005–1011.
17. Thomas, A.E., III, in *Bailey's Industrial Oil and Fat Products*, edited by T.H. Applewhite, Wiley-Interscience, New York, 1985, Vol. 3, pp. 1–22.
18. Paulicka, F.R., *J. Amer. Oil Chem. Soc. 53:* 423 (1976).
19. Kreulen, H.P., *J. Amer. Oil Chem. Soc. 53*: 393 (1976).

Chapter 16

Interesterification

Michael D. Erickson

Kraft Food Ingredients Technology Center
8000 Horizon Blvd.
Memphis, TN

Introduction

During the synthesis of fats and oils in maturing plants and animals, enzymes attach free fatty acids to glycerol in a specific order. Interesterification changes this orderly distribution to a random distribution, provided the temperature of the reaction is above the melting point of the oil. There are two exceptions to the rule of randomized distribution: *directed* interesterification and *enzymatic* interesterification. Although the emphasis of this chapter is on chemical interesterification, both alternatives are discussed in following sections.

Interesterification, often referred to by the appropriately descriptive terms *randomization* or *rearrangement*, offers an important alternative for modifying the behavior of fats and oils. The reaction begins when an appropriate catalyst is added to the oil. The "active form" of the catalyst then forms, which promotes the detachment of fatty acids from the glycerol "backbone." As the reaction proceeds, fatty acids detach and reattach simultaneously at open positions within the same glyceride and at vacant positions on adjacent glycerides. Thus, when the reaction achieves equilibrium, the fatty acids have formed a new mixture of triglycerides that no longer reflects their initial orderly distribution.

The finished product performance of hydrogenated fats and oils is due in large part to physical changes in the fatty acids (*trans* isomers). Interesterification, however, does not change the fatty acid profile of the starting material. The changes in melting and solidification properties of interesterified fats and oils are due to the relative proportions of component glycerides after rearrangement of the fatty acids (1). Consequently, the inherent stability of an interesterified oil, or oil blend, remains predictable. The exact effect of interesterification on melting and solidification properties ultimately depends on the type and mixture of starting materials.

Interesterification also affects the crystallization tendencies of fats and oils. This was used on a commercial scale before vegetable-based shortening displaced "modified" lard as the preferred household shortening. Before interesterification, (unmodified) lard has a strong β crystallization tendency (see Chapter 15). After interesterification, its most stable crystal form is β′. The result of this change in crystal habit increased lard's plastic range and ability to incorporate air, thus increasing its utility.

Theoretical Triglyceride Composition

The triglyceride composition resulting from interesterification can be calculated from probability considerations. For n fatty acids A, B, C, D, ... in amounts of a, b, c, d, the types of triglycerides can be predicted, as shown in Table 16.1.

If, for example, the starting material has six different fatty acids participating in the interesterification reaction, the following calculation can be made:

Simple triglyceride (all fatty acids are the same): $n = 6$
Triglycerides with two of the same fatty acids: $n(n - 1) = 6(6 - 1) = 30$
Totally mixed triglycerides: $\frac{1}{6}n(n - 1)(n - 2) = 1(6 - 1)(6 - 2) = (5)(4) = 20$
Total obtainable triglycerides: $6 + 30 + 20 = 56$

Chemical Description of Interesterification

A detailed investigation of interesterification mechanisms is beyond the scope of this chapter. However, a basic description of how the reaction proceeds may help anticipate where and why problems arise during production.

Rozendaal (2) gives an excellent description of catalyst activation and the resulting mechanism of interesterification, as depicted in Figs. 16.1 and 16.2. Both figures show that the "true" (or active) catalyst is sodium diglyceride, which forms after the sodium-containing compounds contact a triglyceride.

The figures also show the opportunity to form soaps (sodium salts of fatty acids) and mono- and diglycerides. Nearly all of the mono- and diglycerides are removed during deodorization. However, soaps require removal by other physical or chemical means. If not dealt with properly, these factors will lead to significant emulsion problems during processing. Therefore, adding the appropriate amount of catalyst is a critical control point during processing.

Fig. 16.3 shows the generally accepted reaction mechanism for interesterification. The first step is migration of fatty acids within the same triglyceride. This is followed by participation in the entire fatty acid pool until equilibrium is achieved.

TABLE 16.1 Theoretical Triglyceride Compositions after Interesterification (2)

Type	Number	Amounts
Simple triglyceride (AAA, BBB, ...)	n	a^3, b^3, c^3, ...
Triglycerides with two of the same fatty acids (AAB, AAC, ...)	n(n−1)	$3a^2b$, $3ab^2$, $3a^2c$, ...
Mixed triglyceride (ABC, BCD, ...)	$\frac{1}{6}$n(n−1)(n−2)	$6abc$, $6acd$, ...

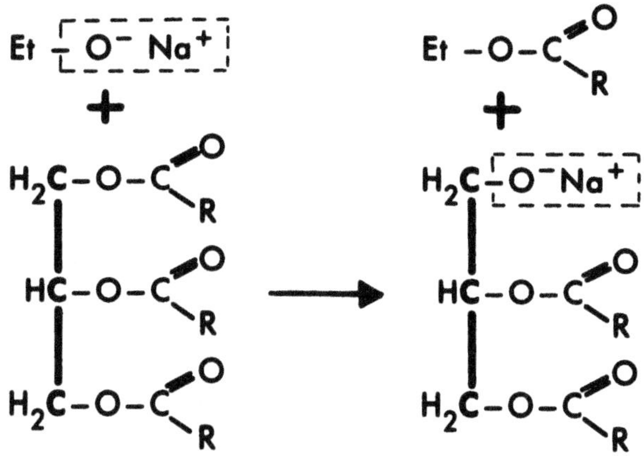

Fig. 16.1. Activation of sodium ethylate. *Source:* Rozendaal, A., in *World Conference Proceedings, Edible Fats and Oils Processing: Basic Principles and Modern Practices,* edited by D.R. Erickson, American Oil Chemists' Society, Champaign, IL, 1989, pp. 152–157.

Enzymatic Interesterification

Enzymes, specifically lipases, are naturally occurring biological catalysts. These "biocatalysts" are excreted by microorganisms into their growth medium to aid in the digestion of fats and oils. Lipases hydrolyze (detach) fatty acids from triglycerides. Although this is the opposite of triglyceride synthesis, the reaction is reversible. When water availability is properly manipulated, simultaneous hydrolysis and synthesis occur. The process can be further controlled by the type of lipase used. Fig. 16.4 shows the effect of specific and nonspecific enzymes.

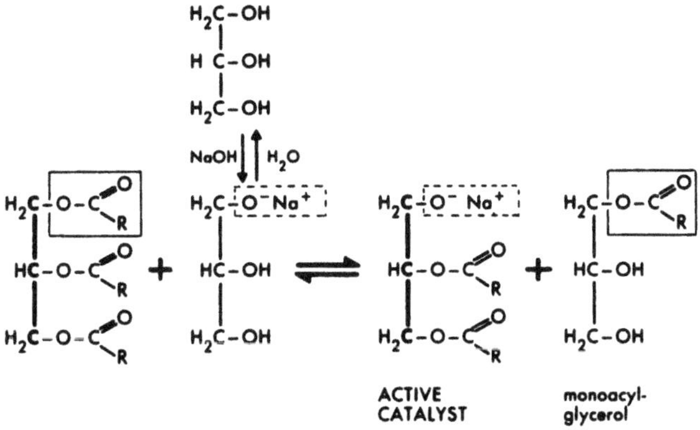

Fig. 16.2. Formation of active catalyst from sodium hydroxide, glycerol, and water. *Source:* Rozendaal, A., in *World Conference Proceedings, Edible Fats and Oils Processing: Basic Principles and Modern Practices,* edited by D.R. Erickson, American Oil Chemists' Society, Champaign, IL, 1989, pp. 152–157.

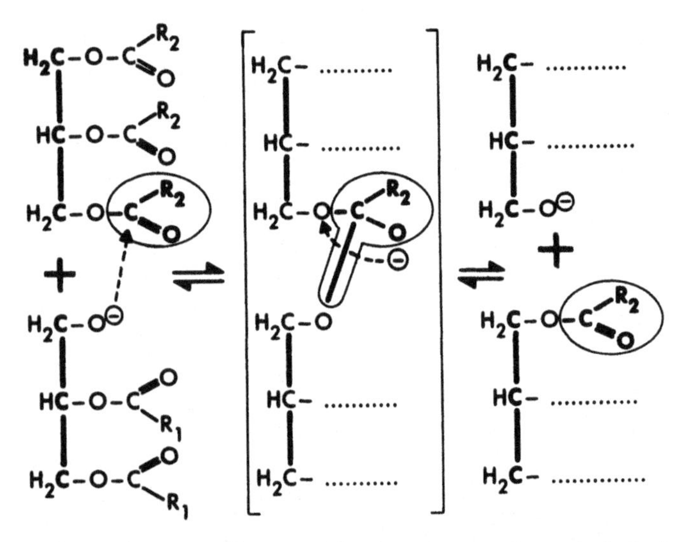

Fig. 16.3. The mechanism of interesterification. *Source:* Rozendaal, A., in *World Conference Proceedings, Edible Fats and Oils Processing: Basic Principles and Modern Practices,* edited by D.R. Erickson, American Oil Chemists' Society, Champaign, IL, 1989, pp. 152–157.

When nonspecific lipases are used, the reaction is analogous to random chemical interesterification. Specific lipases make it possible to produce fats and oils with a customized array of triglycerides, which is not an option with conventional chemical interesterification. This process has great utility when enrichment of a specific fatty acid in a specific position is desired.

(I) Nonspecific lipase:

```
RCOOCH2                                                      CH2OH
  |                                                            |
R'COOCH2  ⇌  RCOOH  +  R'COOH  +  R''COOH  +                 CHOH
  |                                                            |
R''COOCH2                                                    CH2OH
```

(II) 1,3-specific lipase:

```
RCOOCH2              CH2OH                  CH2OH
  \                    \                      \
R'COOCH      ⇌      R'COOCH        ⇌      R'COOCH
  |                    \                      /
R''COOCH2           R''COOCH2                CH2OH
                     + RCOOH               + R''COOH
```

(III) Fatty acid specific lipase:

```
RCOOCH2         R'COOCH2                    CH2OH   R'COOCH2
  |                /                           \       /
R'COOCH    +    RCOOCH ⇌ 2RCOOH + R'COOCH       +   CHOH
  |                \                           /       \
R''COOCH2       R''COOCH2            R''COOCH2      R''COOCH2
```

Fig. 16.4. Products formed from lipase-catalyzed hydrolysis of triglycerides. *Source:* Rozendaal, A., in *World Conference Proceedings, Edible Fats and Oils Processing: Basic Principles and Modern Practices,* edited by D.R. Erickson, American Oil Chemists' Society, Champaign, IL, 1989, pp. 152–157.

Typical batch processing consists of coating a charged inorganic substrate with the enzymes. The catalyst is activated with small amounts of water before the oil–oil or oil–free fatty acid mixtures, which are dissolved in an organic solvent such as hexane or petroleum ether, are added. After reacting, the particles are filtered out, and the interesterified mixture is desolventized.

Continuous enzymatic interesterification requires the reaction mixture to pass through a "fixed-bed" reactor. Enzymes are immobilized on a substrate such as diatomaceous earth or anion exchange resin. Water is mixed with a feedstock, (usually) dissolved in an organic solvent. This results in simultaneous enzyme activation and interesterification as the mixture moves through the bed.

Directed Interesterification

The other exception to random interesterification is called directed interesterification. E.W. Eckey of the Procter & Gamble Co., Cincinnati, OH, is credited with developing the process and was granted several patents for this and related processes (4). Directed interesterification, also known as the "Eckey Process," achieves the desired array of triglycerides by temperature manipulation during the reaction.

If the interesterification reaction is allowed to proceed at lower temperatures, the fatty acids will continue to migrate until triglycerides are formed that solidify, which effectively takes them out of further reaction. The triglycerides solidify because their melting point is above the reaction temperature. This provides the opportunity to fractionate a mixture selectively to obtain desired functional properties.

Before animal fats were displaced by vegetable oils, directed interesterification was a significant modification process. Because of relatively poor yields compared with conventional processing, economic justification of this process must be supported in one of two ways: production of higher-margin products such as cocoa butter substitutes, or low cost and high availability of the raw material. Additionally, the flexibility of hydrogenation has significantly diminished the utility of this process. An example of a directed interesterification process flow is shown in Fig. 16.5.

Since the original patents, several variations of this process have been reported and patented. The major variations are temperature cycling, solvent fractionation, and combinations of both (6,7).

Survey of Interesterification Catalysts

Sreenivasan (5) points out several catalysts that have been used for interesterification (Table 16.2). Metal alkylates, specifically sodium methylate and sodium ethylate, are the most widely used. Their advantages include high activity, low-cost and (high) availability, and *relative* ease of handling, with appropriate protective clothing and equipment. However, they are ***explosive*** if allowed to contact water. The physical and chemical properties of the common alkylates are shown in Table 16.3.

In the past, to make catalyst handling safer, some catalysts were suspended in a nonpolar solvent, such as xylene. The mixture was then added to the oil. This practice is not recommended, nor is it necessary (see the section on "Batch Process Interesterification").

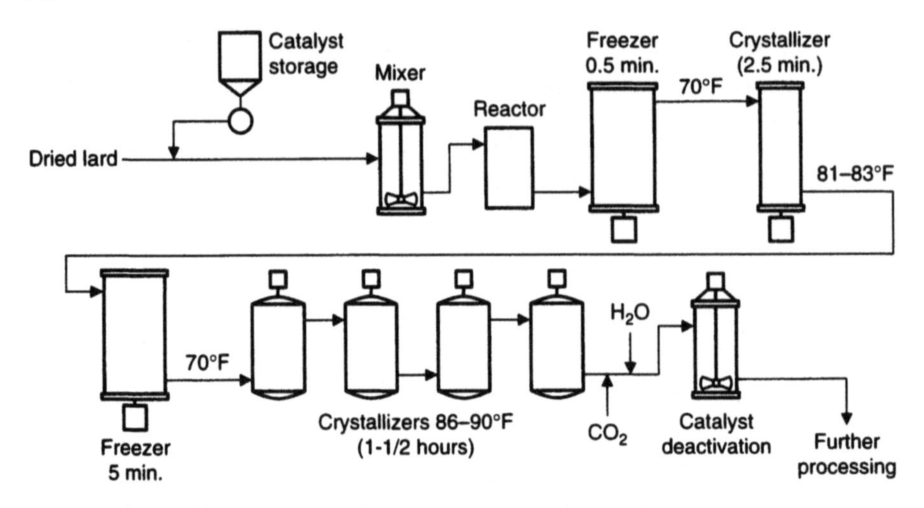

Fig. 16.5. Directed interesterification process flow. *Source:* Sreenivasan, B., *J. Amer. Oil Chem. Soc. 55:* 796 (1978).

TABLE 16.2 Interesterification Catalysts (5)

Catalysts for interesterification	%Treat	Temp°C	Time
Metal alkylates:			
Sodium methylate (methoxide)			
Ethylate (ethoxide), t-butylate, etc.	0.2–2	50–120	5–120 min
Alkali metals:			
Na, K, Na/K alloy	0.1–1	25–270	3–120 min
Alkali hydroxides:			1.5 hr
NaOH, KOH, LiOH	0.5–2	250	under vacuum
Alkali hydroxide	0.05–0.1		
+	+	30–45 min	
Glycerol	0.1–0.2	60–160	under vacuum
Metal soaps:			
Sodium stearate			1 hr
Glyceride	0.5–1	250	under vacuum
Li Al stearate			1 hr
Na Ti stearate	0.2	250	under vacuum
Metal salts:			
Acetates, carbonates, chlorides, nitrates,			
oxides of Sn, Zn, Fe, Co, and Pb			0.5–6 hr
Pb	0.1–2	120–260	under vacuum
Metal hydrides:			
Sodium hydride	0.2–2	170	3–120 min
Metal amide:			
Sodium amide	0.1–1.2	80–120	10–60 min

Sodium, potassium, and sodium/potassium alloy metals are the most efficient catalysts. They also represent the highest risk of explosion if exposed to trace amounts of water.

The most economical catalysts are sodium or potassium hydroxide and glycerol. However, this method can generate excessive amounts of soap and mono- and diglyceride. Both factors contribute to lower yields.

Laboratory Experimentation

Justifying production-scale interesterification begins in the laboratory. It is important to understand that results obtained in the laboratory will never be duplicated exactly in the plant. However, for directional purposes, experimentation is both necessary and prudent.

The following methodology is an example of an economical way of doing laboratory-scale atmospheric chemical interesterification of 500-g batches. The feedstock requirements, apparatus and chemicals are shown in Table 16.4.

Apparatus/Glassware

- Variable speed agitator
- Glass sparge tube
- Thermometer—260°C (500°F)
- Ring stand and ring for separatory funnel
- 2 ea. 1000 mL beakers
- Büchner funnel
- 2000 mL separatory funnel
- Hot plate
- Filter paper
- 2000 mL vacuum filter flask

TABLE 16.3 Physical and Chemical Properties of Common Alkylates

Catalyst	Synonym	Formula	Formula weight	Bulk density (kg/L)	Ave. Particle size (mm)	Shelf life (mo)	Sol/100 Parts	
							H_2O	other
Sodium methylate	Sodium methoxide	CH_3ONa	54.03	0.45–0.6	0.07	3–6	Decom	slt. alc.
Sodium ethylate	Sodium ethoxide	C_2H_5ONa	68.06	0.2–0.3	0.01–0.3	2–3	Decom	slt. abs alc.

TABLE 16.4 Feedstock Requirements (2)

Requirement	Value
Free fatty acid (%)	< 0.1
Peroxide value (meq/kg)	< 10
Moisture (%)	< 0.1
Soap (%)	< 0.1

Chemicals

Sodium methoxide (methylate): Catalyst (see notes 1 and 2 below)
Nitrogen gas (≥98% pure): N_2
Citric acid: CA
Distilled water: D H_2O
Filter aid (diatomaceous earth): DE
Acid-activated bleaching earth: Bleaching clay

1. Sodium methylate is classified by the U.S. Department of Transportation (DOT) as a flammable solid. It reacts violently with water and decomposes into sodium hydroxide (strong base) and methanol.
2. Using fresh or well-preserved catalysts cannot be emphasized enough. An unopened bottle, if old, is no guarantee that the catalyst still has the desired activity. Catalyst with diminished activity will cause incomplete rearrangement and lead to unbreakable emulsions (resulting from the temptation to overtreat). One simple, qualitative way to reduce the risk of using diminished activity catalysts is to rotate the bottle slowly. If the methylate is still very free-flowing, chances are that the catalyst still has the desired activity. If the methylate is caked, has sizable lumps, or does not flow at all, *do not use the catalyst.*

A diagram of the suggested laboratory apparatus is shown in Fig. 16.6.

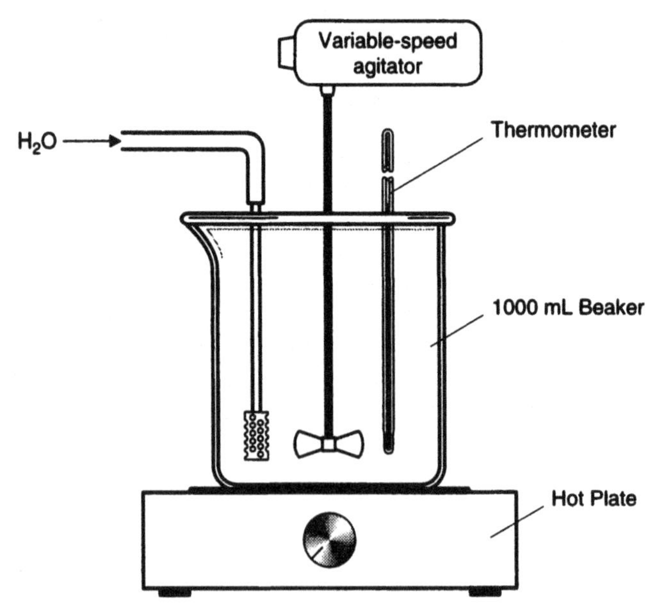

Fig. 16.6. Apparatus for laboratory-scale interesterification.

Procedure

The following is a detailed step-by-step procedure:

1. Ensure that the sample meets the criteria outlined in the feedstock requirements.
2. Place beaker with sample on hot plate.
3. Begin N_2 sparge; adjust for moderate sparge.
4. Turn agitator on to achieve moderate agitation without incorporating air.
5. Adjust hot plate to achieve 110 to 113°C (230 to 235°F).
6. After achieving 110 to 113°C, hold for ~45 min.
7. After having made sure that the catalyst is fresh, *carefully* weigh 0.1% catalyst onto a piece of filter paper (see notes above).
8. Carefully flush the headspace of the catalyst container with dry N_2 and close tightly.
9. After the sample has been held at temperature for ~45 min, carefully sprinkle the catalyst into the sample.
10. Reduce the N_2 sparge to a very slight trickle. *Note:* Excessive N_2 bubbling will cause the mixture to "foam" over the top of the beaker.
11. Reduce the heat to 90 to 93°C (195 to 200°F) while the reaction continues.
12. React for 20 to 30 min using moderate agitation. With experience, this time may be reduced depending on the oil or oil blends. *Note:* A color change should occur within 1 to 3 min after adding the catalyst. The color of the mixture should appear moderate to dark brown. Be prepared to add more methylate if a color change does not occur or the color change is only slight (amber). Add more methylate in 0.1% increments, waiting 4 to 5 min between additions. If 0.5% fails to cause a color change, the activity of the catalyst or the condition of the feedstock may be in question.
13. After 20 to 30 min, and achieving 90 to 93°C, carefully add the correct amount on a mole/mole basis of 50:50 citric acid in distilled water:
    FW sodium methylate: 54.03
    FW citric acid: 192.14

For a 500-g batch:

$$500 \text{ g} \cdot (0.1 \text{ g } CH_3ONa/100 \text{ g batch}) \cdot (\text{mole}/54.03 \text{ g } CH_3ONa) \cdot (192.14 \text{ g CA/mole}) = 1.78 \text{ g CA}$$

From a 50:50 CA soln (D H_2O = 1):

$$50 \text{ g CA}/100 \text{ g soln} = 1.78 \text{ g}/x \text{ g soln}$$
$$x = 1.78\,(100)/50 = 3.56 \text{ g CA soln}$$

Note: This calculation will change proportionately if more than 0.1% methylate is used.

14. Mix CA solution for 1 min. The mixture should become lighter in color. *Note:* Ensure sufficient agitation to keep the CA solution from settling to the bottom. Head will also help.
15. Turn off N_2 sparge.
16. Transfer mixture to the 2000-mL separatory funnel and allow to separate. *Do not shake.*
17. Decant the aqueous layer.
18. Add another 3 to 4 g of the 50% CA solution, gently invert 6 or 7 times, and allow to separate.
19. If separation does not occur (emulsion), sprinkle 2 to 3 g of solid CA into the mixture and gently invert 2 to 3 more times. *Note:* This step provides the greatest opportunity for emulsification.
20. Decant the aqueous layer.
21. Add 400 to 500 mL of hot tap water and gently invert 6 to 8 times. Allow the mixture to separate, and decant the aqueous layer.
22. Repeat water wash step 4 to 5 more times, shaking vigorously, or until the aqueous layer is relatively clear (slightly hazy).
23. Transfer the oil to a clean 1000-mL beaker.
24. Place the beaker on the hot plate. *Note:* Immediately begin vigorous agitation and heavy N_2 sparge.
25. Achieve 104°C (220°F) and hold until the oil turns clear.
26. After the oil turns clear, reduce the temperature to 82°C (180°F) while maintaining vigorous agitation and N_2 sparge.
27. Add 0.5% diatomaceous earth.
28. Add 1.0% bleaching earth.
29. Bring the temperature back to 104°C (220°F) and hold for 20 min.
30. Vacuum filter through Büchner funnel and filter paper.

The oil should now be interesterified. A quick means of determining whether the reaction has occurred is to drop a small amount on a cool surface next to a small amount of melted feedstock at the same temperature. A difference in the solidification rates suggests that the reaction has taken place. The appropriate analyses for verification are melting point (Mettler dropping point) and solid fat index/content.

If application testing is to be done, the oil should be deodorized. However, if laboratory-scale deodorization is not available, applications testing can still occur, because undeodorized oil will not interfere with performance testing. It will, however, interfere with flavor evaluations.

If possible, the drying and bleaching procedure (steps 24–30) should be done under vacuum. Vacuum has the advantage of efficient moisture removal and helps protect the color of the oil.

Pilot Plant Product Development

Once laboratory experimentation shows that an interesterified formula is suitable for a desired application, the next step is scale-up. Typically, the basic procedure will be consistent with the one outlined in the preceding section.

For atmospheric interesterification, the following considerations for 180 kg (400 lb drum) batches should be noted:

- Ideally, the reaction kettle should be stainless steel. A carbon steel kettle is acceptable but requires special attention, because it will rust after washing; requires "sweetening" (coating the inside of the kettle with flush oil from the feedstock) before the start of a batch; and, if not used routinely for this process, it can contribute to color problems.
- Center-mounted variable-speed agitation is highly recommended. Side-mounted agitation tends to create a vortex, which increases the risk of emulsifying at critical times.
- Steam-jacketed kettles are highly recommended. External heat exchangers require more movement of the oil and frequent cleaning.
- The nitrogen sparge line does not require a gas sparger at the end, as recommended for a laboratory setup. However, small-diameter stainless steel tubing that reaches to within a few centimeters of the discharge valve is desirable. Occasionally, when results from a test are not ideal, filtering material may fall to the bottom. A sparge line will help clear obstructions from discharge valve openings.
- A pump and filter press, resembling what would be available on a production scale, is preferred.

Attempting to duplicate laboratory observations in the pilot plant requires additional precautions.

- After drying the oil, turn off the N_2 sparge before adding the catalyst.
- Leaving the sparge on after adding the catalyst will cause excessive foaming. This will significantly increase the risk of creating an unbreakable emulsion.
- The time for complete interesterification is typically longer in the pilot plant than in the laboratory. However, the same color change consideration applies.
- The reaction termination step is the same; however, it is extremely important to turn the agitation off within 30 sec to 1 min after addition of the CA solution.
- Settling usually takes approximately one hour. This is also true for the washing steps.
- *Do not agitate during the first two water washes.*
- When emulsification is no longer a risk, moderate agitation and N_2 sparging is recommended.
- Ensure the N_2 sparge is vigorous during the drying step.

Every operation has unique capabilities and limitations. The foregoing guidelines are intended to advise on points in the process that anyone wishing to do pilot plant–scale development must address.

Batch Process Interesterification

Process Equipment Design Considerations

Fig. 16.7 shows a standard batch reactor for interesterification. One discharge option is to go to the refinery, where a centrifuge will remove the deactivated catalyst and soap after the main washing steps.

The other option is to bleach the batch immediately after the last drying step. This prevents color development and ensures maximum finished product quality. "House water" (water from wells or local utilities) will introduce trace metals and various oxygen-containing compounds into the oil. Therefore, to ensure the highest possible quality, bleaching immediately after the last drying step is the only effective means of removing secondary oxidation products.

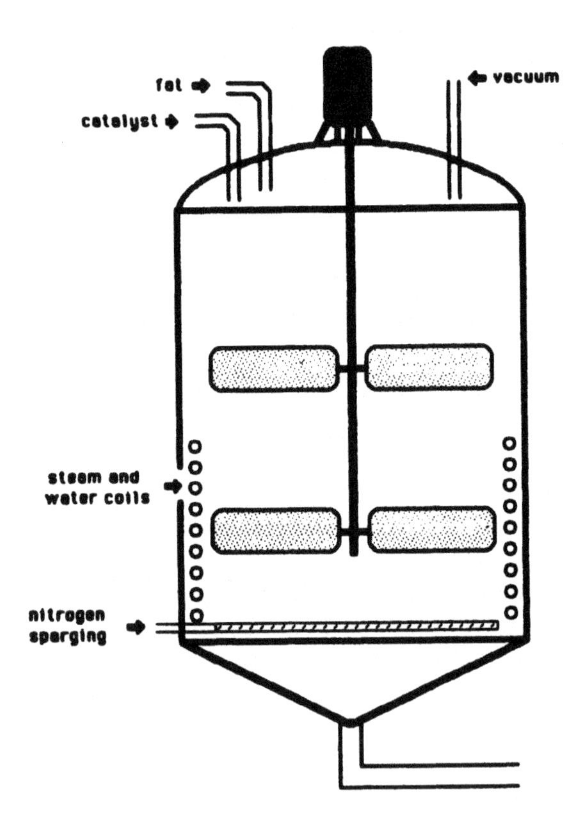

Fig. 16.7. Typical batch interesterification vessel. *Source:* Laning, S.J., *J. Amer. Oil Chem. Soc. 62:* 400 (1985).

Processing Vacuum

Reacting under a vacuum is the most desirable method to do batch interesterification. The major advantages are maximum exclusion of oxygen and increased drying efficiency. However, a suitable alternative is nitrogen sparging during the initial and final drying steps. This alternative requires extra attention during the initial reaction step, because continual sparging at this point will promote emulsification and excessive foaming, which could also cause safety concerns.

Meeting the feedstock requirements (Table 16.4) is critical. Crude oil is not suitable for use in interesterification reactions because of the presence of free fatty acids, moisture, and unsaponifiable matter. Table 16.5 shows the amount of catalyst inactivation by these "poisons."

Conventional interesterification catalysts will react with free fatty acids, reducing the amount of catalyst available. Additionally, large-scale production of interesterified oil blends typically have an optimum "treat (or dose) window," above and below which catalysts become inefficient. This is why a "stubborn" batch should never be overtreated with catalyst in an attempt to force it; an extremely tight emulsion will result.

The following are two methods of introducing the catalyst safely.

- Make a slurry with cold oil from the feedstock. If the feedstock is a blend that does not remain fluid at room temperature or below, then refined, bleached, and deodorized salad oil, such as soybean oil, can be used. If this is the desired practice, a dedicated premix tank must be considered in the initial design.
- If the reactor is equipped with a recirculation line with a separate pump to circulate from the bottom to the top of the reactor, a stainless steel flex hose connected to a side valve on the recirculation line will provide a means of "vacuuming" the catalyst directly from the shipping container, since the reactor is under vacuum. This would eliminate the need to handle the catalyst physically, provided a scale is available to weigh by difference. Furthermore, the side valve for the flex hose should also be tied into the nitrogen line to prevent incorporation of air during the catalyst addition step. Having a recirculation line facilitates fast, intimate mixing.

TABLE 16.5 Inactivation of Catalysts by Poisons (5)

Poison		Catalyst inactivated (lbs/1000 lbs of oil)		
Type	Level	Sodium	Sodium methylate	Sodium hydroxide
Water	0.01%	0.13	0.3	—
Fatty acid	0.1 unit AV	0.04	0.1	0.07
Peroxide	1.0 unit PV	0.023	0.054	0.04
Total catalyst inactivated:		0.193	0.454	0.11

Reaction Step

The following is a general description of production-scale batch interesterification under vacuum, using sodium methylate as the catalyst, without the benefit of a refining centrifuge. Critical control points are marked with an asterisk (*). *Note:* The catalyst needs to be visually inspected for the "flowability" first. If the catalyst is caked, has an appreciable amount of lumps, or does not flow, do not use.

1. *Fill the tank with oil or oil blend that meets the feedstock criteria.
2. Turn on the agitator and adjust for gentle to moderate agitation without incorporating air.
3. After achieving 225°F (107°C) and an absolute pressure of 54 mm Hg (0.07 Atm), allow the oil to dry for 1 hr.
4. After 1 hr, obtain a sample for determination of melting point, such as Mettler dropping point. Dropping point analysis is recommended because of the time it takes for the analysis.
5. *Add the appropriate amount of catalyst. The usual amount of sodium methylate added is 0.1% to 0.5%, depending on the quality of the feedstock. More than 0.5% should be used only with great discretion, because difficulties may be encountered in the washing steps.
6. If appropriate, circulate during the entire reaction time with agitators on.
7. After 30 min, take another sample for dropping point.
8. If the reaction has reached equilibrium turn on cold water to the coils and cool to 200°F (93°C).
9. Turn off circulating pump; break the vacuum; leave the agitator on.

Washing Steps

1. *Add 5 to 10% hot water at 82 to 88°C (180 to 190°F).
2. Immediately turn the agitator off. Overagitation at this stage will cause emulsification.
3. Let the batch sit undisturbed for 1 to 1.5 hr. The degree of separation can be checked by inspecting the discharge after opening the bottom valve.
4. *"Still wash" (no agitation) with another 5 to 10% *hot* water.
5. After another 1 to 1.5 hr, begin dropping (draining) the aqueous layer. If there are signs that the oil may be emulsified, let the mixture sit for another 30 to 45 min.
6. Repeat the "still wash" step again if the discharge is still extremely milky-looking.
7. When the final "still wash" water is no longer milky when compared to the first still wash, turn the agitator on.
8. *Turn on water. After adding approximately 2 to 3% hot water, *turn the agitator off.* Continue filling until approximately 10% treat is achieved.
9. Let the batch sit for 1 to 1.5 hr and drain the aqueous layer.
10. The aqueous layer should now be relatively clear. If not, repeat steps #8 and #9 until the decanted water is clear or slightly hazy.

Drying Step

1. Turn on agitator.
2. *Close the vessel and achieve 107°C (225°F) and 54 mm Hg absolute pressure (29″ of vacuum). *Note:* If operating a system without vacuum, heavy nitrogen sparge provides protection against oxidation, helps color preservation, and speeds up moisture removal.
3. Allow the batch to dry for 45 min to 1 hr.
4. When oil is dry (check by sampling), turn on cold water to coils and reduce batch temperature to 60 to 63°C (140 to 145°F).
5. After achieving temperature, turn off agitator and pump to bleaching.

Use of citric acid. Before the water washing step, addition of a citric acid solution with a concentration that reflects the correct molar amount based on the amount of catalyst used (see step 13 of the procedure in the section on "Laboratory Experimentation") offers the following potential benefits: emulsion control, increased efficiency of soap removal, yield improvement, and trace metal removal. Since different oils and oil blends tend to behave differently during interesterification, citric acid may have utility, especially if emulsion control is a consistent problem.

Use of phosphoric acid or carbon dioxide. Catalyst deactivation with carbon dioxide or phosphoric acid before the water washing step has been reported (8). Sodium methylate will decompose into sodium hydroxide and methanol upon contact with water. Both can convert neutral oil into soaps and mono- and diglycerides, reducing yields.

Emulsion Management

Should the batch refuse to separate because of a stubborn emulsion, there are two ways to effect separation: add sodium chloride (salt) or lower the pH. Changing the pH can be done with either citric acid or phosphoric acid.

Salt, preferably in solution, can be added in an amount equal to the amount of catalyst used. Usually one to three treatments will break the emulsion at least enough to manage further separation with either water or aqueous citric acid.

Lowering the pH with citric or phosphoric acid usually has a greater effect than sodium chloride. An aqueous solution with a concentration that reflects the correct amount (see step 13 of the procedure in the section on "Laboratory Experimentation") is required. Multiple treatments may be necessary to effect separation. Citric acid is preferable to phosphoric acid because of safety considerations.

End Point Detection

The most convenient method for detecting the end point of production-scale volumes is determination of melting, dropping, or "slip" point. The methods, however, become less effective when there is little difference before and after interesterification, as is often the case with some animal fats.

Another method that may be more conclusive for production monitoring is a variation of the solid fat index (SFI) method. Selection of an appropriate SFI temperature (bath) may allow obtaining a direct reading of the dilatometer to decide whether the reaction has reached equilibrium.

Glyceride analysis by thin layer chromatography (TLC) by silver nitrate complexing has also been reported (9). A blend of highly polyunsaturated liquid vegetable oil and fully hydrogenated hardfat was monitored for the disappearance of trisaturated (S_3) glycerides. When the S_3 reached a minimum, the reaction was concluded to be at equilibrium.

Gas chromatography (GC) of triglycerides is a fairly rapid method and is good for oils or oil blends that contain shorter-chain fatty acid glycerides. Profiles before and after interesterification will have different "fingerprints," which will show when equilibrium is achieved.

Other, less practical methods for production monitoring include mass spectrometry, pancreatic lipase hydrolysis, X-ray diffraction, differential scanning calorimetry, and pulsed nuclear magnetic resonance (NMR).

Continuous Interesterification

Keulemans, et al. (10) describe a process for continuous interesterification in their patent. The process first homogenizes a single stream of catalyst solution (aqueous sodium hydroxide), glycerol, and the oil or oil blends. The mixture then passes through a vacuum drier to concentrate the sodium hydroxide and form active catalyst.

After the aqueous phase of the catalyst solution is reduced, the mixture proceeds to a "reactor coil," whose length dictates the residence time. Water is then injected into the stream on the discharge side of the reactor coil to stop the reaction. The interesterified oil can now go on to refining for primary centrifugation (initial soap removal); secondary centrifugation (water wash), bleaching, and deodorization.

As mentioned earlier in the chapter, water is generally detrimental to the interesterification process, in terms of safety and catalyst inactivation. However, part of the elegance of this process lies in the necessary presence of water to effect the introduction of finely divided catalyst. The subsequent vacuum drying step results in catalyst activation through a sodium glycerolate intermediate due to the presence of glycerol.

Glycerol has the additional benefit of promoting preferential reactivity. This allows increased control of competing reactions that, in turn, can reduce oil loss due to soap formation. Also, if the sodium is not bound to a fatty acid, it is available for catalyst activation. The presence of glycerol also promotes the formation of mono- and diglycerides before the drying step. The mixture then has the benefits of emulsifiers offering intimacy of mixing beyond what the homogenizer alone can provide.

Another potential benefit of this process lies in relaxing the criteria previously emphasized for the feedstock. The sodium hydroxide solution will neutralize free fatty acids in crude oil. However, the oil should be free of particulates, because they will offer other active sites for other reactions to occur. This also suggests that upper limits on the amounts of unsaponifiable matter be part of the criteria if crude oil is going to be used (that is, degummed oil).

Applications

Recent reports suggest that *trans* fatty acids may increase the risk of coronary heart disease if certain levels are consumed routinely (11) (see Chapter 23). This has generated renewed interest in developing fats with plastic properties suitable for shortening and margarine applications, because the melting and solidification characteristics unique to hydrogenated vegetable fats are primarily due to *trans* fatty acids. The only alternative for achieving comparable performance is to interesterify blends of liquid (unhydrogenated) oil with fully hydrogenated hardfat. Straight blends of liquid oils and fully hydrogenated hardfat will have an SFI profile with a slope equal to zero, because the solid phase of the blend melts well above 40°C. After interesterification the SFI will shift to show higher solids at 10°C than at 40°C.

Significant amounts of fully hydrogenated hardfat are required to begin approaching the necessary physical properties for some applications. Product developers who are considering interesterification as an alternative to hydrogenation should decide whether a significant increase in saturated fatty acid content is an appropriate compromise for the particular application.

Although interesterification will increase the plastic range of an oil blend, the analytical data from hydrogenated oils should not be simply translated to interesterified oils. Interesterified oils tend to be more "forgiving" than hydrogenated counterparts with comparable melting point and SFI. Even though the melting point and SFI of an interesterified blend may be higher than desired, the mouthfeel is not necessarily unacceptable, even though unacceptable mouthfeel (such as "waxy," "grainy" or "sandy") can usually be predicted from the high solids and melting point of hydrogenated oils.

The most important part of any strategy directed toward development of a new product is a genuine understanding of the customer's desires. Whether the customer is interested in "no hydrogenation" or "no *trans* acids" defines what the capabilities and limitations are, in terms of providing a suitable *trans*-free replacement.

It should be pointed out that detectable levels of *trans* acids can be found in unhydrogenated oils. This has been attributed to high-temperature deodorization or bleaching conditions. These acids also occur naturally in animal fats from ruminants.

As mentioned in the introduction, changes in physical properties of an interesterified oil or oil blend ultimately depend on the type and amount of oils used. This is evidenced in Tables 16.6 and 16.7. Note that in Table 16.6 there is little difference in melting points. However, Table 16.7 clearly shows significant differences in the melting performance of the same blends. Furthermore, comparison of Tables 16.6 and 16.7 with Table 16.8 affords the product developer the ability to relate melting and solidification properties of various blends to the types of triglycerides found after interesterification. Efficiencies in applications development can be gained because of the data contained in these tables.

Frying

Cottonseed, peanut, high-oleic sunflower, palm, and coconut oils are alternatives to hydrogenated products for frying applications. Interesterified blends of highly polyunsaturated oils and fully hydrogenated hardfat have little or no benefit for developing a no-*trans* solid frying fat, because interesterification maintains the initial polyunsaturation of the starting material.

Margarine and Spreads

Fats and oils used for the production of margarine and spreads are required to meet specific performance criteria. Specifically, they must maintain their integrity at room temperature (for a reasonable amount of time), be relatively unaffected by temperature cycling, melt cleanly at or about body temperature, and contribute to acceptable flavor release. Additionally, they must resist becoming "grainy" or "sandy" over time. These criteria are met by using β′-stable fats. Table 16.9 shows that oils that

TABLE 16.6 Fatty Acid Composition (FAC) and Melting Points of Interesterified Blends (1 : 1, wt%) of Fully Hydrogenated Soybean Oil with Nine Vegetable Oils (12)

Vegetable oil in blend	Fatty acid composition (area %)									Melting point, °C[a]
	10:0	12:0	11:0	16:0	18:0	18:1	18:2	18:3	20:1	
Palm	—	—	—	26.6	46.3	22.4	4.7	—	—	48.2
Coconut	1.6	19.2	9.0	10.8	59.0	2.4	—	—	—	41.0
Cottonseed	—	—	—	17.3	46.1	10.7	25.9	—	—	50.0
Peanut	—	—	—	11.4	46.0	25.8	15.5	0.8	0.5	51.0
Soybean	—	—	—	10.2	46.9	12.3	26.6	4.0	—	52.0
Corn	—	—	—	10.7	45.0	14.1	29.5	0.7	—	52.0
Sunflower	—	—	—	8.2	46.9	10.0	34.9	—	—	51.0
Safflower	—	—	—	8.4	45.3	8.8	37.2	0.3	—	51.0
Canola	—	—	—	6.8	42.8	34.6	10.7	4.4	0.7	50.0

[a]Melting point determined by capillary tube method.

TABLE 16.7 Solid Fat Content[a] of Interesterified Blends of Vegetable Oil with Fully Hydrogenated Soybean Oil (1 : 1, wt%) (12)

Vegetable oil in blend	% Solid fat at °C										
	0	10	21	27	33	38	40	43	46	49	54
Palm 3 : 1[b]	63.4	55. 9	43.2	38.3	23.0	17.2	13.3	7.1	2.2	1.1	0.0
Palm	86.2	80.5	73.3	71.5	54.2	44.8	37.2	28.9	21.2	14.0	0.0
Coconut	97.0	89.6	73.4	65.4	40.6	28.1	19.6	11.9	7.5	4.6	0.3
Cottonseed	69.0	58.8	43.0	30.4	24.8	22.0	20.4	15.6	10.6	5.7	1.9
Peanut	65.1	55.7	40.6	38.3	26.9	21.0	17.2	12.2	7.4	6.3	1.0
Soybean	63.0	53.3	38.5	33.8	23.8	19.4	16.3	13.7	9.8	7.5	3.8
Corn	59.7	48.9	33.1	30.4	19.7	17.5	14.0	10.5	8.2	6.2	1.7
Sunflower	60.9	51.0	36.3	30.1	21.3	17.6	14.9	11.9	9.4	7.2	3.0
Safflower	58.7	48.8	34.8	28.0	19.5	16.2	13.5	10.6	8.4	6.2	2.5
Canola	53.8	43.2	32.1	30.1	17.9	14.9	10.8	8.0	6.8	5.6	1.1

[a]Determined by pulsed NMR spectrometry.
[b]Blend of palm oil (75%) and fully hydrogenated soybean oil (25%).

normally have β crystallization tendencies shift significantly toward β′ after interesterification. This suggests that oils once considered inappropriate for certain applications may now be worth evaluating in an interesterified system.

List, et al. (1) reported a method for the preparation of a "zero" trans tub-type margarine oil by interesterifying a blend of 80% refined, bleached, and deodorized (RBD) unhydrogenated soybean oil and 20% RBD fully hydrogenated soybean hardfat. The resulting SFIs are compared to those of conventional tub and stick margarine oils in Table 16.10.

TABLE 16.8 Triacylglycerol Composition[a] of Blends of Vegetable Oil with Fully Hydrogenated Soybean Oil (1 : 1, wt%), Before (B) and After (I) Interesterification (12)

Vegetable oil in blend	Triacylglycerol[b] (area %)							
	U_3		U_2S		US_2		S_3	
	B	I	B	I	B	I	B	I
Palm	9.5	2.3	20.5	13.7	13.3	43.4	56.7	40.6
Cottonseed	15.7	9.8	24.3	31.2	8.5	33.9	51.5	25.1
Peanut	29.0	6.2	15.3	31.1	2.4	47.3	53.3	15.4
Soybean	31.0	9.2	17.1	32.1	1.5	46.4	50.4	12.3
Corn	31.8	9.0	15.3	35.2	1.1	45.9	51.8	9.9
Sunflower	37.6	8.3	11.3	34.3	0.3	46.9	50.8	10.5
Safflower	37.4	17.2	11.5	38.9	0.4	34.0	50.7	9.9
Canola	39.1	9.2	8.5	37.1	0.7	44.2	51.7	9.5

[a]Determined by reverse-phase high-performance liquid chromatography with flame-ionization detector.
[b]Fatty acids in triacylglycerol: U = unsaturated. S = saturated.

TABLE 16.9 β and β′ Crystallization Form Content[a] in Blends of Vegetable Oil with Fully Hydrogenated Soybean Oil (1 : 1, wt%) Before and After Interesterification (12)

Vegetable oil	Before (%)		After (%)	
	β	β′	β	β′
Palm	55.0	45.0	28.0	72.0
Coconut	53.0	47.0	25.3	74.7
Cottonseed	62.4	37.6	41.0	59.0
Peanut	59.0	41.0	48.0	52.0
Soybean	75.3	24.7	54.0	46.0
Corn	77.0	23.0	52.0	48.0
Sunflower	100.0	0.0	63.0	37.0
Safflower	81.2	18.8	62.0	38.0
Canola	77.4	22.6	55.5	44.5

[a]Determined by x-ray diffraction at 23°C; β′ strong short spacing of 4.6Å; β′ strong short spacing of 3.8 and 4.2Å.

TABLE 16.10 Composition and Properties of Hydrogenated and Interesterified Margarine Oils (1)

		Solid fat index[a]			Melting point	Fatty acid composition (wt%)[b]				
Oil	Type	10	21.1	33.3	(C)[c]	S	M	D	T	P:S ratio
A. Hydrogenated	Stick margarine	28.6	18.9	5.3	46	23.1	49.9	24.4	2.6	1.17
B. Hydrogenated	Tub margarine	15.6	8.8	1.3	46	18.8	42.9	33.8	4.8	2.05
C. Hydrogenated	Tub margarine	7.1	4.5	2.0	46	17.9	30.1	45.5	6.5	2.90
D. Interesterified	90:10[d,e]	1.7	1.3	0.2	40	23.2	18.4	51.0	7.5	2.52
E. Interesterified	85:15	4.3	2.2	0.9	46	27.6	17.3	48.0	7.1	2.00
F. Interesterified	80:20	8.0	3.5	2.2	47	31.7	16.6	44.8	6.7	1.62

Temperatures measured in Celsius.
[a]By dilatometry. Temperatures measured in Celsius.
[b]S = saturated, M = monoene, D = diene, T = triene.
[c]By differential scanning calorimetry.
[d]Parts soybean oil: parts soy trisaturate by weight.
[e]As simple mixtures the soy-soy trisaturate blends contained 1.5% *trans* by gas–liquid chromatography.

The investigations reported that by increasing the soybean oil hardfat component, the SFI values at 10, 21.1, and 33.3°C increase, and that for each 5% increase, the SFI values roughly double.

Labeling

The United States currently has no labeling guidelines for interesterified fats and oils. However, if a blend contains a mixture of interesterified unhydrogenated oil and fully hydrogenated hardfat, the hydrogenated oil must be declared.

References

1. List, G.R., E.A. Emken, W.F. Kwolek, T.D. Simpson, and H.J. Dutton, *J. Amer. Oil Chem. Soc. 54:* 408 (1977).
2. Rozendaal, A., in *World Conference Proceedings, Edible Fats and Oils Processing: Basic Principles and Modern Practices*, edited by D.R. Erickson, American Oil Chemists' Society, Champaign, IL, 1989, pp. 152–157.
3. Macrae, A.R., *J. Amer. Oil Chem. Soc. 60:* 243A (1983).
4. Eckey, E.W., *Ind. Eng. Chem. 40:* 1183 (1948).
5. Sreenivasan, B., *J. Amer. Oil Chem. Soc. 55:* 796 (1978).
6. De Lathauwer, et al., U.S. Patent 4,284,578 (1981).
7. Klein, et al., U.S. Patent 4,243,603 (1981).
8. Laning, S.J., *J. Amer. Oil Chem. Soc. 62:* 400 (1985).
9. Freeman, I.P., *J. Amer. Oil Chem. Soc. 45:* 456 (1968).
10. Keulemans, et al., U.S. Patent 4,585,593 (1986).
11. Mensink, R.P., and M.B. Katan, *N. Eng. J. Med. 323:* 439 (1990).
12. Zeitoun, M.A.M., W.E. Neff, G.R. List, and T.L. Mounts, *J. Amer. Oil Chem. Soc. 70:* 467 (1993).

Chapter 17

Soybean Oil Processing Byproducts and Their Utilization

John B. Woerfel

Consultant
Tucson, AZ

Introduction

Refining of soybean oil to make a neutral, bland-flavored, and light-colored oil results in several byproducts. Fig. 17.1 diagrams the conventional refining process and shows the byproducts that are generated at each stage.

In addition, waste oils are generated accidentally by spills and incidentally by operations such as cleaning equipment. Good practice dictates that these fatty materials be controlled at the source or recovered as saleable material to minimize waste treatment costs. Witte (1) discusses control of wastewater loads and estimates treatment costs at $0.20/lb ($0.44/kg) of oil or $0.10/lb ($0.22/kg) of organic material in water measured as biochemical oxygen demand (BOD).

These byproducts consist of various mixtures of phosphatides, unsaponifiables, glycerides, free fatty acids, and soap. The amounts of these materials present in the crude and refined oil are illustrated in Table 17.1 (2). The value of each byproduct depends on its composition.

Lecithin contains mostly hydratable phosphatides together with some free fatty acids and neutral oil (glycerides). It is used, in various modifications, for many food and industrial purposes. The amount recovered by water degumming varies depending upon the oil quality but will usually be 2.5 to 3.0% of the crude soybean oil degummed. Lecithin processing and utilization are covered in detail in Chapter 10.

Soapstock, or *refining byproduct lipids*, consists of soap, neutral oil, phosphatides, and some unsaponifiables. It is valuable as an animal feed additive and as a source of fatty acids for soap and oleochemicals. The amount of soapstock will vary depending on the quality of the crude oil and whether or not the oil is degummed before refining. The amount of soapstock recovered, as measured by Total Fatty Acid (TFA), may be less than 1% for high-quality degummed oil or may be very much higher for poor-quality oil or nondegummed oil.

If oil is physically refined rather than caustic-refined, no soapstock is generated. Instead, the fatty acids are recovered as distillate in the final physical refining.

Spent bleaching clay is not generally considered a byproduct, and much is simply disposed of in a landfill. However, it contains 20 to 45% oil, which can be utilized. In addition to recovering the value of the oil, processing the clay reduces fire hazards and pollution. Although various practices have been used over the years, growing environmental concerns have increased emphasis on finding improved methods of disposal and utilization.

Deodorizer distillate amounts to about 0.25 to 0.50% of the feed to the deodorizer. It consists primarily of unsaponifiables plus some fatty acids. It is valuable as a source of tocopherols and sterols, which are raw materials for vitamin E and pharmaceuticals (3,4).

Refining Byproduct Lipid (Soapstock)

Soapstock is the traditional name for the byproduct resulting from caustic refining of soybean and other oils. In 1992 the NOPA (National Oilseed Processors Association) *Trading Rules* (5) were changed to substitute the name *refining byproduct lipid* for *soapstock* to relate more closely to the Association of American Feed Control Officials (AAFCO) definition of a feed ingredient. This change reflects the impor-

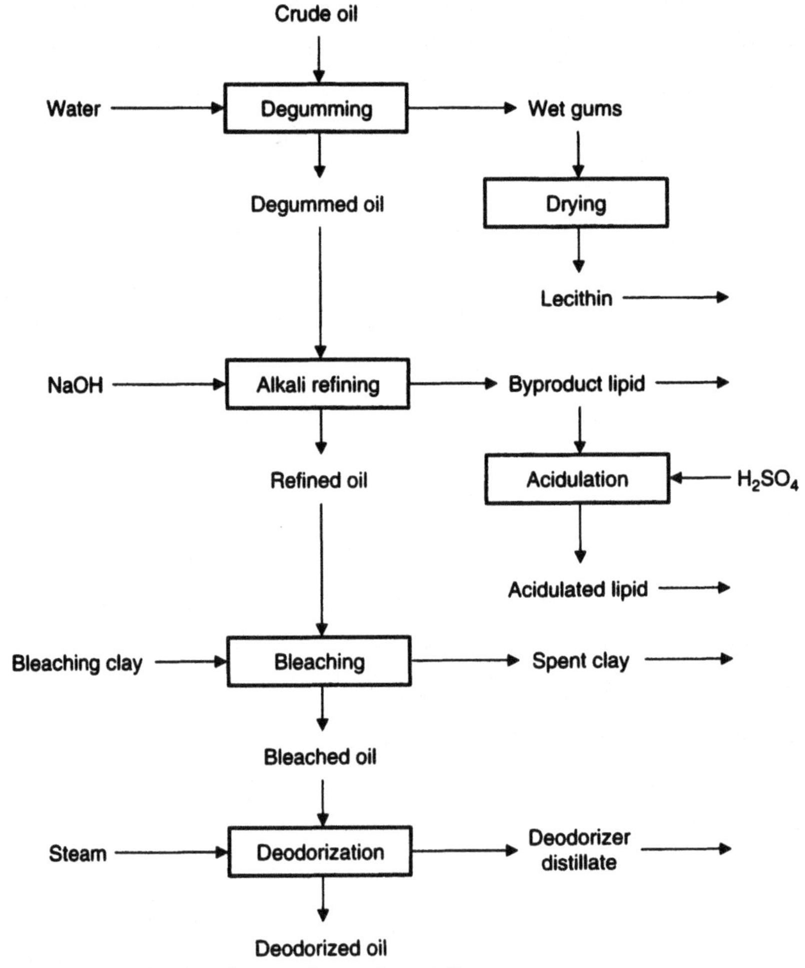

Fig. 17.1. Processing diagram for soybean oil.

tance of the product as an animal feed ingredient. In this chapter both terms are used depending on context.

NOPA contract rules specify that the "product must not be adulterated with any other oil or refining byproduct lipid made therefrom without consent of the purchaser, and nature and amount of the same must be declared in the contract" (5). This provision is significant, because soapstock from other oils may differ in properties essential to the buyer. For example, cottonseed oil derivatives are not used in poultry feed, because of concerns about gossypol, and the oleochemical industry may require raw materials with a defined fatty acid content.

The price of the raw product directly from refining is based on 50% total fatty acid (TFA) content, and the product is not merchantable if less than 30% TFA. The price of an acidulated product is based on 95% TFA, and it is not merchantable below 85% TFA.

Degumming is not essential for the refining process, and it is common practice for many processors to refine crude soybean oil without degumming (2). Soapstock from refining nondegummed oil is much greater in quantity than that from degummed oil and has a higher content of phosphatides, which would have been removed in degumming. The higher levels of phosphatides complicate acidulation and subsequent wastewater treatment.

Crude soybean byproduct lipids consist of soaps of soybean fatty acids, triglycerides, phosphatides, and degradation products as well as small amounts of sterols, tocopherols, pigments, proteins, and carbohydrates. The exact composition will depend on the composition of the feed to the refining process and refining conditions. Refining conditions are adjusted to remove free fatty acids and phosphatides with minimum hydrolysis or removal of neutral oil. Good practice is to keep the refining loss to less than 0.5% above the neutral oil loss (NOL) (6); see Chapter 11.

The principal material of value is the soybean fatty acids which are present in the form of soaps, triglycerides, and phosphatides.

Raw soapstock is difficult to handle. It solidifies readily when cooled and heated tanks and lines are necessary to maintain the temperature above 60°C. On standing, it may separate into two phases. If heated to boiling, it has a tendency to foam, especially if high in phosphatides. It contains large amounts of water and organic compounds and ferments readily unless maintained at a high temperature or treated with a preservative.

TABLE 17.1 Average Compositions for Crude and Refined Soybean Oil (2)

Average composition (%)	Crude oil	Refined oil
Triglycerides	95–97	>99
Phosphatides	1.5–2.5	0.003–0.045
Unsaponifiable matter	1.6	0.3
Plant sterols	0.33	0.13
Tocopherols	0.15–0.21	0.11–0.18
Hydrocarbons (squalene)	0.014	0.01
Free fatty acids	0.3–0.7	<0.05
Trace metals (ppm)		
Iron	1–3	0.1–0.3
Copper	0.03–0.05	0.02–0.06

To reduce weight and stabilize the product for storing and handling, raw soapstock is commonly acidulated with sulfuric acid. Traditionally, soapstock acidulation has been done in a batch process, where soapstock is charged to a corrosion-resistant tank. Wooden tanks fitted with copper or bronze coils are still used, although more recent installations are likely to be *monel metal*, *Carpenter 20 CB*, stainless steel, or fiberglass-reinforced plastic (7,8,9). Typically, sulfuric acid diluted to *ca.* 10% is added in excess to the soapstock charge and the mass is boiled with sparge steam for 2 to 4 h. The tank is then settled, and the acid water layer drawn off. The acid oil is water-washed by adding 25 to 50% water, boiling for a short time, and settling thoroughly. After the water layer is drawn off, the acidulated soapstock will be stored or shipped in steel tanks. Detailed practices may vary considerably from plant to plant and even from charge to charge.

Continuous Acidulation

Several processes have been developed for continuous acidulation of soapstock. One such process (Fig. 17.2) that is currently being practiced on soapstock from degummed soybean oil has been patented by Bloomberg and Hutchins (10).

A mixture of soapstock and centrifuge flush water is delivered at a controlled rate into an acidulation vessel, together with a metered stream of sulfuric acid, to bring the pH to 1.5 to 2.0. The vessel is designed so that the contents are both mixed and heated with sparge steam to approximately 90°C. The mixture constantly overflows into a settling basin. In the settling basin the acid oil floats to the top and the acid water settles to the bottom. A series of valves at different levels are provided for drawing off the acid oil. The acid water overflows through a standpipe, which maintains the level in the settling basin. This process has been in successful use for several years. It is limited to use on soapstock from oil that has been thoroughly degummed, before even small amounts of gums create emulsions that will not separate.

De Smet (Brussels, Belgium) offers a process (Fig. 17.3) that is made continuous by providing three reactor tanks in parallel. Each tank is charged successively. The sequence of acidulation, water washing, and decanting is performed in each vessel. The acid layer from one vessel may be mixed into the soapstock charged to the next vessel to economize on acid usage.

The continuous acidulation process (11) described by Braae (Fig. 17.4) removes the bulk of the acid water by decanting. Wash water is added to the acid oil, and the water and solids are removed by centrifuging in a self-cleaning separator. The use of a surfactant, ethylhydroxyethylcellulose (50 to 100 ppm), to break emulsions is also described.

Another continuous acidulation process has been described by Crauer (12). In this process, continuous centrifuging is used to separate the acid water and acid oil. It is claimed that this reduces the fat content of the acid water to less than 0.4%, and the acid water stream is improved over the batch process in having a higher pH and lower fat and BOD content. A further process step, to neutralize the acid water continuously with lime and clarify it centrifugally, is said to reduce the BOD by 62 to 76% and to remove 80 to 95% of invert sugars and all of the fat.

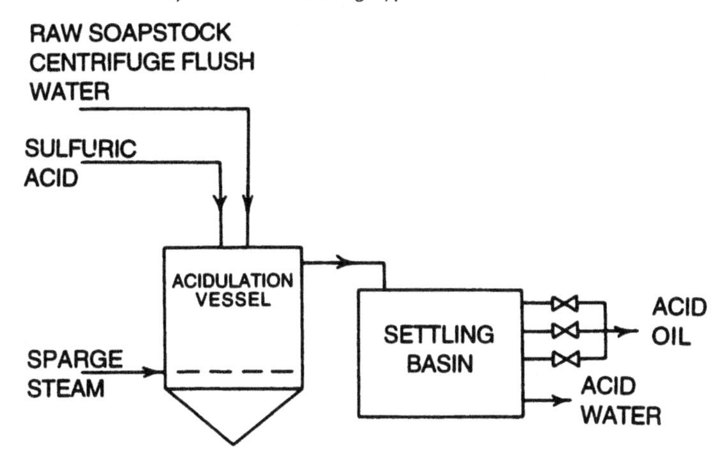

Fig. 17.2. Continuous acidulation of degummed soy oil soapstock. *Source:* Woerfel, J.B., *J. Amer. Oil Chem. Soc. 60(2):* 310 (1983).

Morren describes continuous acidulation (13,14), as shown here in Fig. 17.5 using a Podbielniak centrifuge. The residual TFA in the acid water is reported as 0.03 to 0.74% for different types of oils.

The use of centrifuges for separating acid oil from acid water in continuous soapstock systems has not been widely accepted in the United States, apparently because of the expense of the equipment and the severe corrosion problems inherent in the process.

A particularly interesting development was described by Mag et al. (15) and is shown in Fig. 17.6. In this work, they adopted a practice from petroleum oil/water separation technology. After decanting, the acid water is passed through a fiberglass bed coalescer and then to a smaller decanter tank. Fatty material concentrations of

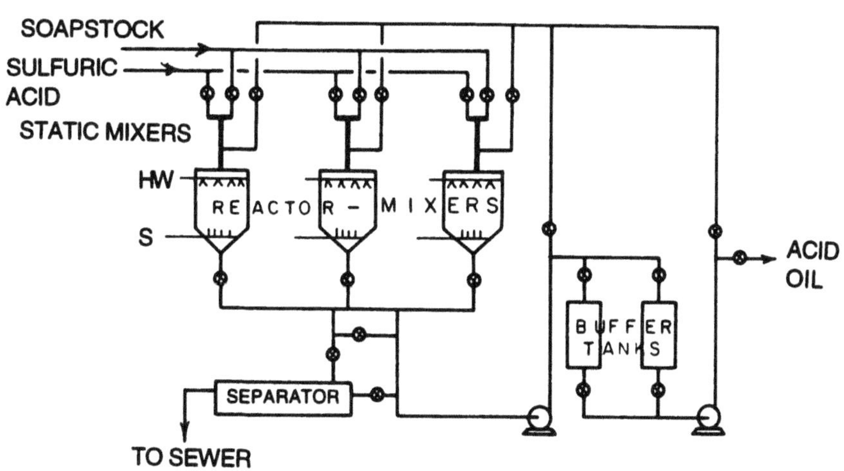

Fig. 17.3. Continuous soapstock acidulation. *Source:* DeSmet, Brussels, Belgium.

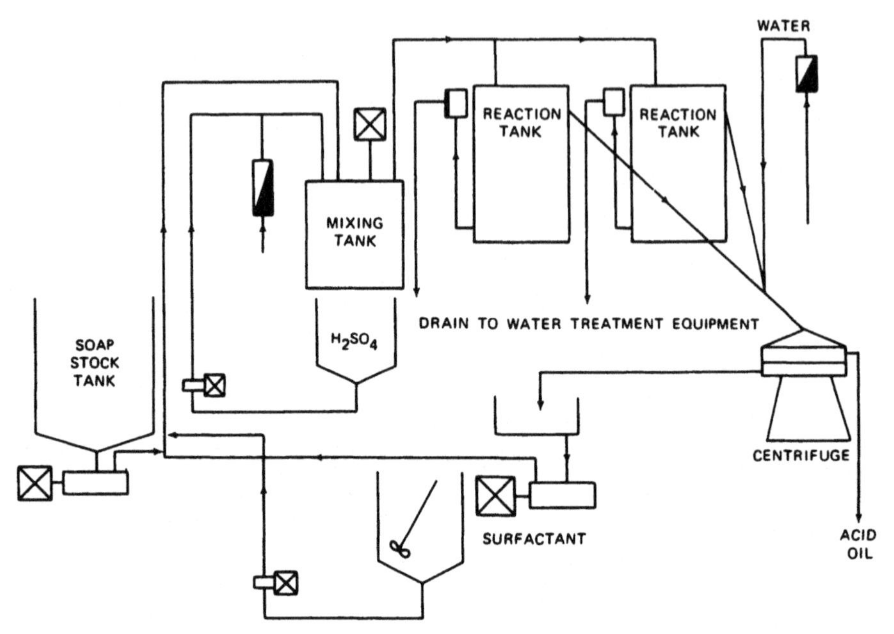

Fig. 17.4. Continuous acidulation. *Source:* Braae, B., *J. Amer. Oil Chem. Soc. 53:* 353 (1976).

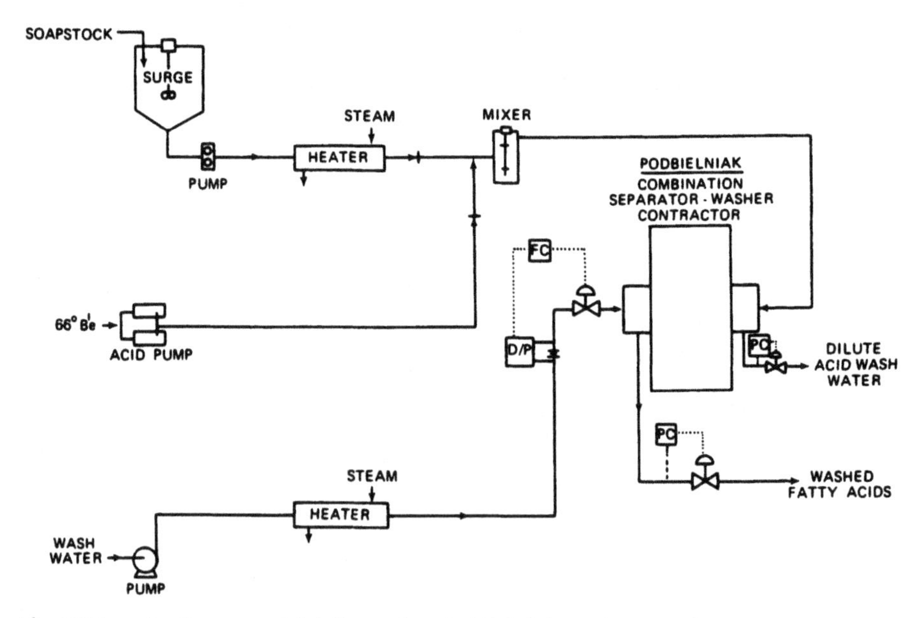

Fig. 17.5. Continuous acidulation using Podbielniak equipment. *Source:* Woerfel, J.B., *J. Amer. Oil Chem. Soc. 60:* 310 (1983).

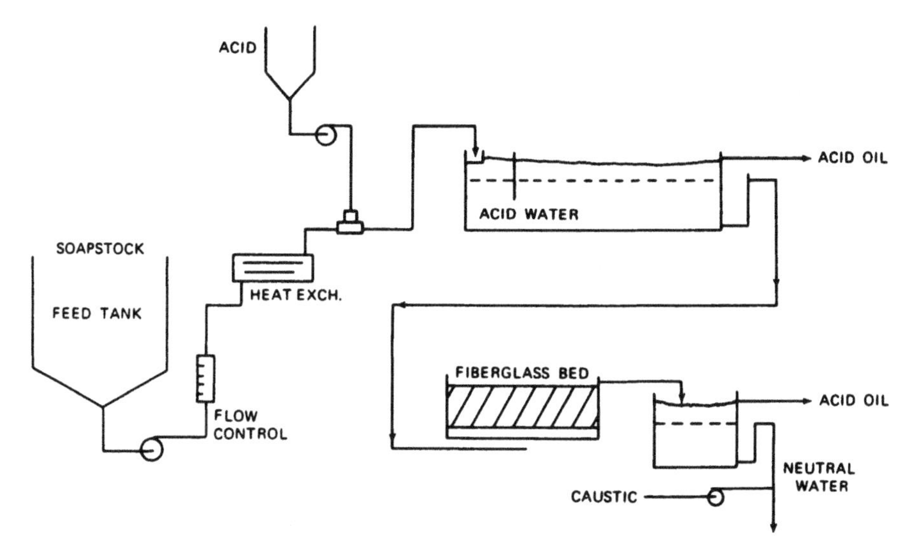

Fig. 17.6. Continuous acidulation and recovery of acid oil. *Source:* Mag, T.K., D.H. Green, and A.T. Kwong, *J. Amer. Oil Chem. Soc. 60:* 1008 (1983).

the acid water of less than 150 ppm can be achieved. The acid water is then neutralized for discharge to municipal sewers.

Acidulated soapstocks are largely water-free and contain approximately 20 to 30% triglyceride oil, 65 to 70% total fatty acids, and about 5% tocopherols, sterols, degraded oxidized components, pigments, salts, and color bodies (16). Some refineries preboil soapstocks with excess caustic to saponify the neutral oil components completely before acidulation. Acidulated preboiled soapstocks will have mostly free fatty acids and only small amounts of glycerides and other fatty acid compounds.

Witte (1) states that the problem of oily materials in acid water is associated with phosphatides, which are not broken down in the acidulation step and act as oil emulsifiers in the decantation step. These materials may be more effectively broken down in acidulation by the use of more severe conditions of pH and temperature. Pressure acidulation allows temperatures in excess of the boiling point of the aqueous mixture. Operation of the pressure process in one plant is said to give improved product yield and lower wastewater loads.

Acid water is high in BOD and low in pH. When properly treated to adjust pH and remove immiscible materials, it is readily biodegradable. Table 17.2 shows estimates by Boyer (17) of the wastewater loads from different operations for a plant that extracts, refines, and produces margarine, mayonnaise, and salad dressings. Acidulation is the largest source of BOD, amounting to 45% overall, or 70% if the margarine and salad oil departments are excluded.

An alternative to acidulation for processing refining lipids at the refinery location is drying the crude material. The dried material is suitable for animal feeding, and the treatment of wastewater is eliminated. If the refinery is adjacent to an extrac-

TABLE 17.2 Edible Oil Processes and Waste Loads (17)

	Waste load				
	Flow, gpd	BOD, lb/d		FOG, lb/d	
Process	average	Average	Maximum	Average	Maximum
Milling and extraction	75,000	370	600	25	65
Caustic refining	11,000	220	1,000	115	400
Further processing	5,000	150	300	75	150
Deodorization	5,000	40	100	20	50
Acidulation	19,000	3,200	5,000	25	800
Tank car washing	5,000	250	1,500	125	250
Packaging	10,000	250	1,000	125	500
Subtotal	130,000	4,480	8,500	510	2,215
Margarine production	70,000	600	1,000	300	500
Salad dressing/mayonnaise	50,000	2,000	3,500	1,000	1,700
Total	750,000	7,080	13,000	1,810	4,415

FOG, fat, oil, grease; BOD, biochemical oxygen demand.

tion plant, the raw soapstock may be neutralized and added to the soybean meal in the desolventizer-toaster (DT), where it is mixed and subsequently dried with the meal. If only the output of the extraction plant is being refined, the fat content of the meal will typically be increased by about 0.4%. This approach is limited to refineries close to an extraction plant.

Another approach is to dry the soapstock, which makes it a stable product that can be stored and transported.

Patents for drum drying of soapstock were issued to workers at Central Soya (Ft. Wayne, IN) (18,19). The dried product contains 0.8% moisture, 67% fatty acid and 608 ppm (0.068%) xanthophyll and is inhibited against hygroscopicity by use of calcium phosphate. It is suited for use in poultry feed.

Beal and Sohns (20) of the U.S. Department of Agriculture (USDA) laboratory developed a process of neutralizing with sulfuric acid and drying under vacuum in either a natural-circulation evaporator or scraped-film evaporator. The product is a waxlike solid at room temperature, containing between 54 and 71% TFA. When made from nondegummed soybean oil, a high level of carotene and xanthophyll was retained. Important advantages were improved pigmentation in shanks of broilers and in egg yolks.

Although drying produces a product useful for animal feeding, the product is not suitable as a raw material for other industrial uses such as soap and oleochemicals.

Utilization of Soybean Refining Lipids

The primary use is for animal feeds, with lesser use in soap and oleochemical manufacture. Table 17.3 shows a projection of 1995 usage for several purposes (16).

For feed, the acidulated dried, or even raw, product may be shipped and used directly. For other purposes, the raw or acidulated material is a raw material for more sophisticated processing.

Individual U.S. refineries rarely process refining lipids beyond acidulation or drying. Some refineries, especially those that refine nondegummed oil, prefer to ship raw soapstock rather than to deal with the problems and costs of acidulation and treatment of acid water. The value of the soapstock represents only 1 or 2% of the total product value from a refinery, and acidulation and wastewater treatment may require an inordinate investment and commitment of management time (21).

In general, the processing to finished soaps, distilled fatty acids, and oleochemicals has been done in specialized plants that draw raw materials from a number of sources. Such plants may be part of organizations that include refineries, or they may purchase on contract from others. In regions where there is a concentration of refineries, a contract buyer can operate trucks to collect raw soapstock on a regular basis. With appropriate traffic management, rail shipments of raw soapstock are feasible.

Soybean oil soapstock–derived fatty acids have the same fatty acid composition as soybean oil–derived fatty acids and are used for similar purposes by the oleochemical and detergent industries. The value of the soapstock-derived acids is less than the crude oil value by the difference in glycerol credit and because of quality differences, such as higher color or degree of oxidation, which limit usage in some higher-grade products.

Industrial uses include ore flotation, alkyd resins, foundry products, medium-grade industrial soaps and as a fatty acid feedstock for nitrogen derivatives and other oleochemicals (see Chapter 21) (22).

Sonntag (16) reviewed a number of processes that are used or proposed for processing soapstocks. Of particular interest are processes that convert the acids to methyl esters. He points out that esters have been replacing corresponding fatty acids for a number of reasons. They are more easily produced, more economically fractionally distilled, more reactive, more thermally stable, less corrosive, and basically cheaper.

In addition to fatty acids, soapstocks contain glycerol (combined as glycerides) and unsaponifiables. Although these are recovered in some operations, it is not always economical to do so because of the small quantities.

TABLE 17.3 Future Prospects for U.S. Soybean Soapstocks (16)

Product	1987 U.S. volumes lbs	Growth potential %/yr	1995 U.S. volume lbs product
Dimer acid manufacture	3 MM	small	3.1 MM
Low-grade oleochemicals	10 MM	5	15 MM
Pet food and feed fortification	260 MM (1.75 MMM all fats)	3	329 MM
Low-grade cleaning, soap, and other surfactant products	20 MM	5	31 MM

Deodorizer Distillate Recovery

Deodorizer distillate is the volatile organic material recovered as a valuable byproduct in the deodorization of fats and oils. Deodorization, a high-temperature, high-vacuum steam distillation, discussed in Chapter 14 is the last step in processing, which improves taste, odor, color, and stability by removing undesirable volatiles (such as free fatty acids, aldehydes, ketones, and other compounds formed by the decomposition of peroxides) and pigments.

Winters (4) discusses deodorization as related specifically to recovery of deodorizer distillate. In particular he presents data on relative volatility (Table 17.4) and vapor–pressure temperature (Fig. 17.7) relationship for key components.

The factors affecting volatile removal are the following:

1. High temperature increases vaporization by increasing pressure of the key components.
2. Lower absolute pressure (high vacuum) decreases the weight of steam needed by increasing its volume. At 4 mm Hg, 33% more steam is required than at 3 mm Hg.
3. Increased volume of steam increases rate of volatilization and decreases time required.
4. Different conditions will affect the composition of the distillate because of the differences in relative volatility.
5. A shortened time at higher temperature decreases hydrolysis and other undesirable reactions.

Deodorizer operating conditions are based on oil quality rather than distillate concerns. In the course of deodorization, a portion of the tocopherols and sterols are vaporized. Complete removal is not necessary, and residual tocopherols are desirable for their antioxidant effect in the finished oil.

Distillate was originally recovered from the *hotwell oil* collected from deodorizer barometric condenser systems. About 1960, wet scrubbing systems using aqueous solutions were introduced. These were soon replaced by the dry recovery systems used today, which condense by recirculated cooled distillate, thus producing a dry product.

Fig. 17.8 (22) is a schematic flow sheet for an EMI distillate recovery system. Steam and distillate vapors coming from the deodorizer at a pressure of 2 to 5 mm Hg are compressed to 50 to 60 mm Hg through one or two booster stages. The vapors pass upward through a packed tower, where they are contacted with liquid distillate cooled to approximately 60°C (140°F) by circulating through a heat exchanger. A

TABLE 17.4 Relative Volatility of Some Key Components of Vegetable Oil (4)

Component	Molecular weight	Relative volatility
Fatty acid	280	2.5
Squalene	411	5.0
Tocopherol	415	1.0
Sterol	410	0.6
Sterol ester	675	0.038
Oil	885	small

level controller in the bottom of the tower maintains a constant volume of condensed distillate in the system; the excess is continuously drawn off.

Packing in the tower is an open grid of stainless steel, to minimize the pressure drop while providing good liquid-vapor contact. Demisters are installed to minimize entrainment losses. It is very important that these be of proper design and kept in place to achieve efficient removal of droplets being carried in the high-velocity vapor stream.

Efficiency of the deodorizer distillate removal is quite high. In tests, tocopherols and sterols were recovered from the distillate and from the fatty residue in the barometric condenser water. These tests indicated more than 98% recovery of tocopherols and more than 95% recovery of sterols in the deodorizer distillate.

Fig. 17.9 (22) shows a deodorizer distillate recovery system by Elliott. This is similar in principle, but employs a scrub cooler rather than a packed tower. Also, the

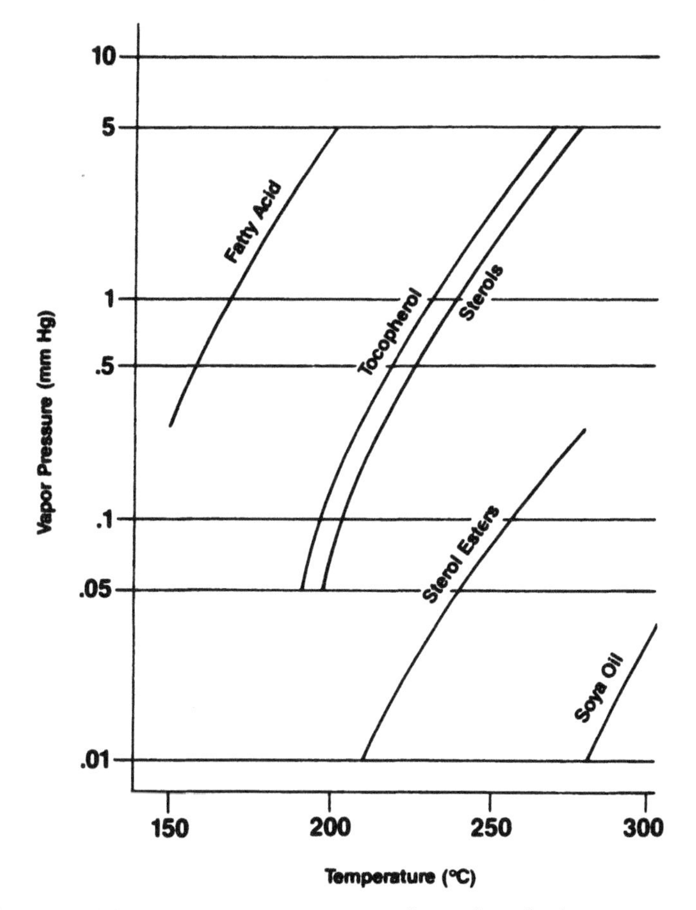

Fig. 17.7. Vapor pressure–temperature relationships for key components of vegetable oil. *Source:* Winters, Robert L., in *Proceedings: World Conference on Emerging Technologies in the Fats and Oils Industry,* edited by A.R. Baldwin, American Oil Chemists' Society, Champaign, IL, 1986, p. 186.

water cooler is an integral part of the scrub cooler. This sketch also shows that the scrub cooler is located immediately after the deodorizer and before the ejector boosters. This configuration is said to minimize steam consumption to the booster and conserve energy.

Where a distillate recovery system of this type is not installed, the deodorizer distillate is condensed in the barometric cooling water. This distillate may then be recovered as hotwell skimmings. Certain anionic polymers can be used to coagulate and float the greasy material to improve its separation from the water.

Recovery of the distillate as hotwell skimmings is not nearly as satisfactory as recovering it dry, because the skimmings contain a large amount of water, which must be removed. Yields of tocopherol may be reduced because of oxidation occurring in the barometric water cooling tower or during the drying.

Handling of Deodorizer Distillate

Steel tanks can be used for collection, storage, and transport of deodorizer distillates, providing that low temperatures are maintained and storage time is minimal. Distillate may be collected in unagitated tanks and allowed to cool, then heated only enough for loading. Protection from oxidation can be provided by nitrogen blanketing. Since the percentage of material is low, a period of several weeks may be required to collect a full tank car or truck load for bulk shipment.

Utilization of Deodorizer Distillate: Tocopherols

Processing of deodorizer distillate consists of highly sophisticated techniques to separate the components including molecular and fractional distillation and solvent separa-

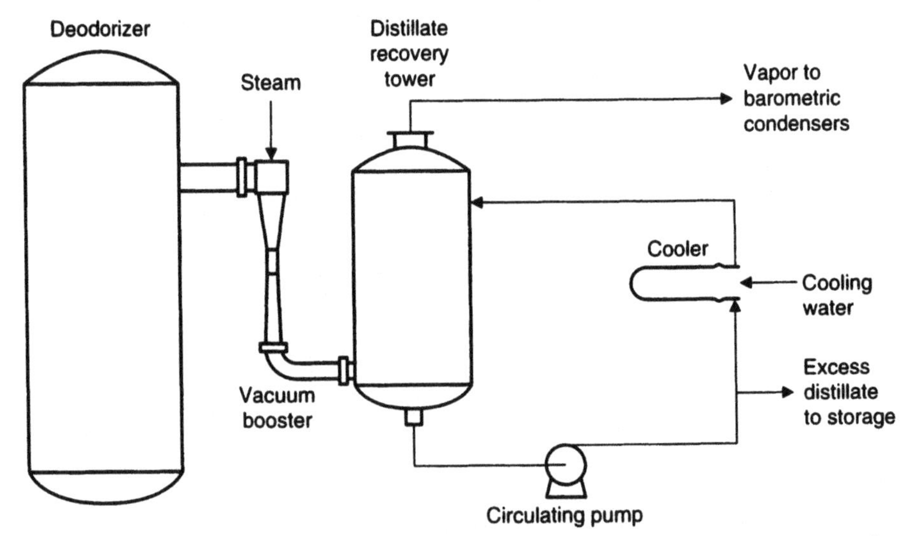

Fig. 17.8. Deodorizer distillate recovery system. *Source:* Woerfel, J.B., *J. Amer. Oil Chem. Soc. 583:* 188 (1981).

tion. In the United States there are two companies that purchase and process distillates, Henkel Corporation (LaGrange, IL) and Eastman Chemical Products (Kingsport, TN).

The primary products from distillate are vitamin E (D-α-tocopherol), mixed tocopherols used as antioxidants, and sterols used for manufacture of pharmaceuticals. A byproduct is the fatty acids remaining after removal of the sterols and tocopherols. These fatty acids are of low quality because of the extensive processing and are limited to nonfood, nonfeed, low-cost applications (3).

Table 17.5 (3) shows analysis of deodorizer distillates from several vegetable oils. Although distillate from other oils may have higher levels of α-tocopherol, the larger volume of soybean oil processed makes soybean distillate an important source of natural vitamin E.

Dougherty (23) and Buford (24) discuss the use of tocopherols as food antioxidants and comparisons with other natural and synthetic products.

The α-tocopherols have some antioxidant activity, but δ- and γ-tocopherols are more effective antioxidants and will usually constitute a minimum of 80% of the total tocopherol content of mixed tocopherols. Mixed tocopherols are used where synthetic antioxidants are not permitted or where there is a preference for a natural product.

Natural source tocopherols are widely accepted for use in foods and are regulated in the United States by the Food and Drug Administration (FDA). They are permitted by the USDA at levels of 0.03% (300 ppm) in animal fats and at 0.03% (300 ppm) or 0.02% (200 ppm) in combination with other antioxidants in poultry products. Many foreign countries permit the use of tocopherols in foods, including Canada, Japan, Korea, Australia, and all countries of the European Economic Community.

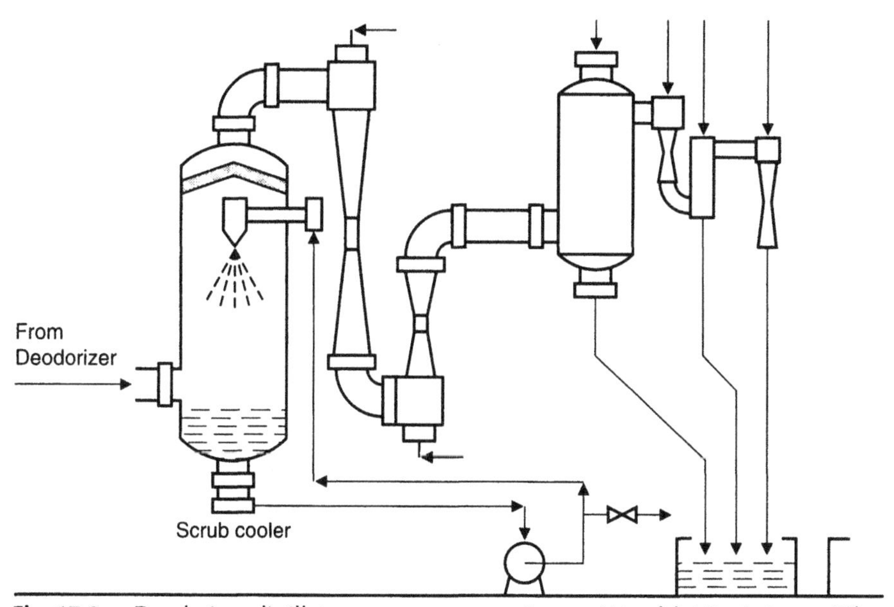

Fig. 17.9. Deodorizer distillate recovery system. *Source:* Woerfel, J.B., *J. Amer. Oil Chem. Soc. 583:* 188 (1981).

Antioxidant preparations containing mixed tocopherols, at various concentrations and combinations with other antioxidants, synergists, and carriers, are distributed by several U.S. companies, including Eastman Chemical Co., Kingsport, TN, (25); U.O.P. Food Products and Processing Dept., Des Plaines, IL (26); and Kalsec, Kalamazoo, MI (27).

Utilization of Deodorizer Distillate: Sterols

Table 17.5 (3) shows soybean distillate to be 18% sterols, of which 4.4% is stigmasterol. Distribution of sterol components for soybean oil is reported to approximate 20% campesterol, 20% stigmasterol, 53% β-sitosterol, 4% δ-avenasterol, and 3% δ-stigmasterol (28). This indicates soybean oil to be a good source of sterols, stigmasterol in particular.

Sterols are used in the manufacture of pharmaceuticals. Stigmasterol from soybean oil is used in the manufacture of progesterone and corticoids, whereas β-sitosterol is used to produce estrogens, contraceptives, diuretics, and male hormones (28).

Value of Deodorizer Distillate

Value of deodorizer distillate depends upon content of unsaponifiables. Table 17.6 (3) shows a typical specification. Pricing is based on tocopherol content and stigmasterol content, or both, depending on market demand for each ingredient, and is subject to wide fluctuations.

Winters (3) lists values and pricing basis over the years for a typical distillate, composed of 10% tocopherol, 14% sterol, and 2.9% stigmasterol (Table 17.7). Prices from 1961 to 1989 varied from no market in February, 1975, to $1433 per MT in December, 1983. The tocopherol market appears to have a greater effect in determining price than the sterol market.

TABLE 17.5 Deodorizer Distillate from Different Oils (3)

Item	Sunflower	Cotton	Soybean	Rapeseed
%Unsaponifiable	39	42	33	35
%Tocopherol	9.3	11.4	11.1	8.2
%α	5.7	6.3	0.9	1.4
%Sterol	18	20	18	14.8
%Stigmasterol	2.9	0.3	4.4	1.8

TABLE 17.6 Typical Specification for Deodorizer Distillate (3)

Item	Specification
Stigmasterol	2.5% min.
Tocopherol	6.0% min.
Water	2.0% max.

TABLE 17.7. Value and Pricing Basis for Typical Distillate (3)

Typical composition: 10.0% Tocopherol
14.0% Sterol
2.9% Stigmasterol

Date	Price basis	Value $ (USD)/MT
06/16/61	Sterol	386
01/01/68	Toco	441
04/05/72	Mixed	649
04/08/74	Toco	897
09/03/74	Toco	1190
02/21/75	No market	N/A
04/01/75	Mixed	162
03/18/76	Mixed	311
10/27/78	Toco	661
12/31/83	Toco	1433
05/01/85	Toco	772
09/16/86	Toco	507
01/01/89	Toco	397
1989	Toco	243–397

The distillate from physical refining is lower in value than distillate from caustic-refined oil. Even though equal amounts of sterol and tocopherol are recovered, the distillate is less concentrated because of higher fatty acid content. Additionally, the oxidative quality of the physical refinery distillate is lower, as measured by carbonyl and peroxide values. Both of these factors affect the cost of handling and processing and reduce the price and total return from sale of the distillate.

Spent Bleaching Earth Utilization

Spent bleaching earth is difficult to handle, primarily because of its tendency to catch fire. This is more pronounced with highly unsaturated oils such as soybean. If spent earth is left uncovered on a warm day, ignition can occur within an hour or two. Several precautions are helpful in addressing this problem. Spraying with water and covering bins in which bleaching earth is collected and transported are simple and helpful in reducing heating. Spraying with an antioxidant has been used; while this has some effect, it does not provide complete protection and is expensive (Smith, S.J., personal communication, 1993).

It is obviously desirable to reduce the amount of oil lost in the clay. This is controlled by several factors. Well-refined oil, especially one low in soap and phosphatides, requires less clay. More efficient clay and other bleaching agents reduce the amount of clay required; even though more active clay may have higher oil retention, the smaller amount required will result in lower total amount of oil lost in the clay. The third factor is operation of the filter press. Control of pressure and flow rates and proper use of filter aid as precoat and body feed can reduce the amount of clay needed by getting maximum utilization of the clay.

Blowing the press with nitrogen results in recovery of a higher-quality oil than does the use of air or steam. Sampling each press and analyzing composite samples for oil retention is an important control practice.

A number of methods are available for treating spent clay to obtain usable oil and defatted clay which can be disposed of more easily. Klein (29) describes methods for solvent extraction with polar and nonpolar solvents, aqueous extraction, and pressure extraction. He cites the importance of processing the spent earth promptly to obtain better-quality oil and points out that distilled fatty acids from clay and from soapstocks were quite similar. Flammability tests indicated that below 5% residual fat, risk of fire is eliminated.

Economic calculations comparing solvent extraction, pressure processing, burning and returning the clay to the meal stream for a 620 MT/yr plant led to the conclusion that a refiner should either sell the spent clay or incorporate the clay into meal cake at the maximum amount allowed by law.

One plant in the United States was established to process spent clay some years ago in an area where there were several refiners to supply raw material. It closed after several years, apparently because it was not profitable (Hastert, R.J., personal communication, 1993).

A number of U.S. processors add spent clay to animal feeds, especially for cattle. This provides energy and is said to reduce caking. Burning the clay in solid fuel boilers has been practiced for more than 50 years and is still in use. Fuel content of the oil is recovered and clay is disposed of in the ash.

One interesting development is incorporating spent bleaching clay in compost consisting of wood and other waste organic materials. At least one plant is reporting using this practice (Smith, S., personal communication, 1993).

Disposal of spent clay in a landfill is not desirable if there are other materials that might be flammable. One practice being used is to spread the clay on the ground and control the heating by a water spray until it stabilizes, and then put it on a dedicated landfill.

References

1. Witte, N., in *Proceedings: World Conference on Emerging Technologies in the Fats and Oils Industry*, edited by A.R. Baldwin, American Oil Chemists' Society, Champaign, IL, 1986, pp. 146–148.
2. Erickson, D.R., and L.H. Wiedermann, in *Proceedings World Conference Edible Fats and Oils Processing, Basic Principles and Modern Practices*, edited by D.R. Erickson, American Oil Chemists' Society Champaign, IL, 1989, pp. 275–283.
3. Winters, Robert L., in *Proceedings: World Conference Edible Fats and Oils Processing, Basic Principles and Modern Practices*, edited by D.R. Erickson, American Oil Chemists Society, Champaign, IL, 1989, pp. 402–405.
4. Winters, Robert L., in *Proceedings: World Conference on Emerging Technologies in the Fats and Oils Industry*, edited by A.R. Baldwin, American Oil Chemists Society, Champaign, IL, 1986, p. 186.
5. *1992–93 Yearbook and Trading Rules*, National Oilseed Producers Association, Washington, DC, 1992.

6. Method Ca 9f-57, *Official and Tentative Methods*, American Oil Chemists' Society, Champaign, IL.
7. Woerfel, J.B., *J. Amer. Oil Chem. Soc. 60:* 310 (1983).
8. Rice, E.E., *J. Amer. Oil Chem. Soc. 65:* 754A (1979).
9. Duff, A.J., *J. Amer. Oil Chem. Soc. 53:* 370 (1976).
10. Bloomberg, F.M., and T.W. Hutchins, U.S. Patent 3,425,938 (1969).
11. Braae, B., *J. Amer. Oil Chem. Soc. 53:* 353 (1976).
12. Crauer, L.S., *J. Amer. Oil Chem. Soc. 47:* 210A (1970).
13. Todd, D.B., and Morren, J.E., *J. Amer. Oil Chem. Soc. 42:* 172A (1964).
14. Morren, J.E., U.S. Patent 3,428,660 (1981).
15. Mag, T.K., D.H. Green, and A.T. Kwong, *J. Amer. Oil Chem. Soc. 60:* 1008 (1983).
16. Sonntag, N. O.V., in *World Conference Proceedings Edible Oils and Fats Processing, Basic Principles and Modern Practices*, edited by D.R. Erickson, American Oil Chemists' Society, 1989, pp. 406–411.
17. Boyer, M.J., in *Proceedings: World Conference on Emerging Technologies in the Fats and Oils Industry*, edited by A.R. Baldwin, American Oil Chemists' Society, Champaign, IL, 1986, pp. 149–154.
18. Witte, N.H., and E. Sipos, U.S. Patent 2,929,714 (1958).
19. Kruse, N.F., and W.W. Cravens, U.S. Patent 2,929,252 (1960).
20. Beal, R.E., and V.E. Sohns, *J. Amer. Oil Chem. Soc. 49:* 447 (1972).
21. Woerfel, J., in *Proceedings: World Conference on Emerging Technologies in the Fats and Oils Industry*, edited by A.R. Baldwin, American Oil Chemists' Society, Champaign, IL, 1985, pp. 165–168.
22. Woerfel, J.B., *J. Amer. Oil Chem. Soc. 58:* 188 (1981).
23. Dougherty, M.E. Jr., *Cereal Foods World 33:* 222 (1988).
24. Buford, R.J., in "A Review of Natural and Synthetic Food Antioxidants," presented at American Oil Chemists' Society Meeting, Anaheim, CA, March 27, 1993.
25. Anonymous, *Tenox GT-1 and Tenox GT-2*, Publication ZG-263C, Eastman Chemical Co., Kingsport, TN, 1993.
26. Anonymous, *Pristine® Natural Based Food-Grade Antioxidants and Flavors*, Publication UOP 17-5073, UOP Food Products and Processes Dept., Des Plaines, IL, 1991.
27. Anonymous, *Duralox™ Oxidation Management Blends*, Publication D-21, Kalsec, Inc., Kalamazoo, MI, 1993.
28. Sonntag, N.O.V., *J. Amer. Oil Chem. Soc. 62:* 928 (1985).
29. Klein, J.M., in *Proceedings of the World Conference on Emerging Technologies in the Fats and Oils Industry*, edited by A.R. Baldwin, American Oil Chemists' Society, Champaign, IL, 1986, pp. 169–171.

Chapter 18

Salad Oil, Mayonnaise, and Salad Dressings

Ahmad Moustafa

Consultant
Cincinnati, OH

History of Mayonnaise and Salad Dressing

It has been stated that dressings go back to the days of Louis XIV of France (1). This period is considered by the French people to be the golden age of French cooking. The French have always excelled in preparing fine foods, and they have been credited with introducing mayonnaise to the Americans during the 19th century. For years, the making of mayonnaise was limited to the housewife, who produced a dressing that not only was tasty but possessed body that would not break after standing for a short period of time.

Mayonnaise was first manufactured commercially during the early part of the 20th century. Several years later the first salad dressing was commercially manufactured. Today, mayonnaise and salad dressing are manufactured under U.S. Standards of Identity for mayonnaise, salad dressing, and French dressing (2). The defining of government standards for these three products in 1928 was instrumental in relieving the early chaos caused by product adulteration. The Standards have been revised from time to time and were finally promulgated by the Food and Drug Administration (FDA) and adopted August 12, 1950. Under the provisions of these Standards, each of these three products must contain certain minimum quantities of vegetable oil and eggs and may contain only certain commonly used spices, flavors, and condiments.

Government Regulations

Mayonnaise and mayonnaise-type dressings are some of the oldest emulsions known to humans and, according to the U.S. Standard of Identity (2), are emulsified semi-solid food preparations containing:

- A minimum of 65% of one, or a blend of two, or more, edible vegetable oils
- Acidifying agents, consisting of vinegar, calculated as acetic acid, not less than 2.5% by weight, optionally mixed with citric acid; in any such mixture the weight of the citric acid must not be greater than 25% of the weight of the acids of vinegar calculated as acetic acid. Lemon juice, lime juice, or both, fresh or dried, frozen, canned, concentrated, or diluted with water to an acidity, calculated as citric acid of not less than 2.5% by weight may be used.

- Egg yolks, liquid, frozen, or dried, (whole eggs liquid, frozen, or dried), or any one or more of the foregoing with liquid egg white or frozen egg white may be used. No minimum requirements are stated for egg products.

The mayonnaise or mayonnaise-type dressing may contain salt, sugar, corn syrups, honey or other syrups, dextrose; or other seasonings commonly used in its preparation, such as mustard, paprika, other spices or spice oils; monosodium glutamate; and any suitable harmless food seasoning or flavoring, provided they do not simulate egg yolk color. Mayonnaise may contain ethylenediaminetetraacetate (EDTA) as the calcium disodium salt, disodium dihydrogen salt, or combination thereof. The added EDTA will not only sequester the heavy metals but will synergize the activity of antioxidants such as butylated hydroxytoluene (BHT), butylated hydroxyanisole (BHA) and tert-butylhydroquinone (TBHQ). The quantity of such added EDTA should not exceed 75 ppm by weight of the finished food.

Theory of Emulsion Stability

Emulsification is the formation of an emulsion from two immiscible liquid phases. An *emulsion* is a "significantly stable suspension" of particles of a liquid of a certain size within a second, immiscible liquid. Two types of emulsions are recognized, based on the size of the dispersed particles: macroemulsions and microemulsions. The particles of the former range from 0.2 to 50 micrometers (μm) in size and are easily visible under the microscope. The latter have a particle size ranging from 0.01 to 0.2 μm.

The size of the dispersed particles in an emulsion determines its appearance and degree of stability. If the diameter of the dispersed particles is 1 to 10 μm, the emulsion is milky white in appearance; 0.1 to 1 μm, blue-white; 0.05 to 0.1 μm, gray and semitransparent; and less than 0.05 μm, transparent as shown in Table 18.1 (3,4).

When preparing mayonnaise for commercial purposes, it is absolutely essential that exact production techniques be followed to guarantee maximum product stability in the market place. Since mayonnaise is an oil-in-water (O/W) emulsion in which the oil portion (the discontinuous phase) constitutes from 65 to 80% of the formula, it is evident that dispersion of this large amount of oil in the relatively small aqueous phase requires the utmost care in choosing the type and quantity of emulsifier, the mixing method, and the emulsification equipment.

TABLE 18.1 Relationship of Particle Size to Emulsion Appearance and Stability

Particle size	Appearance	Stability
μm		
<0.05	Transparent	Extremely stable
0.05–0.10	Translucent	Excellent stability
0.10–1.00	Blue-white	Good stability
1.00–10.00	Milky white	Tendency to cream
>10.00	Coarse	Quick-breaking

Ingredients

Vegetable Salad Oils

Among the vegetable salad oils used in commercial production of mayonnaise and salad dressings, soybean oil is the major oil used in the United States, followed by cottonseed and corn oils. Alkali-refined, bleached, and deodorized soybean oil (RBDSBO) is used with very acceptable results by almost all mayonnaise and salad dressing processors. Partially hydrogenated, winterized, and deodorized soybean oil (HWSBO) is an optional oil used alone or in combination with RBDSBO by some who feel that it improves stability and shelf life. HWSBO costs $0.044 to 0.066/kg (2 to 3¢/lb) more than RBD SBO. Both oils must not cloud, solidify, or deposit any crystalline solid fraction at refrigerator temperature [4°C (40°F)] and must pass the American Oil Chemists' Society (AOCS) cold test (AOCS Method Cc 11-53). The test sample should stay clear after immersion in an ice bath at 0°C (32°F) for 5.5 hr.

Great strides have been made during the last two decades through the basic and applied research efforts of government research centers and the industry in developing today's high-quality soybean oil (see Chapter 5). When properly processed, the fresh oil is bland, but it will deteriorate organoleptically when exposed to light or stored at elevated temperatures. The change in flavor will develop at low levels of oxidation because of the high polyunsaturation of the oil. The peroxide value, if measurable, is 1 to 2 meq/kg or less. The early stages of oxidation may give flavors described as "beany" and "grassy" and more typically oxidized at the more advanced stages (5). It has been stated that this potential defect is the major problem in relation to the use of vegetable oils in mayonnaise and salad dressings (4). This defect is affected by time, temperature, light, air, exposed surface, moisture, nitrogenous organic material, and trace metals. All of these extraneous factors are present during the production of mayonnaise and salad dressings.

In practice, the oil is normally delivered at a temperature ranging from 32 to 38°C (90 to 100°F). It is usually stored in carbon steel tanks or epoxy-lined carbon steel or stainless steel tanks and is utilized within three days to a few weeks, depending on the plant capacity and the amount of oil delivered. More modern plants use stainless steel for both transport and storage. It is important that storage temperatures be kept as low as possible to enhance the keeping quality of the oil. On the other hand, control over the storage and display conditions of the mayonnaise and salad dressings is lost once they leave the processing plants. The product may be stored or displayed at elevated temperatures. It may be three months or more on the shelf before it is sold.

In view of these concerns, it is evident that the quality of the finished mayonnaise and salad dressing is influenced by the quality of the oil used in its production. Oil quality is perhaps the foundation for continued success in the marketplace.

Oil-purchasing agents normally buy oils in accordance with specifications adopted by the National Oilseed Processors Association (Washington, DC) (6) or those recommended by the American Oilseed Association (ASA) (Erickson, D.R., private communication, 1993) and shown in Table 18.2, or a modification of their own to these specifications.

TABLE 18.2 Specifications for Soybean Salad Oil

Factor tested for	NOPA specs (2)	ASA specs	AOCS test method
Flavor	Bland	Bland	——
Odor	Odorless	Odorless	——
Free fatty acids,			
% as oleic acid	0.05	0.02–0.03	Ca 5a-40
Peroxide value, meq/Kg	2.0	<1.0	Cd 8-53
Color, Lovibond			Cc 8b-52
Red	2.0	<1.0	Cc 8e-63
Yellow	20.0	10.0	Cc 13b-45
Rancimat stability, hrs	——	5.0	——
AOM fat stability test,			
hr to reach 35 P/V	8.0	15–18	Cd 12-57
Cold test, hr	+5.5	5.5	Cc 11-53
Phosphorus, ppm,			
Colorimetric	——	<1	Ca 12-55
Atomic absorption			Ca 12b-87
Iodine value units	——	125-140	Cd 1-25

Additional testing may be done, such as heating an oil sample to 176°C (350°F in a glass beaker to determine the presence of foreign odors at that temperature anc for odor after it cools to 37°C (99°F). A second test is the Kreis test (or ε modification thereof), which entails the addition of 10 mL of concentrated HCl to 1C g of an oil plus 10 mL of 0.1% phloroglucinol in diethyl ether, followed by shaking for 30 sec and allowing to stand for color development. The degree of pink to red color in the bottom layer indicates potential onset of rancidity (pink) or actual rancidity (red).

These two tests are quick indicators of potential or actual oxidation, but the crucial test is the determination of flavor by organoleptic means. This should be required for both fresh and aged incoming and outgoing products. Such testing should be routinely done as part of an ongoing quality program even though results from the aged samples are after the fact. Such data are invaluable in tracking quality over time for both oil receipts and outgoing products (see Chapter 24).

Under normal operating conditions and for maximum emulsion stability, either the processor transfers the daily oil requirement to another tank, where it is cooled to 7°C (45°F) or lower before use, or the oil temperature is reduced to 7°C (45°F) or lower in a plate heat exchanger prior to production.

Eggs

In all edible emulsions, including mayonnaise, the choice of emulsifying agent is limited. From the standpoint of phase-volume theory of emulsion stability, this represents a considerable challenge. It is a fact of solid geometry that an assembly of spheres of equal radii can be placed in a position of densest packing in two ways. In either case the spheres are found to occupy 74.02% of the total volume, and the remaining 25.98% consists of empty space. This means that any attempt to exceed a

TABLE 18.3 Typical Formulas for Commercial Mayonnaise

	Content (%)			
Ingredient	Industry standard	Light	Medium	Heavy
Oil	75.0	77.3	78.8	81.2
Salt	1.5	[a]	[a]	[a]
Egg yolk	8.0	9.0	8.5	6.3
Mustard	1.0	[a]	[a]	[a]
Water	3.5	[b]	[b]	[b]
Vinegar[c]	11.0	[b]	[b]	[b]
Spice mix	—	3.2	3.2	3.2
Vinegar and water	—	10.5	9.5	9.3

[a]Combined with spice mix.
[b]Combined with vinegar water mix.
[c]6% Vinegar (60 grain).

phase volume of 0.74 for the internal phase must result in either inversion or breaking (7,8). For a given system, both oil-in-water (O/W) and water-in-oil (W/O) emulsions are possible between the phase volume concentrations 0.26 and 0.74; below the one and above the other only one system form can exist (9).

Typical formulas for commercial mayonnaise are shown in Table 18.3. The formulas given in the table and the various factors that influence the stability of the mayonnaise emulsion were critically scrutinized by Corran (10). From many points of view, the most critical ingredient is the egg yolk, because it is far from being a satisfactory emulsifier in this particular system. The surface-active components of egg yolk are lecithin and cholesterol. Lecithin, the major surface-active component, is known to be a good O/W emulsifier, but cholesterol is an efficient W/O emulsifying agent and thus exerts an effect antagonistic to lecithin (10,11).

The average composition of eggs is 9.5% shell, 63% albumin, and 27.5% yolk (12). Egg yolks are defined as the yolks of domestic hens separated from whites and containing not less than 43% total egg solids. Almost all mayonnaise manufacturers use frozen and salted egg yolks. The approximate compositions of various types of frozen eggs used in the manufacture of mayonnaise and salad dressings are shown in Table 18.4 (12,13).

The total solids of albumin, yolk, and whole egg are around 12, 52, and 24% respectively. The lipid composition of egg yolk varies between 32 and 36 and is composed of 65.5% triglycerides, 28.3% phospholipid, and 5.2% cholesterol (14). The composition of the phospholipid fraction is shown in Table 18.5 (15). As previously stated, the surface-active components of the egg yolk are lecithin and cholesterol. Lecithin in its natural state is known to be a very powerful emulsifying agent in promoting the O/W emulsions, whereas cholesterol exerts the reverse effect. The effect of the lecithin-cholesterol ratio on the type of emulsion produced has been investigated and it was found that an approximate ratio of 50:50 oil to water or an 8:1 lecithin to cholesterol ratio will produce an O/W emulsion, whereas a lower ratio will invert the emulsion, with oil becoming the continuous phase. In naturally occurring fresh egg yolk, the lecithin:cholesterol ratio is 6.7:1.0, which favors a W/O

TABLE 18.4 Approximate Composition of Frozen Eggs (%)

Component	Reference (10)	Whole egg (12)	Egg white (12)	Egg yolk (12)
Fat	22.5	12.3	0.2	27.1
Protein	16.0	13.3	12.0	15.2
Lecithin	10.0	—	—	—
Cholesterol	1.5	—	—	—
Salt	2.0	—	—	—
Water	48.0	73.7	87.2	56.7
Ash	—	0.7	0.6	1.0

emulsion. Hence, the use of fresh egg yolks is recommended for the production of stable mayonnaise, because lecithin suffers some breakdown upon storage, which reduces its emulsifying effectiveness.

A stable emulsion is the suspension of particles of two immiscible liquids, one into the other, with an emulsifying agent or a surface-active agent at the interface, oriented in a monomolecular fashion with the polar groups oriented toward the water phase and the nonpolar hydrocarbon chain oriented toward the oil (see Fig. 18.1). Albumin and lecithin in eggs are classified as O/W emulsifiers.

It is estimated that about one-third of the total production of eggs is frozen (16). When raw egg yolk is frozen and stored below –6°C (21°F), the viscosity increases, gelation occurs, and all fluidity is lost. This is easily controlled by the addition of either sodium chloride or sucrose, which are commonly added to frozen egg products at a level of 10% to control gelation. Other additives include glycerine, syrups, gums, and sodium metaphosphate. The addition of phosphates makes it possible to pasteurize the egg product at a lower temperature, so as not to denature proteins.

Egg yolk gelation is affected by the rate and temperature of freezing, thawing conditions, and storage time and temperature. The effects of these conditions are so interrelated that they are difficult to quantitate individually.

In industrial practice, egg yolks are frozen to –29°C (–20°F) for 72 hr and then transferred to –18°C (0°F) for the balance of the storage period (17). However, care must be taken when thawing frozen eggs for production purposes. The frozen egg containers are moved from the freezer, where they are normally stored at –18 to –12°C (0 to 10°F), to a walk-in cooler the day before production, kept there at 7 to 10°C (45 to 50°F), and used at that temperature.

TABLE 18.5 Composition of Egg Yolk Phospholipid Fraction

Lipid Fraction	Composition, %
Phosphatidyl choline	73.0
Phosphatidyl ethanolamine	15.0
Lysophosphatidyl choline	5.8
Sphingomyelin	2.5
Lisophosphatidyl ethanolamine	2.1
Inositol phospholipid	0.6

The functional properties of plain egg yolks are little affected by freezing, and mayonnaise made with frozen egg yolk is more stiff but slightly less stable than that made from unfrozen yolks. The addition of sodium chloride increases the emulsifying ability of the defrosted yolks.

Frozen egg yolks are packaged in a variety of containers, and until recently the 13.6 Kg (30 lb) tin can predominated. Cartons and plastic bags are also used, and bulk shipments are made in 208 L (55 gal) steel drums with plastic bag liners.

Mustard

From the above discussion on mayonnaise stability, it can be deduced that the only other reason a stable emulsion is produced may be attributed to the presence of the mustard powder. It has been suggested (18) that the fundamental condition for the formation of O/W emulsions with solids is that the solids must be more easily wetted by the water than by the oil phase. As a general rule, the phase in which the emulsifying agent is more soluble will be the external phase. The fineness of the mustard flour also influences its emulsifying activity; the finer the particle size, the more effective it is. Consequently, the concentration of mustard flour solid particles at the interface represents an interfacial film of considerable strength and stability.

Mustard flour has a definite effect on the character of mayonnaise in storage. It is responsible for the sensation of heat and flavor. In about two weeks after production, the former character will disappear and the latter will change to an egglike flavor (19). Mustard is marketed as black or brown (*Brassica nigra*) or white or yellow (*B. alba*) flours. The two species are blended together to achieve a desired balance between flavor and pungency. Black or brown mustard flour has a sharp odor, whereas white or yellow mustard flour is hot to the taste but is odorless.

According to Weiss, wetting mustard flour with water activates the glucosidase enzyme which releases the oil of mustard responsible for the two characteristics (17) just mentioned. However, some manufacturers elect to use oil of mustard because it retains its original pungency and flavor for a longer time, perhaps as long as three months, and is speck-free. The specks may originate from the black coat of rapeseeds, which are a common contaminant with mustard seed. Hence, extreme care must be exercised in the selection, cleaning and preparation to achieve an almost speck-free mustard flour.

Due to the hazardous nature of mustard oil, its dilution in salad oil is perhaps the safest way to handle it (17). Based on an oil content of between 0.5 and 1.0% mustard oil in the flour, a 0.25 to 0.5% master mix solution of mustard oil in vegetable salad oil can be added safely to the mayonnaise formulation at the same level as would be used for mustard flour.

Acidulants

The acidifying ingredients allowed for use in the production of mayonnaise and salad dressings are vinegar (acetic acid), citric acid, and malic acid, singularly or in combination.

Vinegar and Acetic Acid. Vinegar is one of the oldest fermentation products known to man. The formation of acetic acid from ethanol by acetic acid bacteria was first noted by Pasteur. The reaction has since been termed "acetic acid fermentation" and requires the presence of oxygen (20). The manufacture of vinegar saw substantial change in the 20th century with the development of pure- and submerged-culture techniques; the latter introduced in the late 1940s.

The highest ethanol concentration that allows the formation of acetic acid varies with the *Acetobacter* species and ranges between 5 and 11% by volume. The optimum temperature for acetic acid production lies between 20 and 30°C (68 to 86°F) at pH 5 to 6.

Very little pure acetic acid, as such, is used in foods, although it is classified by the U.S. Food and Drug Administration (FDA) as Generally Recognized as Safe (GRAS) material. Consequently, it may be employed in products that are not covered by definition and standards of identity. As the major component of vinegars, acetic acid has been known for centuries and is one of the earliest flavoring agents.

Pure acetic acid is produced commercially by a number of processes. As a dilute solution, it is obtained from alcohol by the *quick-vinegar process*. It is manufactured synthetically in high yields by the oxidation of acetaldehyde and of butane, and as the reaction product of methanol and carbon monoxide (21).

Vinegars are produced from cider, grapes or wine, sucrose, glucose, malt, or any sugar-containing material that can be converted by successive alcoholic and acetous fermentation (22). Such materials include molasses, sorghum syrup, honey, fruits, maple syrup, potatoes, beets, malt, grain, and whey.

In the United States the term *vinegar* without qualifying adjectives implies only *cider vinegar*. This stems from definitions drawn up by the FDA while it was part of the U.S. Department of Agriculture. A 4 to 8% solution of pure acetic acid would not qualify as vinegar because it lacks the other minor components shown in Table 18.6.

The following is a list of vinegars commonly available for use by mayonnaise and salad dressing manufacturers:

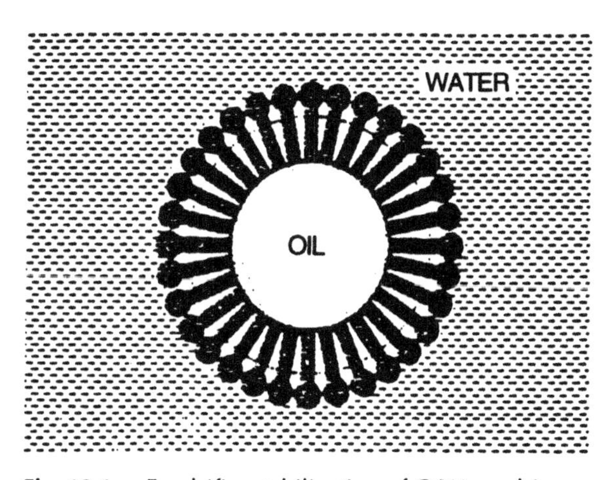

Fig. 18.1. Emulsifier stabilization of O/W emulsion.

TABLE 18.6 Composition of Cider Vinegar (15)

Component	Composition, g/100 mL
Acidity as acetic acid	3.24–9.96
Nonvolatile acids as lactic acid	0.05–0.30
Alcohol	0.03–2.00
Glycerine	0.23–0.46
Total solids	1.20–4.45
Reducing sugars (as invert sugar)	0.11–1.12
Volatile reducing substances (as invert sugar)	0.14–0.34
Pentosans	0.08–0.22
Volatile esters (as ethyl acetate)	0.30–0.91
Total ash	0.20–0.57
Soluble ash	0.17–0.51
Phosphoric acid in soluble ash	0.007–0.040
Sugars in total solids	5.30–43.30
Total ash in nonsugar solids	12.30–30.00

1. Plain white distilled vinegar
2. Apple cider vinegar
3. Red and white wine vinegars
4. Malt vinegar
5. Barley vinegar
6. Rice vinegar
7. Balsamic vinegar (imported from Italy)

In the United States wine vinegar is the most common variety in use.

The strength of vinegar is very important. It is measured by *grain*, 100 grain equaling 10% acetic acid. The acetic acid content varies from 4 to 7%, with the standard being 5% (50 grain). On the other hand, industrial vinegar is usually available as 100 and 120 grain.

White distilled vinegar has hardly any taste other than a sour note. On the other hand, top-quality wines are used for the production of wine vinegars. Cider vinegar lends a strong fruit taste to any product, and rice vinegar is quite sweet. Balsamic vinegar is expensive as it is imported from Italy. Malt and barley vinegars impart their own flavor and strong nature to dressings.

Citric Acid

Citric acid, a tricarboxylic acid, has been in use in foods in the United States for more than 100 years. The acid and its potassium and sodium salts are classified by the FDA as GRAS for miscellaneous and general use in foods. The acid may be used in preparing mayonnaise, salad dressings, and French dressing.

In addition to its acidulating properties, citric acid sequesters trace metals, which can act as prooxidant catalysts in mayonnaise and salad dressings. It also acts as a synergist for antioxidants employed in inhibiting rancidity of the oil phase.

Citric acid is produced commercially by the fermentation of corn syrups that are inoculated with various pure strains of *Aspergillus niger* (23). The acid is also obtained from lemons and from pineapple cannery waste. The acid is also produced in deep-vat fermentation using fungal mycelia, a process that requires careful control of the pH of the sugar solution and the rate of nutrient addition.

The important function in the use of acidulants by the formulator is the skillful incorporation with other adjuncts, such as sugar and spices, in the food preparation to impart specific desirable physiological and psychological effects. Citric acid is often used to create a rapid build-up or "burst" in taste (24).

Malic Acid. Malic acid is one of the general-purpose additives in the FDA list of GRAS substances. It is commercially synthesized as a racemic mixture of D- and L-isomers from maleic and fumaric acids in the presence of a suitable catalyst. However, it is found in nature as the L-isomer in many fruits and vegetables and is the second most prevalent acid in citrus fruits.

Malic acid has a smooth tart taste resembling that of citric acid, and although its degree of ionization in water is similar to that of citric acid, malic acid has a much stronger apparent acidic taste (25). Another function of malic acid is its synergistic effect with antioxidants to prevent fats and oils from becoming rancid (26).

Hence, the addition of small amounts of malic acid in the production of mayonnaise and salad dressings will not only extend shelf life but will also give a smooth and tart taste. Most importantly it will act as a flavor potentiator with many of the flavoring materials used.

Starches

Starch is the chief source of carbohydrate in the human diet. Common starches found in products such as corn, wheat, sorghum, and tapioca are heterogeneous complex polymers and are characterized by their amylose-amylopectin ratios. Amylose has a linear chain containing from 400 to 1200 monomer units of glucose molecules (27), whereas amylopectin is highly branched, as shown in Fig. 18.2.

The molecular orientation of these two compounds plays an important role in the character of starch-thickened salad dressings. The amylose exists in a helical form as a 1-butanol complex and can readily assume other configurations. Freshly prepared starch solutions are strongly opalescent. On standing the opalescence increases until finally a portion of the starch precipitates. This is the phenomenon of *retrogradation*. Retrograded amylopectin can return to its original dispersed state by heating to 50 to 60°C, but retrogradation of amylose cannot be reversed (28). In dilute solution, the polysaccharide molecules gradually align in linear forms, and on slow cooling the solution clouds as these bundles start to grow and precipitate. In a more concentrated form, such as is found in hot corn starch paste, rapid cooling to room temperature will cause the paste to set into a rigid gel by hydrogen-type bonding, as shown in Fig. 18.3.

In contrast, the branched amylopectin is quite stable and does not show this tendency to retrograde to an insoluble precipitate or to form a gel (29). Retrogradation is rather complicated and depends on factors such as type of amylose, molecular size, concentration, and pH of the dispersion.

The starch granule is insoluble in cold water but will absorb about 25 to 30% by weight of water without appreciable swelling. As the temperature of the starch slurry approaches the gelatinization temperature, the starch molecules swell and their

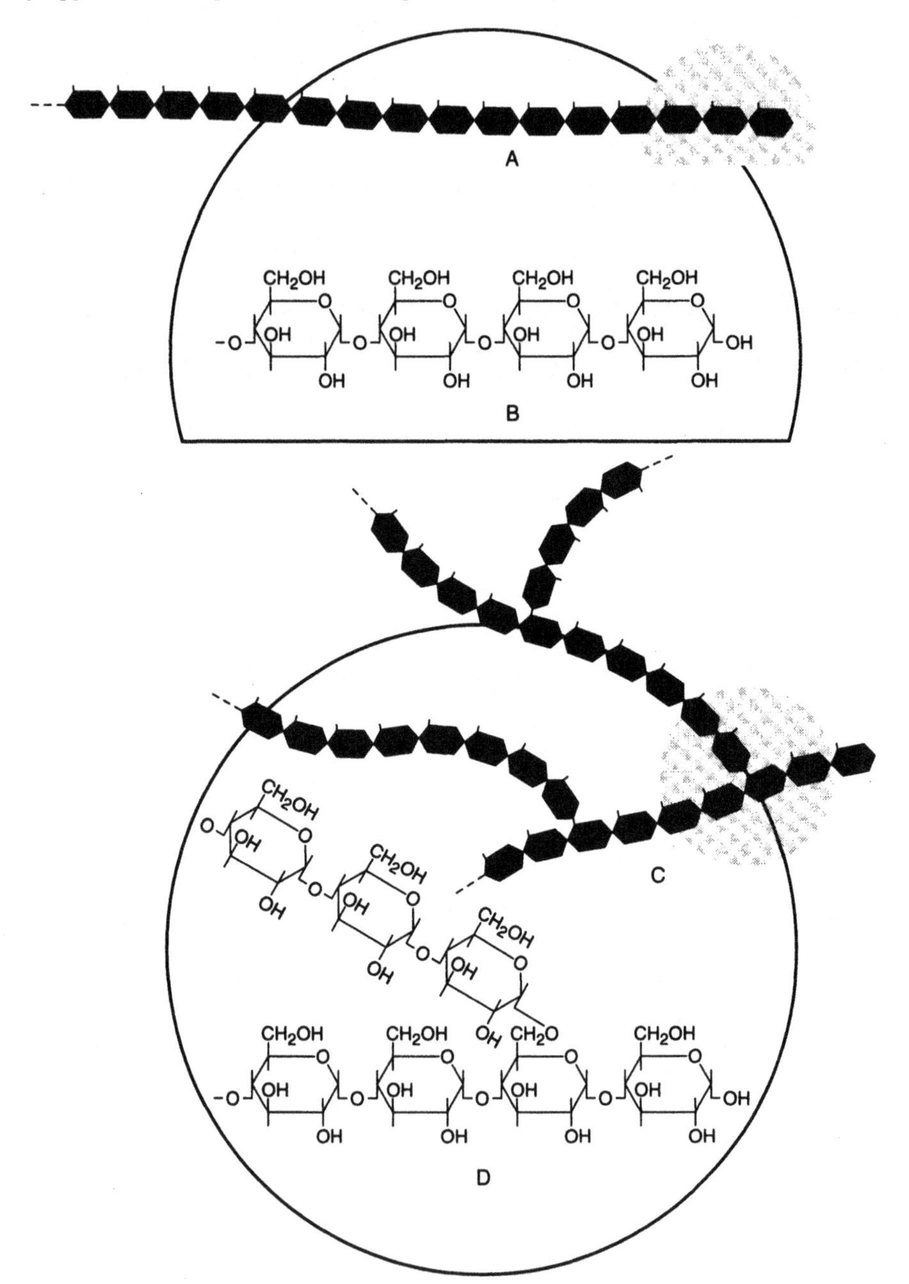

Fig. 18.2. Structures of the amylose (A,B) and amylopectin (C,D) components of starch.

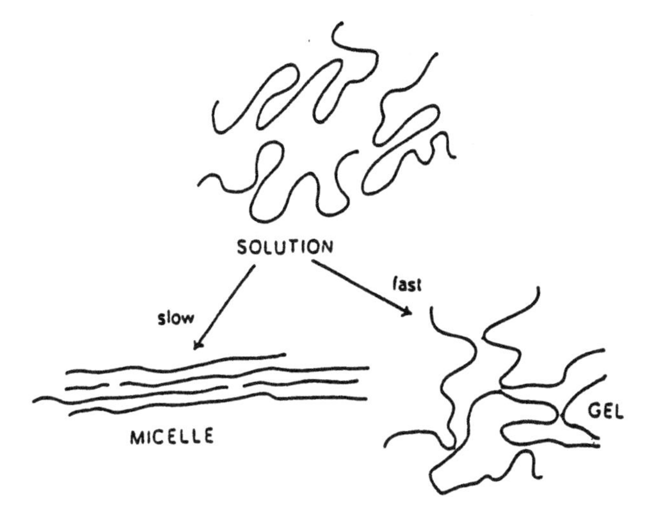

Fig. 18.3. Effect of cooling rate on starch retrogradation.

water absorptive capacity increases 100-fold, their swelling becomes irreversible, and the crystallinity of the granule is destroyed.

Of all the food applications of starch, its use in salad dressings is one of the most rigorous and demanding. The starch used must meet the following requirements:

1. The starch should withstand the highly acid medium of the salad dressing and must not thin out or lose its water-holding capacity on storage.
2. It must have a high and stable viscosity capable of withstanding the high colloid-mill shearing action and not degrade in consistency.
3. The type of starch paste needed is termed "short" in texture, as opposed to the "long" and "stringy and cohesive" types.
4. It should impart gloss and surface sheen to the salad dressing.
5. The salad dressing should be smooth and not appear grainy or discontinuous in texture.
6. The starch should maintain the dispersed oil phase in a stable emulsion and not permit the oil droplets to coalesce.
7. Last but not least, the starch should be on the U.S. GRAS list of food ingredients for human consumption.

The starch used in the manufacture of salad dressings must be resistant to the high acidity of the medium, which is usually between pH 3.0 and 3.5. Unmodified corn starch is affected by this low pH, and its paste undergoes thinning. However, modifying the associative forces of either corn or waxy starches by cross-linking or chemical substitution stabilizes them, and minimal acid hydrolysis and loss of viscosity takes place. The extent of cross-linking is controlled to produce starches showing varying degrees of swelling and pasting properties. Cross-linking the granules yield starches whose cooking characteristics are short or salvelike.

It is common practice in formulating salad dressings to use a combination of starches, frequently a blend of cross-linked waxy corn starch and cross-linked normal corn starch.

Spices and Seasonings

The use of any of the permitted spices, herbs, natural spice oil, spice extract, or any combination thereof for seasoning purposes will impart a specific flavor character and quality to the mayonnaise or salad dressing. Such use is not permitted in either product if it results in a color simulating the color imparted by eggs.

Even though mustard is included in the Standard of Identity as an optional ingredient, black pepper oleoresin or oil may be used to replace mustard as a source of pungency (heat) (30).

Also, paprika is normally used in the production of mayonnaise and salad dressing as the oleoresin because of its distinct odor, color, and flavor. Garlic powder is frequently used in mayonnaise, because its characteristic flavor seems to be quite stable on long-term storage. In contrast, onion is milder in flavor but less stable (17). Spices such as ginger, mace, cloves, tarragon, cinnamon, lemon peel, and celery seed are used to a limited degree (4).

Sweeteners, Salt, and Water

Sweeteners play an important role in mayonnaise and salad dressings. Originally only cane sugar was used, but now a greater variety of sweeteners are employed. The U.S. Standard of Identity provides that nutritive carbohydrate sweeteners such as cane or beet sugar (granulated or in liquid form) corn syrup, dextrose, invert sugar syrup, glucose syrup, nondiastatic malt syrup, and honey may be used as sweeteners in the production of mayonnaise and salad dressings.

Corn syrups are produced from refined corn starch by subjecting the wet starch to acid or enzymatic hydrolysis, which convert the starch to dextrose, maltose, and dextrins. Partial acid conversion of corn starch followed by enzyme conversion yields a syrup with increased sweetness, lower viscosity, and better flavor than that made by acid hydrolysis alone. It has been found that corn syrups may replace a part of the sucrose pound for pound on a solids basis without affecting the sweetness of the product (31).

An important aspect of flavor acceptance by the consumer is the correct vinegar to sugar balance. The product should not be excessively tart nor excessively sweet. The use of a 1:1 sugar to 50-grain vinegar is a good starting point.

Salt is used for its stabilizing action as well as for flavor enhancement. Most formulas contain from 1.0 to 1.5% salt on an "as is" basis or 5.0 to 7.5% in the mayonnaise continuous water phase. At this concentration, the salt will exert its protective effect by lowering the water activity, thus inhibiting and retarding any further bacterial growth. Fine flake or granulated 99.5% pure sodium chloride salt is generally used. It must be free from impurities and contain no added calcium or magnesium salts, copper, or iron, because those accelerate oil rancidity.

Water used in manufacturing mayonnaise and salad dressings must be potable and of no more than moderate hardness, because the presence of calcium or magnesium salts will accelerate oxidation of the oil phase. Also, if the water is on the hard side, the alkalinity should be determined and additional amount of vinegar used to balance the formula.

Since the Standards of Identity for mayonnaise and salad dressings do not allow the use of antioxidants, the use of an inert gas, such as carbon dioxide or nitrogen, should be used. However, nitrogen is preferred. It is inert, has low solubility in fats and oils, and has no effect on product flavor. The gas is introduced into the processing system just ahead of the colloid mill. Since mayonnaise and salad dressing are sold by volume rather than weight, the amount of nitrogen gas injected (12% maximum) is usually adjusted to reduce the product's specific gravity from 0.94 to between 0.90 and 0.92. Injecting higher levels of gas is not recommended, because the product loses gas to the head space during shipping and standing in the store, resulting in low product volume in the container and frequent rejection by the consumer.

Manufacture of Mayonnaise and Salad Dressing

Preparing a stable mayonnaise emulsion is not an easy task, because an emulsion tends to form with the major component in the continuous phase and the minor component as the dispersed phase, which is the opposite of a mayonnaise emulsion. In the process of mixing the product's ingredients not only can the order of adding the ingredients be varied, but, in addition, both time and degree of agitation exert a profound influence on the stability and taste of the final emulsion. The temperature of the oil and other ingredients during mixing also influences the viscosity and stability of the product. A thin product results if the operation is carried out with materials that are too warm. The starting ingredient temperatures should be 10 to 15°C (50 to 60°F), and the final product temperature should be 15 to 21°C (60 to 70°F), because emulsion failure will normally occur at temperatures above 24°C (75°F) (17,32).

In small-scale operations the initial mixing of mayonnaise is carried out in ordinary planetary-type mixers equipped with high-speed beaters or paddles. The Hobart (Troy, OH) mixer is an example of this type. The egg yolks are placed in the bowl with a small portion of the oil and beaten, and then sugar, salt, dry spices, and one-half of the vinegar are added and beaten. The oil is gradually beaten in and finished with the remainder of the vinegar.

Batch Methods

Commercial methods of mayonnaise production vary with the size of operation and type of equipment. The Dixie-Charlotte system (Chemicolloid Laboratories, Inc., New York, NY) is probably the one most widely used. The system can be operated with one tank as a batch or two or more tanks in a semicontinuous fashion utilizing staggered production procedures. It can produce from 230 to 4600 L (60 to 1200 gal) per hour. In normal operation, the premix tanks used are 380 to 950 L (100 to 250 gal) in capacity and are equipped with horizontal high-shear turbine-wheel agitators driven with a variable-speed drive having pneumatic drive adjustments (Fig. 18.4). Eggs, oil, vinegar, and possibly water and a liquid sugar solution or liquid sweetener are metered in. The flow is programmed as to both sequence and rate of flow. The agitation is varied during the mixing to suit the product state and to enhance the building of the emulsion (33). Spices and other dry ingredients are carefully weighed and hand-fed to the batch at the desired time. When the batch is completed, it is pumped through a colloid mill set at 0.02 cm (0.05 in) clearance to the filler capper. It is common practice to prime the mixer at startup with approximately 50 to 60 L (15 gal) of mayonnaise to cover the agitators, and on subsequent batches the premix is emptied, leaving 50 to 60 L of mayonnaise in the bottom of the tank.

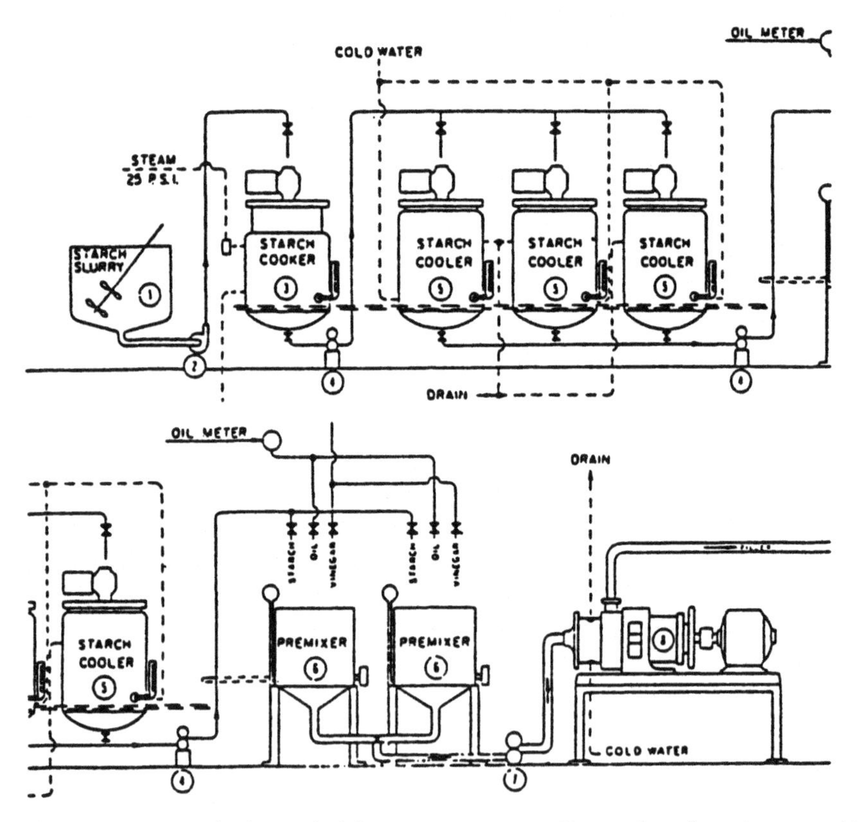

Fig. 18.4. Dixie-Charlotte salad dressing system (top). Preparation of starch paste, which is added to the mayonnaise unit (bottom). *Source:* Chemicolloid Laboratories, Inc.

Fryma (Switzerland) is another system that can be operated on a batch or semi-continuous basis. Three 75 L (20 gal) stainless steel tanks equipped with upper and lower control sensors are supplied continuously with oil, egg, vinegar, water, sweetener, and a dry ingredients slurry from large bulk storage tanks. The flows from the three tanks are programmed both as to sequence and rate of flow using proportioning pumps through a static mixer to a premix holding tank of 380 to 1200 L (100 to 300 gal) capacity, maintained under 300 to 400 mbar vacuum. At the same time, the coarse emulsion is withdrawn from the bottom of the premix tank and circulated back to the top of the premix through a colloid mill set at 1.25 mm (0.05 in) clear-

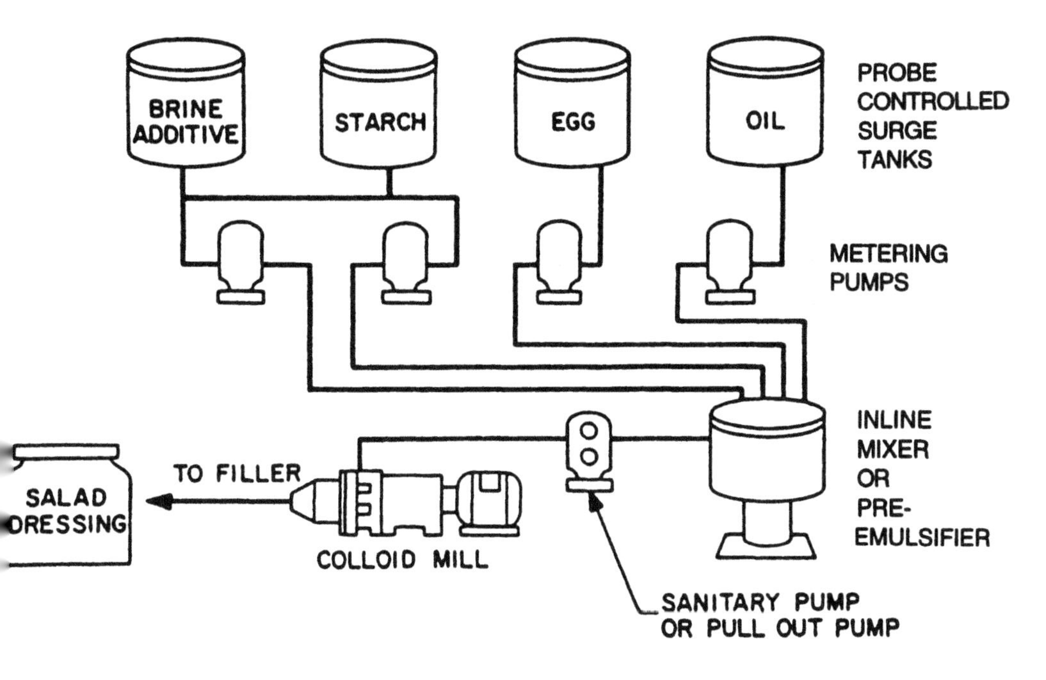

Fig. 18.5. Flow diagram for continuous manufacture of mayonnaise and salad dressings (Emulsol System). *Source:* Used with permission of CTI Publications, Inc., from the 12th edition of *A Complete Cource in Canning,* Book III.

ance. Preparing a 950 L (250 gal) mix takes about 15 min and 45 passes through the colloid mill before the mayonnaise is diverted to the filler.

Continuous Methods

The Emulsol System (Chicago, Illinois) (34) provides continuous automatic manufacture of mayonnaise (Fig. 18.5) or salad dressing from a central process center with an accuracy of 0.5%. The system includes four surge tanks, equipped with upper and lower control limit sensors, and four proportioning pumps for the oil, vinegar and/or eggs, starch paste, and slurry of water, sweeteners and dry ingredients. The pump settings are preprogrammed for speed and sequence of addition. The metered quantities are fed to an in-line mixer, or *preemulsifier*, where a coarse pre-emulsion is formed before being metered to a colloid mill set at the proper setting for either mayonnaise 1.25 mm (0.05 in), or salad dressing 2.5 to 3.75 mm (0.1 to 0.15 in). The system is capable of producing from 1500 to 7500 L (400 to 2000 gal) or more per hour and is monitored by one operator.

The Bran & Luebbe System (The Hartel Corp., Fort Atkinson, WI) (35) (Fig. 18.6) is very much like the Emulsol system where all oil, egg, vinegar, starch paste and water/sweetener/dry ingredient slurry tanks are maintained at a constant level to feed the ingredients head of the proportional metering pump. The capacities

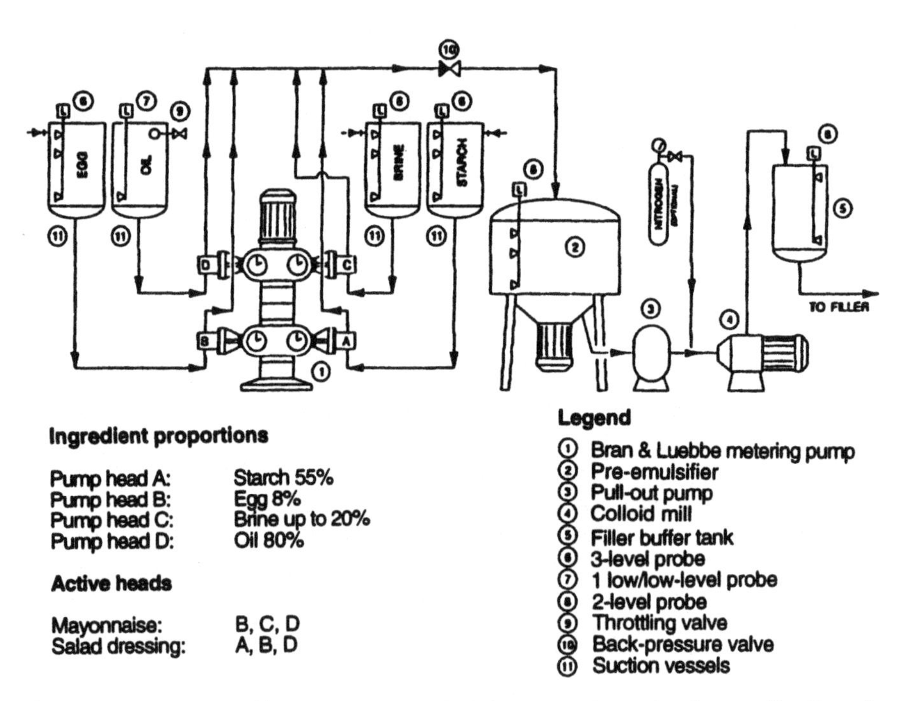

Fig. 18.6. Bran & Luebbe mayonnaise/salad dressing schematic. *Source:* The Bran & Luebbe System, Bran & Luebbe, Inc., Buffalo Grove, IL.

of these tanks vary from as small as 38 L (10 gal) to as large as 280 L (75 gal). All vessels include manual outlet-isolating valving. A manual back pressure valve is located on the discharge of the system to maintain system back pressure. The Bran & Luebbe proportional pump controls the ingredient injection flow rates. Each head has a manual stroke length adjustment. The flow rate of the system is controlled by manually varying the speed of an alternating current (AC) three-phase motor. Once all ingredients are injected into the common header, they are premixed through an in-line static mixer discharging into a 200 L (50 gal) preemulsifier tank equipped with a high-shear agitator. The coarse emulsion is mixed to the proper consistency before milling in the colloid mill; if the proper consistency is not reached, the emulsion is diverted to another tank, where it is corrected and then sent to the colloid mill and filling equipment.

Salad Dressing

Salad dressing is defined as the emulsified semisolid food made from edible vegetable oil, acidifying ingredients, and egg yolks. It contains not less than 30% by weight of vegetable oil and not less than 4% liquid egg yolk. It also contains a paste made from a food starch, modified food starch, tapioca flour, rye flour, wheat flour, or any two or more of these ingredients. Optional emulsifying ingredients are gum acacia (gum arabic), carob bean gum (locust bean gum) (36), guar gum, gum karaya, gum tragacanth, carrageenan (extract of Irish moss), pectin, methyl cellulose, propylene glycol ester of alginic acid, sodium carboxymethylcellulose (cellulose gum), hydroxypropyl methylcellulose, or any mixture of two or more of these. The quantity used of any such emulsifying agent mixture amounts to not more than 0.75% by weight of the finished salad dressing.

Starch-Cooking and -Cooling Equipment

The first step in manufacturing salad dressing is the formation of the starch paste. The following types of equipment are available to choose from, depending on plant capacity and individual preference.

A *batch-type cooker/cooler* is essentially a jacketed, 470 to 1200 L (125 to 300 gal) tank made of 316 stainless steel, equipped with a single- or double-action special agitator mounted on a removable bridge assembly (33). The starch slurry ingredients, consisting of starch, vinegar, water, and sugar, are mixed and then heated to 85°C (186°F) using low-pressure steam (about 1.7 atm, 25 psig), held for 8 to 10 min to ensure sufficient cooking, and then cooled to 27 to 32°C (80 to 90°F) in the same tank using either cold water or ice water or is transferred to another similar tank for cooling.

In larger plants, the starch paste is cooked automatically on a continuous basis using a *plate heat exchanger cooker* (37), the Girdler-Votator *tube cooker* (Cherry-Burrell, Louisville, KY) (38), or the Emulsol *steam infusion system* (Chicago, IL) (39).

In the first system, the starch slurry is cooked at about 88 to 93°C (190 to 200°F) and cooled to 21 to 32°C (70 to 90°F) in a plate heat exchanger equipped with a regeneration section for energy savings.

In the Votator method the slurry is metered under continuous pressure by the feed pump to a four- or five-tube scraped-surface heat exchanger. The slurry is cooked at 90°C (195°F) with steam in the first two tubes and cooled with domestic water or ice water in the second two or three cylinders to 27 to 32°C (80 to 90°F) before discharging to the paste holding tank(s).

The Emulsol System claims to be the most efficient method of processing starch paste, because it saves about 60% of the energy input in the regeneration section. The starch slurry at 21°C (70°F) is fed through the upstream regenerator section of the plate heat exchanger, leaving it at 63°C (145°F). The hot slurry then passes through a special stainless steel steam valve injector, where the proper amount of steam is injected to raise the temperature to 90°C (195°F). The discharged cooked starch is collected in a receiving funnel and then cooled, first to 44°C (120°F) in the regeneration section of the heat exchanger and finally to 27 to 30°C (80 to 85°F) using cold domestic water.

French Dressing

Production of French dressing is governed by a Standard of Identity that defines it as the separable liquid food, or the emulsified viscous fluid food, prepared from not less than 35% edible vegetable oil and vinegar, which may be diluted with water or with citric acid, provided the citric acid is not greater than 25% by weight of the total acid in the final product.

French dressing may be seasoned or flavored by:

1. Salt, sugar, dextrose, corn syrup, invert sugar syrup, nondiastatic maltose syrup, glucose syrup, or honey.
2. Mustard, paprika, other spices or spice oil or spice extract, or any harmless food seasoning or flavoring.
3. Tomato paste, tomato puree, catsup, or sherry wine.

The optional stabilizers and/or emulsifying agents used are generally alginates, xanthan gum, and vegetable gums. Egg yolk is also permitted but is seldom used. The total amount of emulsifying agents is limited to a maximum of 0.75% by weight of the finished product, but it normally ranges from 0.15 to 0.75%. In the production of emulsified French dressing, two methods are used for incorporating the stabilizing agents. In the first method the dry emulsifiers are mixed with other dry ingredients, added to water with sufficient agitation, and allowed to hydrate properly before the oil is added. In the second method, the stabilizer is suspended in a small portion of the oil and then metered into the remainder of the oil phase before the water phase is added, mixed and emulsified in the colloid mill.

Separating, or nonemulsified, French dressing is made by a single- or two-stage filling operation. In the single-stage operation all dry ingredients are thoroughly mixed, added to the oil phase, followed by the water phase, and then metered into the containers and capped. In the two-stage option, all dry ingredients are thoroughly mixed, added to the water phase with sufficient agitation to allow rehydration, and then metered into the appropriate containers to a predetermined volume; finally the oil is added and the container capped.

Nonstandard Salad Dressings

A large variety of nonstandard spoonable and pourable salad dressings are offered to the consumer in supermarkets. Also, there is a host of low-calorie dressings with reduced oil content, as well as some with no oil at all. Less oil emphasizes the gummy, slimy, or stringy character of the gum stabilizer(s) used. In the no-oil dressings, a few ingredients impart richness and excellent oil-like mouthfeel to the dressings.

In the spoonable category, Thousand Island dressing, tartar sauce, sandwich spreads, and a variety of cheese dressings are examples. They contain either mayonnaise or salad dressing as a base, plus drained sweet relish, dill relish, chili sauce, chopped onion, chopped capers, pepper flakes, blue cheese, or other cheeses. Vegetable gum stabilizers may be incorporated with the base to compensate for loss of viscosity and water retention control.

Pourable or liquid salad dressings include a great variety of products with unlimited formula variations. The classic old-fashioned example is the oil- and vinegar-type dressing.

Production Cleanup Guidelines in the Manufacture of Dressings and Sauces

Mayonnaise and salad dressings commercially produced in the United States are defined in accordance with the FDA Standards of Identity. Food has been preserved since early history by acid (40), and the microbiological content of these products are dictated primarily by the high acetic acid concentration found in their aqueous phases. The overall microbiological contents of these products are low with very low incidences of spoilage. Yeasts and bacilli are the organisms commonly found.

The major preservative effect is therefore from the acetic acid content, with minor influences from salt and sugar concentrations. Mayonnaise and salad dressings produced in the United States are antagonistic to bacteria, especially pathogens. The acetic acid levels used by the major producers, 0.31 to 0.32% for mayonnaise and 0.90 to 0.928% for salad dressing, are effective in destroying *Salmonella* and *Staphylococcus* (41). The length of time required for the destruction of *Salmonella* depends on the pH, the number and type of organisms present, and the temperature of storage. The lower the pH of a mayonnaise, the sooner the organism will be killed. The pH should be below 4.0 (42).

The following guidelines were prepared by the Association for Dressings and Sauces to assist manufacturers in developing internal specifications for individual companies. The applicability of specifics of these guidelines may differ; however, the primary concern is the prevention of product contamination from critical areas within the manufacturing system. Bad products on the market create prejudices among customers that discredit not only the particular brand but the entire industry, and this prejudice is exceedingly hard to overcome.

Essentially, a cleanup system involves the following four basic steps:

1. Rinsing to remove all remaining product as soon as possible at the end of a day's production; promptness of cleaning saves effort over the long run
2. Cleaning the system using safe and suitable cleaners
3. Rinsing the cleaning solution thoroughly from equipment surfaces
4. Holding the system in a clean, sanitary state ready for production

All equipment surfaces in contact with the product should be constructed of 316 stainless steel.

All procedures, tests, checks, temperatures, and other parameters should be documented. Such documentation helps eliminate errors and gives valuable information if microbiological problems do occur in the products. Equipment from each finished product should have written cleanup procedures to be followed exactly. Cleanup procedures should never be done from memory.

All cleaning and sanitizing agents should be food-approved by the FDA and should be stored in a dry, segregated area separate from the production area. They should be used as specified by the manufacturer with respect to concentration and temperature. Only potable certified domestic or well water should be used in the food plant for food preparation and for cleaning and sanitizing of equipment.

Cleaning-in-place (CIP) system(s) should never be taken for granted. All parts of the system should be torn down and inspected routinely to ensure adequate performance.

Drain all water from the system after cleaning and let it air-dry. If the acid sanitizer is not of the "no-rinse" type, rinsing should be done immediately before startup, preferably with hot water giving a minimum of 80°C (175°F) surface temperature.

All gaskets should be of a sanitary material and should be replaced routinely.

The quality of the cleanup is directly related to the quality of the people doing that cleanup. The importance of these people to the company should be reflected in their qualifications, training, and pay scale.

Every cleanup procedure should have some means by which the adequacy of that procedure is checked routinely by quality control personnel. This may entail swab tests, visual teardown of critical control points, pH checks, and other measures.

Quality Control and Quality Assurance

Commercial mayonnaise and salad dressings are prepared under strict quality controls instituted by industry. Both products are processed to conform mainly to the Federal Standards of Identity for mayonnaise and salad dressings or from specifications originating from the customer's requirements which may be the same or an extension of the standard specifications or may be unique to a particular end use (43).

Almost all large mayonnaise and salad dressing manufacturing plants have established product quality and product safety control procedures. A "Quality Assurance Program" or "Quality Control Program" is a written statement, or protocol, identifying the procedures, control points, tests and related activities by which a plant protects the quality and safety of the food item it produces and of the ingredients used, and the appropriate records of such procedures and tests.

Normally, a comprehensive quality program covers the raw materials and the finished goods. All raw materials intended for ingredient use in the manufacture of mayonnaise and salad dressings should be procured according to written specifications prepared by those in charge of quality assurance. These specifications may require that certificates of analysis, showing the supplier's results for certain tests, accompany the shipment to be given to the quality assurance personnel. Also, there should be an established protocol for adequate sampling and testing procedures. All raw materials must be stored using good storage practices applicable to each type of raw material thereby ensuring that the integrity of the raw materials will be maintained.

Since vegetable salad oils, such as soybean, cottonseed, corn, and possibly canola, are the predominant ingredients used in mayonnaise and salad dressings, careful specifications should be set up for their purchase. Random samples for laboratory analysis should be taken before unloading. The tests may include some or all of those listed in Table 18.7.

As far as finished products are concerned, the first and most obvious characteristic of a finished product is its appearance. The whiteness of the mayonnaise or dressing depends on the color, amount, and kind of eggs used; color of oil used; the presence of mustard, paprika or other spices; and the processing equipment used to prepare the final emulsion. The whiteness of the product may be improved to some extent by the amount and degree of dispersion of the injected nitrogen gas that is used to control the final viscosity of the product.

TABLE 18.7 Quality Control Laboratory Tests Used in The Manufacture of Mayonnaise and Salad Dressings

Test	Method
Color (Lovibond)	AOCS Cc 13b-45 89
Free fatty acids	AOCS Ca 5a-40 89
Peroxide value	AOCS Cd 8-53 89
Impurities	AOCS Ca 3a-46 89
Moisture (hot plate)	AOCS Ca 2b-38 89
Refractive index	AOCS Cc 7-25 89
Iodine value	AOCS Cd 1-25 89
Kreis test	(see Text)
Cold test	AOCS Cc 11-53 89
Fat stability test	AOPCS Cd 12-57 89
Flavor and odor	AOCS Cg 2-83 89
Microbiological tests:	
Standard plate count	
Yeast and mold	
E. coli	
Salmonella	

The next characteristic after appearance that is observed is the flavor and odor of the finished product. A trained panel usually determines how close the finished product is to the standard in a triangular test. All samples must never be tested off the line. They must be stored at room temperature overnight or for a specific number of hours in order to let the flavors reach equilibrium in the system.

Microbiological testing of finished products is required for safety as well as quality assurance. Such testing ensures that proper sanitation procedures have been followed, that the product is acceptable from the standpoint of public health and laws and regulations relating to healthful foods, and that the product will have the keeping qualities necessary to distribute the product safely with minimum risk of spoilage or deterioration. Testing should be performed on random samples using standard microbiological procedures listed in Table 18.7.

Another function of quality control is returned product disposal. The quality control laboratory should inspect all returned product and order its appropriate disposal.

Finally, any good quality control and quality assurance program is not complete without a recall procedure. Manufacturers should have established recall procedures or guidelines to regulate recalls of the product(s). Recalls may be of two kinds:

1. Market withdrawal, by which the manufacturer initiates a recall-of-product procedure, for any reason;
2. Recall, done at the insistence of a regulatory agency.

Although the foregoing material has been outlined as a guide to those engaged in the production of mayonnaise and salad dressings, constant observations of the following factors will ensure continued customer acceptance, satisfaction, and increased sales:

1. Adherence to specifications
2. Quality control of raw materials
3. Selection and use of the proper equipment in relation to the formulas
4. Following good manufacturing practices and adherence to strict sanitation procedures

References

1. *The History and Manufacture of Mayonnaise and Salad Dressing*, J.H. Filbert, Inc., Baltimore, MD, 1965.
2. *United States Standard of Identity, Code of Federal Regulations*, 21, Chapter 1 (4-1-90 edition), Subpart B–Requirements for Specific Standardized Food Dressings and Flavoring.
3. Rosen, M.J., *Surfactants and Interfacial Phenomena*, InterScience Publications, Wiley, New York, 1978, p. 224.
4. Lipschultz, M., *A Complete Course in Canning*, The Canning Trade, Inc., Baltimore, MD, 1987.

5. Frankel, E.N., in *Handbook of Soybean Oil Processing and Utilization*, edited by D. Erickson, et al., American Oil Chemists' Society, Champaign, IL, 1980, p. 230.
6. Brekke, O.L., in *Handbook of Soybean Oil Processing and Utilization*, edited by D. Erickson, et al., American Oil Chemists' Society, Champaign, IL, 1980, p. 379.
7. Ostwald, W., *Kolloid-Z. 6:* 103 (1910).
8. Ostwald, W., *Kolloid-Z. 7:* 64 (1910).
9. Becher, P., *Emulsions, Theory and Practice*, Reinhold, Brooklyn, New York, 1946, p. 100.
10. Corran, J.W., *Emulsion Technology*, Chemical Publishing Company, Brooklyn, New York, 1946, p. 178.
11. Becher, P., *Emulsions, Theory and Practice*, 2nd edn., Reinhold, New York, 1965, p. 345.
12. Cotterill, O.J., and G.S. Geiger, *Poultry Sci. 56:* 1027 (1977).
13. Finberg, A.J., *Food Eng. (2):* 83 (1955).
14. Privett, O.S., M.L. Bland, and J.A. Schmidt, *J. Food Sci. 27:* 463 (1962).
15. Rhodes, D.N., and C.H. Lea, *Biochem. J. 65:* 526 (1957).
16. Powrie, W.D., and S. Nakai, *The Chemistry of Eggs and Egg Products*. Egg Science and Technology, 3rd edn., AVI Publishing Company, Westport, CN, 1986.
17. Weiss, T.J., *Food Oils and Their Uses*, 2nd edn., AVI Publishing Company, Westport, CN, 1983, pp. 211–230.
18. Pickering, S.U., *J. Soc. Chem. Ind. 13:* 1008 (1921).
19. Bice, S., *Spices and Mustard in Today's Dressings, Mayonnaise, and Salad Dressing*, Mayonnaise and Salad Dressing Institute, Chicago, IL, 1965.
20. Allgeier, R.L., G.B. Nickol, and H.A. Conner, *Food Product Development Magazine*, June–July, Part 1, 1974.
21. Faith, W.L., D.B. Keyes, and R.L. Clark, in *Industrial Chemicals*, 2nd edn., Wiley, Inc., New York, 1957, pp. 11–26.
22. Jacobs, M.B., in *Chemical Analysis of Foods and Food Products*, 3rd edn., Van Nostrand, Princeton, NJ, 1958, pp. 614–616.
23. Lockwood, L.B., and W.E. Irwin, in *Kirk-Othmer Encyclopedia of Chemical Technology*, Vol. 5, edited by A. Standen, Interscience, New York, 1964, pp. 528–531.
24. Gardner, W.H., *CRC Handbook of Food Additives*, 2nd edn., edited by Thomas E. Furia, CRC Press, Cleveland, OH, 1972, pp. 225–270.
25. Gardner, W.H., *Baked Goods, Food Acidulants*, Allied Chemical Corporation, New York, 1956, p. 156.
26. Gardner, W.H., *Meat, Fish, and Oil*, Allied Chemical Corporation, New York, 1966, p. 173.
27. Schoch, T.L., *Glass Packer*, p. 69, (January, 1967).
28. Osman, E.M. in *Starch: Chemistry and Technology, Vol. II*, Industrial Aspects, edited by R.L. Whistler and E.F. Paschall, p. 169.
29. Greenwood, D., *The Carbohydrates, Chemistry, and Biochemistry*, 2nd edn., edited by W. Pigman and D. Horton, 1970, p. 471.
30. Cumming, D., *Food Technol. 18:* 1901 (1964).
31. A.E. Staley Manufacturing Company, *Industrial Sales*. Technical Bulletin, Decatur, IL, 1966.
32. Mattil, K.E., in *Bailey's Industrial Oil and Fat Products*, 3rd edn., edited by D. Swern, Interscience, New York, 1964, p. 260.

33. Haycock, R., *Mayonnaise and Salad Dressing Systems*, Cherry-Burrell, Louisville, KY, Process Equipment Literature, November, 1993.
34. *Emulsol Mayonnaise and Salad Dressing Batch Systems*, Emulsol Equipment, Inc., Chicago, IL.
35. *The Bran & Luebbe System*, Bran & Luebbe, Inc., Buffalo Grove, IL.
36. Anderson, D.M.W., and S.A. Andon, *Cereal Food World Vol. 33:* 844 (1988).
37. Ziemba, J.V., *Food Ind. 3:* 124 and 200 (1949).
38. *Starch Paste Cooker (SPC) Engineering Drawings #C-02224O*, Cherry-Burrell Process Equipment, Louisville, KY.
39. *Emulsol Continuous Cooker-Cooler System Sales Bulletin*, Emulsol Equipment, Inc., Chicago, IL.
40. Weiser, H.H., *Practical Food Microb. Techn. 16:* 238 (1962).
41. Smittle, R.B., *J. Food Protection 40:* 415 (1977).
42. Lerche, M., *Wierner Tierarztliche Monatsschrift. 48:* 348 (1961).
43. Erickson, D.R., *J. Amer. Oil Chem. Soc. 44:* 534A (1967).

Chapter 19

Consumer and Industrial Margarines

Ahmad Moustafa

Consultant
Cincinnati, OH

Introduction

Margarine is an engineered product invented in 1869 (1) because of a butter shortage in Europe. Its evolution to a highly accepted spread is a prime example of technological advancement made through the combined efforts of food technologists, oil chemists, nutritionists, and chemical engineers. Margarine has taken its place worldwide as an excellent nutritive food because it is a concentrated source of food energy, it can be a uniform supplement of vitamins A and D, it can be a source of polyunsaturated essential fatty acids, it contributes satiety, it contributes appetizing flavor and it has complementary effects on other foods (2).

History

A number of factors were responsible for stimulating inventors to seek a substitute for butter in Europe during the mid-1800s. Increasing pressure of the population on the food supply and a growing industrial base, as well as the rise of French nationalism, required a more secure food supply. The invention of margarine occurred, then, at a time of need in Europe for a cheap substitute for butter. It was the direct result of an offer by Napoleon III of a handsome prize to develop an economic and yet nutritious and appetizing substitute for butter. The pharmacist Hippolyte Mege Mouries won the prize and in 1869 applied for French and English patents.

Attempts to create butterlike substances had been made for years previous. The name *margarine* was originally coined from a fatty acid component isolated in 1813 by the oil chemist Michel Eugene Chevreul that became known as margaric acid. Even though Mouries' product did not resemble today's margarine, it was nevertheless made from naturally-occurring products (beef tallow) and possessed all the physical attributes and food values of butter. Mouries attached the word *oleo* for beef tallow to Chevreul's *margaric* naming his new product *Oleo-Margarine*, a name that is still used today. On the American scene, margarine production began in 1873 with the issuance of a patent to the originator of the "Oleomargarine" (3).

Like most European countries, the United States had a large and powerful dairy industry, whose lobbying hindered the growth of the margarine industry through legislative restrictions. Therefore, real growth in the margarine industry did not come about until after 1941, when the Federal Standard of Identity for margarine was promulgated, and the following repeal of many state tax laws on colored margarine in the 1950s. By 1956 the per capita consumption of margarine surpassed that of butter for the first time and has never relinquished that position since.

Major Developments in Margarine Technology

The original production process by Mouries (3,4) consisted of the following four basic steps:

1. Mincing and washing of fresh fat
2. Digestion of the minced fat with artificial gastric juice
3. Expression of the softer portion of the fat from the harder fat
4. Digestion and agitation of the soft fat with milk and mammary-gland tissue extract

From such an unappetizing beginning, the margarine industry has seen many technological changes in attempts to improve and refine the product. The following is a summary of the major developments during the last 125 years:

1. In 1917 coconut oil replaced tallow in the oil phase as the preferred fat.
2. In 1923 the addition of Vitamin A brought margarine nutritionally closer to butter.
3. One of the most significant developments contributing to the rapid acceptance of margarine worldwide was the advancement in vegetable oil refining technology, followed by the development of the hydrogenation process in 1909, which allowed the use of refined and partially hydrogenated vegetable oils in the manufacture of margarine (see Chapter 13).
4. In 1934 partially hydrogenated cottonseed oil replaced coconut oil and became the major component of all vegetable oil margarines.
5. Equally important was the development, in the United States in 1937, of the brine-cooled revolving drum for the rapid chilling of the fat emulsion, and the ultimate development of the closed continuous internal chiller and plasticizer, or scraped surface heat exchanger (SSHE).
6. Another breakthrough in the development in 1947 of today's margarine was the use of β-carotene (also a vitamin A source) and annatto as replacements for coal tar–derived dye colors.
7. By 1948 cottonseed oil was the major component of the oil in 62% of all vegetable oil–based margarines in the United States, but it soon relinquished its position to soybean oil by 1956, when 68% of all vegetable oil–based margarines were composed of soybean oil. Today, soybean oil amounts to 90% of the oil used in margarines sold in the United States.
8. The year 1957 saw the introduction of soft whipped margarine in stick form.
9. In 1958 the adaptability of the margarine formulation to meeting nutritional needs was demonstrated by the use of liquid corn oil in the production of the first margarine that had a high ratio of polyunsaturated fatty acids to saturated fatty acids (P/S) (2.1:1.0) in response to the perceived need for additional polyunsaturated fatty acids in the diet.

10. The following year saw the introduction of a grocery store margarine made with liquid corn oil and partially hydrogenated corn oil, with a P/S ratio of 1.55:1.00.
11. Pourable, nonseparating margarines, in which the emulsion remained uniform and pourable under refrigeration and at room temperature, were introduced in 1967.
12. Restrictions on the use of emulsifiers were lifted in 1992.
13. The flexibility of the margarine industry and its willingness to adapt to changing times was demonstrated in 1993 by the latest amendment to the Standard of Identity, allowing the addition of refined and partially hydrogenated menhaden fish oil to the list of oils that can be used in margarine.
14. All the foregoing developments were paralleled with contributions of microbiologists and flavor chemists whose cultures and synthetic flavors brought the flavor of today's margarine very close to that of butter.

Of all of these developments, the three most significant developments contributing to the rapid worldwide spread and acceptance of margarine were (i) advances in oil refining technology; (ii) development of hydrogenation, which allowed the use of partially hydrogenated vegetable oils in margarines; and (iii) development in the United States of the brine-cooled revolving drum for rapid chilling of the fat emulsion and the subsequent development in 1937 of the closed continuous internal chiller-plasticizer or scraped-surface heat exchanger (SSHE) that became universally known as the *Votator* (Cherry-Burrell, Louisville, KY) (5–8).

Today's margarine is neither an imitation of, nor a substitute for, butter, even though the spread is made from naturally occurring products and possesses all the physical, sensory, and nutritional attributes of butter. It is also fair to say that margarine is a mature food product in demand by consumers worldwide. Margarine is manufactured in different varieties and styles and is priced within the means of most potential consumers. Per capita consumption of margarine in the United States is shown in Table 19.1. Also, margarine accounts for approximately 20% of the total fat intake, as shown in Table 19.2 (8).

TABLE 19.1 Per Capita Consumption of Margarine and Butter in the United States

	Margarine		Butter	
Year	lb	kg	lb	kg
1900	1.4	0.64	19.6	8.89
1920	3.4	1.54	14.9	6.76
1930	2.6	1.18	17.6	7.98
1950	6.1	2.77	10.7	4.85
1956	8.2	3.72	8.7	3.95
1957	8.6	3.91	8.3	3.77
1970 (8)	11.0	4.99	5.3	2.40
1975 (8)	11.0	4.99	4.7	2.13
1980 (8)	11.3	5.12	4.5	2.04
1985 (8)	11.8	5.82	4.9	2.42
1990 (8)	10.9	5.37	4.4	2.17

Source: Historical Statistics of the USA. Colonial times to 1970, U.S. Bureau of Census. 1989.

TABLE 19.2 U.S. Per Capita Consumption of Edible Fats and Oils (in lbs) (8)

Year	Total fat intake	Margarine	Butter	Shortening	Salad oil, cooking oil	Other uses
1970	52.6	10.8	5.40	17.3	15.4	6.9
1975	52.6	11.0	4.70	17.0	17.9	5.2
1980	57.2	11.3	4.50	18.2	21.2	5.2
1985	64.3	11.8	4.90	22.9	23.5	5.3
1990	62.7	10.9	4.40	22.2	24.2	4.2

Formulation of Margarine Oil Blends

The subjects of margarine and margarine oil formulation and their control are covered in several articles in the literature (9,10,11).

To produce the proper consistency in the various types of margarines and spreads, a thorough understanding of the theory and of the practical application of emulsion technology is needed. A formulator must focus on the following basic principles:

1. For each margarine style to be produced, the desired spreadability at refrigerator temperature and acceptable mouth feel or "getaway" requires that a specific balance be maintained between the liquid and solid oil fractions of the continuous phase.
2. Crystallization of the high-melting triglycerides, or the solid phase, must be accomplished quickly so that almost all of the crystal nuclei are formed in the desired beta prime (β') form (1 μm or less crystal size).
3. The low-melting, or liquid, triglycerides must be entrapped within the crystal lattice of the solidified high melting fat as honey is held in a honeycomb.
4. The finely formed crystals must form a lattice strong enough to hold the finely divided water droplets, which range from 2 to 20 μm, with the majority being 2 to 5 μm, as the discontinuous phase.

Margarine is often produced by manufacturers independent from oil refiners. Hence, the former, in consultation with their oil supplier(s), must write meaningful specifications for their purchased base stocks and oil as well as for their finished margarine products. Detailed processing specifications must also be issued for plant operators to follow. These measures are very important in maintaining product integrity with respect to plasticity and spreadability.

A thorough description of what takes place during cooling and supercooling of the polymorphic fat blend in the chilling equipment has been published (11); see also Chapter 15. It must be emphasized that close control of the processing temperature of the emulsion during chilling in the SSHE (plasticizing) is as important as is specifying a narrow range for the solid fat index (SFI) or solid fat content (SFC) curve and melting properties of the incoming oil.

Formulating fats and oils for all applications with satisfactory functionality using a single source of oil is probably impossible (12). This presents the shortening and margarine oil manufacturer with a challenge when it comes to developing a practical base stock system. The versatility and other desirable characteristics of soybean oil, some of which are shown in the following list, have made it the dominant edible oil base stock material (13) (see Chapter 13):

1. Plentiful, dependable supplies available at competitive price
2. High content of essential fatty acids
3. A high iodine value, which permits hydrogenation and blending for the production of a large number of finished products
4. Low refining losses

With this in mind, an example of a practical base stock system, in which a limited number of hydrogenated soybean base stocks are blended to meet the requirements of many finished products is shown in Table 19.3 (14). The advantages of a base stock program are improved product uniformity, elimination of rework and heels, accommodation to normal variation in hydrogenation, and increased plant efficiency (see Chapter 13).

Typical soybean oil base stock margarine blends can be prepared from the stocks described in Table 19.3; see Table 19.4.

However, the indicated values are not absolute and may vary slightly from batch to batch to take into account the expected, but minor, variability in hydrogenation end points.

The following hydrogenation process controls are used to determine base stock end points are (13,14):

1. *High IV* (+90): refractive index or butyro scale number (AOCS Method Cc 7-25)

TABLE 19.3 Hydrogenation Conditions and Properties of Partially Hydrogenated Soybean Oil Base Stocks for Margarine Production (14)

Stock Number	1	2	3	4	5
Hydrogenation conditions:					
Initial temp, °F	300	300	300	300	300
°C	150	150	150	150	150
Final temp, °F	330	350	425	425	330
°C	165	177	218	218	165
Pressure, psig	15	15	15	5	15
Nickel catalyst[a],%	0.02	0.02	0.02	0.05[b]	0.02
Oil Properties:					
Final IV	80–82	106–108	73–76	64–68	70–72
Congeal point, °F	—	—	75–76	91.4–92.3	77.9–78.8
°C	—	—	23.9–24.4	33.0–33.5	25.5–26.0
Solid fat index					
10.0°C (50°F)	19–21	4 max	36–38	58–61	40–43
21.1°C (70°F)	11–13	2 max	19–21	42–46	27–29
33.3°C (92°F)	0.0	0.0	2 max	21 max	9–11

[a]Based on weight of oil.
[b]Very selective catalyst needed.

TABLE 19.4 Typical Margarine Formulas Using Soybean Oil Base Stocks (13,14)

Formula number	1	2	3	4
Margarine type	Soft stick	Stick	Stick	Tub
SFI				
10.0°C (50°F)	20–24	27–30	28–32	10–14
21.1°C (70°F)	12–15	17.5 min.	16–18	6–9
33.3°C (92°F)	2–4	2.5–3.5	1–2	2–4
Stock #		Percentage Composition		
1	—	—	60	—
2	—	42	—	80
3	—	20	25	—
4	50	38	15	20
Soybean salad oil	50	—	—	—

2. *Intermediate IV* (55 to 90): refractive index or butyro scale number (AOCS Method Cc 7-25) plus congeal point (AOCS Method Cc 14-59) or Mettler dropping point (AOCS Method Cc 18-80)
3. *Low IV* (10 or less): quick titer, which is a modification of the standard AOCS Method Cc 12-59.

The quick titer test involves dipping the bulb of a glass thermometer into liquid fat and then rotating the thermometer stem between the fingers to cool the fat. The end point is, the temperature reading when the fat on the bulb clouds. Constants have been determined to add to the quick titer reading to approximate the real titer as determined by the standard method (15).

In addition to the oil base stocks just described, the refiner probably will have on hand one or more of the following oil stocks in order to formulate a wide range of consumer table-grade margarines and spreads, industrial bakery margarines, and frying shortenings:

1. Refined, bleached, and deodorized (RBD) soybean, corn, sunflower, safflower, or canola oils
2. RBD palm oil
3. Partially hydrogenated cottonseed oil (IV 30)
4. Partially hydrogenated cottonseed flakes (IV 5)
5. Partially hydrogenated palm oil flakes (5 IV).

However, in most countries other than the United States, the oils and fats used for food purposes depend on local availability and imports from all edible oil sources giving rise to the multi-feedstock situation (16), where one oil type is substituted for another. Table 19.5 lists a number of oils and fats used in Europe.

The polymorphism and crystal habit characteristics of the partially hydrogenated soybean base stock may cause consistency problems in the finished margarine because of the formation of large β crystals. The addition of one of the aforementioned β′ crystal promoter fats or the addition of a crystal-modifying additive is required to offset the tendency to form the larger undesirable β crystals (17).

TABLE 19.5 Refined, Bleached Oils and Fats Used in Europe

Vegetable	Animal	Marine
Coconut	Lard	Sardine
Cottonseed	Beef tallow	Anchovy
Groundnut	Anhydrous butter oil	Menhaden
Maize germ	Crystal modified lard	Herring
Palm		Cephalin
Palm kernel		Whale
Canola		
Safflower		
Soybean		
Sunflower		

Despite the fact that hydrogenated soybean oil is a β-tending triglyceride, it is used in the manufacture of 90% of all table-grade margarines and spreads produced in the United States. It is theorized that the more a triglyceride approaches a pure chemical compound, the more it will tend to crystallize in the β; and the more it deviates from a pure compound, the more it tends to be β′. Therefore, although soybean oil is roughly 10% C16 and 90% C18 chain lengths when it is hydrogenated, the formation of *trans* isomers of the C18 fatty acids results in a greater mixture of different fatty acids, making it less of a "pure" compound (18). Thus, blending of any three partially hydrogenated soybean stocks listed in Table 19.3 will protect the finished margarine spread against graininess, as long as the product is maintained under refrigeration throughout its distribution and marketing cycle.

The use of a blend of three base stocks increases the heterogeneity of the triglycerides and prompts the fat phase to crystallize in the β′ form, or small needle-shaped crystals, giving the final margarine spread a smooth, pleasing mouthfeel and good spreadability.

Since 1962 most major premium and private label margarine and spread brands available in U.S. grocery stores have been produced using blends of two or three soybean base stocks. These products have shelf lives of 9 mon or more with no sign of sandiness or graining.

It is very important to avoid temperature cycling during in-plant storage, during shipping, and at retail outlets. Melting and solidification as a continuing process could result in shifting of the β′ crystals to the larger β type. This condition is usually accompanied by a reduction in the specific area of the crystal mass and leads to oiling out. This will become more apparent in blends of two base stocks than in a product made from three oil base stocks.

Formation of the Margarine Emulsion

Oil and water are immiscible liquids. With vigorous mechanical agitation they can be emulsified, but the emulsion will separate rapidly on cessation of agitation unless a third component—an emulsifier—is placed at the interface between the two immiscible phases. Also, when the two are mixed, two types of emulsions are possible. In the oil-in-water type (O/W), the oil is the *dispersed, discontinuous*, or *internal* phase, surrounded by the *continuous* or *external* water phase. In the water-in-oil type (W/O), the opposite is true.

An emulsion is a "significantly stable" suspension of droplets of a liquid of a certain size within a second, immiscible liquid. Two types of emulsions based on the sizes of the dispersed particles are recognized: macroemulsions and microemulsions. The particles of macroemulsions range from 0.2 to 50 μm and are easily observed under a microscope, whereas a microemulsion has particles from 0.01 to 0.2 μm. Standard margarine and spread emulsions are macroemulsions.

In margarine manufacture, stirring milk or water into the bulk oil favors the formation of a water-in-oil emulsion (W/O). On the other hand, if the oil is run into the water phase, the opposite type of emulsion is formed (O/W) (19). Initially, the water phase is dispersed in the continuous oil phase by means of a good shear agitator mixer, and final emulsification is achieved in the positive-displacement pump that feeds the SSHE and in the SSHE itself. Rapid solidification of the fat phase restricts the separation of the two phases by immobilizing the fat phase as well as the fine water droplets in the plastic fat crystal mass.

The ratio of the two phases, oil and water, influences the type of emulsion formed. In the production of a standard 80% fat margarine with 17 to 20% aqueous phase, the emulsion type is predominantly, or entirely, W/O. On the other hand, with the introduction of the 60 to 40% fat margarine spreads in recent years, strict processing controls must be practiced to guard against the formation of a mixture of W/O and O/W emulsions or a cream-type (O/W) emulsion.

The degree of dispersion and stabilization that is considered optimal depends on the melting points of the glyceride fractions and their crystal structures. A uniform structure of small crystals reduces the plasticity of the solid margarine and gives a "heavy" sensation on the palate. It has been shown that if 80 to 85% of the globules have diameters of <1 μm and the remainder not above 3 μm, a very dense, "fatty" product is obtained. Margarine with 95% globules of diameter 1 to 5 μm, 4% of 5 to 10 μm, and 1% 10 to 20 μm is usually "light" on the palate.

The average droplet size and distribution of the dispersed water phase plays an important role in the flavor release of the water-soluble flavor ingredients and the keeping quality of the finished margarine. As previously mentioned, the majority of the water droplets range in size from 2 to 5 μm; however, for good flavor release, some droplets should remain close to the upper limits (20 μm).

A very fine dispersion of the aqueous phase helps to protect the margarine against microbial deterioration, because many of these droplets are smaller than many of the undesirable microorganisms such as yeasts and molds.

Experience has shown that as the moisture content of the spread increases, taste panelists experience difficulty in flavor perception. This may be attributed to the fact that in making these low-calorie spreads, higher percentages of strongly lipophilic emulsifiers, with an HLB of 3 to 6, are used, creating a strong and stable interfacial film between the oil and water phases. This will prevent phase reversion of the W/O emulsion to O/W in the mouth, facilitating the release of the flavor additives. In other words, the spread does not stay in the mouth long enough for the fat to melt and release the flavor-laden water phase. Hence, rapid melting of part of the fat at temperatures well below that of the body gives the palate immediate access to the water droplets and the release of flavor ingredients.

Margarine Production Practices

Margarine Products

As mentioned above, U.S. margarine production is governed by a Federal Standard of Identity regulation that specifies fat content limits. In a previous publication (20), two products were listed:

1. Standard, 80% minimum fat content margarine
2. Diet or imitation margarine, 40% fat max.

That left a wide gap between the two limits, allowing for the introduction of the present-day "spreads." In 1973, the foregoing standards were amended by the U.S. Food and Drug Administration (FDA) to bring them into line with those published by *Codex Alimentarius*, which provided increased flexibility in product formulation, especially for that of the water phase, and delineated the use of the required 10% of nonfat dry milk solids (NFDMS) in formulating the water-phase.

After 1975, when the Kraft Foods Company (Memphis, TN) introduced its first 60% fat content spread (21), a host of other spreads between 60 and 70% fat appeared on the market. These were produced in stick or tub form according to one of the SFI curves shown in Table 19.6.

These substitutes did not conform to the Federal Standard of Identity for margarine in that they did not contain a minimum 80% fat by weight, but they resembled margarine in both taste and appearance. The products were not nutritionally identical to margarine, prompting the FDA to propose the establishment of a descriptive common name for these substitutes in order to signal to the consumer that the products were "imitations" and "inferior" to the standard margarine (22).

TABLE 19.6 Solid Fat Index and Fatty Acids Types of Margarines and Spreads Sold in the United States (21)

	Stick		Soft tub
Temperature	Regular	Hi-Li[a]	(70% liq. oil)
10.0°C (50°F)	25–30	16–24	8–14
21.1°C (70°F)	14–18	10–15	5–8
33.3°C (92°F)	2–4	1.5–4	0.5–2.5
Polyunsaturates	3–10	20–40	30–60
Monounsaturates	50–70	20–50	15–42
Saturates	16–25	13–23	10–20

[a]High-linoleic acid content oil blends.

In the meantime, whipped margarine appeared on the market during 1978, followed by another spread containing 53.3% fat, or one-third less fat than standard margarine. This apparently was done in response to the American Heart Association's dietary recommendation that total dietary fat should comprise no more than 30% of the daily caloric intake.

Since vitamin A, and sometimes vitamins D and E, are present in significant quantities in standard margarines, manufacturers of the new spreads fortified their products to the same levels as in standard margarine in order to maintain nutritional equivalence for these vitamins.

Since then, the Standard has been amended to include only two classes of compounds: the Standard 80% category and the Spread, which is defined as containing less than 80% fat. The popularity of this new product is shown in Table 19.7.

The latest amendment, allowing the use of marine oils, became effective September 27, 1993 by the revision of section 166.110, paragraph (a)(i) as follows:

> *Edible fats and/or oils, or mixtures of these, whose origin is vegetable or rendered animal carcass fat, or any form of oil from a marine species that has been affirmed as Generally Recognized as Safe (GRAS) or listed as a food additive for this use, any or all of which may have been subjected to an accepted process of physico-chemical modification.*

TABLE 19.7 Production of Margarine and Spread in the United States[a] in Millions of Pounds

Year	Margarine	Spread	Spread% of Total[c]
1979[b]	2553.2	195.3	7.65
1980	2553.2	248.4	9.73
1981	2576.4	293.6	11.98
1982	2596.0	320.9	12.36
1983	2451.2	376.1	15.34
1984	2480.9	520.5	20.96
1985	2603.3	581.7	22.34
1986	2808.8	611.4	21.77
1987	2685.2	607.3	22.62
1988	2627.2	648.5	24.68
1989	2799.4	823.3	29.41
1990	2767.1	856.7	30.96
1991	2791.7	898.0	32.17
1992	2755.2	991.9	36.00
1993	2750.0	1193.8	43.41

[a]National Association of Margarine Manufacturers (NAMM).
[b]First year production data reported to NAMM.
[c]Includes spread quarters, solids, and soft.

Oil Supply Considerations

As previously mentioned, the majority of U.S. margarine manufacturers do not own their oil-refining facilities and must rely on two or three different oil refineries to supply them with oils according to written specifications that outline SFI or SFC curves, melting point, peroxide value, color, and flavor. Since oil constitutes between 50 and 80% of the product by weight, and follows labor as the second most important "variable cost" factor, close attention to daily oil prices and the logistics of production scheduling and future oil deliveries become essential considerations of any profitable margarine operation.

Large companies with nationally advertised brands can well afford to contract to purchase large quantities of oil to cover their projected sales requirements. On the other hand, small companies specializing in the disappearing "private label" or store brands usually operate within limited market areas and may overlap each other's sales territory; these companies have limited cash flows and cannot afford to speculate on oil futures. They can only commit themselves to known sales promotion target dates, because they know that the quantities needed and the selling price have been determined before production starts.

With competition as keen as it is today, product formulation and control become essential steps in costing. Margarine is a product of combination between an oil phase and a water phase, each of which must be accurately formulated in order to meet both government regulation and formula specifications. In the United States, almost all margarine manufacturing companies rely on weighing each of the oil and water ingredients (gravimetric) rather than dosing (volumetric) them as practiced in Europe, thus eliminating the need for strict temperature control of oils and water phases and continued monitoring of the displacement accuracy of the dosing equipment. However, with the introduction of computer control, this system may appeal to some operators.

Steps in Margarine Processing

The essential steps for processing margarines and spreads are somewhat similar to those for shortenings except that the first contains a two-phase system—oil phase and water phase—and the second is a 100% fat phase. Also, the processing equipment of all margarines must be 100% stainless steel whereas that for shortenings can be fabricated of carbon steel, like oil storage tanks and pipes, because there is no water in the system to cause rust development. Most modern SSHE equipment is fabricated from stainless steel. In either case, the process may require all the following steps (23):

1. Batching, weighing, or proportioning of the essential ingredients
2. Mixing and creation of a coarse emulsion
3. Cooling and plasticizing
4. Working to complete crystallization and create the proper filling characteristics
5. Resting
6. Packaging

Margarine Production Equipment

Fig. 19.1 shows a simplified flow diagram for a semicontinuous margarine production batching system, where oil-soluble ingredients (such as lecithin, mono- and diglycerides; vitamins A, D, or E; and, to some extent, oil-soluble flavors and colors) are added to the margarine oil blend and kept moderately heated and mildly agitated in order not to affect adversely either the oil quality or the color of the blend. The water phase, which can range from 100% fresh skim milk to reconstituted nonfat dry milk solids (NFDMS) and water, is pasteurized and stored at 4.4°C (40°F) until used. Some of that milk may be cultured with special lactic acid bacteria for the proper development or the typical buttery flavor when used in the formulation of margarines. A typical milk blend will be made of 70 to 75% uncultured pasteurized milk and 25 to 30% cultured milk, to which the water-soluble ingredients (such as salt, preservatives and water-soluble flavors) are added. First the oil phase and then the water phase is weighed into another tank, which is mounted on a beam scale and provided with an appropriately sized mixer to initiate the formation of the W/O emulsion; then the mixture is transferred to an emulsion tank, which feeds the proper chilling and packaging equipment. The system works well in plants that produce a limited number of formulas that require the use of the same oil blends and water phase.

However, if the plant is producing a variety of formulations that differ in both oil blends and water phases as well as color, flavor, emulsifier, salt content, vitamin potency, and preservatives, then the system shown in Fig. 19.2 is recommended. It depends on a load cell (strain gauge) suspended from an appropriate part in the pro-

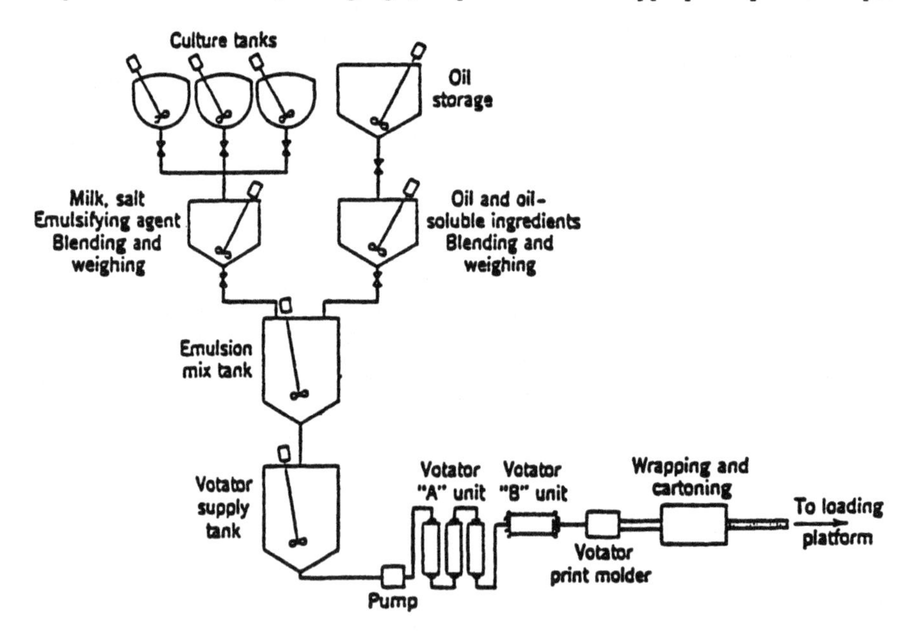

Fig. 19.1. Simplified flow diagram for continuous margarine solidification. Courtesy of Cherry-Burrell Process Equipment.

duction department or from a specially designed cradle and connected to the formulating computer. The computer feeds the necessary commands to the different oil and water phase pumps and air actuated delivery valves, which are mounted on separate manifolds above the weigh tank (one for the oil phase and the other for the water phase) to open and close as needed as shown in Fig. 19.3.

Another important part of this system is an adequate oil tank storage facility for storing all the different types of oil. Depending upon the particular plant operation, one may find:

1. Vegetable base stocks, such as refined, bleached, and deodorized (RBD) soybean oil; partially hydrogenated soybean oil; RBD palm oil; cottonseed oil; or canola oil
2. Animal fats, such as lard or tallow

The same is found for the water phase, where each type of milk or water is stored in its own tank with its own delivery pump.

When a standard 80% fat margarine formula is called for, the proper soft oil valve will open, and the oil will be delivered to the weigh scale tank. At the same time the proper amount of lecithin or mono- and diglyceride is metered in with this first oil from a calibrated positive-displacement pump. This will be followed by other oils called for in the formula as needed. In each case, the delivery valve at the

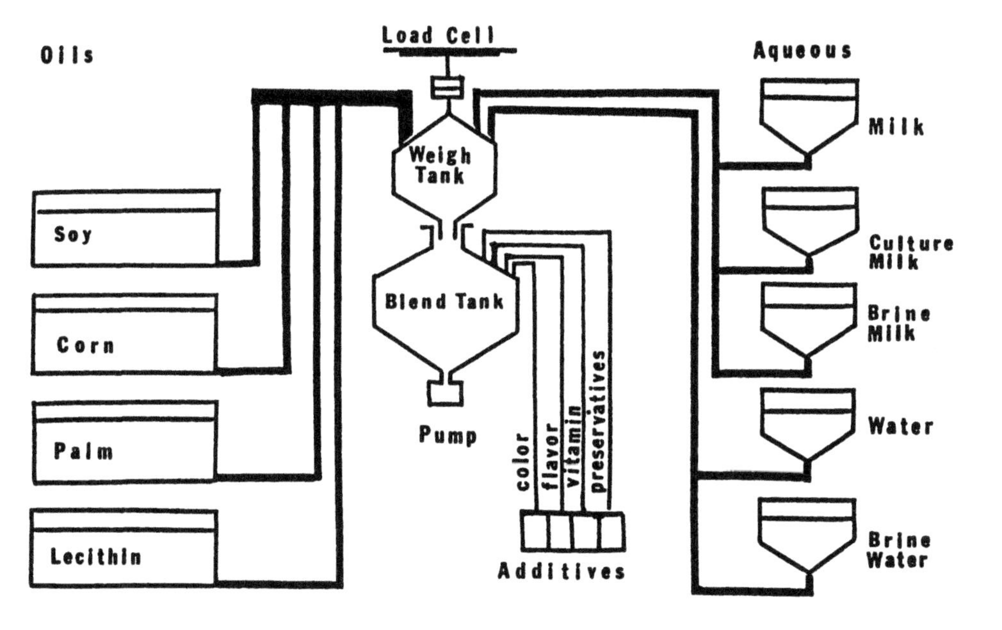

Fig. 19.2. Margarine Batching.

end of the manifold will partially close toward the end of the addition until the load cell registers the proper weight for that particular oil, and so on.

The same thing will take place with the water phase, and a final correction of the percentage of oil in the final emulsion is made by adding either water or oil to the top scale, before the emulsion is dropped to the lower mix tank. Then the rest of the minor ingredients, such as color, flavor, preservatives, and sometimes mono- and diglycerides, are added from calibrated stainless steel positive-displacement pumps. High-shear agitation and mixing for a specific length of time is an essential requirement for the proper formation of the W/O emulsion before the mixed emulsion is sent to a holding tank for processing in the SSHE equipment.

Such a system will work very well with margarine formulas containing from 53 to 80% fat. However, as the fat content decreases below 53%, strict processing con-

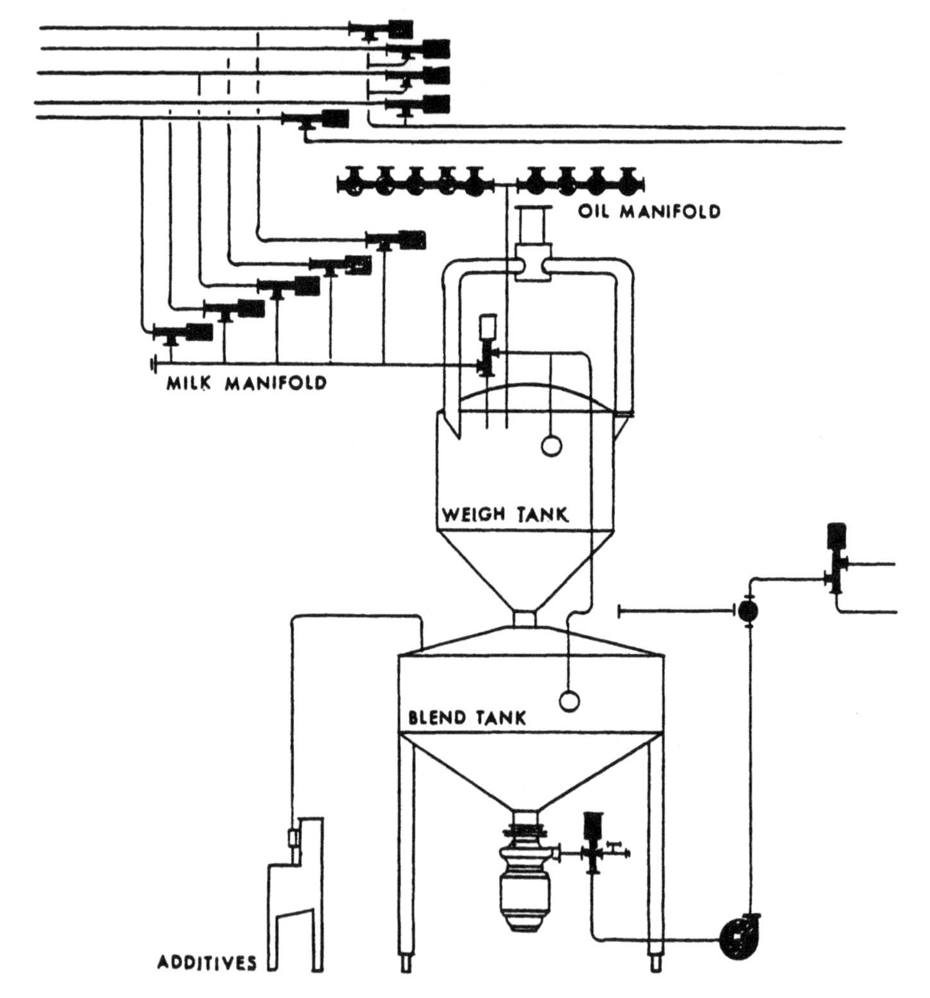

Fig. 19.3. Auto Batch Weigh System.

trols, such as oil and water temperatures and the rate of water addition to the oil, must be employed to guard against the formation of a mixture of the W/O and O/W emulsions or even the formation of a 100% cream-type O/W emulsion. In such a case, the system just described can be modified to accommodate these high-water spreads by either of the following:

1. Having two load cells and two weigh tanks of the appropriate size, one for weighing the oil phase and the other for weighing the water phase; the final oil blend will be dropped into the blend tank and the water phase will be discharged slowly to the lower emulsion tank through a partially opened valve, thus allowing for the proper mixing and emulsion formation below
2. Modifying the mode of addition in the single-load-cell system, where the oil phase is first weighed and then dropped into the mixing tank, after which the water phase is dropped slowly and carefully with mixing

The system as described saves floor space. However, if floor space is available, two or three load cells per weighing tank, with a totalizer, can be mounted under the tank legs. Fig. 19.4 is a simplified flow chart for production of stick margarine, and Fig. 19.5 is a typical flow chart for the production of soft and spread margarines.

As the ratio of water to oil increases, reduced-fat spread emulsions begin to increase in viscosity, requiring additional pumping capacity to maintain production rates. Also, very close control of crystallization rate and temperature must be ob-

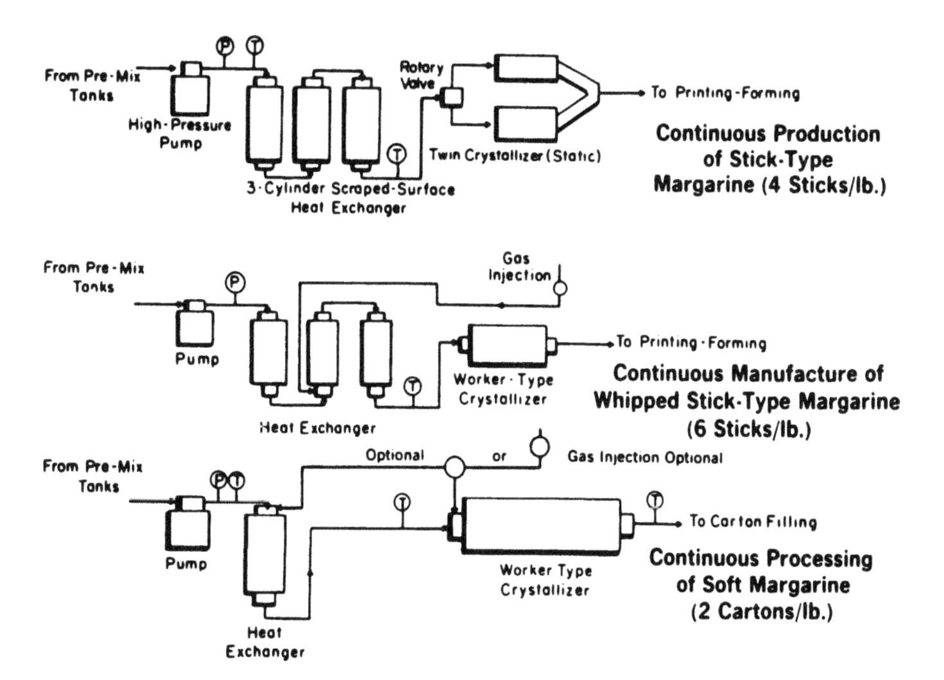

Fig. 19.4. Flow diagram of the Votator Stick margarine process.

served in order to guard against emulsion breakage or inversion of phases, which invariably requires a higher fill temperature.

Very Low Fat Spreads

In 1989 a new margarine spread, containing less than 20% fat as the continuous phase, appeared on the market. It required the use of special processing equipment to form the W/O emulsion. However, O/W emulsion spreads containing approximately 20% fat have been prepared using homogenization prior to plasticizing in the SSHE (24).

As recently as 1967 it was believed that emulsions containing more than 75% by volume of an internal or dispersed phase could not be prepared, due to theoretical considerations of structural packing densities; however, it was discovered that these emulsions could be prepared by using special mixing techniques. They are "structured" systems and are referred to as *high-internal-phase-ratio emulsions* (HIPREs) (25). The basic property that distinguishes them from conventional emulsions is behavior like that of a solid when at rest yet like that of a liquid when force is applied.

Present margarine-processing equipment does not apply enough shear to very-low-fat margarine spread emulsion to achieve an apparent viscosity near those of the external and internal phases without going above the shear stability point of the emulsion. A volume of 74% is the limit for packing monodispersed hard spheres. To attain internal phase loadings greater than 74%, the particles must be nonspherical, or "polyhedra," with 14 faces on the average. The processing apparatus is substantially

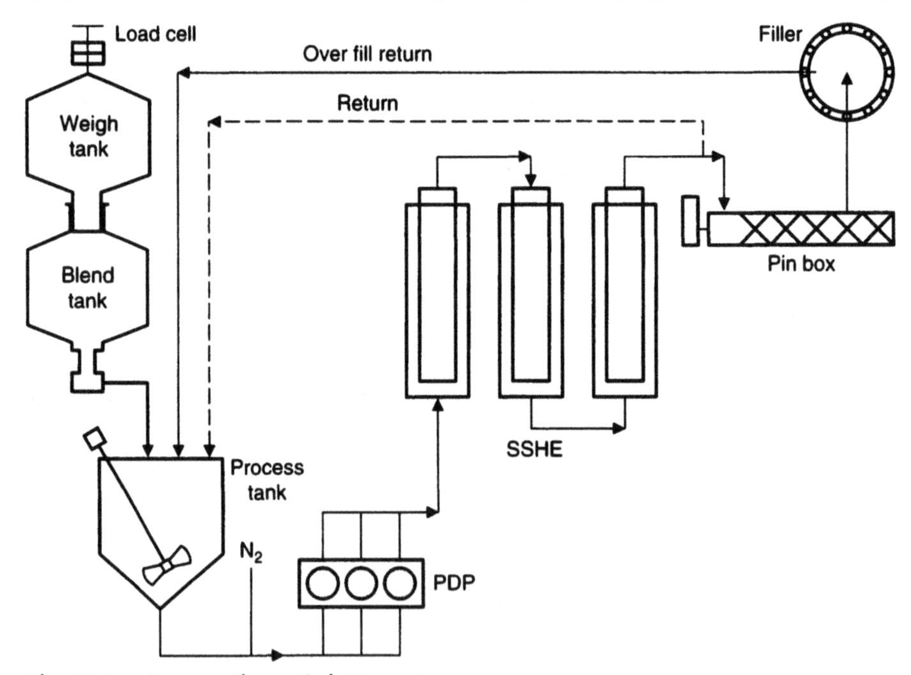

Fig. 19.5. Process Flow—Soft Margarine.

a cylindrical mixing chamber fitted with a specially constructed shaft and discs, which rotate at 1200 to 3000 rpm, and an inlet and an outlet port plus two positive-displacement pumps, one for the oil phase and the other for the water phase, operating at a flow rate of 20:80 oil to water (see Figs. 19.6 and 19.7) (25).

The two phases are prepared in two separate tanks and two feed pumps supplying the processing positive displacement pumps. Emulsification takes place in the specially designed cylinder followed by crystallization in a single SSHE tube at a flow rate of 1600 kg/hr (3600 lb/h).

Chilling, Forming, and Tempering

Stick margarine emulsions are quickly chilled in the SSHE equipment. In a typical plant operation using the American closed-system Lynch (Anderson, IN) "Morpack" packaging equipment, it takes 5.44 atm gauge (80 psig) from the positive-displacement pump (PDP) to deliver the chilled mass at 15°C (55°F) to the "B" unit or resting tube of the packaging machine, where it moves slowly and undisturbed in order to set before it enters the mold block cavity, forming the "print." Under normal operating conditions, 6.8 atm (100 psig) is set for the pump and 8.16 atm (120 psig) for the cavity filler compensating piston to push the margarine into the cavity

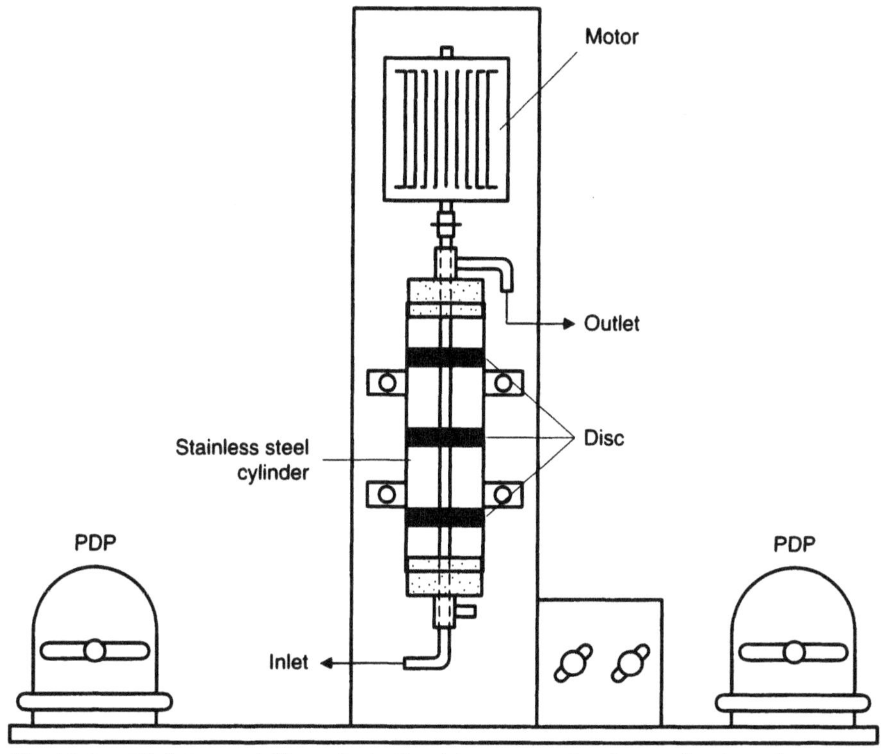

Fig. 19.6. High-Internal-Phase-Ratio-Emulsion system for the production of margarine spread.

and send any excess back to the return or remelt line. With European-type packaging machinery such as the Benhil (Dusseldorf, Germany), where the paper inner wrap is preformed into the cavity and the plastic product is dispensed into it, only 1.4 to 2.4 atm (20 to 35 psig) is needed on the filling piston.

Soft tub margarines and spreads, for which spreadability at refrigerator temperatures is a measure of good quality, will be very firm and brittle and exhibit a narrow plastic range if the supercooled emulsion is allowed to crystallize into the tub without mixing in an agitated "B" unit (pin box). Under normal plant production conditions, an appropriately sized agitated "B" unit attached to a variable-speed drive is placed in-line after the "SSHE," where most of the crystallization and dissipation of the heat of crystallization (see Chapter 15) takes place before the plastic mass is metered into the cup or tub. This treatment is essential in preventing the previously mentioned defect and is instrumental in extending the plastic range of the spread. Many types of American- as well as European-made filling equipment are in use for this purpose.

After chilling in the SSHE, industrial plastic margarines are treated much like a shortening, where much emphasis is placed on consistency because of its importance to the baker from the standpoint of handling and performance in bakery goods and icings.

The main purpose for quick-chilling a plastic fat is to obtain a fine crystal structure with high- and low-melting triglycerides. Subsequent heat treatment (*tempering*) of the fat mass may have a large effect on consistency at a given temperature. Commercial bakery margarines and shortenings are commonly tempered to improve their consistencies at 27 to 30°C (80 to 85°F) for from 24 to 72 h depending on type, processing conditions, container size, stacking arrangement, and customer specifications.

Tempering changes the character of the crystals as well as their amounts. Tempering causes a shift in the solid fat index curve as evident by an increase in the amount of solid fat above the tempering temperature and a decrease below that temperature (26). In other words, the plastic or working range of the product will be extended. Tempering is essentially an "unmixing" process, proceeding through par-

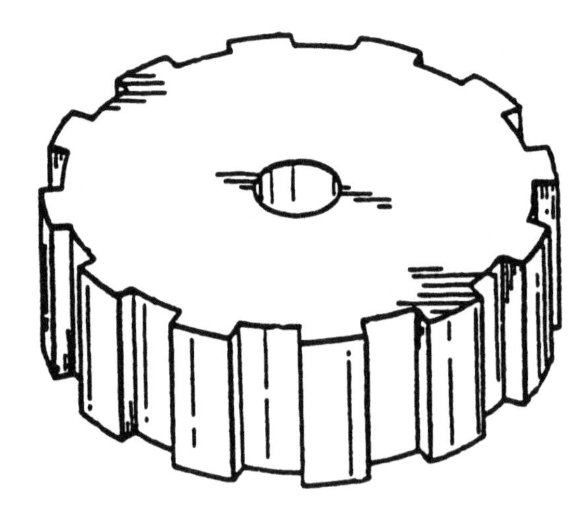

Fig. 19.7. Specially designed rotating disc (3 to 12 discs).

tial melting followed by slow recrystallization. This presumably eliminates low-melting glycerides from the crystals and produces a more homogeneous and mechanically stronger crystal lattice. The constant melting and recrystallization that take place as the temperature of the fat follows that of its environment does not seem to cause new crystals to appear or disappear but is merely a process of growth or diminution of preexisting crystals (27).

Packaging and Marketing of Margarine

With the recent changes in the Standard of Identity for margarine and packaging regulations, the American consumer shopping at the local supermarket will find the following types or forms of margarine and spreads displayed in the dairy section of the store:

1. Standard 80% fat stick margarine usually available in four 114 g (4 oz) sticks in one package (quarters) or 0.45 kg (1 lb) solid prints
2. Light margarine, containing 53.3% fat sold in either four 114 g (4 oz) sticks to a package, two 227 g (8 oz) plastic tubs per sleeve, or one 0.45 kg (1 lb) tub package
3. Standard soft margarine, containing 80% fat, sold either in two 227 g (8 oz) plastic tubs per one-sleeve package or in a single plastic 454 g (1 lb) bowl
4. Liquid margarine in a 0.45 kg (1 lb) squeezeable plastic dispenser
5. Spreads containing from 75% to 18% fat, sold in stick form in two 227 g (8 oz) plastic tubs to a sleeve, or in 0.45, 0.90, or 1.4 kg (1, 2, or 3 lb) plastic bowls

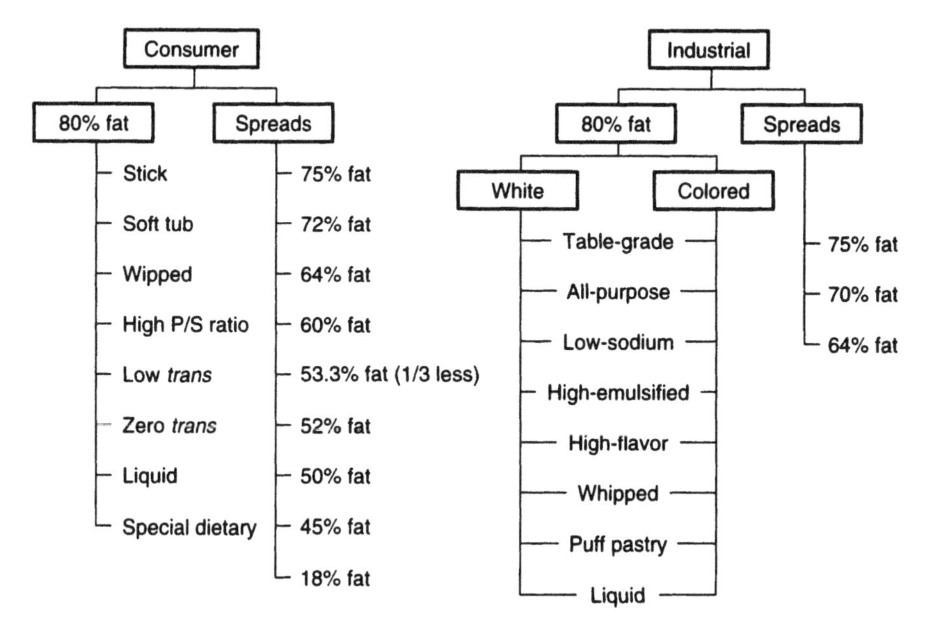

Fig. 19.8. Margarine Types and Styles.

Whipped margarine is no longer sold in six 75.7 g sticks to a 0.45 kg (1 lb) package, and diet margarine containing 40% fat, if produced, is included in the spread category.

On the other hand, industrial 80% fat, nonaerated margarines and spreads are sold in 454 g (1 lb) solids, 15.9 kg (35 lb) plastic pails or 22.7 kg (50 lb) polyethylene-lined cardboard boxes, depending on the formula. Soft table-grade and whipped margarines, containing 33% injected nitrogen gas (volume basis), are packed in either 1.6 or 9.1 kg (3.5 or 20 lb) plastic pails. Liquid pourable margarine is sold in two 2.5 gal (9.5 L, holding 9.1 kg) plastic jugs or one 15.9 kg (35 lb) narrow-mouthed plastic container for ease of handling. The range of margarine products is depicted in Fig. 19.8.

The marketing of margarine in the 1980s and 1990s is markedly different from the situation in the 1960s and 1970s. Once the federal tax on yellow margarine was repealed, there was an explosive growth in per capita consumption of margarine in the United States (Table 19.1). Equally important was the removal of all state restrictions on the production and sales of margarine which helped margarine to establish itself as an important and indispensable food item competing for consumer dollars.

With product developments such as soft margarine and the appearance of innovative packaging, the pressure for advertising intensified as the large producers of advertised brands sought to increase their market shares. However, because of a substantial price differential, private-label or store brands still maintained 50% or more of the market. By the mid-1970s, competition intensified and advertised brands began selling at about private-label prices. In addition, with the introduction of spreads, these large producers found a means to increase their market share at the expense of the private-label category. All of these efforts have seriously affected the private-label or store brand segment. It is estimated that spreads account for more than 45% of the total margarine production volume, whereas private-label margarines plummeted to less than 20%.

Conclusions

Margarine is only 125 years old. It has become an integral part of human diets worldwide and is by far the preferred choice of the consuming public to butter. National trends in amounts and types of fat consumed have not alleviated our problems in nutrition and health despite the efforts of The American Heart Association to educate the American public about the relationship between diet and incidence of coronary heart disease. Although margarine consumption has increased and that of butter has decreased, the per capita consumption of both seems to have reached a static state, despite the fact that the per capita consumption of edible fats and oils has increased (Table 19.2).

Unlike butter, margarine is a product of combination. It is an excellent source of calories per unit weight, suggesting a complementary role with high-protein foods (i.e. "protein sparing") when formulating special diets high in protein content or fortified with certain amino acids, such as lysine, cysteine, and methionine for geriatrics and the underfed (28). In this type of product, energy will be supplied by the margarine, and both protein and amino acids will be utilized more efficiently (2).

As an engineered food product, margarine is an excellent vehicle for assisting research investigative teams in all scientific disciplines in supplying the specific type of fats for their experimental work. Spreads can be formulated with any ratio of saturated : monounsaturated : polyunsaturated fatty acids from any combination of natural and synthetic sources, as well as spreads with zero *trans* fatty acids or reduced calories to satisfy the needs of those on restricted fatty acid diets and those under the supervision of nutritionists and dietitians. For more information, refer to Chapter 23.

The consumer has been well-educated about margarine and realizes that almost all margarines are comparable in functionality, taste and texture. Therefore, marketing strategies must follow the latest nutritional findings and formulate accordingly to remain competitive.

Recent New Developments

Since the development of the scraped surface heat exchanger in the late 1930s and its application in the continuous process for the manufacture of margarine, the conventional process has been to create water-in-oil (W/O) emulsion by thoroughly mixing the added water phase to the oil phase and cooling to a temperature sufficient to effect substantial solidification of the hard fat in the fat phase. The cooled dispersion is either subjected to shearing force or left undisturbed to attain maximum fat crystallization, depending on final product application.

With the introduction of the low-fat spread category, close attention of water-phase-rate addition to the oil phase was necessary to avoid phase reversion, especially when the aqueous phase was more than 50% of the formula. Initial water and oil phase temperatures, agitation type and efficiency also played a significant role in the success or failure of the process.

The Mouries process (29) is a new processing technology for the production of conventional margarine and low-fat spreads. Basically, the new process separates the water and oil phases into two discrete operations. The oil phase is cooled and crystallized in a newly designed scraped surface heat exchanger, then held in a holding tube that provides the required residence time for maximum fat phase crystallization. On the other hand, the water phase is pasteurized and cooled in another heat exchanger. A suitable proportioning pump is used to supply the proper amount of each phase to the emulsification stage. The emulsification device controls the size of water droplets and so the texture and flavor of the finished product (Fig 19.9).

In another modification of the above process where hydrocolloids are not included in the formulation (30), scraped surface heat exchangers for cooling are not used. Rather, the formed oil-in-water emulsion is inverted in the emulsification device to the desired water-in-oil type.

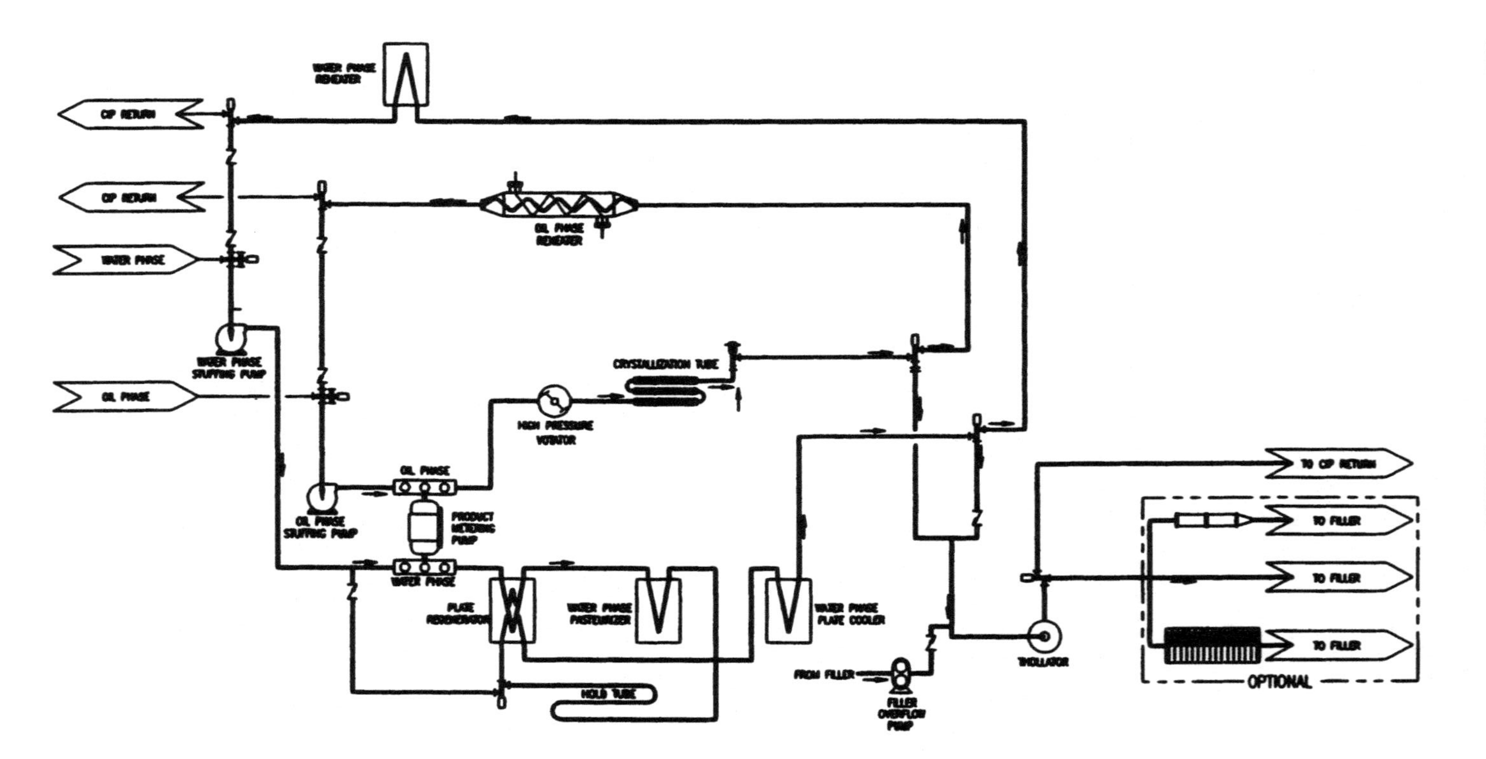

Fig. 19.9. Votator "Mouries" System. Courtesy of Cherry-Burrell Process Equipment.

References

1. Kenny, A.H., *The American Mercury 45(284):* 201 (1947).
2. Moustafa, A., in *Fats Production and Composition, Technologies and Nutritional Implications*, NATO ASI Series A: Life and Sciences, vol. 131, edited by C. Gali and E. Fedelli, 1986, p. 215.
3. Riepman, S.F., *The Story of Margarine*, Public Affairs Press, Washington DC, 1987, p. 108.
4. Snodgrass, K., *Margarine as a Food Substitute*, Food Research Institute, Stanford University, 1930, p. 121.
5. Slaughter, J.E., and C.E. McMichael, *J. Amer. Oil Chem. Soc. 26:* 623 (1949).
6. Joyner, N.T., *J. Amer. Oil Chem. Soc. 30:* 526 (1953).
7. Mattil, K.F., F.A. Morris, and A.J. Stirton, in *Bailey's Industrial Oil and Fat Products*, 3rd edn., edited by D. Swern, John Wiley & Sons, New York, 1964, p. 1063.
8. U.S. Dept. of Commerce, Economics and Statistics Administration, Bureau of Census, 112th edn., 1992, p. 132.
9. Wiedermann, L.H., *J. Amer. Oil Chem. Soc. 49:* 478 (1972).
10. Wiedermann, L.H., *J. Amer. Oil Chem. Soc. 45:* 515A (1968).
11. Wiedermann, L.H., *Amer. Oil Chem. Soc. 55:* 823 (1978).
12. O'Brien, R.D., in *Hydrogenation: Proceedings of an AOCS Colloquium*, edited by R. Hastert, American Oil Chemists' Society, Champaign, IL, 1987, p. 153.
13. Latondress, E.G., in *Handbook of Soy Oil Processing and Utilization*, edited by D.R. Erickson, et al., American Oil Chemists' Society, Champaign, IL, 1980, pp. 145–154.
14. Latondress, E.G., *J. Amer. Oil Chem. Soc. 58:* 185 (1981).
15. Latondress, E.G., *Hydrogenation and Bleaching Control Procedures*, presented at the AOCS Short Course, East Lansing, MI, August, 1966.
16. Young, F.V.K., in *Hydrogenation: Proceedings of an AOCS Colloquium*, edited by R. Hastert, American Oil Chemists' Society, Champaign, IL, 1987, p. 171.
17. Krog, N.J., in *Encyclopedia of Emulsion Technology*, vol. 2, ed. by P. Becher, 1985, p. 321.
18. Erickson, D.R., in *Proceedings of Soybean Extraction and Oil Processing Short Course*, edited by E.W. Lusas, L.R. Watkins, and S.S. Koseoglu, Food Protein Research and Development Center, Texas A & M University, College Station, TX, 1988, Ch. 79.
19. Clayton, W., in *Emulsion Technology*, Chemical Publishing Co., Brooklyn, NY, 1946, p. 181.
20. Moustafa, A., in *Proceedings of World Conference, Edible Fats and Oils Processing: Basic Principles and Modern Practices*, edited by D.R. Erickson, American Oil Chemists' Society, Champaign, IL, 1990, p. 214.
21. Chrysam, M.M., in *Bailey's Industrial Oil and Fat Products*, edited by T.H. Applewhite, John Wiley and Sons, New York, 1985, p. 56.
22. *Federal Register 41* (169): 36509 (August 23, 1976).
23. Chrysam, M.M., in *Bailey's Industrial Oil and Fat Products*, 4th edn., vol. 3, edited by T.H. Applewhite, John Wiley and Sons, New York, 1985, pp. 41–111.
24. Miller, D.E., and C.E. Werstack, U.S. Patent 4,238,520 (1980).
25. Bradley, G.M., and T.D. Stone, U.S. Patent 5,147,134 (1992).
26. Lutton, E.S., Crystallization of Fats in Margarine, 5th Margarine Research Symposium, Oct. 17, 1958.

27. Bailey, A.E., *Melting and Solidification of Fats*, Interscience Publishers, New York, 1950, pp. 304–305.
28. Bonnell, R.H., B. Bornstein, and G.W. Schutt, *J. Amer. Oil Chem. Soc. 48:* 175A (1971).
29. Tholl, G.W., U.S. Patent 5,352,475 (1995).
30. Tholl, G.W., U.S. Patent and Trademark Office. Application serial number 08/006, 656, January 22, 1995.

Chapter 20

Soybean Oil Products Utilization: Shortenings

R.D. O'Brien

Introduction

Shortening is an American invention, growing out of the cotton industry, and was perfected for soybean oil utilization. Cotton acreage expansion between the Civil War and the close of the nineteenth century resulted in large quantities of cottonseed oil. Shortenings were developed as an outlet for this oil. The first cottonseed oil shortenings developed as lard substitutes for a market created by classical supply and demand; the pork industry could not satisfy all of the plasticized shortening requirements (l). Supply and demand plus technology led to the utilization of soybean oil as the major oil source for shortenings in the United States. Increased shortening demand during and after World War II (1940 to 1946) promoted the development of technology to overcome the flavor and stability problems attributed to soybean oil shortenings when compared with cottonseed oil products.

The fats and oils industry has undergone tremendous change since the introduction of vegetable oil–based shortenings. The changes have involved new manufacturing techniques and facilities and the number of products produced. Advances in equipment and processing techniques changes were so rapid for a time that the industry was hard pressed to modify its facilities fast enough to take advantage of the new developments. During this period such developments as continuous vacuum bleaching, continuous and semi-continuous deodorization, and improved hydrogenation processes were introduced, which are taken for granted today.

Technological advances in laboratory methods, equipment, and techniques have also contributed to the progress of the fats and oils industry. Fats and oils laboratories have seen the perfection of many different analytical methods for studying the characteristics of fats as well as controlling the manufacturing process. Some of the analytical tools used today that were still in the development stage just a few years ago are (also see Chapter 24):

- *The solid fat index* (*SFI*) measures the ratio of solids to liquid in a fat at several standard temperatures.
- *The fatty acid composition* (*FAC*) identifies the composition of a fat or blend of fats in terms of the amount of each fatty acid.
- *Melting points* can be determined by several analytical methods available to the fat and oil chemist. Probably the most widely used procedure today is the Mettler dropping point, which observes the first drop of oil to break a light beam.
- Minute quantities of various *trace metals* that can be detrimental to fats and oils can be determined by several different methods.

- Accelerated rancidity determinations have been developed that predict the flavor life of a shortening. The *accelerated oxygen method* (AOM) has been accepted by many fat and oil producers as well as food manufacturers. Some less time-consuming techniques and equipment are also currently available.

These analytical tools have not replaced many of the methods used previously but have enhanced chemists' ability to characterize a fat or shortening product. Some of these time-proven techniques are

- The *flavor* of a fat can be organoleptically evaluated. Most fats are processed to be as bland as possible. However, some products will have faint characteristic flavors or mouth feel. Despite improved analytical techniques, flavor remains the prime indicator of product quality.
- The *free fatty acid* (FFA) content measures the uncombined fatty acids in a shortening. It can reflect how well a product is processed, unless the finished product contains an emulsifier, lecithin, or some other agent that titrates as free fatty acid.
- The *peroxide value* (PV) measures the degree of oxidation of a shortening. This value must be used with caution, because peroxides are both reactive and volatile and are a measure of a dynamic oxidation reaction. Therefore, a shortening could have a low peroxide value yet have a poor flavor and vice versa.
- The *iodine value* (IV) measures the unsaturation of a fat. Fatty acid composition also determines unsaturation by identifying the individual fatty acids. Iodine value is a titration method that usually requires less time to perform than FAC and does not need a gas chromatograph.
- *Moisture* determinations range from very simple effective evaluations to quite detailed procedures requiring exacting techniques. The simple effective methods are preferred because of speed, reproducibility, and confidence that the tests can be performed unless an exacting level is necessary.

The advances in technology have contributed greatly to the storehouse of fats and oils information, resulting in many new specialized shortening products. These specialized shortening developments have been responsible for many advances throughout the food industry. Likewise, new food developments have caused the development of new specialty shortenings. It has become increasingly important for shortening developers to understand the functionalities of their products in individual food application, down to the particulars of the individual producers' requirements for processing, packaging, distribution, product claims, and so forth.

Shortening Development

The development of a shortening product for a food application depends on many factors that differ from customer to customer, such as equipment, processing limitations, product preference, and customer base. Fat and oil products are now being designed to satisfy individual specific requirements as well as offering products with broad general appeal. The design criteria for a general-appeal product must be of a broader nature than those for a specific product and process.

The important attributes of a shortening in a food product vary considerably. In some food items, the flavor contribution of the shortening is of minor importance; however, it does contribute beneficial effects to the eating quality of the finished product. This fact is evident with the recent introductions of fat-free products. Most of these products lack the eating characteristics contributed by shortenings. In many products, such as cakes, pie crusts, icings, cookies, and certain pastries, shortening is a major contributor to the characteristic structure of the product and has a significant effect on the finished product quality.

Satisfactory shortening performance depends on many factors; three of the most important are:

1. *Flavor.* Generally the flavor of shortening should be completely bland so that it can enhance a food product's flavor rather than contribute a flavor. In some specific cases, the shortening flavor should be typical of lard or a butter-flavored product. The bland or typical flavor must also be stable throughout the life of the food product. Therefore, oxidative stability must be considered. The oxidation rate of a shortening is directly related to the type and amount of unsaturated fatty acids present. The relative oxidation rates for the three most common unsaturated fatty acids are listed in Table 20.1. Oxidation of deodorized shortenings brings on the flavor usually characteristic of the original source oil. Shortenings must be designed with flavor stabilities suitable for the finished product requirements. Reduction of unsaturated fatty acids can be accomplished with hydrogenation, fractionation, or both to increase flavor stability.
2. *Physical Characteristics.* The characteristics of the fats and oils utilized for a shortening are of primary importance in designing a shortening for a specific use. Oils can be modified through various processes to produce the desired properties. Hydrogenation has been the primary process used to change the physical characteristics of oils (see Chapter 13). The melting point and solids profile of an oil can be completely altered with this process and the changes controlled by the conditions used to hydrogenate the oil. In the hydrogenation process, hydrogen gas is reacted with oil at a suitable temperature and pressure with agitation in the presence of a catalyst. Control of these conditions and the end point enables the processor to meet the physical characteristics desired for shortening products. Some typical shortening SFI curves, presented in Fig. 20.1, illustrate how the hydrogenation process can be manipulated to produce physical characteristics suitable for the performance desired.

TABLE 20.1 Relative Oxidation Rates of Fatty Acids (2)

Fatty acid	Relative oxidation rate
Oleic C18:1	1
Linoleic C18:2	10
Linolenic C18:3	25

3. *Crystal habits.* Fats can exist in three crystal forms: α, β, and β′. Large β crystals can produce shortenings with sandy, brittle consistencies that result in poor baking performance where creaming properties are important. The small, uniform, tightly knit β′ crystals produce shortenings with good plasticity, heat resistance, and creaming properties. However, the large β crystals are not detrimental but desirable for some applications like frying or pie crusts. Crystal habits can be controlled by the selection of oil sources in formulation, plasticization conditions employed, and tempering after packaging.

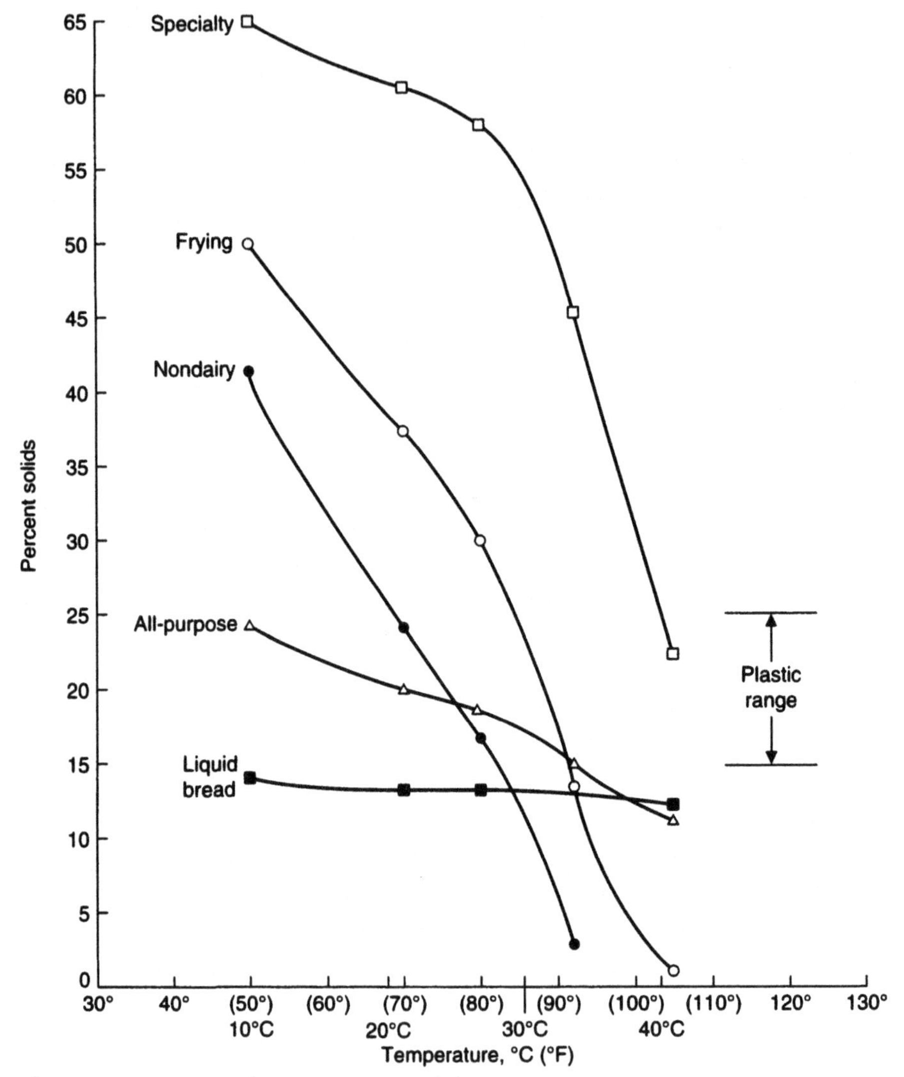

Fig. 20.1. SFI curves for various types of shortenings.

Shortening Formulation

Most shortenings are identified and formulated according to usage. Fig. 20.1 illustrates the diverse SFI–melting point relationships among five different shortening products. This chart indicates the differences in plastic range necessary to perform the desired function in the finished products. Shortenings with the flattest SFI curves have the widest plastic range for workability at cool temperatures as well as elevated temperatures. All-purpose shortenings have the widest plastic range. Nondairy and solid frying shortenings have relatively steep SFI curves, and so will provide firm, brittle consistency at room temperature but become practically fluid at only slightly elevated temperatures.

The product with a very flat SFI slope is a fluid or pumpable shortening. These types of products have become popular because of convenience, lower handling costs in some situations, and lower saturate levels. In these systems, the stable β crystal form is necessary to produce a fluid product (3).

Wide-Plastic-Range Shortenings

The basic all-purpose shortening has been the building block for shortenings where creaming properties, plasticity, and heat tolerance are important. Initially, an unhydrogenated oil was blended with a hardfat to make a compound shortening that had a very flat SFI curve, which provided an excellent plastic range. However, the low oxidative stability of these shortenings precludes their use today for most products. Currently, these products are formulated with a partially hydrogenated soybean oil base stock and a low-IV cottonseed or palm oil hardstock. The β′ crystal form hardstock functions as the shortening plasticizer for improved creaming properties (see Chapter 13).

Hydrogenating the shortening base stock increases the oxidative stability. As a rule, the lower the IV of the base stock, the greater the AOM stability. However, as the hardness of the base stock increases, the level of hardstock required to reach a desired consistency decreases. Reducing the amount of hardstock reduces the plastic range and heat tolerance. Therefore, oxidative stability improvements are achieved at the expense of plasticity. The extent that one attribute can be compromised to improve another must be determined by the requirements of the intended food product.

The plastic range is important for bakery shortenings intended for roll-in and creaming applications alike, because the consistency changes with temperature. Shortenings become brittle above the plastic range and soft below the range; both conditions adversely affect creaming and workability alike. The normal plastic range for a general purpose shortening is produced with SFI values between 15 and 25 (4). In other words, the steeper the SFI curve, the narrower the working range. The all-purpose shortening SFI plotted in Fig. 20.1 illustrates a wide-plastic-range shortening, while very narrow plastic ranges are illustrated by nondairy and frying shortenings, as shown in Table 20.2.

All-purpose shortenings must be formulated carefully and quick-chilled or plasticized properly to start the crystallization process, and finally tempered adequately for crystal stabilization.

Narrow-Plastic-Range Shortenings

Plasticity is of minor importance and can be a detriment for products requiring high oxidative stability or sharp melting characteristics. Shortenings specially designed for specific frying situations, nondairy systems, cookie fillers, confectionery fats, and spray oils require flavor stability and eating characteristics not possible with blends of moderately hydrogenated oils and hardfat. These products require lower-IV shortenings, for oxidative stability, and steeper SFI slopes and a lower melting point (below 37°C or 98°F) for good eating characteristics. The frying and nondairy shortenings plotted in Fig. 20.1 illustrate the steep SFI slopes.

The steep SFI products are generally produced with selectively hydrogenated base stocks that have high solids at lower temperatures but decrease rapidly to a low melting point for desired eating or mouth feel characteristics and excellent oxidative stability. These products are usually composed of a single straight hardened base, or possibly two selectively hydrogenated bases, for slightly different slopes than possible with a single base.

Fluid Shortenings

Liquid or pumpable shortenings are flowable suspensions of solid fat in liquid oil. The liquid oil phase may or may not be hydrogenated, depending on the finished product's consistency and the oxidative stability required. Low-IV soybean oil hardfat seeds crystallization. It can vary from as low as 1% to higher levels as required to produce the desired finished product viscosity. The ease in which soybean oil converts to the stable β crystal form makes it ideal for fluid shortenings (3).

Fluid shortenings have been developed for food products where pourability or pumpability at room temperature is important. The major uses of liquid shortenings are

- *Frying*. Liquid frying shortening provides a convenience to the operator; it can be poured into the fryer and quickly heated to frying temperature.
- *Bread and cake*. Liquid shortening systems can be pumped at room temperature, eliminating the need for heated tanks or handling packaged shortenings.

TABLE 20.2 Shortening Plastic Ranges

	Working range		Total	
Shortening type	°C	°F	°C	°F
All-purpose	<10–33	<50–92	23	42
Nondairy	20–28	68–82	8	14
Frying	29–33	84.5–91.5	4	7

- *Nondairy.* Liquid shortening systems offer the room temperature pumping convenience and high polyunsaturates which are attractive for some nondairy products like liquid creamers, filled milks, and toppings.

Some typical compositions and SFI results for the different types of fluid shortenings are compared in Table 20.3. These products must contain high levels of liquid or only lightly hydrogenated oils for fluidity. Therefore, additives must contribute heavily to the product performance of liquid shortenings. The major additives for the shortening types are listed in Table 20.4.

Bread shortenings are the only product in which an additive is not mandatory for liquid shortening performance. Lubrication is the primary function of a bread shortening. However, the emulsifiers that improve bread shelf life are usually produced with fully saturated fats, which can be difficult to add to doughs. Liquid shortening is an ideal carrier for these emulsifiers, providing another convenience for the user.

Bakery Products Applications

Fats and oils are major ingredients in most baked products. The flat-SFI shortening made by blending a nonselectively hydrogenated, 80 to 90-IV soybean oil base with a β′-crystal, habit hardstock has been the building block for many of the general and specialty bakery shortenings. Fig. 20.2 shows the progression of bakery shortening development from an all-purpose shortening to specialty baking shortenings for cakes, icings, puff pastry, roll-in products, and so forth.

Mono- and diglyceride emulsifiers added to an all-purpose shortening changed baking technology in the early 1930s. The Hi Ratio all-purpose, emulsified shortening allowed the production of high-sugar and -moisture cakes with improved eating characteristics, a finer grain and texture, and extended keeping qualities. This shortening also improved the aeration and eating qualities of icings. Performance improvements were also found with many of other bakery products in which cream-

TABLE 20.3 Typical Fluid Shortening Compositions and SFI Values

Soybean oil base stock	Frying	Bread	Cake	Nondairy
IV 135	—	90.0	—	—
IV 110	—	—	—	98.0
IV 104	98.0	—	99.0	—
IV 5	2.0	10.0	1.0	2.0
Additives				
Methyl silicone, ppm	1.0	—	—	—
Emulsifiers, %	—	—	11.0	6.0
Solid Fat Index				
At 50°F (10°C)	4.0	13.0	7.0	5.5
At 80°F (27°C)	2.0	12.0	5.5	2.5
At 104°F (40°C)	0.5	10.0	1.0	1.0

ing properties are important. The discovery that mono- and diglycerides improved creaming properties, which resulted in superior baked products, opened the door for developing other emulsifiers and emulsifier combinations to produce specialty baking shortenings. This technology has helped to develop new consumer products for the baking industry. The mix industry utilizes specialty shortenings for cakes, icings, and yeast-raised products that are formulated with an all-purpose shortening base with emulsifier systems developed for the specific product.

The all-purpose shortening formulation technique of producing a flat SFI curve by blending a nonselectively hydrogenated base with a β′ hardstock has also been adapted to many specialty purpose products. Fig. 20.2 identifies the following product types among the nonemulsified shortening:

Roll-In Shortenings. These are utilized in puff pastry and Danish pastry. The characteristic features of a roll-in shortening are plasticity and firmness. Plasticity is necessary because the shortening should remain as unbroken layers between dough layers during the repeated folding and rolling operations. Firmness is equally important, because soft or oily shortening would be partly absorbed by the dough and its role as a barrier between the dough layers would be greatly reduced. The wide workable range for the firm shortening is usually developed with a nonselectively hydrogenated, 80 to 90-IV soybean oil base stabilized with a β′ hardstock at approximately 15 to 20%, followed by careful chilling and tempering to produce a smooth and workable but firm product.

Yeast-Raised Shortenings. Bread shortenings are selected for their ability to provide lubricity and structure and for the beneficial effects on flour starch and protein. General-purpose shortenings provide lubricity and structure, and can serve as the carrier for emulsifiers that complex the starch for anti-staling properties.

Cookie Shortenings. Cookies are made in many varieties, shapes, sizes, and flavors but only two basic types: hard and soft. All-purpose shortening formulations with partially hydrogenated soybean oil bases coupled with a β′ crystal habit hardstock provide the creaming properties and oxidative stability required for soft cookies. The shortening functions in hard cookies are lubrication and enhancement of the eating character. Sharp–melting point products with high AOM stability ratings are usually more functional for the hard varieties. The steep SFI soybean oil based shortenings meet these requirements.

TABLE 20.4 Additives Used in Liquid Shortenings

Liquid shortening type	Additive
Frying	Methyl silicone
Cake	Emulsifier system
Bread	None, or emulsifier
Nondairy	Flavors and/or emulsifier

Biscuit Shortenings. Biscuits should be flaky and tender, by Southern United States tradition. A firm shortening that does not completely mix into the biscuit dough but can still be handled by the operator are required and filled by a firm version of the all-purpose shortening.

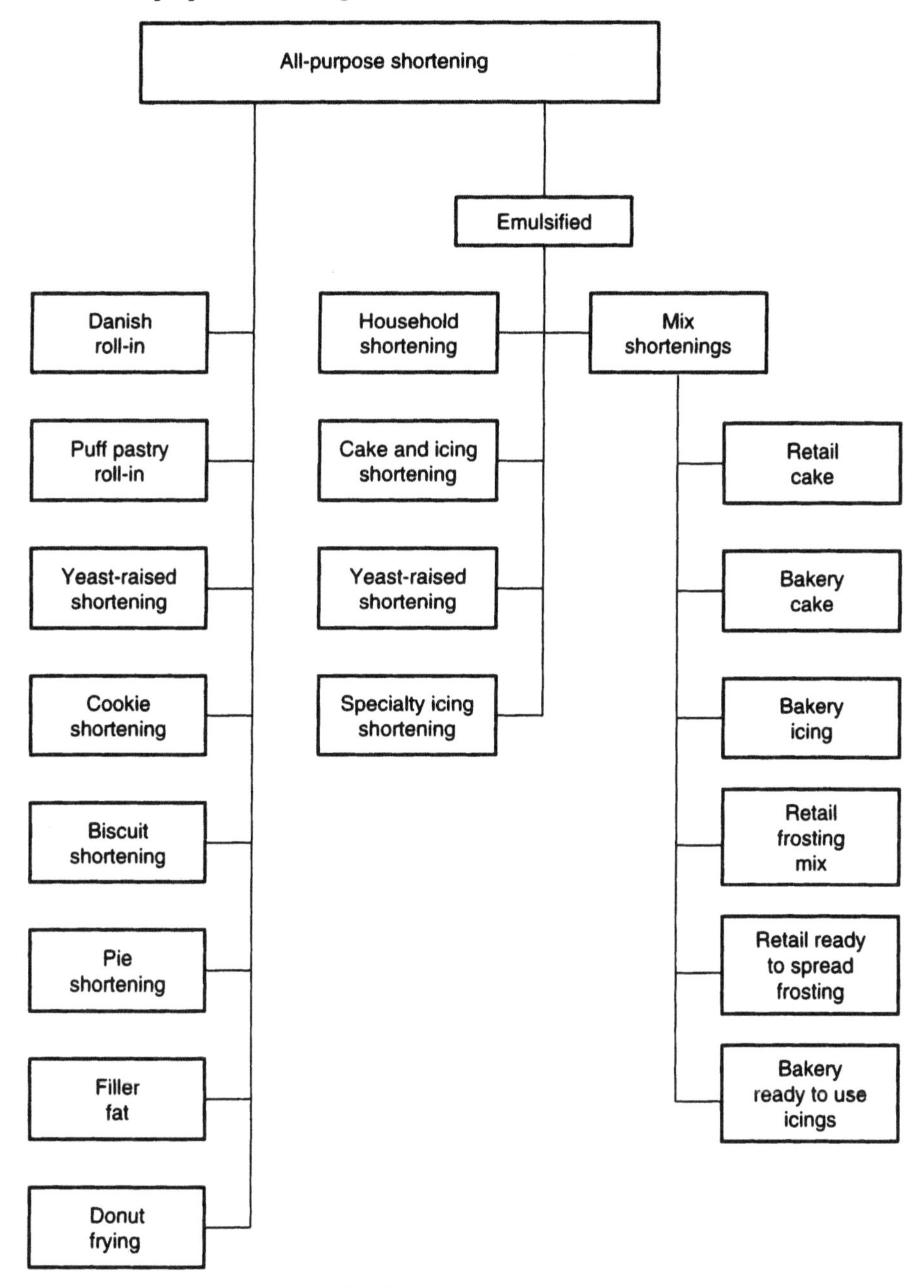

Fig. 20.2. Bakery shortening development.

Pie Shortenings. Lard has been the preferred pie shortening because of the eating characteristics provided by the grainy β crystal habit animal fat. A hydrogenated soybean oil base stock, stabilized with a soybean hardstock all-purpose shortening, has been accepted by many specialty pie bakers.

Filler Fats. Some fillers are best when produced with the standard all-purpose shortening formulation. Cookie fillers are formulated with high fat levels for eating characteristics and creaming properties for some filler types. The all-purpose shortening formulation forced to a β′ crystal habit performs well in these applications. Other cookie fillers will require nonaerating shortenings to perform properly in packaging equipment.

Doughnut Frying. All-purpose type shortenings are still the preferred shortening for doughnut frying and other fried bakery products with high fat absorptions where a dry fried appearance is desired.

Frying Shortenings

Deep fat frying is more efficient than dry oven heat and faster than boiling in water, because shortenings are very effective heat transfer agents. The high temperatures used cause rapid heat penetration for short cooking times. Also, unlike boiling or baking, in frying, the heat transfer medium becomes a part of the food. The frying shortening reacts with protein and carbohydrates in the food to develop unique flavors and odors.

Frying has long been the preferred or the only preparation method for potato chips, doughnuts, and other snack food products. Also, most restaurants use frying to prepare either all or a high portion of the food served. Shortening formulations for the restaurant and snack food industries have very different requirements.

Restaurant Frying

Frying shortenings command the largest share of the foodservice fats and oils requirements. Frying is the only food preparation method used in many fast-food operations and the major one in many others. Frying stability is a major economic and quality factor that restaurant operators should consider when choosing a frying shortening. The factors that affect frying stability or frying life are:

- *Polymerization* is the combining of triglyceride molecules to form three-dimensional polymers. Polymerization causes increased viscosity, which results in foaming. The foam develops when the oil will not release moisture but keeps it trapped as steam vapor. Unsaturated fatty acids have the least resistance to polymerization.
- *Oxidation* is the combination of oxygen with the unsaturated fatty acids, causing off-flavor and odor development with color darkening.

- *Hydrolysis* is the breakdown of the triglyceride molecule by chemical reaction with water, leading to off-flavors and darkening.

The various menus and operators' preferences require five different foodservice soybean oil frying types:

- Refined, bleached, deodorized soybean oil
- Hydrogenated, winterized soybean salad oil
- Liquid frying shortening
- All-purpose shortening
- Hydrogenated soybean oil frying shortening

The characteristics that contribute to the restaurant operator's choice of frying shortening to use are shown in Table 20.5. Hydrogenated soybean oil frying shortenings were among the first shortenings specifically formulated for a definite function. These solid shortenings were developed for maximum frying stability by reducing the polyunsaturate content with selective hydrogenation conditions. The shortenings appealed to the foodservice operator over the all-purpose shortenings because of their longer frying life and lower melting point, which improved the fried food's eating characteristics.

A frying additive, identified after hydrogenated frying shortenings had begun to enjoy good success, increased frying stability dramatically by reducing foaming. Methyl silicone at 0.5 to 2.0 parts per million effectively retards oxidation and polymerization, thus reducing foaming. Frying stability increases of 3- to 10-fold with the antifoamer have been confirmed in controlled laboratory frying evaluations. The degree of frying stability increase depends on the original stability of the frying shortening before the antifoamer is added; the more stable original products show a greater increase in stability with the additive.

TABLE 20.5 Foodservice Frying Shortenings

Frying shortening properly	RBD soybean oil	Hydrogenated, winterized soybean oil (HWSBO)	All-purpose shortening	Opaque liquid shortening	Hydrogenated soybean oil shortening
Consistency	Clear liquid	Clear liquid	Plastic solid	Opaque pourable	Solid
Typical analysis:					
Melting point, °F	Liquid	Liquid	110–120	92–98	105–110
°C			43–49	33–37	41–43
AOM stability, h	10–25	16–20	40+	35+	200+
Polyunsaturates, %	34–61	54–60	15–20	35–40	4–8
Saturates, %	13–27	14–21	20–30	15–20	22–25
Iodine value	130–135	108–112	80–85	102–108	65–75
Selection criteria	Pourable Oily fry Nutrition Price	Pourable Oily fry Nutrition	All-purpose Dry fry High melt pt.	Pourable Semi-oily fry Nutrition Stability	Stability Dry fry Nutrition Economy

Liquid shortenings containing the methyl silicone antifoamer were introduced with claims of the convenience of a pourable product and longer frying stabilities than the specially hydrogenated solid frying shortenings. Frying tests confirmed these claims until the antifoamer was also incorporated into the specially hydrogenated frying shortening. However, the convenience of the liquid shortening, the long frying stability, and the slightly oilier food appearance have made the liquid frying shortening type a major foodservice ingredient.

Refined, bleached, and deodorized (RBD) and hydrogenated and winterized soybean oils (HWSBO) are used by some foodservice operators who have a rapid turnover of their frying oil and desire an oily appearance. HWSBO is slightly more stable than RBD oil, because of a reduction of the polyunsaturates. Frying life can be extended with the addition of methyl silicone in these products too. However, the frying stability will not approach that of the aforementioned (fractionated) liquid or specially hydrogenated frying shortenings.

Before the addition of antifoamers to frying shortenings, foaming was the indicator used by foodservice operators to decide when to replace their frying shortenings. Now, a frying fat that reaches the persistent foaming stage may have gone beyond the discard point.

Currently, foodservice operators use several different shortening discard indicators which include the following (5):

- Some operators rely on *color*, discarding the oil when it reaches a certain darkness or when visibility is impaired at a definite distance under the oil surface.
- *Smoke* is caused by hydrolysis and oxidation, which lower the smoke point, so some operators replace their shortening when smoking reaches a certain intensity.
- Some operators replace their frying shortening after a prescribed *time* has elapsed or a specified quantity of food has been fried.
- A number of *test kits* have been introduced to measure color, conductivity, free fatty acid, and other factors.
- *Foaming* is still used by some operators and is acceptable if the amount of foaming is not excessive.

Pan and Grill Shortenings

Foodservice operators have readily accepted the pan and grill shortenings, which are butter-flavored, and colored products with lecithin as an anti-sticking agent. These products are replacements for butter, margarine, or cooking oils previously used for pan or grill frying. Table 20.6 reviews the three general types of pan and grill shortenings available: two liquids and a solid. The distinguishing difference between the two liquid types is the addition of salt. Salt enhances the butter flavor to give this product a more pleasing flavor, making it more closely duplicate the flavor of clarified butter or margarine. The finely milled salt can only be dispersed, because it is not soluble in oil. Lecithin is added to all of these products to provide antisticking properties during pan and grill use. Both liquid products are formulated like liquid opaque frying shortenings with soft soybean oil base stock and hardstock. Coconut oil is usually added to solid pan and grill products to achieve rapid melting characteristics and stability. The soybean oil base stock used is selectively hydrogenated for stability and melting properties (5).

Snack Food Frying

Frying shortenings function as an ingredient and the cooking medium for chips of all kinds. Unlike restaurant fried foods, these products are fried and packaged for distribution and consumption at a later time. Most fried snack foods have high fat absorptions, which generate high usage or turnover rates of the frying shortening. Therefore, frying stability is not a normal concern, but flavor stability of the snack after frying determines shelf life. The high fat absorption rate makes the frying shortening a major contributor to flavor stability (6,7). Typical snack frying shortenings are shown in Table 20.7.

The selection of the frying shortening for chip snacks depends on two major factors:

- *Chip finish.* Liquid or lightly hydrogenated oils are preferred where a bright oily finish is desired. Higher-melting shortenings produce chips with a dry finish.
- *Flavor.* Bland initial oil flavors, with a long induction period before oxidation develops objectionable flavors, are preferred for most products. Some producers use particular sources of oil or blends of oils for a less objectionable oxidized flavor.

TABLE 20.6 Foodservice Pan and Grill Products

	Pourable	Salted/pourable	Solid
Source oil	Soybean oil	Soybean oil	Coconut and soybean oil
Other ingredients	Lecithin, color, Butter flavor, Methyl silicone	Salt, lecithin, color, butter flavor	Lecithin, color, butter flavor, methyl silicone
Solid fat index			
at 50°F (10°C)	2–10	2–10	35–45
at 92°F (33°C)	0–8	0–8	2–5
Applications	Pan frying Grilling Soups, sauces, and gravies Basting Brush-on Dressing Corn on the cob	Pan frying Grilling Seasonings Soups, sauces, and gravies Brush-on Dressing Basting Popcorn Seafood dip Corn on the cob	Pan frying Grilling Bun dressing Bun toasters

Nondairy Shortenings

Butterfat has been replaced with a shortening in two basic types of dairylike products, called *filled* and *imitation*. Filled products are essentially made from defatted milk with a nondairy shortening replacing the butterfat content. Imitation product formulations usually replace milk with protein, stabilizers, emulsifiers, and other ingredients as well as the nondairy fats.

For the initial developments of many nondairy products, all-purpose shortenings were used to replace butterfat. Mouth feel and associated deficiencies contributed by the all-purpose shortenings hastened the development of nondairy shortenings more closely duplicating the melting characteristics of butterfat. Developers have now taken the next steps for specialty products and improved the functionality of the shortenings specifically for each application and process. The melting and SFI properties of some selected nondairy shortenings are compared to butterfat and an all-purpose shortening in Table 20.8 and briefly reviewed as follows:

Beverage fats are liquid shortening systems used in producing filled or imitation milks, liquid coffee creamers, whipping products, dips, and other dairylike products. These fluid shortening systems were designed to provide high levels of polyunsaturated fatty acids. Refrigerated or frozen distribution of these products allows the use of shortenings with relatively low AOM stabilities.

Frozen dessert or *mellorine shortenings* still closely resemble butterfat characteristics to achieve the overrun, smooth texture, proper meltdown, freezer stability, and eating characteristics associated with quality ice cream.

Coffee whitener shortenings have been improved by increasing the melting point to help provide powders with creaminess, whitening, and flavor stability for extended shelf life.

Imitation cheeses have in most cases utilized the butterfat replacement–type shortenings, but liquid oils have been used to provide a desired nutritional claim for polyunsaturates.

For most *whipped toppings*, coconut or palm kernel oils were preferred initially because of their sharp melting points and flavor stabilities. Specialty hydrogenated soybean oil shortenings have been adopted by some producers because of the adverse publicity received by tropical saturated fats high in lauric acid.

TABLE 20.7 Typical Snack Frying Shortenings

	A	B	C	D	E
Source oil:					
Soybean oil	100	100	75	50	100
Cottonseed oil	—	—	25	50	—
Analytical characteristics:					
Melting point, °C	Liquid	24	33	33	35
°F		75	92	92	95
AOM stability, h	10	30	60	60	200
Iodine value	110	94	80	82	74

Margarines were probably the first nondairy product. The fat source for these products has seen many changes over the years. See Chapter 19 for more detailed information.

Specialty Products

Today, most shortenings are tailor-made for specific food products and, in many cases, for a specific manufacturer's process. Specialty fat development has been responsible for many advances throughout the food industry. Many of the specialty shortening–type products developed do not conveniently fit into the broad baking, frying, or nondairy classifications. Soybean oil has been the choice domestic oil for many of the specialty products because of its availability, economics, and product characteristics.

Shortening Chips

These are large, thick flakes made from selectively hydrogenated soybean oil, typically with a 43 to 46°C (110 to 115°F) melting point. The chips are available unflavored, butter-flavored and colored, and with sugar and cinnamon added. The predominant application for shortening chips is in baked products to produce a flaky, tender product with pockets of color and flavor after baking. Biscuits, cookies, pizza crusts, breads, and sweet rolls have been some of the products benefiting from the use of shortening chips.

TABLE 20.8 Selected Nondairy Shortenings Compared with Butterfat

Typical Characteristics		Butterfat	All-purpose shortening	Selectively hydrogenated	Beverage fat	Coffee whitener
Solid Fat Index						
% Solids at	50°F (10°C)	33	24	41	5.5	64
	70°F (21°C)	14	20	24	—	54
	80°F (27°C)	10	19	16	2.5	49
	92°F (33°C)	3	15	3	—	32
	104°F (40°C)	—	11	—	1.0	12
Melting point, °F		95	115	95	Liquid	111
°C		35	46	35		44
AOM stability, h			40	200	20	200+
Dairylike products		All original products		Mellorine Margarine Imitation cheese Liquid creamer Whipped topping Dips Filled milk	Filled products Liquid whitener Dips	Powdered coffee creamers

Icing Stabilizers

Selectively hydrogenated soybean oils with melting points centered at 45°C (113°F), 47°C (117°F), and 52°C (125°F) are flaked with lecithin for use as a stabilizer for roll icings and glazes. Addition of only 1 to 5% of these products improves warm weather stability by preventing moisture bleeding; contributes to smooth surface appearance with improved gloss retention; and finally promotes a drier icing that will not stick to the package but is still pliable.

Emulsifier Bases

Mono- and diglyceride emulsifiers are made by reacting glycerine with hydrogenated oils of varying hardness. The base hardness, as indicated by IV, determines the functionality of the mono- and diglyceride emulsifiers. Table 20.9 identifies the functionality of the mono and diglyceride emulsifiers produced with hard, soft, and intermediate soybean oil base stocks (3). Other fat-based emulsifiers can be prepared with the same hydrogenated soybean oil base stocks but different reactants.

High-Stability Oils

Fractionation of hydrogenated soybean oils has produced oils that are liquid at room temperature but possess very high stabilities. AOM stabilities of 300+ h have been recorded with one of these specially processed oils. This flavor stability, which

TABLE 20.9 Mono- and Diglyceride Functionality

Base stock hardness	Functionality
Hard (5 IV)	Fine cake grain and texture Baked product moisture retention Bread crumb softener Yeast raised anti-staling Chewing gum antisticking agent Peanut butter stabilizer Dehydrated potato processing Pasta antisticking properties Margarine emulsion stability Yeast-raised doughnuts Nondairy emulsion stability Frozen dessert freeze/thaw stability Prepared mixes Coffee whiteners
Soft (90 IV or higher)	Aeration for batters, icings, and fillings Nondairy whipping properties Gravy and sauce texture improvement Pet foods Loose margarine emulsion stability Cake doughnuts
Intermediate, plastic shortening consistency	Compromise functionality of both hard and soft

equates to 20+ times that of typical hydrogenated winterized salad oil, has many specialty applications where an oil appearance or eating characteristic is preferred but a long shelf life is necessary, including use in frying, spray oils, and dressings; as a coating agent, flavor base, spice carrier, release agent, or wetting agent; and for viscosity control or dust control.

Confectioners' Fats

The confectionery coating industry has desired a fat that is compatible with chocolate liquor; resists coating bloom (formation of a discolored separated layer on the surface); is easily tempered; and is produced from domestically grown oilseeds for better economics. Soybean oil–based confectioners' fats, produced with two different processes, have had limited success:

1. Solvent fractionation of a cottonseed–soybean oil blend has some compatibility with chocolate liquor, has a melting point slightly higher than that of cocoa butter, and requires no critical tempering procedure to produce chocolate coatings.
2. Hydrogenation with a sulfur-treated catalyst has produced a soybean oil product that has a steep SFI slope and a sharp melting point, also requires no critical tempering, and is moderately compatible with chocolate liquor.

References

1. Bell, R.J., in *Introduction to Fats and Oils Technology*, edited by Peter J. Wan, American Oil Chemists' Society, Champaign, IL, 1992, pp. 187–188.
2. Gunstone, F.D., and T.P. Hilditch, *J. Amer. Oil Chem. Soc.* 836 (1945).
3. O'Brien, R.D., in *Hydrogenation: Proceedings of an AOCS Colloquium*, edited by R. Hastert, American Oil Chemists' Society, Champaign, IL, 1987, pp. 157–165.
4. Latondress, E.G., in *Handbook of Soy Oil Processing and Utilization*, American Soybean Association, St. Louis, MO, and American Oil Chemists' Society, Champaign, IL, edited by D.R. Erickson, 1980, pp. 145–152.
5. O'Brien, R.D., *INFORM 4*: 913 (1993).
6. Weiss, T.J., *Food Oils and Their Uses*, AVI Publishing Co., Inc., Westport, CT, 1983, pp. 170–174.
7. Brekke, O.L., *Handbook of Soy Oil Processing and Utilization*, American Soybean Association, St. Louis, MO, and American Oil Chemists' Society, Champaign, IL, edited by D.R. Erickson, 1980, pp. 426–429.

Chapter 21

Industrial Uses for Soybeans

Lawrence A. Johnson and Deland J. Myers

Center for Crops Utilization Research and
Department of Food Science and Human Nutrition
Iowa State University, Ames, IA

Introduction

Today, industrial (nonfood, nonfeed) uses for soybeans comprise no more than 0.5% of the protein (1) and 2.6% of the oil (2) produced from soybeans grown in the United States. There are no current industrial uses for whole soybeans, although some have humorously speculated about their values as projectiles for pea shooters and for filling bean bag chairs. There is at least one instance when soybeans were used for ship ballast (3), and powdered soybeans were patented as a floor covering (4). The economically viable industrial uses today are paper coatings and wood veneer adhesives (protein), and alkyd resins, printing ink, and oleochemicals (oil).

Any complete discussion of industrial uses for soybeans must contain a historical perspective, as the existing industrial uses of soybeans have considerable historical significance and provide insight into future uses, especially when nonrenewable resources become limiting. Soybeans, originating as an industrial crop in the United States, experienced considerable progress and excitement in providing consumer goods and helping the United States to emerge from the Great Depression. Today, imminent growth of soybean usage in fuel, adhesives, plastics, and construction materials is predicted.

Early Industrial Uses in the Orient

Soybeans were first domesticated in northeastern China around the 11th century B.C. (5), and it is not surprising that the Chinese were the first to crush soybeans into oil and cake using mechanical presses. In addition to using the oil for cooking, they also used the oil for lubricating fluids, lamp oils, coatings, and marine caulking materials (5). Reportedly, the oldest known industrial use of soybeans dates from A.D. 980, when soybean oil was first used in caulking compounds for boats (6). Up until the 20th century, the Chinese used the cake for another important industrial purpose: fertilizer and soil amendment, often referred to in early literature as "green manure" or "bean cake manure" (7,8).

Early Industrial Uses in the West

It has been reported that soybeans were first introduced into North America in 1804 as the ballast of a Yankee sailing ship involved in the China trade (3); thus, ship ballast became the first industrial use for soybeans in the West. But it was not until the late 19th century that soybeans began to attract the serious attention of Western scientists, farmers, and businesses. The first commercial uses for soybeans in the United States were industrial, because soybean oil was regarded as inferior to alternative food oils because of its flavor instability. In 1910 flax (linseed), the primary industrial oil crop at that time, escalated in price, and soybean oil began to be used as a substitute or extender for linseed oil, widely used in paints and varnishes.

Early Industrial Uses for Soybean Oil. In 1908 some European countries, particularly Great Britain, began to import soybeans from Manchuria to supplement short supplies of cottonseed and flax to meet a growing demand for edible oils, soaps, and glycerine. At that time, glycerine was in great demand for manufacturing explosives used in mining and in constructing the Panama Canal (9). Soybeans were pressed into oil and meal; the oil was chiefly used in soaps, and the meal was fed to dairy cattle. Soybean oil was used as a partial substitute for cottonseed oil in hard soap. For soft soaps, soybean oil could completely replace cottonseed oil and partially replace linseed oil (10). The glycerine, a by-product of soap and candle making, was distilled for explosives (dynamite, blasting gelatin, and cordite) and for manufacturing printing inks and printers' rollers (11). In 1909 the first of several patents was issued for the use of soybean oil in rubber substitutes (12), and in 1910 the first use of soybean oil in linoleum flooring was reported (13).

In the United States during World War I (1914–1918), the largest market for soybean oil was in the soap industry. Lesser amounts were used in paint, varnish, enamel, linoleum, oilcloth, asphalt, and other waterproofing materials.

As early as 1855, the potential for using soybean oil to replace linseed oil as a drying oil in wood finishes was recognized (7), and as early as 1880, the feasibility of using soybean oil in paints was recognized (14). Replacing linseed oil was particularly important in the United States in 1910, when the flax crop failed, and during World War I, when shortages of linseed oil were severe due to military demand. This contributed immensely to the demand for soybeans, which in turn led to rapid growth in the number of acres that U.S. farmers devoted to soybeans.

Although soybeans typically contain only 5 to 11% linolenic acid compared with 35 to 60% for linseed, the amount in soybeans is sufficient for soybean oil to be classed as a "semi-drying" oil. Soybean oil does not have drying properties as good as those of the "drying oil" linseed oil, which in those early formulations would dry to a nontacky film within several days: 4.5 days with a linseed oil base (15) but 6 days for a soybean oil base (16). Adding up to 25% soybean oil reportedly extended drying time by only several hours and produced marginally acceptable paint films (15). By 1919 it was learned that heating and blowing air through soybean oil, which gave rise to the term "blown oil," would increase viscosity by initiating oxidation and partially polymerizing the oil (17). Blowing soybean oil also improved its properties in printing inks. Soybean oil also was used to replace linseed oil as a binder in foundry cores (18). As early as 1926 R. Ditmar used soybean oil as an agent for plasticizing and increasing elongation of rubber (19).

Soybean oil was not widely used for food until the 1930s, when oil processing technologies advanced sufficiently to produce hydrogenated soybean oils with acceptable stability and flavor. Today, less than 168 million kg (369 million lb), equivalent to 2.6% of the soybean oil produced in the United States, is used in industrial applications, and the remainder is used primarily for food (although much smaller, but ever-increasing, amounts are also used in feeds). The 2.6% usage level of soybean oil in industrial products is much less than the 16.5% level for the total of all fats and oils (2). Thus, most industrial products made from fats and oils are supplied by other sources. Major industrial uses for oil today include paint and varnish, resins and plastics, and as a source of fatty acids. The latter is often derived from refining by-products (sometimes known as foots).

Early Industrial Uses for Soybean Protein. The crudest form of soy protein, namely, ground cake or meal, had limited industrial uses in adhesives and paper sizings (20), for binding charcoal briquettes (21), as an additive in wall plaster (1), and as a spray emulsifier for dormant fruit trees (22).

In 1883 the German scientists E. Meissl and F. Boecker published the first compositional analysis of soybeans and introduced the terms "soy casein" and "soy albumin" to describe soy protein fractions (23). The term "soy casein" was used because soy proteins were similar to milk casein in solubility, viscosity, and other properties. The term "soy albumin" was also used because soy protein had many of the same properties as egg albumin. O. Nagel in 1903 described how to produce soy casein and discussed its applications in commercial products (24). The general approach of his protein isolation procedure became the basis for today's commercial soy protein isolate production (see Chapter 8). The oil was first extracted with benzene and the defatted meal was extracted with alkaline sodium carbonate solution. The resulting soluble extract was then separated from the insoluble fibrous residue, and the protein was precipitated with rennet or dilute hydrochloric acid.

Alternatively, industrial-grade soy casein was prepared from oil-free meal by grinding with cold water, filtering, and treating with powdered gypsum (25). The mixture was boiled to precipitate the protein, which was collected and washed on filters. The soy casein was dissolved in dilute soda, filtered, precipitated with acetic acid, washed, collected by filtering, and dried. This soy casein was white and quite pure and was used to replace milk casein in paper sizing. Other early uses for soy casein were in the preparation of silk and artificial textiles, rubber, leathers, plastic materials, films, photographic emulsions, and paints.

In 1913 S. Satow (26) obtained a U.S. patent for immersing coagulated (acid-treated) or "glutenized" soy proteins in a formaldehyde solution to produce a moldable plastic. This rigid, semitransparent plastic was claimed to be a good electric insulator and a substitute for ebonite, celluloid, Bakelite, ivory, and marble. In 1923 the first of many patents was issued for a glue based on soybean meal (27), and in 1926 the I.F. Laucks Co. began marketing a "soybean-oil-meal" glue to the plywood industry (28).

The Chemurgic Movement

The early successes of soy-based plastics and other industrial products made from soybeans and the economic stresses in the agricultural sector during the Great Depression of the 1930s promoted the establishment of various organizations to foster new industrial uses for agricultural products. This was the goal of the National Farm Chemurgic Council, formed during the early 1930s with the noted industrialist Francis P. Garvan as its first president. Many now-famous persons were involved in chemurgy, such as the scientists George Washington Carver, Leo M. Christiansen, Thomas A. Edison, and Percy Julian; the noted industrialist Henry Ford; and the journalist Wheeler McMillen. Henry Ford founded the Edison Institute in Dearborn, MI, in honor of his close friend Thomas A. Edison for the purpose, among others, of finding uses for soybeans in manufacturing automobiles. Nearly 200 industrial uses for soybeans have been attributed to this movement (29).

Ford not only was a major force in the chemurgic movement but also is credited with being one of the few successful commercializers of many of the chemurgic innovations. His interest in soybean-derived industrial products was based not on altruism but on his commitment to raising farm-sector income so that farmers could purchase his automobiles, trucks, and farm tractors. For 1935 Ford automobiles as much as a 3.8 ML (1 million gal) of soybean oil were used in enamel paint; 2 ML (540,000 gal) were made into glycerine for use in shock absorbers; and 0.75 ML (200,000 gal) were used in its engine foundries as core sand bond (30). In 1935 Ford constructed his own soybean extraction and processing mill at his River Rouge, MI, automobile plant to process soybeans into industrial products; at one time he had three such plants. In 1936 the Ford Motor Company planted over 4,900 hectares (12,000 acres) of soybeans to support its industrial soybean interests. The 1937 automobiles reportedly used 180,000 kg (400,000 lb) of soybean meal in plastic parts, and by this time development of soy fiber for automobile upholstery was underway (31).

On the national level, the chemurgic movement had largely ended shortly after World War II (1939–1945), and by the mid-1950s programs continued at significant levels only at the U.S. Department of Agriculture's Northern Regional Research Center (USDA, NRRC, now known as the National Center for Agricultural Utilization Research). In 1949 Arthur D. Little, Inc., estimated that 23.4 million kg (51.5 million lb) of soy flour and 12.25 million kg (27 million lb) of soy protein isolate were annually used for industrial products in the United States (32). Based on the total meal produced, industrial uses of soy protein amounted to approximately 3.6% of the 1949 domestic supply. This was a much greater proportion than today.

Chemurgic interests are resurfacing today; the New Uses Council, St. Louis, MO, has been established to deal with recent surpluses in U.S. agricultural commodities and rising petroleum prices. In 1991, Congress passed legislation creating the Alternative Agricultural Research and Commercialization Center (AARCC), a program designed to foster commercialization of industrial uses for agricultural materials.

Industrial Uses for Soybean Protein

Wood Adhesives

History and Background. In the past, plywood adhesive has been one of the major industrial uses for soybean products. Patents obtained by O. Johnson (27) and by I. Laucks and G. Davidson (33) formed the basis for the use of soybean meal and protein in adhesives for the Douglas fir plywood industry, which became economically viable in the late 1920s (34). Laucks and his company hired Johnson, purchased the rights to his 1923 patent, and developed additional improvements. Consequently, Laucks is often credited with founding the soy glue industry with a product his company developed in 1926 (28). Laucks' adhesive, often referred to as "Laucks' Bean Soup," was first formulated with soy flour made from soybean press cake imported from Manchuria.

Although the technical development of soy protein isolate was first reported as early as 1919, it did not become a commercial product until 1937, when the Glidden Company, Chicago, IL, put its first plant into operation after purchasing the patent rights to a process developed by Laucks for improving soy adhesives (35). This plant produced "Alpha Protein," an industrial-grade protein isolate. The Glidden Company, prompted by its interest in oil for its Glidden paint operations and Durkee Famous Foods Division, invested heavily in soybean processing during the 1930s. They soon became quite interested in soy protein to replace milk casein in water-based paints and in adhesives for high-grade printing paper and wallpaper. This first plant produced 4,500 kg (10,000 lb) per day of industrial-grade soy protein isolate. However, by the mid-1960s, petroleum-derived glues displaced soy glues and continue to dominate the adhesives market. The soy products of the time simply did not measure up, lacking water and microbial resistance.

Performance Properties. To produce a good glue joint, the glue must bond chemically to the wood, and in order to do so, the molecular structure of the adhesive must be highly polar to bond to polar cellulose wood fibers (34). It is also important that the glue form large polymers to reduce water solubility and increase strength.

Factors important to making soy flour a good adhesive are not well understood. The major wood-bonding component of soy flour is believed to be protein, although the 35% carbohydrates present may also contribute some adhesion. By comparison with other proteins, soy protein has a large number of polar amino acids, particularly ionic amino acids such as glutamic and aspartic acids. The defatted soy flour should not be exposed to temperatures above 71°C (160°F) so that the protein is not denatured. For unknown reasons, wide variations were experienced in performance characteristics, and to reduce the risk of inadequate performance, soy flour from several sources was often blended (34). In theory, one would expect that a higher-protein-content ingredient than soy flour (44 to 52% protein), such as a protein concentrate (>70% protein) or a protein isolate (>90% protein), would give much greater bonding; however, research has shown that the use of a soy protein isolate did not improve the adhesive enough to offset its higher cost (34,36).

The cited advantages of soy glue were soy flour's low cost and plentiful supply compared with casein; its relatively strong and water-resistant (albeit not waterproof) bond; its lack of tackiness, making the glue-coated surfaces and materials easier to handle and improving manufacturing efficiencies; its ability to be applied hot or cold and spread or sprayed; and its compatibility with high-moisture-containing veneer without surface splitting (37).

Soy adhesives were unique in that they could be applied using either hot- or cold-pressing procedures. This property helped soy adhesives remain competitive until the early 1960s. Most of the original processes for bonding wood laminates were cold-pressing processes, requiring clamping for hours during curing and extensive labor to clamp and unclamp the bonded laminates. When the industry moved toward more waterproof phenolic adhesives, the faster hot-pressing method became the preferred method. This change was responsible for temporarily reducing the usage of soy flour in plywood adhesives (37). When a successful hot-pressing method was developed for soybean glue and the Galber and Golick "No-Clamp" cold-press plywood method was developed, the use of soy flour steadily increased again (34,37).

Although the use of soybean adhesives for Douglas fir plywood was relatively successful, there were several factors that limited more widespread usage and still restrict its competitiveness today. The solution used to disperse the protein is highly caustic and can discolor wood; therefore, it has never been used in decorative wood applications, such as hardwood furniture, where discoloring is a concern (32). Another perceived limitation was its sensitivity to microbial degradation. Soy adhesives were susceptible to mold, particularly in humid environments (38). Most industry experts now believe that this problem can be solved by using phenols, chlorinated phenolic compounds, and their salts.

Probably the most significant factor limiting the growth of soy adhesives in wood products has been their lack of water resistance compared with phenol-formaldehyde and other protein adhesives such as blood (38). For this reason, soy adhesives have been limited to interior plywood applications. This lack of water resistance severely limited usage of soy adhesives when the industry switched to producing indoor-outdoor plywood in the early 1960s (39). Although there have been many attempts to improve moisture resistance, using crosslinking agents, tanning agents, formation of calcium proteinates, and blending with other proteins, none have been completely satisfactory in preventing the weakening of the bond in humid conditions (38).

Technology. A key step in preparing the adhesive is to disperse the protein in aqueous alkali (pH ≥11) to solubilize the protein and unfold peptide chains, providing maximum exposure of amino acids for bonding. The flour is first added to water and mixed. Sodium hydroxide is then added to raise the pH, and calcium hydroxide is also added to form proteinates that control glue consistency, improve water resistance, and decrease assembly time. Other "minor" ingredients are added to improve the adhesive properties of the glue. Sodium silicate improves water-holding capacity and viscosity and prolongs the life of the glue. Protein denaturants and crosslinking agents increase water resistance, control viscosity, and stabilize the protein against hydrolysis. Preservatives are added to protect the adhesive against molds. A typical soy adhesive formula is shown in Table 21.1.

Glues based on soy flour have been formulated with blends of other adhesives including casein, blood, urea-formaldehyde, and phenol-formaldehyde (36,40). Soy flour reduces the cost of the adhesive and imparts unique properties. Soy flour gives excellent consistency when blended with blood to improve assembly time, and promotes rapid water loss in blends with casein (42). Today, as in 1947, the major use of soy flour in wood adhesive is in blends with other proteins for specialized uses (36,38).

Soy glue can be applied by a variety of different methods including dry mixing, spraying, extruding, and rolling (34). Viscosity determines the type of application as well as end use and other properties of the adhesive. The viscosity of the soy protein adhesive for wood lamination should be 5,000 to 25,000 cP (centipoise) and over 50,000 cP for mastic-consistency adhesives, which are typically extruded (34).

Markets. In the 1930s the plywood industry, desperate for adhesives to keep pace with automobile industry demand for wood-laminate running boards, adopted glues made with finely ground soy flour. By 1942 virtually every plywood plant on the West Coast was using soy glue exclusively, and soy glue temporarily captured 85% of the U.S. plywood glue market (28). The amount of soy flour used in plywood glue reached 27 million kg (60 million lb) in 1942 (37), dropped to 19 million kg (42 million lb) in 1947 due to the advent of synthetic resin adhesives (32), but increased again, peaking at 25 million kg (56 million lb) in 1956 with the widespread adoption of the cold-press/no-clamp process (32,34). Thereafter, the levels used steadily declined because of the increased competitiveness of petroleum-based substitutes. Today, the wood adhesives market for soy flour is relegated to specialty uses (i.e., high-quality veneers) and is estimated to be approximately 91,000 kg (200,000 lb).

Current Interest. Two factors provide opportunities for soy adhesives to be competitive in the wood products market again: cost and environmental regulation. The demand for phenol is increasing, escalating the prices of phenol-formaldehyde resins. Old-growth forest products are not able to keep up with demand for building materials, so demand is being met with new products such as oriented strandboard, waferboard, chipboard, and particleboard; these products require large amounts of

TABLE 21.1 Basic Soybean Adhesive Formulation (34)

Ingredient	Amount (%)
Soy protein	19.82
Water	68.90
50% NaOH solution	2.86
Ca(OH)2	2.45
Sodium silicate	5.10
Antifoam	0.61
Carbon disulfide	0.20
Carbon tetrachloride	0.06

adhesive. Manufacturers are interested in less expensive adhesive ingredients for replacing a significant portion of the resin to reduce cost without compromising quality standards.

There is increasing concern about emissions ("gassing off") of formaldehyde (a probable carcinogen) from interior-grade wood products, particularly those products bonded with urea-formaldehyde. The industry is interested in alternatives to replace all or a significant amount of the urea-formaldehyde in interior-grade wood products without significantly increasing costs. In 1985 the U.S. Department of Housing and Urban Development specified formaldehyde emission standards for particleboard and hardwood plywood panels used in mobile homes (41). The U.S. Environmental Protection Agency (EPA) and the U.S. Consumer Product Safety Commission have also been considering indoor formaldehyde emission standards. Many feel that more general restrictions are required and that formaldehyde emission problems are not limited to mobile homes. In fact, a total ban of urea-formaldehyde resins, coupled with mandated use of phenolic resins, is under consideration by the EPA. Several states are considering their own regulations for formaldehyde emissions in buildings; however, to date, only Minnesota has done so (42). The future of soy protein in the wood adhesives market will be determined by price, ability to reduce or eliminate fugitive formaldehyde, and the advantages of cold setting, rapid water loss during bonding and improved water resistance of the adhesive bond.

Plastics

History and Background. In 1913 patents were issued in France (43) and Great Britain (44) for preparing semiplastic materials from soybean protein. However, it was S. Satow's patent of 1919 (26,45) that stimulated widespread interest. Isolated soy protein was glutenized with acid and treated with the crosslinking agent formaldehyde. These plastics flowed and molded well but were prone to cracking and shattering with aging.

By 1930 the impending Great Depression forced Henry Ford to look for ways to increase farm income to protect farm-related purchases of his automobiles, trucks, and farm tractors. Ford and his team at the Edison Institute, headed by R. Boyer, developed a process using phenol-formaldehyde resin composed of approximately 30% soybean meal in the production of plastics (46). Soybean meal was mixed in a steam-jacketed mixer with phenol, formaldehyde, and an accelerator and allowed to react for 15 to 20 min. Then a filler (wood flour) was added, along with clay and stearic acid to promote smooth molding (30).

In 1940 *Time* magazine published an article about Ford's soy plastics project (49). The story included a photograph of Ford swinging an axe into the plastic trunk lid of one of his "collision-proof" Ford automobiles (Fig. 21.1). It was never publicly admitted that Ford used the blunt side of the ax, which was also covered with a plastic guard to cushion the blow (6). Reportedly, the new plastic did not dent, nor was the finish chipped or abraded. In pursuit of an "all agricultural" car, Ford's soy plastic was used for gear shift knobs, horn buttons, window frames, and other parts. Ford even had a prototype complete car body made of soy plastic and planned to mass-produce his soy-based plastic car bodies (48). Then in 1941, responding to metal shortages caused by the war effort, Ford ordered his engineers to build soy plastic prototypes of refrigerators, bathtubs, sinks, and household fixtures (49). Despite Ford's vision and commitment, soy plastic could not compete in these applications.

Fig. 21.1. Henry Ford slamming an ax into a 1940 automobile trunk lid made of soy protein plastic. *Source:* From the collections of the Henry Ford Museum and Greenfield Village, Dearborn, MI. Used by permission.

In 1942 the USDA NRRC, working on approaches similar to those of Ford, found that it was possible to use 20% heat-denatured soy protein concentrate (acid-washed meal) with 40% phenol-formaldehyde resin and 40% wood flour to make a very acceptable molding powder for plastics (50). That year, more than 68 million kg (150 million lb) of soy concentrate extruded plastic was produced. It was used in buttons, plastic rods, beads, buckles, frames of eyeglasses, and plastic novelties. USDA scientists also developed a plastic construction helmet out of heavy cotton cloth and soybean protein that would reportedly withstand 18-kg (40-lb) blows (51).

Performance. Water absorption of the protein plastics could not be reduced to the very low level that could be obtained with synthetic, petroleum-derived resin products (52). Soybean meal was also tried, and dyeing properties were improved. These plastics had good color and strength and were relatively low in cost, but they also showed poor water resistance, lacked permanence when subjected to water, and experienced forming difficulties except with water plasticization (32).

G. Brother and L. McKinney led the USDA effort and found that if soy isolate was hardened with an aldehyde at or near its isoelectric point, a material that was thermoplastic and "minimally" absorbed water (10% moisture in 24 hr) could be produced (53). The flow of this formaldehyde-treated protein could be improved with ethylene glycol as the plasticizer and adding oleanolic acid (3,β-hydroxyolean-12-en-28-oic acid) and aluminum stearate to improve water resistance (54). An aqueous suspension of the formaldehyde-hardened soy isolate could be used as a laminating material for kraft paper (55). The laminating material had the same properties as the more expensive phenol- and urea-laminating materials; however, it was less water-resistant.

Work with soy protein isolate was discontinued in favor of modified soybean meals as modifiers and extenders in phenol-formaldehyde plastics (32). A suitable plasticizer was never found for soy protein isolate to work in injection dies (56).

Brother and McKinney (56) improved the thermosetting properties of the formaldehyde-hardened protein by combining it with phenolic resins. They found that formaldehyde-hardened soybean meal combined with phenolic resins in a 50:50 ratio produced a plastic that had higher water absorption rates and longer cure times than the resin alone; however, when blended with 25% wood filler, the plastic had excellent strength, good cure time, and fine finish. Hardened soybean meal imparted more flow to the phenolic resin than wood flour, and it was possible to use the formaldehyde-hardened soybean meal as an extender for phenolic resin (57). McKinney (50) developed a process to remove soluble sugars from soybean meal and treat it to improve water resistance and reduce blistering; to heat-denature the protein to reduce solubility; and "harden" the protein with formaldehyde. Up to 20% of this treated soy meal could be used with 40% phenol-formaldehyde resin and 40% wood flour to produce a high-quality molding powder. In addition to saving valuable phenolic resin, the plastic had improved dyeing properties. However, it cured more slowly than phenolic resin and wood flour alone; therefore, the material never successfully competed (32).

Market. Despite all the early enthusiasm and fanfare, soy plastics did not measure up to alternative materials. Inexpensive and adequate supplies of petroleum and better-performing synthetic materials discouraged serious attempts to make continued improvements. For a short period (1935–1943), Ford used soy plastics in his popular, inexpensive automobiles, but his plan to use over 1.8 million kg (4 million lb) of meal for every million cars produced (46) never materialized.

Current Interest. Environmental concerns and increased polymer prices relative to agricultural products are renewing interest in soy plastics. Today's 30-million-MT (33-million-short-ton) plastics market is dominated by five petroleum-derived polymers: high-density polyethylene (16% of market), low-density polyethylene (19%), polyvinyl chloride (15%), polypropylene (13%), and polystyrene (8%); more than 18 other polymers make up the other 29% (58). Experts envision the greatest market for soy-based plastics to be polystyrene. Polystyrene is widely used in foam cups, foam food packaging trays ("clamshells"), molded table serviceware, and other items used by fast-food restaurants.

Mounting environmental concerns have focused attention on the concept of designing materials from "cradle to grave." In Europe, there is growing legislative concern about introducing new products without having environmentally sound means of disposal in place. Most believe that the solution to many of our waste problems lies in composting and recycling. Despite the recent emphasis on conventional recycling, less than 2% of our plastic wastes are recycled, because the sorting and reclamation costs are quite high compared with the low values of reclaimed plastics. The new soy protein plastics offer new possibilities in both composting strategies and strategies to reclaim resources as fertilizer and livestock feed.

The International Maritime Pollution Treaty, which will take effect January 1, 1999 (recently granted a five-year extension from the previous January 1, 1994 implementation date), prevents the continued practice of ocean dumping of plastic wastes. Military and merchant marine ships and cruise liners have storage capacity for three to five days (sufficient time to enter or leave port), after which wastes are dumped at sea. Plastic, being less dense than seawater, rises to the surface and ultimately fouls beaches and threatens wildlife, especially sea mammals and shore birds. Fish do not possess amylases to break down starch, so protein may be a more suitable degradable polymer in the marine environment.

Today's soy plastics vary from 40 to 90% soy protein isolate, the balance being corn starch, depending on the desired properties (Fig. 21.2). It is possible to make protein/starch plastics with all edible ingredients, and formaldehyde is not required to achieve acceptable properties for many applications. Some estimate a realistic U.S. market potential for soy meal to be equivalent to about 1.6 million MT (60 million bu), or 3% of the U.S. soybean production. Considering that this is equivalent to the total current usage of soybean meal for all food and industrial uses for soy meal, soy plastics represent a huge potential market for soybean producers.

Fig. 21.2. Newly developed plastics using soy protein, corn starch, and other food-grade ingredients. *Source:* J. Jane, Center for Crops Utilization Research and the Department of Food Science & Human Nutrition, Iowa State University, Ames, IA.

Textiles Fibers

History and Background. Soy protein was one of a number of proteins that were used to produce "regenerated" protein textile fibers in the late 1930s. Although the discovery of regenerated protein fibers from casein is attributed to Todtenhaupt in 1904, it was Farretti in 1935 who successfully developed, patented, and produced a fiber with wool-like properties (59,60). In 1939 the Japanese reportedly produced about 450,000 kg (1 million lb) of soy protein fiber (38). In short order the Germans also began producing commercial quantities of soy fiber. Likewise, spurred by the war effort, use of soy protein as textile and felting fibers was explored during World War II in the United States to replace more expensive wool, felt for hats, and rabbit fur no longer available from Europe. The first U.S. patents for soy fibers were granted to T. Kajita and R. Inoue in 1940 (61). In the United States, research by O. Huppert of the Glidden Company resulted in a number of U.S. patents in the years 1942 through 1945 (60,62).

In 1945 R. Boyer of Ford's Edison Institute was awarded an important patent (63) for producing textile fibers from soybean meal for use in automobile upholstery. In addition to soy protein and casein, other proteins, such as corn zein and peanut, were also used to produce regenerated protein fibers. Ford wore a suit made from soy fiber (Fig. 21.3), which was reportedly quite itchy when dry and odoriferous when wet. Soy fiber technology did not reach commercial textile production in the United States. Pilot plant production of soy fiber by the Ford Motor Company reportedly reached 2,300 kg (5,000 lb) per day in 1940 (37).

Technology. The basis for producing these fibers is the fact that silk and wool are naturally occurring fibrous proteins. If other proteins could be "regenerated" into a fibrous form using the principles of the viscose process for making rayon from cellulose, then protein fibers could be produced that would have similar or improved properties, enabling them to compete with wool and silk (60,64). However, this was never realized because the important molecular structure that was required to produce good fiber qualities was lost due to protein denaturation; the relatively long repeat distance and bulky side chains prohibited protein chain interaction; and the denatured protein was extremely flexible, allowing internal rotational motion resulting in random coil configurations (64).

Soy protein isolate was used to produce regenerated protein fibers. According to Boyer's process (65), soluble soy protein isolate was obtained from defatted meal that had not been heat-treated. The meal was then treated with a weakly alkaline solvent (0.1% sodium sulfite) for 30 min. The solution was then clarified by filtering or centrifuging. The protein was precipitated with acid, and the curd was washed and dried. The protein was then dissolved to form a viscous "stringy" solution with greater than 12% solids (65). The protein solution needed to be highly alkaline, pH 12.5, to unfold the protein. Although a solution of 12% solids was difficult to obtain because of the tendency of the solution to gel, solutions as high as 20% solids were reported (37). The solution was filtered, deaerated, and aged to allow the solution to become more viscous and develop stringiness for spinning (37,65). The use of detergents rather than alkali to unfold the protein was rejected because of the large amounts required and the lack of adequate recovery processes. Heat, water, and mechanical shear were moderately successful.

The textile fibers were formed by a wet-spinning process which forced the solubilized protein through a spinneret into an acid bath to recoagulate the protein. The bath typically consisted of a solution of sulfuric acid, formaldehyde (to harden the

Fig. 21.3. Henry Ford wearing a soy protein suit, 1941. *Source:* From the collections of the Henry Ford Museum and Greenfield Village, Dearborn, MI. Used by permission.

fiber), and a salt (to accelerate drying) (37,65). The fibers were collected on a reel and stretched to orient them, improving their strength and elasticity. The fibers were then immersed in a formaldehyde bath to crosslink the protein (to improve the resistance of the threads to attack by water and dilute acids); cut into desired lengths; and dried.

Performance Properties. Soy fibers were white to tan in color and had a warm, soft feel; natural crimp; and high resilience (60,65). They had more elongation than wool and 80% of wool's dry strength (66). Soy fiber blended with wool and cotton and could be processed using worsted textile equipment. However, these textile efforts were not as successful as those with milk casein fibers and failed, largely because soy fibers had poor wet strength and could not withstand repeated washing or dry cleaning. The wet fibers also had a characteristic "wet dog" odor. Early soy fibers resembled casein fibers, and later fibers even resembled rayon (67,68). Later improvements increased resistance to carbonizing, reduced sensitivity to hot alkali, increased resistance to high temperature treatments in aqueous acidic solutions, reduced dry-cleaning damage, and improved dyeing with the common chrome and acid dyes applied to wool.

Soy fiber had one serious problem that was never improved: its low tenacity, particularly when wet. Compared with wool, the fiber was 45% weaker when dry and 75% weaker when wet (68). Because of its low tenacity, the best application of the fiber was found to be in blends with other fibers including wool, rayon, nylon, and cotton.

Market. The Federal Trade Commission adopted the generic name "Azlon" for soy fibers, fashioned after "rayon" for synthetic cellulosic fibers. The original intent of producing these fibers was to compete with the natural protein fibers wool and silk (32,59). The advantage that soy protein isolate had compared with other protein sources was its relatively low price and high protein content. Unfortunately, unlike the relatively more successful commercialization of fibers from casein ("Lanital"), zein ("Vicara"), and peanut ("Ardil"), soy protein fibers were never commercialized in the United States. In Japan, soy fibers were produced under the name "Silkool." The Ford Motor Company produced soy fibers on pilot plant scale to test its use in car upholstery. Ford sold his soy fiber interests to the Drackett Company, and they were unable to commercialize the technology because of problems of quality control and high production costs (69). By the time Ford died in 1947, the Ford Motor Company had divested its industrial soybean interests.

Current Interest. Work on soy fibers is being reestablished at several institutions; however, the likelihood of commercial success is quite speculative and risky. Today, the costs of wool and silk are much lower relative to soy protein; environmental regulations are much more restrictive than in the 1940s; and several synthetic fibers have been developed since. To be successful, one will need to identify properties important to consumers that soy fibers can offer at a reasonable price, and this has not yet been accomplished. Additionally, a dry-spinning process will likely be required, as the older wet-spinning process generated large amounts of waterborne wastes.

Paper Coatings

History and Background. The largest industrial use today for soy protein is paper coatings. The use of soy protein isolate in this industry began with the advent of machine-coated paper during the period 1935–1944 (32,37). Prior to the machine process, all paper was coated using the off-machine process. Machine-coating provided an economical way to produce higher-quality halftone reproductions of engravings. Although starches were originally used in the machine process, casein and soy protein produced coatings that had better rheological properties and were more waterproof (32). Also, paper manufacturers were forced to look for casein substitutes during World War II because much of the milk casein supply was diverted to dry milk production for the military.

Performance Properties. Compared with casein, soy protein had other advantages in addition to those mentioned, including being available in a wider viscosity range and the ability to be used in coatings with higher solids (70,71). Both native and partially hydrolyzed proteins were used. Soy protein did not "string out," leaving tracks in the coating, as milk casein did (70). The protein may be hydrolyzed to various levels to tailor viscosities and rheological properties. In addition, hydrolyzed soy protein produced higher-solids and lower-moisture coatings, facilitating higher machine speeds and lower drying costs.

Paper is coated for a variety of reasons, including upgrading the paper, making it more printable, and providing varying degrees of resistance to water or grease (32). Paper coating involves the application of a thin layer of pigments on the paper to provide a different surface with more desirable properties. A binder is used to bind the pigment particles to each other and to the paper. The binder also will affect the size and volume of the pores of the paper, ultimately influencing the printing and gluing properties of the paper or paperboard (70). The binder also influences the runability of the coating during application by controlling rheological properties (70,71,72).

Today, soy protein is used primarily as a co-binder in these systems with latex-protein and starch-latex binding systems (70,73). Co-binders are used in amounts from 1 to 5% of the binding formulation depending on the application. The blend of latex and soy protein isolate produces a coating that imparts higher brightness (74).

Technology. Soy protein isolate continues to be used in paper manufacturing. Under selective conditions of pH and chemical reactants, the protein is modified to improve its influence as a coating modifier and dispersant (71,73). Only modified proteins are used in the industry today, and these modifications have continually improved the functionality of the protein (e.g., viscosity, water-retention and pigment-stabilizing properties) and kept them competitive in the industry (71,72,73).

One major consideration in the binder formulation is the choice of alkali used to disperse the protein. The alkali used will influence the final properties (color, wet rub resistance, and viscosity) of the protein in the binder formulation (71). The other key consideration is the mixing regimen (mixing time, temperature, speed, and order of ingredient addition). The choice of other additives used to control the properties of the dispersion (flow modifiers to get workable viscosities at higher solids; preservatives, foam controllers, and lubricants to help the release of the coating colors from the rollers) are also important to the formulation (72).

Market. In the early 1940s, approximately 3,600 MT (8 million lb) of soy protein isolate were used in paper coatings, compared with 10,200 MT (22.5 million lb) of casein (38). Because of lower prices and more consistent properties, soy protein isolate then began to outcompete casein in this market. By 1949, the amount of soy protein isolate used rose to 6,600 MT (14.5 million lb), while the amount of casein used dropped to 9,000 MT (20 million lb) (32). Current annual consumption of soy protein isolate in this market is about 24,500 MT (27,000 tons) (70).

Current Interests. Soy protein isolates as co-binders in paper coating binder formulations have remained a vital part of the paper industry. Today, the coating is used in a variety of coated paper and paperboard applications. There should be steady growth of soy protein used in this application, provided petroleum prices remain high and soy protein isolate remains price-competitive relative to synthetic latex binders.

Other Industrial Uses for Soy Protein

Paper and Textile Sizing. Laucks is also credited with developing paper sizings from soy flour (28). Paper is sized to make it resistant to the penetration of moisture and liquids (37). Reportedly, soy isolate helped to improve this process by making the rosin used in the process more uniform and stable upon addition of borax and by binding free alkali. "Prosize," an industrial protein isolate of the Glidden Company, was a product in which industrial soy protein isolate was used to stabilize an alkaline resin size dispersion for internal sizing of cellulosic fibers before complete dewatering (75).

Textile fibers are sized for a variety of reasons depending on the fiber, the process, and the end use. Soy protein isolate has been used as a warp sizing to manufacture rayon (32,37,76). The advantages of soy sizing were its ability to penetrate the yarn, heat stability in the operating temperature ranges of the machines, and stability. However, manufacturers had concerns about low resistance to abrasion and off-color (32). These factors, along with the inroads of synthetics, prevented soy protein isolate from capturing greater market share.

Building Materials. Recently, Phenix Composite Inc. developed "Environ," a composite material composed of used newspapers and soybean meal and having a wide range of uses as a construction material (Fig. 21.4). The material uses soy flour in a monolithic resin system to bind together recycled newspapers. "Environ" possesses the easy-working properties of wood and has the appearance and sales appeal of polished granite. It has been introduced for use as decorative surfaces in furniture, cabinets, and recognition plaques; it is also expected to be available soon as a structural building material for both interior and exterior uses. Although the current cost of Environ is relatively high (\$130 to 150/m^2; \$12 to 14/ft^2), its fire resistance is a major factor in considering its use, in addition to its unique appearance and excellent working properties.

Wallpaper. Soy flour, alone or in combination with soy isolate, was used for a number of years as a coating for wallpaper. At concentrations below 16.5%, the alkaline protein dispersion was thixotropic (could be made liquid by shaking) and able to suspend pigments and clays. The pH and viscosity remained fairly stable if preservatives were added to prevent mold and bacterial growth. Water resistance was imparted to the paper coating by applying a tanning solution containing formaldehyde. Soy flour not only was used in the coating formula; it was also in the formulations of water-based inks used for printing. High-grade washable wallpaper could be made with soy flour, and several million kg were used annually up to 1947 (37); however, soy flour was actually used in only low-grade wallpaper, and those industries using soy flour declined significantly by 1949 (32).

Miscellaneous Adhesives. Soy protein was used in adhesive applications other than plywood, although at significantly smaller usage levels (37,77). Soy protein isolate has been used in tacky and remoistening adhesives. Soy protein isolate was also used in the formulation of glue for shotgun shell casings, about 45,000 kg (100,000 lb) in 1949 (32); it was chosen for this application over soy flour because of better initial tack and water resistance. A process was also patented for using an alkaline dispersion of soy flour as an adhesive for charcoal briquets (37). Briquets formulated with a soy flour dispersion along with other chemicals were resistant to weathering and breakage during handling.

Water-Thinned Paints. In 1949 the amount of soy protein isolate and flour used in water-thinned paints amounted to 900,000 kg (2 million lb) (32). Presently, there is no reported use of soy protein in this application.

Powder and Paste Paints. Soy protein isolate was used in paste and powdered cold-water paint formulations because of its adhesive, film-forming, and water-proofing properties. The paste paints were considered to be more durable than the powder paints.

The advantages of soy protein in paint formulations were increased washability of the painted surface, improved hiding power of a single application, increased durability on porous surfaces, rapid drying, and absence of odor (because they contained no organic solvents) (37). The disadvantages were low water resistance and durability, poor initial adhesion, and low resistance to microbial degradation if not properly preserved.

These paints used higher protein concentrations (50 g/L, 0.4 lb/gal) than did the resin-oil emulsion paints (32). Due to the superior properties of the resin-oil paints, the importance and market share of these paints declined.

Resin-Oil Emulsion and Latex Paints. In resin-oil emulsion systems, the addition of soy protein isolate or flour aids in dispersing the resin-oil phase and stabilizing the emulsion. Emulsion paints form hydrophobic films on drying. The advantages of

Fig. 21.4. Wood- and granite-like building products produced with soy flour and recycled newspapers. Materials provided by the Phenix Co., Mankato, MN.

the paint were that they were easy to apply, quick drying, and odorless and had good washability and high hiding power (37). Although casein and soy protein could be used interchangeably in many of these paint formulations, soy protein had some advantages, including improved washability and brushability. However, casein was still preferred over soy protein in spite of the higher cost, because of its better viscosity control and bonding characteristics (32).

Printing Ink. Currently, there is high interest in the use of soy oil in printing inks; however, soy protein has been used in printing ink formulations as well. Soy protein isolate was used in dispersions as vehicles that were set by precipitation. A. Schmutzler and D. Othmer (80) developed a process to disperse soy isolate in polyhydroxy alcohols with other dispersing agents to replace zein as the vehicle. They improved the water resistance of the ink by modifying the dispersion to form complexes with guanidine carbonate. At one time an estimated 45,000 kg (100,000 lb) of soy protein isolate were used annually in this application (32). Today, there is no evidence of protein being used in this application.

Fire-Fighting Foams. During World War II soy foams for fighting shipboard fires were made with lime-hydrolyzed soy protein isolate for the U.S. Navy and were known as "bean soup" to sailors (79). During the war, much of the soy protein isolate manufactured in the United States was hydrolyzed and used to formulate such foams for fighting fires on ships (37). The protein was fed into a water stream, and foam was produced by means of an aerating nozzle. It formed a tight foam blanket to smother flames and check reignition and was adhesive so that it would stick to walls (80). As late as 1949, 113,400 kg (250,000 lb) of soy isolate were used in this application (32). There is no report of soy isolate being used in this application today.

Other Miscellaneous Industrial Uses for Soy Protein. Before World War II substantial tonnage of soy meal was used for mixed fertilizers because of its high nitrogen content. Today, no soy meal is used for this purpose.

In the late 1940s the Glidden Company marketed soy protein as a "sticker and spreader" in agricultural sprays. The product was known as "Spraysoy."

There were numerous other industrial applications for soy flour and protein isolate, albeit in very small quantities. Some of these uses were stabilizers for latex emulsions in the rubber industry, coatings for rubber and tire cord in tire manufacturing, foundry core binders, leather finishes, insecticide emulsifiers, and asphalt emulsions. Soy flour is used as an ingredient in fermentation media and in honeybee diets (81).

Industrial Uses of Soybean Oil

The United States produced over 5 million MT (11 billion lb) of soybean oil in 1990, and of this amount only about 0.13 million MT (300 million lb), or 2.6%, are used for industrial products (Table 21.2). Although soybean oil provides over 50% of the total fats and oils consumed in the United States, soybean oil provides less than 10%

of our industrial needs for fats and oils. Important end-use categories for which there are economic data are fatty acids, paints and varnish, resins and plastics, drying oil products, and other industrial products. Epoxidized oil (resins and plastics) and coating vehicles (paints and varnishes) comprise 50% of the industrial market for soybean oil. The former is about a 45 million kg (100 million lb) market, which has been expanding, but the latter is losing market share to synthetic latex coating vehicles.

Soybean oil is often regarded as being too viscous and reactive to atmospheric oxygen to be used as fuels, cosmetics, lubricants, and chemical additives, but not reactive enough for most paints and coatings applications. However, more stringent environmental standards, rising costs for competing petroleum-derived products, ability to tailor soybean oil for improved performance properties, and more cost-effective chemical conversion processes are leading to increased attention on soybean oil as a feedstock for industrial products.

Paints, Coatings, and Varnishes

History and Background. Paint is an old use for soybean oil and other vegetable oils. Archeological evidence indicates that the ancient Egyptians grew flax and used

TABLE 21.2 Utilization of Soybean Oil and All Fats and Oils in the United States for Industrial Products (million lbs)[a]

Year	Fatty acids	Soap	Paints and varnish	Resins and plastics	Lubri-cants	Drying oil products	Other industrial products	Total industrial utilization	Total utilization
				All fats and oils					
1955	606	1,115	D[b]	D	D	1,104	834	3,659	11,237
1960	1,245	860	D	D	D	821	841	3,767	12,335
1965	1,735	706	D	D	D	895	763	4,099	13,963
1970	2,004	679	D	D	D	620	872	4,175	16,014
1975	1,701	836	227	106	170	D	591	3,631	16,209
1980	2,154	848	190	126	172	D	678	4,168	18,246
1985	1,911	754	221	163	103	D	453	3,605	20,445
1990	1,981	799	99	203	160	D	296	3,538	21,490
				Soybean oil					
1955	D	—	115	71	—	12[c]	39	237	2,539
1960	6	—	96	64	—	5[c]	30	201	3,190
1965	—	—	100	104	—	6	53	263	4,481
1970	3	—	82	65	—	6	49	205	5,985
1975	23	—	94	66	—	3	24	207	7,612
1980	23	—	46	71	—	D	63	202	8,812
1985	27	—	45	100	—	D	80	252	10,256
1990	D	—	50	106	—	D	73	291	11,013

[a]*Source:* U.S. Department of Agriculture, Economic Research Service. Does not include animal feeds or foods and loss data included in non-food utilization data. Method of reporting data was changed in 1972.

[b]D denotes data not available.

[c]Includes linoleum and oilcloth.

linseed oil in decorative coatings. Unless treated, soybean oil is only a semi-drying oil, because it contains insufficient polyunsaturation to dry or cure to a nontacky state. Thus, soybean oil alone made very poor paint and was largely limited to extending linseed oil in periods of linseed shortages.

It was not until the 1930s that soybean oil became widely used in paints, coatings, and varnishes, when R. Kienle and A. Hovey of the General Electric Company (82) developed soybean oil alkyd resins for incorporating into soybean oil to improve drying, adherence, endurance, and color. Soybean oil alkyd resins were often blended with linseed and tung oil alkyd resins to improve performance at acceptable costs. Then Henry Ford used soybean oil and soybean oil derivatives in enamel paints for his automobiles. However, Du Pont's development of the "four-hour enamel" based on soybean alkyds is generally considered to be the most important event in furthering usage of soybean oil in paints.

However, after World War II, oil-based paints lost market share to less expensive and easier-to-use latexes (rubber-based) using water as the solvent and lower levels of soybean oil–derived ingredients.

Technology. Excellent descriptions of technologies used to produce paints, coatings, and resins have been prepared by Formo (82) and Stanton (83), and much of the information in this section comes from Stanton's work. Paint hides, protects, and decorates the surface with a pigment, which is bonded to the surface by a polymer (the resin) (84). Coatings are similar to paints but are used to coat surfaces of consumer goods (e.g., automobile bodies and furniture) and must dry quickly. Varnishes are similar, except that no pigment is used to hide the surface, and they produce clear films. All three products contain solvents, either water or a straight-chain or aromatic petrochemical. The solvent either dissolves or suspends the resin and pigment in the can and during application. Once applied, the solvent evaporates, leaving a thin film of pigment and resin on the surface, and the film polymerizes through oxidation of double bonds (Fig. 21.5). The oxidation or polymerization is often accelerated by heating, by using fatty acids with more polyunsaturation or more conjugated unsaturation, or by air-blowing the oil. Blown soybean oil is widely used in caulks and sealants.

Maleinized oils were developed by E. Clocker (85,86) in the 1930s to improve the drying rate of soybean oil so that it could be used in many of the same applications as linseed oil. By heating soybean oil, normal unconjugated fatty acids of soybean oil are converted to conjugated forms, which react with maleic anhydride (Fig. 21.5) in the Diels-Alder reaction. The derivatives are then esterified to a polyol such as glycerol or pentaerythritol (Fig. 21.5). Maleic adducts of higher levels can be neutralized with ammonia to produce an adduct that is water-dispersible.

By far, the most important application for soybean oil in paints and coatings is as alkyd resins. Alkyd resins are also used in core resins, adhesives, lithographic inks, and plasticizers (87). Originally, alkyd resins were polymers produced by esterifying a dibasic acid (phthalic anhydride or isophthalic acid, Fig. 21.5) with a polyol (glycerol or pentaerythritol, Fig. 21.5); but today, the term also refers to a polymer modified with fatty acids. Normally, the oil and polyol are reacted by a process known as alcoholysis, or transesterification, to transfer some of the fatty acids from the soybean triglycerides to form esters with the polyol. After alcoholysis a dibasic acid is added, and esterification takes place until the proper end point is achieved. When fatty acids are used in place of oil, alcoholysis is eliminated and the fatty acids, dibasic acid, and polyol are reacted in one step. Fatty acids of soybean oil, tung oil, linseed oil, and tall oil are all used.

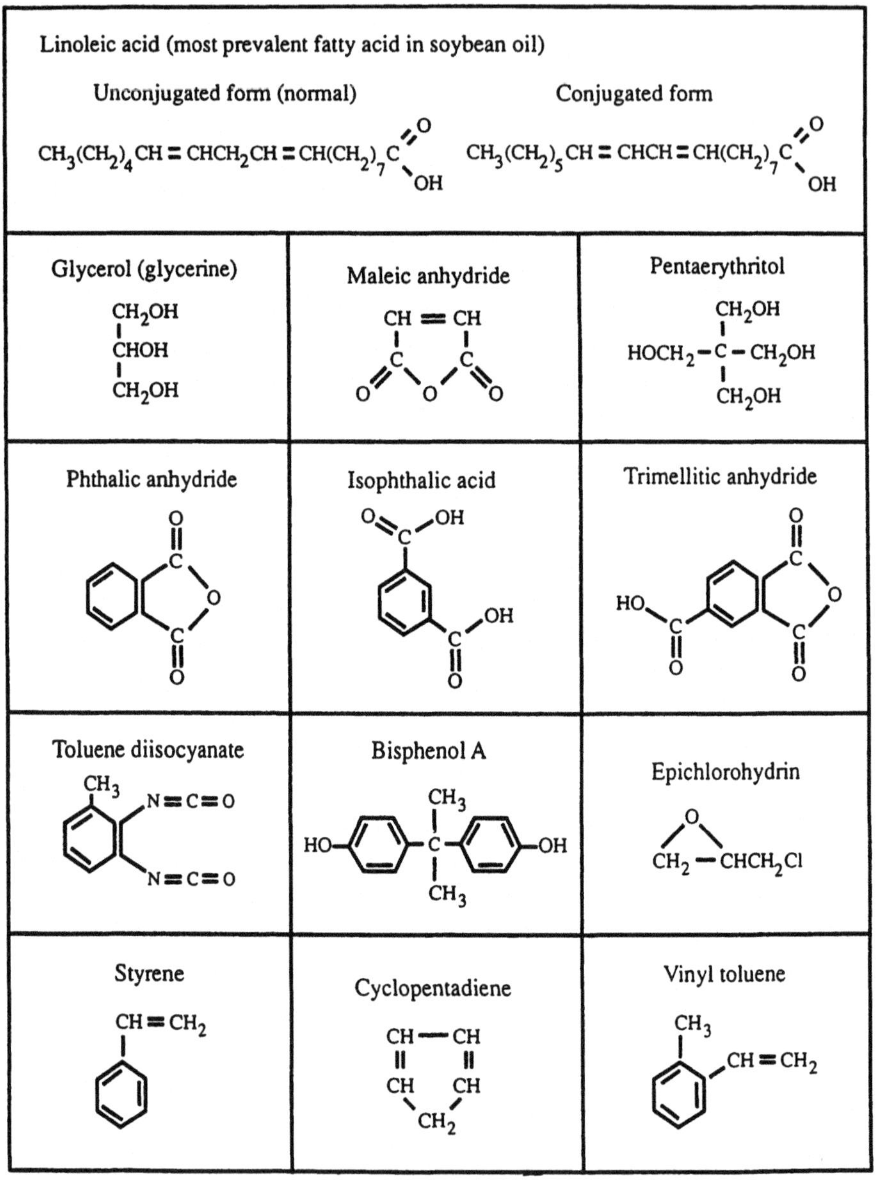

Linoleic acid (most prevalent fatty acid in soybean oil)

Unconjugated form (normal)

$CH_3(CH_2)_4CH{=}CHCH_2CH{=}CH(CH_2)_7C({=}O)OH$

Conjugated form

$CH_3(CH_2)_5CH{=}CHCH{=}CH(CH_2)_7C({=}O)OH$

Glycerol (glycerine)	Maleic anhydride	Pentaerythritol
CH_2OH–$CHOH$–CH_2OH		$HOCH_2$–C(CH_2OH)(CH_2OH)–CH_2OH
Phthalic anhydride	Isophthalic acid	Trimellitic anhydride
Toluene diisocyanate	Bisphenol A	Epichlorohydrin
		CH_2–$CHCH_2Cl$ (epoxide)
Styrene	Cyclopentadiene	Vinyl toluene
$CH{=}CH_2$		CH_3, $CH{=}CH_2$

Fig. 21.5. Chemicals used in converting soybean oil into alkyd resins. Prepared by D.J. Burden, Center for Crops Utilization Research, Iowa State University.

Typical alkyd resins are composed of 60% fatty acids, 20% pentaerythritol, and 20% phthalic anhydride. However, alkyds are classified as long oil, medium oil, or short oil, depending upon the ratio of oil to dibasic acid used.

Long oil alkyds contain 55 to 70% oil and are used in architectural glossy and semigloss enamels in both interior and exterior applications, latex paints, stipple paint, spar varnishes, and fire-retardant paints (87). They are noted for rapid drying, good gloss, good weathering, and low yellowing. Medium oil alkyds contain 45 to 55% oil and are more viscous than long oil alkyds. Medium and long oil alkyds are soluble in aliphatic solvents such as mineral spirits. Paints made from them can be brushed or sprayed and are used in architectural enamels, enamel undercoats, self-sealing flat paints, and wood primers (89). Medium oil alkyds have poorer application properties than those made with long oil alkyds.

Short oil alkyds have the least oil and the highest amount of phthalic anhydride. They are not soluble in aliphatic solvents and require aromatic solvents. These paints do not dry well unless forced-air dried or baked. Short oils are often compounded with urea formaldehyde or melamine formaldehyde to make them thermosetting. These resins make hard, tough, durable coatings and are widely used as finishes for consumer goods, especially as automotive topcoats, dipping enamels, baking enamels, lacquer primers, and mar-resistant finishes (87).

Alkyds can be made water-dispersible by increasing the number of carboxyl groups so that the resin can be neutralized with amines and dispersed in the aqueous phase. Usually, part of the dibasic acid is replaced with trifunctional trimellitic anhydride (Fig. 21.5). Once the desired amount of esterification takes place, the remaining carboxyl groups are neutralized, and the resin is dissolved in a "co-solvent," usually glycol ether. The pigments are dispersed in the resin before combining with water. Soybean oil is the least expensive unsaturated oil, and its high linoleic acid content imparts good resistance to yellowing.

Urethane oils or polyurethanes are similar to alkyds except that the dibasic acid is replaced with a difunctional isocyanate such as toluene diisocyanate (TDI, Fig. 21.5). The process used to make urethanes is also similar to alkyds in that after alcoholysis, TDI is added, and the urethane reaction takes place. Urethane oils, first marketed in 1957, dry faster and harder than alkyds yet retain flexibility. They have better resistance to water, chemicals, and abrasion; however, they will yellow. Urethanes are highly useful in clear finishes for wood floors and cabinets. They are used in about 80% of the clear finishes used by homeowners and painters.

Epoxy resin–based coatings are resins derived from the reaction of bisphenol A and epichlorohydrin (Fig. 21.5). The epoxy groups of the resin are esterified with unsaturated fatty acids, predominantly from soybean and tall oils. These resins make coatings that are extremely chemical-resistant and adhere to a wide variety of materials. They are widely used in can coatings and in primers and underbody coatings for cars and trucks.

Soybean oil will react with various reactive monomers such as cyclopentadiene, styrene, and vinyl toluene (Fig. 21.5). Cyclopentadiene reacts readily with unsaturated fatty acids to produce products that dry hard and fast and are soluble in aliphatic solvents. These copolymers are widely used in aluminum paints due to their good "leafing" properties. Copolymers of oils and vinyl toluene are widely used in fast-drying aerosols and as sanding sealers.

Performance Properties. Durability is important to all paints, but many modern paints last longer than the customer likes the color. Therefore, in many paints, especially interior household paint, durability is less important than color, finish (gloss and texture), quick drying, and easy cleanup. Durability is more important to surfaces in direct contact with the environment, such as exterior household paint. Exterior paint deteriorates because it is subjected to sunlight, moisture, and thermal stress. Most latex paints contain vegetable oil alkyds to improve adhesion. About 25 to 50% of the binder in exterior latex or acrylic house paint is alkyd resin made from vegetable oils and petroleum products.

Market. Although paints, coatings, and varnishes have been the largest industrial market for soybean oil, it is relatively mature and growing only about as fast as the Gross National Product. A few large companies dominate both ingredient manufacturing and mixing/merchandising. Vegetable oils have a declining share of oil-based paints, and oil-based paints are a declining share of paints, coatings, and varnishes. Usage of vegetable oils in paints, coatings, and varnishes peaked around 1950. In 1940 over 80% of paints, coatings, and varnishes were composed of vegetable oils or resins made from vegetable oils. By 1987 that market share had dropped to about 30%, with the remainder coming from petroleum-derived products.

Of the 56 million kg (124 million lb) of vegetable oils used in paints, coatings, and varnishes during 1992, approximately 35% was supplied by soybeans. In recent years, 18 to 23 million kg (40 to 50 million lb) of soybean oil have been used by this industry. Soybean oil is the preferred vegetable oil in alkyd resins and some urethane and epoxy resins. Of the approximate 1 billion kg (2.3 billion lb) resin market today, about 270 million kg (600 million lb) are for alkyd resins (the largest category), 86 million kg (190 million lb) for epoxy resin, and 64 million kg (140 million lb) for urethane resin.

Growth in the latex paint market has slowed in recent years; however, the market for water-based paints is likely to grow because of environmental issues. Therefore, water-dispersible alkyds are playing an increasingly important role. While water-dispersible alkyds have generally been limited to finishes for consumer goods, their use in exterior and interior paints and varnishes will likely grow.

Although the cost of the paint is always of some importance, even more important is performance and total cost per unit surface area including application. In painting houses or other items, the cost of the paint is small relative to the cost of preparing the surface and applying the paint. Only in retail marketing of paints applied by homeowners does cost per gallon seem important. Better durability and lower application costs are more important to manufactured goods and construction trades.

Current Interest. Paints, varnishes, and coatings are being reformulated today to reduce volatile organic compounds (VOCs), and the EPA is becoming increasingly concerned about paint sources of VOCs. EPA concern over VOC emissions is driving the demand for water-based systems. Latex paints now have 50% of the exterior house paint market. Water-dispersible alkyds are becoming more important. The next generation of paints will likely use nonvolatile and perhaps reactive solvents, such as oils high in epoxy fatty acids.

Advances in biotechnology are encouraging molecular biologists and soybean breeders to contemplate redesigning soybeans to produce more useful oils that could expand soybean oil's share of the coatings market. Redesigning soybeans to produce conjugated fatty acids, as does tung, and epoxy fatty acids, as does vernonia, could improve performance of soybean oil in paints, varnishes, and coatings.

Plasticizers

History and Background. Plasticizers are incorporated into plastics or elastomers (rubber and rubberlike materials) and increase flexibility, workability, distensibility, and toughness. An excellent discussion of plasticizers, from which much of the information provided here was obtained, has been published by Formo (88). Plastics such as polyvinyl chloride (PVC) are hard, hornlike materials with limited usefulness unless properly plasticized. Soybean oil products, specifically epoxidized soybean oil, are used extensively to plasticize PVC resins. "Primary plasticizers" are those that can be used over a broad range. Phthalate esters, such as dioctyl phthalate (DOP), are the most important. Other plasticizers, such as epoxidized soybean oil, cannot be used alone over a wide range of plastic compositions, because on aging, the plastic becomes opaque, sticky, and inflexible. These plasticizers are known as "secondary plasticizers" and are often used to improve stability to heat and light (preventing the plastic from becoming brittle) and reduce the cost of the plasticizer.

The use of epoxidized soybean oil as a plasticizer and stabilizer for plastics is based upon the pioneering work of the noted lipid chemist D. Swern and his coworkers during the 1950s. Swern, a USDA scientist, showed that the peroxy acid oxidation reaction could be stopped at the epoxy stage before forming dihydroxy compounds.

Technology and Performance. To have plasticizing properties, the plasticizer must act as a solvent for the polymeric material and be a high-boiling polar liquid having a proper balance of polar functional groups versus nonpolar hydrocarbon structure and proper balance of size of groups. For instance, the compatibility of simple esters and PVC depends on molecular symmetry; that is, the ester functionality should be near the center of the molecule.

Epoxy plasticizers are produced from oils or unsaturated fatty acid esters of monohydric alcohols by oxidizing (Fig. 21.6) with performic acid, peracetic acid or with hydrogen peroxide in the presence of a strong ion-exchange resin (87). The epoxidized fatty acids have oxirane moieties that give the molecule polar character. Most plastic formulations contain about 3% of epoxidized oil (87).

Epoxidized oils also have a stabilizing effect on PVC, extending useful life. As PVC ages, hydrogen chloride is released, accelerating deterioration. Epoxidized fatty acids minimize autocatalytic deterioration by scavenging the acid (Fig. 21.6).

Market. The plasticizer market is dominated by di-2-ethylhexyl (dioctyl) phthalate (DOP). About 10 to 15% of the plasticizer market is derived from vegetable oils. The annual plasticizer market is about 1 billion kg (2 billion lb) and growing. Of the epoxidized vegetable oils, soybean oil is used in about 75% of the production.

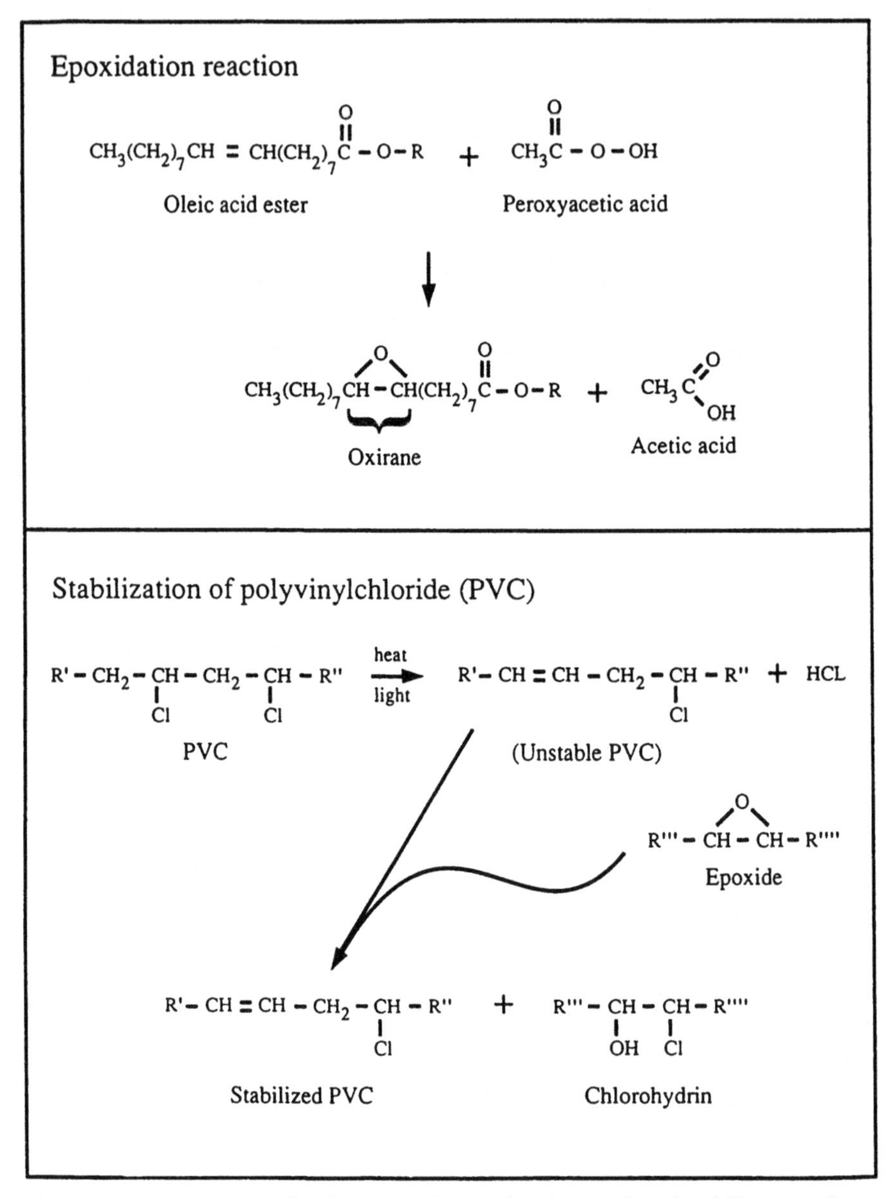

Fig. 21.6. Reactions involved in epoxidation of soybean oil and stabilization of PVC. Prepared by D.J. Burden, Center for Crops Utilization Research, Iowa State University, Ames, IA.

Epoxidized linseed oil is preferred in vinyl liners of bottle caps and medical tubing, where extraction of the plasticizer must be avoided. Sunflower oil, being higher in oleic acid content, produces epoxy plasticizers that are more compatible with vinyl resin. Epoxides of tall oil are also used in some plastic products.

Drying Oil Products

History and Background. Numerous products used the moderate drying properties of soybean oil, especially up until the mid-1950s. Those of historical importance include linoleum, oil cloth, sealing and caulking compounds, rubberlike materials, and core oils. Formo also provides a good treatise on these applications (88). Many of these uses no longer consume significant amounts of soybean oil, and economic and marketing data are difficult to find.

From around the turn of the century until the early 1950s, linoleum was the dominant flexible floor covering. Linoleum was being manufactured as early as 1864 in England and ten years later in the United States. However, floor coverings based on vinyl and phenolic resins or asphalt have now captured this market. Linoleum typically consisted of one-third binder (oxidized soybean, linseed and tung oils, natural gums, and rosin), one-third inorganic fillers (pigments and ground limestone), and one-third organic fillers (ground cork and wood flour). The binder, known as cement, must be durable and resilient, yet thermoplastic. A highly oxidized and polymerizing drying oil worked well when compounded with rosin and other natural resins.

Oils were used in two ways in oilcloth: one in which the entire cloth was impregnated with oil and another in which only one side of the cloth was coated. Impregnated cloth was used in raincoats and coverings for machines and instruments. Coated oilcloth was used as coverings for walls, tables, shelves, and other surfaces. The polymerized oil gave water resistance with flexibility.

Sealing and caulking compounds have used significant quantities of drying and semidrying oils. Blown soybean oil has been used in mastic products or glazing compounds, which have largely replaced putty. Glazing compounds develop heavy surface skins with excellent painting properties, but unlike putty, they remain plastic in the interior for long periods. Other sealing materials have a vegetable oil base to produce rubberlike properties.

Unsaturated fatty acids are capable of polymerizing to form various elastic, rubberlike materials. *Factices* are oils that have been polymerized using sulfur in the same manner as vulcanizing rubber, in which crosslinking is possible through sulfur or disulfide bonds. Factices are somewhat rubberlike, but none have the combination of elasticity and tensile strength of vulcanized natural rubber, because instead of long linear chains with occasional crosslinking, factice has an extensively crosslinked structure. Factice is used in materials for gaskets, stoppers, bumpers, tubing, electrical insulation, and rubberized fabrics.

Soybean oil has been used in core oils for binding sand cores in manufacturing metal castings. Cores are formed by mixing about 2% oil in sand, molding in a wooden form, and baking until the core becomes hard through polymerization of the oil. The core must be sufficiently hard to retain its form during metal casting, but not so hard as to be difficult to break and remove the cooled metal casting. The core must be permeable to permit escape of gases during casting.

Technology and Performance. Preferred drying oils have high levels of polyunsaturated fatty acids such as linolenic acid (18:3) and conjugated unsaturation (without methylene interruption). Soybean oil has a moderate level of polyunsaturation (less than linseed oil) and is naturally unconjugated (unlike tung oil). The drying properties of soybean oil can be improved by maleinizing (see the previous section on paints, coatings, and varnishes), heating, or blowing air through the oil to oxidize it partially and convert unsaturation to the conjugated form. Additionally, the drying properties of soybean oil can be improved by adding metallic driers such as lead, cobalt, and manganese, which accelerate drying (89).

Markets. Markets for these products have been lost to petrochemicals because of their better performance properties. Consumption of soybean oil in all other drying oil products (not including paints and varnishes) reached a maximum in about 1954, when as much as 5.4 million kg (12 million lb) were consumed. In the last year the USDA collected data on this market, 1978, as little as 1.8 million kg (4 million lb) were consumed. Usage of soybean oil in linoleum and oilcloth peaked in 1948, when as much as 14 million kg (30 million lb) of soybean oil were used, equivalent to 20% of the total fats and oils in linoleum and oilcloth; by 1957, less than 450,000 kg (1 million lb) were used.

Lubricants and Fluids

Vegetable oils were once widely used for lubrication fluids; but today, petroleum-derived mineral oils have largely supplanted vegetable oils. The principal advantage of vegetable oils is their superiority in clinging to metal surfaces in very thin films, while they also have problems in that under extreme conditions they can hydrolyze and become acid and corrosive (88). Today, most of the fats and oils used for lubricating oils and greases are animal fats or castor oil (due to its high ricinoleic acid content). Sulfurized fatty products should have good stability to oxidation and heat (90). Less than 1.8 million kg (4 million lb) of soybean oil are used as lubricants and lubricant additives, about 10% of the total fats and oils lubricant market (90). Small quantities of sulfurized soybean oil are used in extreme-pressure and antiwear additives, where they form metal sulfides or chlorides that permit sliding contact and sloughing off of the metal salt rather than the scarring associated with metal-to-metal contact (91).

Soybean oil is being formulated into bar chain oils for chainsaws used in cutting timber and firewood. Soybean oil offers environmental advantages, being completely biodegradable.

Irrigation wells need lubricants dripped down the well shaft, and these lubricants ultimately discharge into the aquifer. Soybean oil is a good, harmless, and biodegradable lubricant in irrigation wells and performs as effectively as petroleum-derived lubricants.

There is interest in using soybean oil as hydraulic fluids to replace petroleum derivatives. In addition to lower cost, soybean oil hydraulic fluids are environmentally benign, should hydraulic lines break or leak and a spill result, and are easily and safely disposable. Lubricating capacity is adequate, and wear of hydraulic pump parts is reportedly the same as for standard products, but oxidative and polymerization stability can be a problem. Soybean oil is also being tested as a quench oil for heat-treating hot metal and for way-oil for lathe lubrication and cooling oil. The market for lubricating liquids is estimated to be 9.5 GL (2.5 billion gal), with 55% of this for automobiles and the remainder for industry uses.

Soybean oil, soybean refining byproducts, and emulsions of soybean methyl esters are being used as slip or release agents for concrete and asphalt and compete against petroleum-based release agents. Spraying of truck beds carrying highway asphalt and of forms for concrete structures greatly improves separation, making handling easier than would otherwise be. Soybean oil products are being promoted for their environmental advantages and low toxicity and skin irritation to workers.

Oleochemicals

History and Background. Soybean oil can be converted into many different organic chemicals. The variety of possible reactions are depicted in Fig. 21.7. Soybean oil triglycerides may be split (hydrolyzed) using the Colgate-Emery process into glycerol and fatty acids, or soybean oil soapstock (often referred to as foots) may be acidified to produce fatty acids that can be recovered. Crude soybean fatty acids are used to make adhesive tape, shaving compounds, textile water repellents, carbon paper, and typewriter ribbons. However, more often the fatty acids are separated into various fractions by distilling, and they are used in candles, crayons, cosmetics, polishes and buffing compounds, and mold lubricants. Purified fatty acids and fractions thereof are converted into a wide variety of oleochemicals, such as dimer and trimer acids, diacids, alcohols, amines, amides, and esters (Fig. 21.8).

Dimer Acids. Soybean fatty acids can be thermally or catalytically polymerized to yield both dimer and trimer polybasic acids (Fig. 21.8). Dimer acids may be hydrogenated to improve color and oxidative stability. The main usage of these materials has been in polyamide resins, paints, plastics, and coatings, especially as bodying, curing, and flexibilizing agents; however, they are also used in manufacturing corrosion inhibitors, lubricant additives, fuel additives, and antiwear agents (92). The annual dimer acid production is about 20 million kg (45 million lb).

Polyamide Resins. Dimer acids can be converted into both reactive and nonreactive polyamide resins (87). In preparing nonreactive polyamides, dimer acids are condensed with ethylenediamine to form low-molecular-weight polyamides (Fig. 1.8). Dimer acids typically contain 10 to 30% trimers which lead to three-dimensional polymers that cause stickiness, narrow melting ranges, cold flow resistance, and rheological properties useful in adhesives. Nonreactive polyamides are used without solvents as hot-melt adhesives in packaging, can seam solders, book bindings, and shoe soles. Nonreactive polyamides are used in flexographic printing inks for food packaging films, where flexibility, adhesiveness, and resistance to cold temperatures and to cracking are important. Nonreactive polyamides are also used in thixotropic alkyd paints which remain homogeneous during storage, are thick to make application less messy, have low drag to make application easier, and are not absorbed into porous surfaces. The annual market for nonreactive polyamides is about 11 million kg (25 million lb).

Reactive polyamide resins are produced by reacting dimer acids with slightly excessive amounts of polyamines, such as diethyl triamine and triethyl tetramine, to produce a low-molecular-weight polyamide resin having a free amino group in the chain (Fig. 21.8). Reactive polyamides are used as curing agents in epoxy resins used in surface coatings, adhesives, potting and casting compounds, and patching and sealing compounds (93). Epoxy adhesives are typically two-component systems including an epoxy resin and a polyamide curing agent. Epoxy coatings inhibit corrosion, are highly flexible, and are highly resistant to impact, abrasion, water, acids, and bases. The two-component epoxy systems are used in toppings for concrete floors, bridge decks, and airport landing strips. The market for reactive polyamides is about 5.4 million kg (12 million lb).

Soaps, Detergents, and Surfactants. Although in earlier years soybean oil was used in toilet bars and other solid soaps, the amount used and its desirability were limited by the high unsaturation and high price of soybean oil compared with alternative animal-derived fats. Today, little soybean oil or soybean oil soapstock is used in soaps. Soybean oil and fatty acids could be hydrogenated to make them more saturated and increase solids, but the cost of doing so would prevent competition with the naturally more saturated and less costly animal fats. Most toilet soaps are about 20% coconut oil–derived and 80% inedible tallow–derived. Today, most soybean soapstock is used in animal feeds and pet food. Small amounts are used as dust suppressants.

Fatty acids recovered from soybean soapstock are used in medium-grade liquid soaps for laundry and dry cleaning, building maintenance and industrial cleaners, car wash liquids, cutting oils, and food and dairy cleaners (94). About 20% of this 113 million kg (250 million lb) market is served by soybean fatty acids.

Soybean oil and soybean fatty acids are used to manufacture surface-active compounds, including sodium and potassium salts, diethanolamine derivatives, and polyethylene glycol esters. Also, a number of soy fatty amines are used as surfactants, including soyamine, disoyamine, N,N-dimethylsoyamine, N-soya-1,3-diaminopropane, trimethylsoyammonium chloride, dimethylsoyammonium chloride, bis(2-hydroxyethyl)soyamine, and polyoxythylenesoyamine. In addition to surfactant properties, some have antimicrobial, pesticidal, and anticorrosion activities.

Diesel Fuel

History and Background. Vegetable oils were early fuels tested in the compression ignition engine invented by Rudolf Diesel in the late 1800s (96). While on display at the 1900 Paris Exposition, Diesel's engine was fueled with peanut oil (97). From

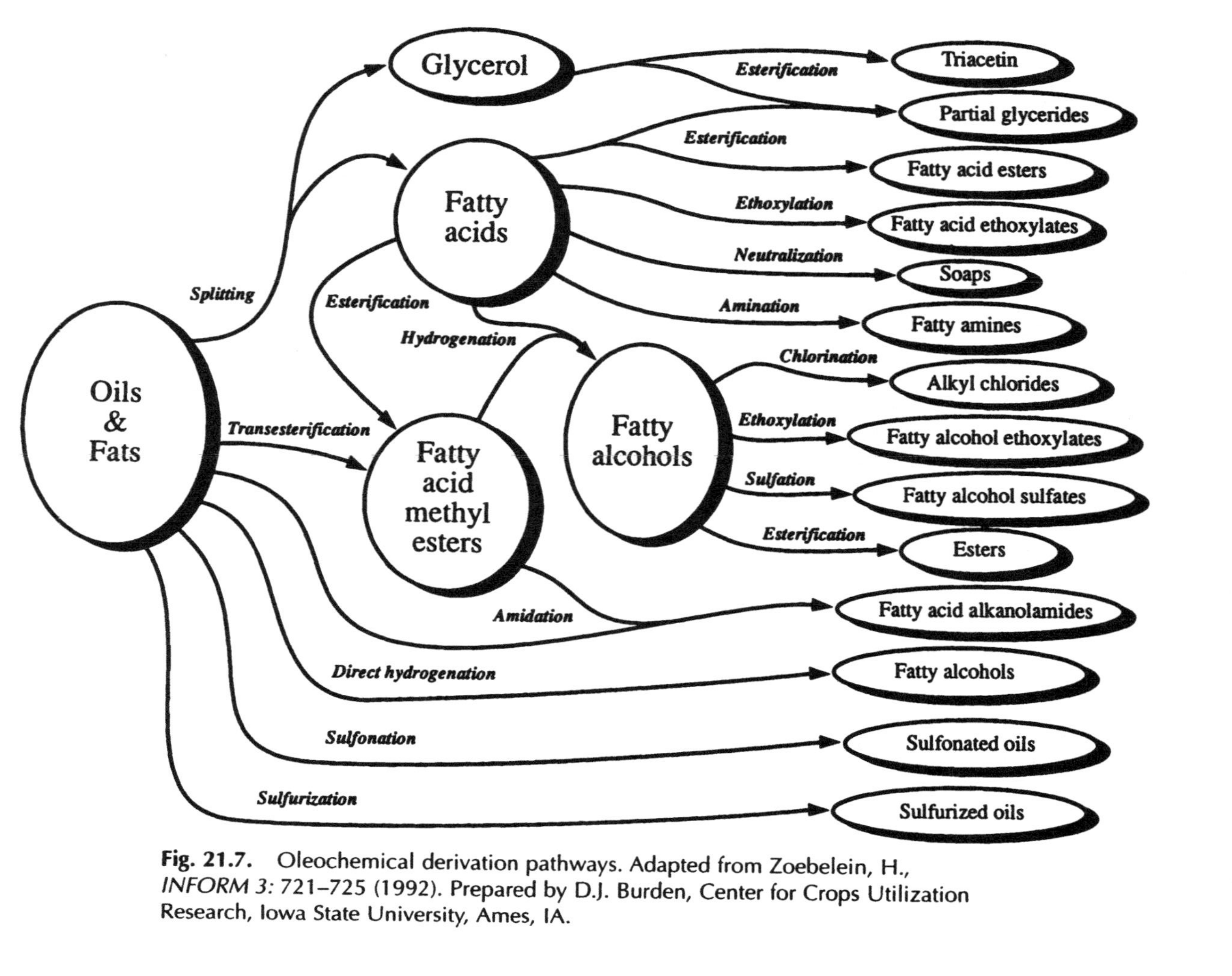

Fig. 21.7. Oleochemical derivation pathways. Adapted from Zoebelein, H., *INFORM 3*: 721–725 (1992). Prepared by D.J. Burden, Center for Crops Utilization Research, Iowa State University, Ames, IA.

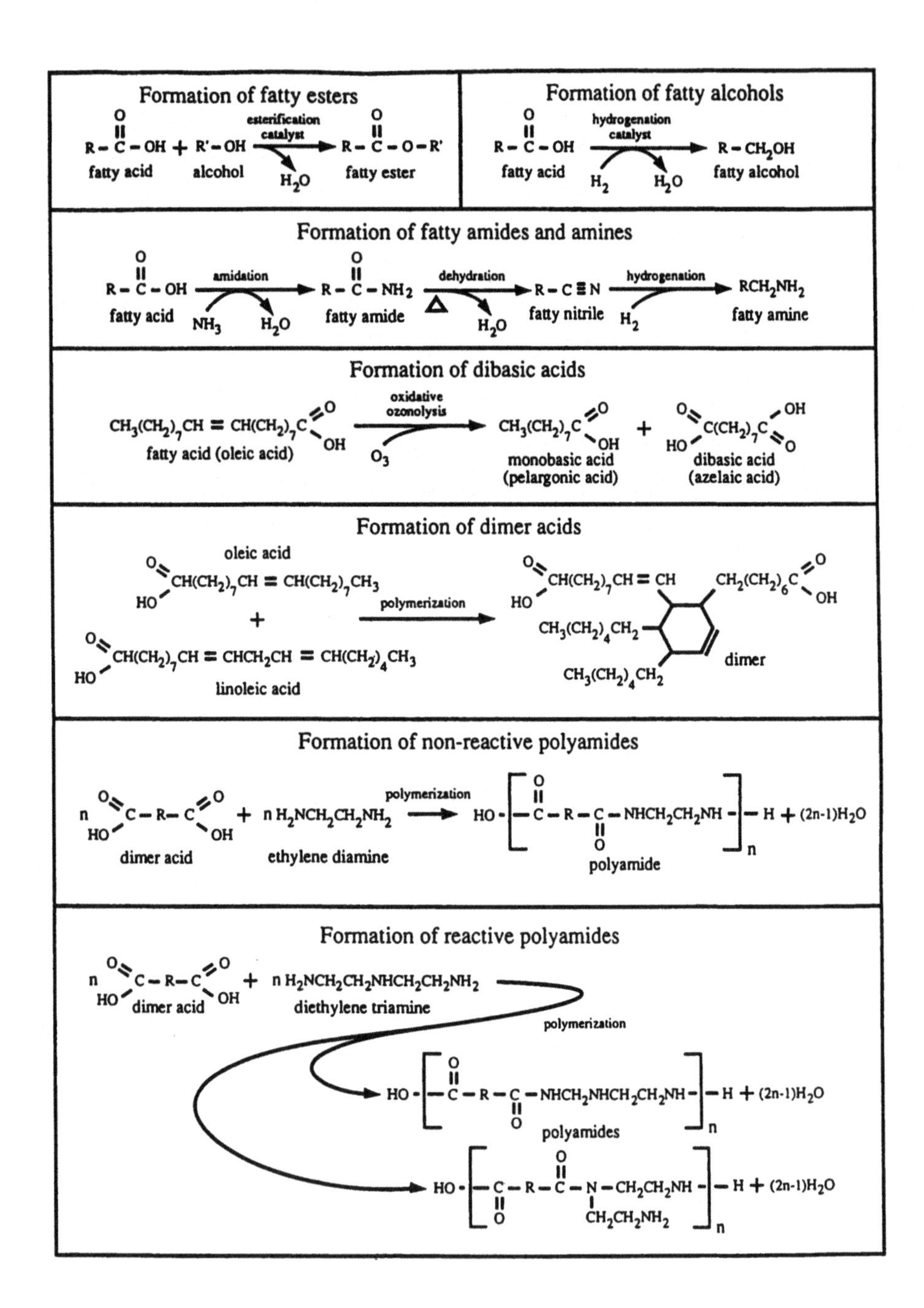

Fig. 21.8. Structures and reactions of common oleochemicals. Prepared by D.J. Burden, Center for Crops Utilization Research, Iowa State University, Ames, IA.

that time until the early 1950s, varying degrees of success were achieved when substituting vegetable oils either completely or partially for diesel fuel (98). The interest in vegetable oil fuels never gained widespread serious attention because of abundant supplies, stable markets, and low prices for petroleum, and so diesel engines were optimally designed to operate on petroleum distillates. However, in the early 1970s the Organization of Petroleum Exporting Countries (OPEC) began to limit supplies (popularly known as the OPEC Oil Embargo), and prices for petroleum rapidly escalated. Interest reemerged in vegetable oils as diesel fuel extenders because of fear of high petroleum prices and unstable Middle East politics. However, high petroleum prices did not persist, and as prices declined in the 1980s, so did interest in using vegetable oils as fuel. In recent years, however, interest has revived, because methyl esters of vegetable oils reduce atmospheric pollution and have other environmental benefits.

Performance Properties. The varying success of vegetable oils as diesel fuel experienced prior to 1970 has been attributed to the lack of recorded data regarding the chemical and physical characteristics of the oils tested to guide others; in most cases, only the type (soybean, peanut, or the like) was reported, not the extent of processing and quality characteristics. In addition, there was great variation in the design and construction of the test engines; instrumentation was less accurate; and operating conditions, such as ambient temperature and barometric pressure, which are now known to affect performance, were not controlled or reported (95). Most frequently, the thermal efficiency and power output relative to diesel fuel were reported, because they could be determined in short-term engine tests. Some reported that engine performance with vegetable oils was equivalent to or better than with diesel fuel; others reported just the opposite. In several cases, severe problems with carbon deposits, corrosion, and wear were reported. No general trends regarding the suitability of vegetable oils as alternative fuels emerged from these studies.

Work conducted during the 1970s was much more systematic, and efforts focused on developing techniques to reduce performance problems by either processing vegetable oils or altering the engine design or operating conditions. The common vegetable oils, including soybean oil (98,99), have been subjected to standard ASTM tests for diesel fuels along with extensive chemical characterizations. The major problems in using unmodified vegetable oils included (100) clogging of fuel lines and filters with fines, phosphatide gums, waxes, or high-melting fats; polymerization or partial oxidation of oil during storage, which increased viscosity or caused clogging problems; thermal or free radical polymerization in the combustion chamber, causing varnish formation and other deposits on cylinder walls and pistons; poor ignition and combustion characteristics caused by improper atomization; increased coke formation; and polymerization of fuel blowby, which caused thickening and/or increased solids contamination of the lubricating oil. In general, these problems were caused by the higher viscosities of vegetable oils, low volatility, and incomplete combustion; by the presence of highly reactive unsaturated fatty acids that led to polymerization; and by the presence of minor components such as phosphatides. Most studies found that vegetable oils must at least be degummed before blending with diesel fuel. Also, indirect combustion engines, which have precombustion chambers for improved atomization, worked better with vegetable oils than direct injection engines. The problems caused by high viscosity could be lessened by heating the oil, transesterifying with low-molecular-weight alcohols, diluting with diesel fuel, pyrolyzing, or microemulsifying.

It is possible to recycle heat from fuel combustion to preheat the fuel prior to injection. However, most do not consider it practical to heat the oil, especially during startup.

Converting the oil (triglycerides) to methyl or ethyl esters by the process of transesterification (also called alcoholysis) has proved to be one successful method of improving combustion characteristics and is widely used. Methyl esters are preferred, because about half as much methanol is required as ethanol to form the esters and, methanol costs about one-third the price of ethanol. When methyl esters are prepared from soybean oil, they are referred to as "methyl soyate." Methyl esters have significantly lower viscosity, because they have about one-third the molecular weight of the original oil. The viscosity of methyl esters is about one-tenth that of triglycerides and nearly equivalent to diesel fuel (at 40°C diesel fuel is 3 mPa versus 30 mPa for soybean oil versus 4 mPa methyl soyate). Triglyceride molecules have molecular weights about four times as large as those of diesel fuel molecules. The heating value of vegetable oil methyl esters is about 90% of that of unblended diesel fuel; thus, the specific fuel consumption is slightly less for methyl esters than for pure diesel fuel. The properties of SoyDiesel (methyl soyate) are compared with No. 2 diesel fuel in Table 21.3 (101).

Diluting the oil or esters with diesel fuel can also reduce viscosity, and today's efforts focus on diluting methyl esters by using blends of 20 to 30 vol% methyl esters in diesel fuel for specific niche markets. The industry has been blending 20 vol% of methyl esters and 80 vol% petrodiesel, having a total oxygen content of 2.2%. This blend is becoming known as BD-20 (BD denotes biodiesel and 20 refers to 20 vol% methyl esters).

Vegetable oils, pyrolyzed by heating in the absence of oxygen, can be used as a diesel fuel replacer (102,103). Pyrolyzed soybean oil has a cetane number of 43, exceeding that of soybean oil and the minimum ASTM value of 40 (104). The viscosity of the distillate is higher than that of diesel fuel but much below that of soybean oil.

Microemulsions are optically isotopic, thermodynamically stable liquid dispersions of two normally immiscible components along with surface-active compounds as dispersing agents (104). A microemulsion of soybean oil:methanol:2-octanol:cetane improver (52.7:13.3:33.3:1.0 by volume) has given reasonably good combustion characteristics (105).

Today, the term "Biodiesel" refers to an ester-based fuel derived from biological sources for use in compression-ignition engines (101) and "SoyDiesel" refers to methyl or ethyl soyate. The neat form (100% esters) is biodegradable, nontoxic, and essentially free of sulfur and aromatics. Although there are conflicting reports, most engine tests have shown methyl soyate to perform nearly equivalent to diesel fuel without any of the engine fouling problems encountered with direct use of soybean oil. The presence of oxygen (as esters) in the fuel gives more complete combustion, which reduces carbon monoxide, unburned hydrocarbons, particulate matter, oxides of nitrogen (NO_x), and smoke opacity (Table 21.4). The most favorable results have been achieved with a combination of BD-20, an injector timing change, and a catalytic converter. Rubber fuel line components are not affected as they are with other alternative fuels (106).

Minor technical problems remain, including the gradual accumulation of ester in the crankcase, which alters the viscosity of the lubricating oil, making more frequent changing of lubricating oil necessary (107), and the high cloud point of methyl soyate (3°C versus −12°C for No. 2 diesel), which can result in fouled fuel lines and filters in winter use. Crystal modifiers and esters of branched-chain alcohols are being examined to lower the crystallization temperature. Usually, crystal modifiers lower pour point but not cloud point, whereas branched-chain esters are more effective and reduce both points but are more expensive.

Technology. The oil should be alkali-refined to remove free fatty acids, which can poison the catalyst, and phosphatides, which can form gums by adsorbing moisture. The equipment needed to convert refined soybean oil into fuel-grade methyl esters is relatively simple (95,108). Typically, 0.1 to 0.5% (based on oil) of sodium methoxide or potassium or sodium hydroxide is dissolved in 99% methanol, and the alcohol mixture is added to the heated oil (80°C). For essentially complete reaction,

TABLE 21.3 Comparative Properties of Diesel and Soy Diesel Fuels (101)

ASTM fuel property	Test method	Soy diesel (methyl soyate)	Diesel (No. 2)	ASTM D 975
Total acid number (mg KOH/g)	D 644	0.45	0	
Water and sediments max. (vol%)	D 2709	<0.005	0.01	0.05
Color	D 1500	0.5	3.5	
Ash, max. (wt%)	D 482	0.001	0.01	0.01
Relative density (@ 15°C)	D 1296	0.883	0.840	
Distillation temp. (°C)	D 86			
Initial point		313	100.9	
50%		335	270.2	
90%		345	351.1	
95%		354	370.2	
End point		357	380.5	
Recovery		96	96.5	
Residue		1	1.5	
Carbon residue max. (wt %)	D 189	0.19	0.08	0.35
Cetane number	D 613-86	51.8	51.0	40 (min)
Phase changes (°C)				
Flash point	D 93	118	64	
Cloud point	D 2500	−3	−21	50 (min)
Viscosity at 37.8°C (min.)	D 445	4.70	3.25	1.9
Sulfur (wt %)	D 1552	<0.01	<0.21	0.50

approximately twice as much alcohol must be added as is stoichiometrically required. The mixture is stirred and allowed to stand. (Compare the interesterification process, described in Chapter 16.) One mole of triglyceride reacts in the presence of the alkaline (or acid) catalyst with three moles of methanol to give three moles of methyl ester and one mole of glycerol; the glycerol begins to separate almost immediately in a lower phase. The ester layer is vacuum-distilled to remove the unreacted alcohol for recycling and washed with water to remove the catalyst and traces of glycerol and soap. The esters are then dried and blended with diesel fuel. Separation of the catalyst from the esters is critical; otherwise the fuel will be slightly alkaline or acidic, depending on the catalyst and may cause deposit formation or corrosion problems in the engine. In addition, during storage the fuel may absorb moisture from the air, which reacts to form soap. Glycerol has been a valuable coproduct, and sales of glycerol often defray the cost of transesterification. As Biodiesel becomes more widely produced, increased production of glycerol will likely drive glycerol prices down unless significant high-value uses are developed.

Market. Soybean oil–based fuels make sense from an energy production viewpoint. When all nonsolar energy required to grow and process soybean oil into Biodiesel is compared with the energy content of the oil, the return is at least 2.0 units of energy for each one used (101). Some have even estimated the return to be far higher—more than 4.5 units of energy produced for each one used (109). It has been estimated that if all of the U.S. soybeans were crushed for fuel, only 5 to 6% of the total U.S. demand for diesel fuel could be met, so the potential market is huge. Unfortunately, the price of methyl soyate is considerably higher than current prices for diesel fuel. In the United States, there are few suppliers of methyl soyate, and it currently commands prices of about $0.55/L ($2.00/gal), so at the present time, soybean oil as a fuel cannot compete with depressed petroleum prices without subsidy or being mandated by legislation. Many have proposed that Biodiesel should have incentives similar to those of alcohol fuels, such as tax credits (about $0.77/gal of Biodiesel). Reportedly, more than 200,000 MT/yr of Biodiesel is being produced in

Table 21.4 Emissions for Diesel Engines Operating on Different Fuels and Engine Configurations[a]

Emissions reduction method: Engine configuration	Particulates, g/bhp-hr	Carbon monoxide, g/bhp-hr	Hydrocarbons, g/bhp-hr	Oxides of nitrogen, g/bhp-hr
Baseline: No. 2 diesel, no engine modifications	0.261	1.67	0.45	4.46
BD-20 with 3 deg. retardation in injector pump timing	0.216	1.50	0.38	4.25
BD-20 with timing change and catalytic converter	0.191	0.45	0.12	4.32

[a]*Source:* Fosseen Mfg. and Development, Ltd., "Evaluation of Methyl Soyate/Diesel Blend in a DDC 6V-92TA Engine," ORTECH Final Report No. 93-E14-21/93-E14-36, ORTECH International, Mississauga, Ontario, Canada, 1993.

Europe (110). The European market is made possible by subsidies in the form of eliminating all or part of the various fuel taxes.

Current Interest. Current interest in Biodiesel is high, especially for fueling urban buses in cities with air pollution problems. Interest in California is especially keen, due to the environmental benefits. A large number of city transit companies have long-term tests and demonstrations underway. Biodiesel blends have powered more than 1500 vehicles for nearly 13 million km (8 million miles) in the United States and several hundred times that in Europe, because of favorable tax policies and much higher fuel prices (101). Most operators have noted no differences in power and fuel efficiency (106).

The National SoyDiesel Development Board (NSDB), Jefferson City, MO, was established in 1992 to foster development of SoyDiesel. Technical issues of interest to the Board include modifying the timing, injectors, injection pressure, and other engine design features to reduce emissions; modifying the fuel to improve engine performance, reduce emissions, and lower crystallization temperature; reducing the cost of producing SoyDiesel; evaluating SoyDiesel in marine applications; demonstrating reduced NOx emissions; and conducting long-term engine durability tests. Additional issues of concern are marketing fuel technology, engine warranty, underground mining applications, fuel certification, evaluating fuel biodegradation and toxicity, compatibility with fuel delivery and storage components, and developing fuel standards.

Further market commercialization depends on EPA recognition in regulations that Biodiesel is "substantially similar" to petrodiesel; full recognition of Biodiesel as a clean-burning fuel under the Clean Air Act Amendments of 1990 and as an alternative fuel under the Energy Policy Act of 1992; full support from engine manufacturers; development of niche markets in ozone nonattainment areas and in environmentally sensitive areas; and establishment of incentives comparable to those already available to fuel ethanol and bio-based methanol (101).

Printing Ink

Background. Newspaper printing primarily uses lithography or offset printing and letterpress printing. Lithographic and letterpress inks are composed of a pigment (such as carbon black for black inks or various organic compounds for colored inks), a vehicle, and other minor proprietary ingredients. The vehicle is used as the carrier for the pigment. Until recent times, the vehicle has been composed of a resin and a petroleum-derived hydrocarbon solvent such as mineral oil. The resin binds the pigment to the paper and provides appropriate viscosity and tackiness to the ink. The solvent is the carrier used to dissolve or disperse the resin and to achieve desired flow properties of the ink important for proper absorption by the paper. The solid ingredients are dispersed in the vehicle to form a paste ink. Lithography and letterpress processes require paste inks. A typical lithographic ink consists of 50 to 70% mineral oil solvent, 15 to 25% hydrocarbon and/or alkyd resins, and 15 to 20% pigments. During the 1970s, petroleum prices escalated rapidly, and the supply of printing inks became unpredictable, stimulating the search for inks based on renewable materials. About the same time, the ink industry was forced to more expensive, highly refined mineral oils because less refined counterparts contained polynuclear aromatics, shown to be carcinogenic.

By the early 1980s, the American Newspaper Publishers' Association (ANPA) developed ink formulations containing tall oil, carbon black, and gilsonite (resinous asphalt), which worked marginally well but were expensive, limited in supply, and difficult to clean up (111,112,113). However, about the same time, tall oil also started to escalate in price, and ANPA then focused on vegetable oil–based inks.

Technology. In 1985 ANPA formulated the "first-generation" soybean oil–based news ink, comprised of alkali-refined soybean oil, a hydrocarbon resin, a pigment, and an antioxidant. Originally, soybean oil was merely substituted for mineral oil. ANPA's technology has been used since 1987 for both black and color printing.

A "second generation" of soybean-oil-based inks was developed at the USDA's NCAUR (114,115,116). They developed ink vehicles totally derived from vegetable oil, thus even eliminating the need for the petroleum-based resin (114). In one process, the vegetable oil was heat-polymerized at a constant temperature to the desired viscosity, while in a second process the oil was heat- polymerized to a gel point and then the gel was mixed with vegetable oils to obtain the desired viscosity. Both methods facilitated tailoring the properties of the vehicle for a variety of news inks.

Performance. Printers are highly concerned about "rub-off," print quality, and readability (contrast between the paper and the print). Newspapers want bolder blacks and more vivid colors (117). Additionally, waste and cleanout inks made with mineral oil products are regarded as hazardous wastes and are becoming increasingly more difficult to dispose of in landfills as has been the practice. Disposal of waste ink can cost as much as $300 per barrel. Properly formulated soybean oil inks are "biodegradable" and cost far less to dispose. The NCAUR group examined the biodegradability of their ink and found that the soybean oil–derived components were almost completely degraded within 25 days; the 100% soy-based vehicle degraded 82 to 92%, compared with 58 to 68% for the ANPA vehicle and 20% for the petroleum-based vehicle.

Industry experience with ANPA inks since 1987 has shown that soybean oil–petroleum hybrid-based inks are as good or better than conventional mineral oil–based inks (118). The major advantages of soybean oil–petroleum-based inks are superior print qualities, brighter colors, lower rub-off, better mileage (spread quality), cleaner press runs, and environmental benefits (121).

The NCAUR ink (118) contains no (nonpigment) petroleum-based component; both solvent and resin have been replaced with soy oil–based materials. Replacing the resin reduced the cost of NCAUR inks so that they are competitive with petroleum-based lithographic news ink. The light color of the NCAUR vehicle allows less pigment to be used, further reducing the cost of colored inks. Their soy oil ink formulations should be cost-competitive with petroleum-based inks of comparable quality and meet or exceed the industry standards in regard to print quality, ease of cleanup, rub-off resistance, viscosity, and tack. Furthermore, they are developing vegetable oil–based lithographic sheetfed and heat-set and flexographic inks.

Market. About 180 million kg (400 million lb) of news ink is produced annually in the United States (120). Thus, as much as 144 million kg (320 million lb) of soybean oil could be used in this ink market alone. In 1992, the printing industry actually used 12.5 million kg (27.5 million lb) of soybean oil, and growth has been projected to go to 40 million kg (89 million lb) by 1997. Soybean oil–petroleum hybrid-based inks have been much more successful in the colored ink market than the black ink market. These inks have already captured about 70% of the color newsprint ink market because of their superior print qualities. This market penetration occurred despite high costs for these inks. Colored ink pigments are more expensive than carbon black, so the added cost of soy oil blends are relatively small for significantly improved print quality. Soon, conventional color inks will be sold in such small volumes that they will undoubtedly become more expensive than their soy counterparts.

On the other hand, soybean oil–petroleum hybrid-based inks have yet to capture more than about 2% of the black newsprint ink market. This black newsprint ink has not yet achieved price-quality parity with conventional inks. Unfortunately, the cost of the oil vehicle is a far more important factor in the pricing of black newsprint inks, and as a result, soybean oil–petroleum hybrid-based inks are currently priced about 30% higher than their petroleum counterparts. Ink represents about 1 to 2% of the cost of printing newspapers; 40% of the cost is for paper. Nevertheless, ink costs are a major driving force in selection of inks unless environmental issues or consumer interest become more important. The American Soybean Association allows the use of its soy ink logo if colored news ink contains 30% soybean oil and if black news ink contains 40% soybean oil.

Current Interest. The optimum vegetable oil–based ink is one that contains no petroleum-derived components; is cost-competitive; has low rub-off; is flexible, so that formulations with a wide range of viscosities can be produced; and is environmentally benign and biodegradable. Soybean oil–petroleum hybrid-based inks, in part, meet some of these criteria, while the NCAUR technology readily meets these criteria. When adopted, market share should continue to increase. The major scientific issues and technical problems related to vegetable oil use in news inks have been solved. Government regulation related to the protection of the environment may help the marketing of soybean oil–based news inks.

Other Industrial Uses for Soybean Oil

Dust Suppressants. Airborne grain dust in a confined space can cause a disastrous explosion when present in sufficient concentration and in the presence of oxygen and an ignition source. Every year about 30 explosions are attributed to grain dust, causing deaths of several workers and major losses of property. In 1978 H. Barham, Jr., and H. Barham, Sr., were awarded a patent for using soybean oil (among other uses) to control dust in grain elevators and to reduce mold growth (121). However, to expand usage, the patent was successfully challenged by the American Soybean Association on the basis of prior widespread use for dust control and lack of evidence supporting some claims.

Soybean oil reduces grain dust in elevators by 94%. In 1987 the U.S. Federal Grain Inspection Service ruled that soybean and other edible oils could be used to control grain dust in elevators (122). Spraying soybean oil is an inexpensive means of reducing the risk of dust explosion, and the investment in equipment to use soybean oil has been estimated to cost about 1% of that needed for dust collection equipment (123).

Incorporating 1 to 2% soybean oil in livestock feeds also greatly reduces hog house dust and often gives 5 to 10% increase in weight gains (124). Normally, degummed oil is used. Controlling dust levels in hog houses leads to healthier pigs, improved weaning rates, and reduced odor (hog house odor is attributed to airborne dust). The market for using soybean oil to control dust in hog houses and in grain elevators has been estimated to be 0.6 billion kg (1.3 billion lb) (125).

Recently, interest has developed in using soybean oil soapstock in similar dust control applications. Excellent control of road dust has been achieved by spraying soybean refinery byproducts onto gravel roads.

Herbicide and Insecticide Carriers. Herbicides and insecticides must be diluted in a carrier so that they can be uniformly applied (126). Water has been almost exclusively used for insecticides, and the relatively large volumes (200 L/ha) used require frequent time-consuming refilling of spray tanks. The advent of rotary atomizing nozzles has facilitated the application of very small carrier volumes (2 to 3 L/ha). This has made it practical to use soybean oil as a carrier. Additional benefits include reduced drift and evaporative losses and increased penetration, which reduce effective application rates.

There is growing widespread use of postemergence herbicides on crops, especially because of the acceptance of no-till and other soil conservation tillage methods. Phytobland petroleum oil is used as the carrier and is used in sprays applied at about 2.5 L/ha with rotary nozzle sprayers. Soybean oil has similar flow properties and, unlike petroleum products, is completely biodegradable. Once-refined soybean oil is used.

Spray equipment may need to be cleaned more frequently, because oil films can build up on equipment, and those applying the spray should wear protective clothing, because soybean oil may increase absorption of active ingredients through the skin (126). Although soybean oil has proven to be a superior carrier for agricultural chemicals with probable environmental benefits, it is not widely used, because its cost is greater than that of the petroleum products it replaces (127).

Miscellaneous Uses

Soybean oil is an antifoam agent in aerated fermentations such as production of penicillin, streptomycin, and tetracycline. Soybean oil also markedly increases yields of antibiotic, presumably by providing important nutrients (80). Soybean oil has also been observed to delay the onset of blooms on fruit trees, reducing susceptibility to frost damage.

Soybean Hulls in Industrial Products

Today, there are no industrial markets for soybean hulls; all hulls are sold for roughage in feeding livestock. However, researchers have recently begun looking at the use of soybean hulls in bioremediation of soil contaminated with diesel fuel and crude oil. One group has reported isolating petroleum-degrading bacteria from soybean hulls (personal communication with Anthony Pometto III, Department of Food Science & Human Nutrition, Iowa State University, Ames, IA). As much as 80 to 90% of spilled petroleum was degraded within eight weeks. Rapid degradation has been attributed to the high population of microorganisms capable of degrading petroleum on hulls, the ability of hulls to absorb fuels or oil, and, effectively, present the contaminants to the microorganism in a highly accessible state, and by providing cosubstrates and nutrients to promote microbial growth.

Potential for Increased Usage of Soybean Products in Industrial Products

There are several factors that convince some in the industry that the time may be right to reconsider many of the early industrial uses and expand currently successful industrial products from soybeans. The first is that petroleum prices may soon reach a level where the cost of soybean-derived products will be able to compete against petroleum-derived products. The second is that environmental concerns are forcing the industry to consider the use of more "environmentally friendly" products and processes. Soybean products may be able to help fill those needs.

Renewable resources provide less than 10% of the world's organic chemicals; petroleum, coal, and natural gas provide the other 90% (128). Fats and oils as a whole contributed about 2% of the total U.S. production of organic chemicals in 1985 (129), but much more could be provided before land availability and marketability of coproducts become limiting. As nonrenewable resources become limiting—and some predict petroleum will be in short supply shortly after the turn of the century—we will need to turn more effectively and extensively to renewable materials. Soybeans are renewable resources that can be produced in abundant supplies.

When more bushels of soybeans are needed to buy a barrel of oil, the market provides more incentive to convert soybeans into industrial products. Since 1974 the price of petroleum has risen at least twofold, while prices for soybeans, soybean oil, and soybean meal have remained largely the same (Fig. 21.9). The price ratios of petroleum to soybeans and soybean products have risen considerably over the past quarter century. In 1974 a farmer could trade a bushel of soybeans for a barrel of oil; by 1982, at the height of the OPEC Oil Embargo, it took almost six bushels of soybeans to trade for one barrel of oil. Only at this time did the soybean oil:diesel fuel price ratio ever approach the ratio of 7 that will be required for methyl soyate to compete without subsidy or legislative action. In recent years (except for 1990 due to the U.S.–Iraq war), prices for petroleum have retrenched, but it still takes about three bushels of soybeans to trade for a barrel of oil.

For soybean products to be adopted, however, the performance characteristics of new soybean products will have to meet the quality standards of the ingredients currently being used, regardless of price and "environmental benefits." Furthermore, the environmental advantage of many soybean products in the formulation or process will have to be clearly established before such claims can be seriously considered.

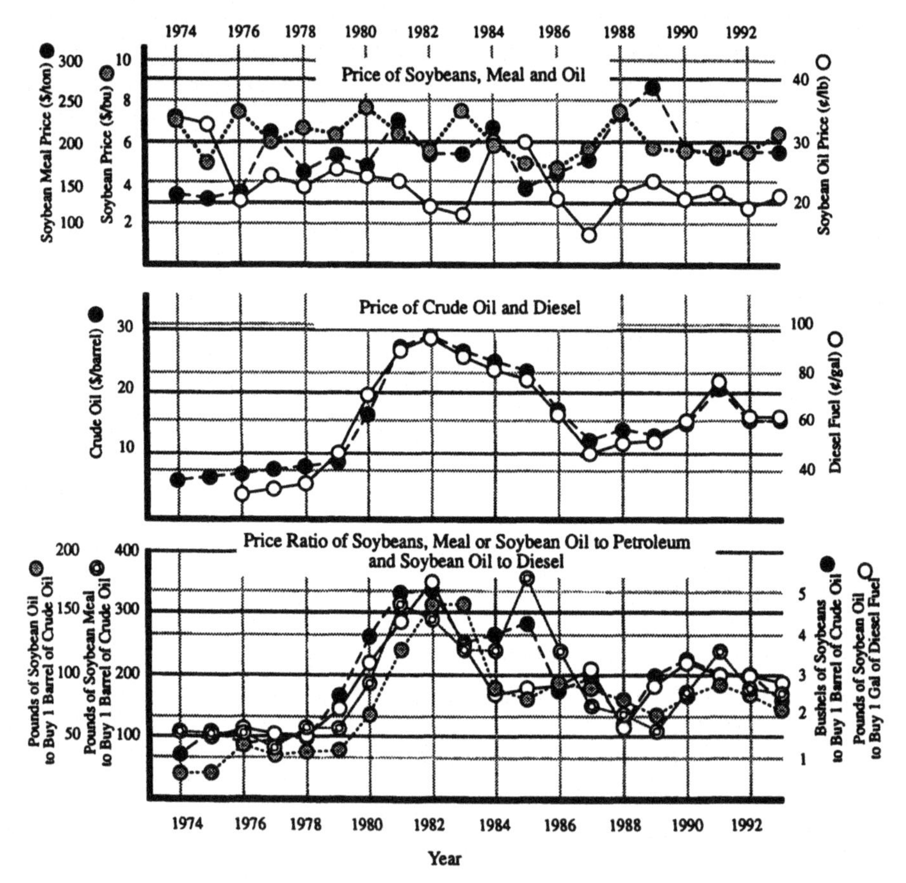

Fig. 21.9. Prices and price ratios between petroleum and soybeans and soybean products. Data provided by A. Paulsen, Department of Economics, Iowa State University. Figure prepared by D.J. Burden, Center for Crops Utilization Research, Iowa State University.

References

1. Johnson, L.A., D.J. Myers, and D.J. Burden, *INFORM 3:* 429 (1992).
2. U.S. Department of Agriculture, Economic Research Service, *Oil Crops Situation and Outlook Reports* and *Fats and Oils Situation and Outlook Reports.*
3. Anonymous, *Time*, Oct. 21, 1935, p. 34.
4. Naemura, T., British Patent 1,466,241 (1923).
5. Spon, W., and F.N. Spon, in *Spon's Encyclopedia of the Industrial Arts, Manufacturers, and Commercial Products*, E. & F.N. Spon, New York, 1980, pp. 1378, 1814.
6. Shurtleff, W., and A. Aoyagi, in *Bibliography of Industrial Utilization of Soybeans*, Soyfoods Center, Lafayette, CA, 1989.
7. Montgaudry, B., *Bulletin de la Societe d'Acclimatation 2:* 16 (1855).
8. Hance, H.F., *Gardeners' Chron. 13:* 209 (1980).
9. Lewkowitsch, J., *Chem. Ind. 33:* 705 (1910).
10. Lewkowitsch, J., *J. Royal Soc. Arts 58:* 519 (1910).
11. Sawer, E.R., *Cedara Memoirs on South African Agriculture, Vol. 2*, Report X, 1911, pp. 183–218.
12. Goessel, F., and A. Sauer, German Patent 228,887 (1909).
13. Goessel, F., and A. Sauer, French Patent 430,183 (1911).
14. Bryan, L.C., *Southern Farmers' Monthly 3:* 170 (1880).
15. Anonymous, *Oil, Paint and Drug Reporter*, Dec. 27, 1909, p. 15.
16. Meister, R., *Farben-Zeitung 15:* 1486 (1910).
17. Gardner, H.A., *Paint Manufacturers' Association of the U.S., Educational Bureau, Science Section, Circular No. 63*, 1919.
18. Sefing, F.G., and M.F. Surls, *Michigan Engineering Experiment Station Bulletin No. 54*, 1933, p. 12.
19. Ditmar, R., *Gummi-Zeitung 41:* 535 (1926).
20. Davidson, G., H.F. Rippey, C.N. Cone, I.F. Laucks, and H.P. Banks, U.S. Patent 1,622,496 (1927).
21. Rippey, H.F., G. Davidson, C.N. Cone, I.F. Laucks, and H.P. Banks, U.S. Patent 1,735,506 (1929).
22. Eddy, C.O., *Trans. Ken. State Hort. Soc.*, 139–141 (1931).
23. Meissl, E., and F. Boecker, *Sitzungsberichte der Akademie der Wissenschaften in Wien, Mathematisch-Naturwissenschaftliche Klasse, 87* (part 1)*:* 371 (1883).
24. Nagel, O., *J. Soc. Chem. Ind. 22:* 1337 (1903).
25. Anonymous, *Sci. Am. Suppl. 72:* 115 (1911).
26. Satow, S., U.S. Patent 1,245,975 (1917).
27. Johnson, O., U.S. Patent 1,460,757 (1923).
28. Anonymous, *Soybean Dig. 2(7):* 6 (1942).
29. Myers, D.J., *Cereal Chem. 38:* 355 (1993).
30. Lougee, E.F., *Modern Plastics 13 (8):* 13 (1936).
31. Anonymous, *Sci. News Let. 33:* 302 (1938).
32. A.D. Little, Inc., in *Marketing Potential for Oilseed Proteins in Industrial Uses*, USDA Technical Bulletin 1043, 1951.
33. Laucks, I.F., and G. Davidson, U.S. Patent No. 1,691,661 (1928).

34. Compiled from information found in Lambuth, A.L., in *Handbook of Adhesives*, 2nd edn., edited by I. Skeist, Van Nostrand Reinhold Co., New York, 1977, pp. 172–180.
35. Johnson, O., U.S. Patent No. 1,680,264 (1928).
36. Babcock, G.E., and A.K. Smith, *Ind. Eng. Chem 39:* 85 (1947).
37. Burnett, R.S., in *Soybeans and Soybean Products, Vol. 2*, edited by K.S. Markley, Interscience Publishers, New York, 1951, pp. 1003–1054.
38. Conner, A.H., in *New Technologies for Value-Added Products from Protein and Co-Products, Symposium Proceedings*, edited by L.A. Johnson, 80th Annual Meeting and Exposition of the American Oil Chemists' Society, Cincinnati, OH, 1989, pp. 15–28.
39. Lambuth, A.L., in *Adhesives from Renewable Resources*, edited by R.W. Hemingway, A.H. Conner, and S.J. Branham, American Chemical Society, Washington DC, 1989, pp. 1–10.
40. Lambuth, A.L., in *Wood Adhesives: Chemistry and Technology, Vol. 2*, edited by A. Pizzi, Marcel Dekker Inc., New York, 1977, pp. 172–180.
41. U.S. Department of Housing and Urban Development, *Manufactured Home Construction and Safety Standard*, 1985.
42. Minnesota Department of Health, *Formaldehyde Product Standard, Minnesota Statutes, Section 144.495*, 1985.
43. Contant, P.J., and J.B.F. Perrot, French Patent No. 461,007 (1913).
44. Dodd, R., and H.B.P. Humphries, British Patent No. 15,316 (1913).
45. Brother, G.H., *Chem. Eng. News 20:* 1511 (1942).
46. Taylor, R.L., *Chem. Metallurgical Eng. 43:* 172 (1936).
47. Anonymous, *Time*, Aug. 25, 1941, p. 63.
48. Anonymous, *Time*, Nov. 11, 1940, p. 65.
49. Anonymous, *Soybean Dig. 1(2):* 11 (1940).
50. McKinney, L.L., *Soybean Dig. 2(8):* 4 (1942).
51. Anonymous, *New York Times*, Dec. 7, 1941.
52. Smith, A.K., and S.J. Circle, in *Soybeans: Chemistry and Technology*, edited by A.K. Smith and A.J. Circle, AVI Publishing Co., Inc., Westport, CN, 1978, pp. 1–26.
53. Brother, G.H., and L.L. McKinney, *Ind. Eng. Chem. 30:* 1236 (1938).
54. Brother, G.H., and L.L. McKinney, *Ind. Eng. Chem. 31:* 84 (1939).
55. Brother, G.H., L.L. McKinney, and W.C. Suttle, *Ind. Eng. Chem. 32:* 1648 (1940).
56. Brother, G.H., and L.L. McKinney, *Ind. Eng. Chem. 32:* 1002 (1940).
57. McKinney, L.L., and G.H. Brother, *Modern Plastics 18(9):* 69 (1941).
58. Beach, E.D., and J.M. Price, *Industrial Uses of Agricultural Materials: Situation and Outlook Report*, USDA, Economic Research Service, June, 1993.
59. Moncrieff, R.W., *Manmade Fibers*, 5th edn., Wiley Interscience Publishers, New York, 1975.
60. Hartsuch, B.E., *Introduction to Textile Chemistry*, John Wiley and Sons, Inc., New York, 1950.
61. Kajita, T., and R. Inoue, U.S. Patent 2,192,194 (1940).
62. Huppert, O., U.S. Patent 2,377,885 (1945).
63. Boyer, R.A., W.T. Atkinson, and C.E. Robinette, U.S. Patent No. 2,377,854 (1945).
64. Anonymous, in *Encyclopedia of Polymer Science and Engineering, Vol. 6*, 2nd edn., edited by J.I. Kroschwitz, John Wiley and Sons, New York, 1985, pp. 647–733.
65. Boyer, R.A, *Ind. Eng. Chem. 32:* 1549 (1940).

66. Boyer, R.A., *Ind. Eng. Chem. 59:* 811 (1940).
67. Von Bergen, W., *Rayon Textile Monthly 20:* 633 (1939).
68. Von Bergen, W., *Rayon Textile Monthly 25:* 225 (1944).
69. Anonymous, *Soybean Dig. 10(7):* 91 (1949).
70. Garey, C.L., in *New Technologies for Value-Added Products from Protein and Co-Products, Symposium Proceedings,* edited by L.A. Johnson, 80th Annual Meeting and Exposition of the American Oil Chemists' Society, Cincinnati, OH, 1989, pp. 10–14.
71. Olson, R.A., and P.T. Hoelderle, in *TAPPI Monograph No. 36.,* edited by R. Strauss, Technical Association of the Pulp and Paper Industry, Atlanta, GA, 1975, pp. 75–96.
72. Skidmore, G., *Paper Trade J. 146(35):* 32 (1962)
73. Coco, C.E., D.R. Dill, and T.L. Krinski., in *Coatings Binders Short Course, Short Course Proceedings,* edited by the Technical Association of the Pulp and Paper Industry, Coating Binders Committee, Boston, MA, 1990, pp. 71–80.
74. Erratt, R.L., R.G. Jahn, and L.H. Silvernail, *TAPPI 42:* 142 (1959).
75. Sprague, P.E., *Soybean Dig. 4(11):* 47 (1944).
76. Sanford, I.G., *Textile World 95:* 127 (1945).
77. Burnett, R.S., *Ind. Eng. Chem. 37:* 861 (1945).
78. Schmutzler, A.F., and D.F. Othmer, *Ind. Eng. Chem. 36:* 847 (1944).
79. Adams, W.H., *Soybean Dig. 4(9):* 8 (1944).
80. Smith, K., and J. Thompson, in *New Technologies for Value-Added Products from Protein and Co-Products, Symposium Proceedings,* edited by L.A. Johnson, 80th Annual Meeting and Exposition of the American Oil Chemists' Society, Cincinnati, OH, 1989, pp. 2–9.
81. Johnson, D.W., in *World Soybean Conference III: Proceedings,* edited by R. Shibles, Westview Press, Boulder, CO, 1985, pp. 175–181.
82. Formo, M.W., in *Bailey's Industrial Oil and Fat Products, Vol. 1,* 4th edn., edited by D. Swern, John Wiley and Sons, Inc., New York, Copyright © 1979, pp. 687–817. Reprinted by permission of John Wiley and Sons, Inc.
83. Stanton, J.M., in *Soybean Utilization Alternatives,* Feb. 16–18, University of Minnesota, St. Paul, MN, 1988, pp. 281–293.
84. Paulsen, A., Paint, Varnish and Coatings, presented at the NCR-159 meeting, Industrial Uses of Soybeans, Sept. 28, St. Louis, MO, 1987.
85. Clocker, E.T., U.S. Patent 2,188,882 (1940).
86. Clocker, E.T., U.S. Patent 2,188,890 (1940).
87. Pryde, E.H., in *Handbook of Soy Oil Processing and Utilization,* edited by D.R. Erickson et al., American Soybean Association, St. Louis, MO, and American Oil Chemists' Society, Champaign, IL, 1980, pp. 459–481.
88. Formo, M.W., in *Bailey's Industrial Oil and Fat Products, Vol. 2,* 4th edn., edited by D. Swern, John Wiley and Sons, Inc., New York, Copyright © 1982, pp. 343–405. Reprinted by permission of John Wiley and Sons, Inc.
89. Gardner, C., *J. Amer. Oil Chem. Soc. 36:* 568 (1959).
90. Kammann, K.P., and A.I. Phillips, *J. Amer. Oil Chem. Soc 62:* 917 (1985).
91. Formo, M.W., in *Soybean Utilization Alternatives,* Feb. 16–18, University of Minnesota, St. Paul, MN, 1988, pp. 35–42.
92. Zilch, K.T., in *World Soybean Research Conference II: Proceedings,* edited by F.T. Corbin, Westview Press, Boulder, CO, 1980, pp. 693–701.

93. Leonard, E.C., in *Fatty Acids*, edited by E.H. Pryde, American Oil Chemists' Society, Champaign, IL, 1980.

94. Sonntag, N.O.V., *J. Amer. Oil Chem. Soc. 62:* 928 (1985).

95. Engler, C.R., L.A. Johnson, W.A. LePori, and M. Yarbrough, in *Biomass Energy—A Monograph*, edited by E.A. Hiler and B.A. Stout, Texas A&M University Press, College Station, TX, 1985, pp. 174–212.

96. Grinaker, G., *Sunflower 7*(1)*:* 28 (1981).

97. Wiebe, R., and J. Nowakowska, *USDA Bibliographical Bulletin 10:* 52–105, 183–195 (1949).

98. Goering, C.E., A.W. Schwab, M.J. Daugherty, E.H. Pryde, and A.J. Heakin, *Trans. ASAE 25:* 1472 (1982).

99. Adams, C., J.F. Peters, M.C. Rand, B.J. Schroer, and M.C. Ziemke, *J. Amer. Oil Chem. Soc. 60:* 1574 (1983).

100. Lipinsky, E.S., S. Kresovich, C.K. Wagner, H.R. Applebaum, T.A. McClure, J.L. Otis, and D.A. Trayser, in *Proceedings of the International Conference on Plant and Vegetable Oils as Fuels*, American Society of Agricultural Engineers, St. Joseph, MI, pp. 1–10.

101. Holmberg, W.C., and J.E. Peeples, *Comprehensive Review of Fuel Performance, Characteristics, and Regulatory Issues for Biodiesel*, American Biofuels Association and Information Resources, Inc., 1994.

102. Dykstra,J.J., A.W. Schwab, S.C. Sorenson, E. Selke, and E.H. Pryde, in *Symposium of Alternative Energy in the Midwest*, Feb. 21–23, Schaumburg, IL, 1985.

103. Niehaus, R.A., C.E. Goering, L.D. Savage, and S.C. Sorenson, Paper #85-1560, *ASAE Winter Meeting, Dec. 19–20*, Chicago, IL, 1985.

104. Bagby, M.O., and B. Freedman, in *New Technologies for Value-Added Products from Protein and Co-Products, Symposium Proceedings*, edited by L.A. Johnson, 80th Annual Meeting and Exposition of the American Oil Chemists' Society, Cincinnati, OH, 1989, pp. 36–43.

105. Goering, C.E., *Final report for the project "Effect of Nonpetroleum Fuels on Durability of Direct-Injection Diesel Engines,"* Grant 59-2171-1-6-057-0, USDA, ARS, Peoria, IL, 1984.

106. Schumacher, L.G., S.C. Borgelt, W.G. Hires, C. Spurling, K. Humphrey, and J. Fink, *Fueling diesel engines with diesel/soydiesel blends*, Paper no. MC93-101, 1993 Mid-Central Conference, April 9–10, American Society of Agricultural Engineers, St. Joseph, MI, 1993.

107. Seikman, R.W., G.H. Pischinger, D. Blackman, and L.D. Carvalo, in *Proceedings of the International Conference on Plant and Vegetable Oil Fuels*, American Society of Agricultural Engineers, Fargo, ND, 1982.

108. Peterson, C.L., M. Feldman, R. Korus, and D.L. Auld, *Trans. ASAE 7:* 711 (1991).

109. Goering, C.E., in *Soybean Utilization Alternatives, Feb. 16–18*, University of Minnesota, St. Paul, MN, 1988, pp. 303–324.

110. Anonymous, *Chem. Eng.* Feb. 1993, pp. 35,37,39.

111. Moynihan, J.T., U.S. Patent 4,419,132 (1983).

112. Moynihan, J.T., U.S. Patent 4,519,841 (1985).

113. Moynihan, J.T., U.S. Patent 4,554,019 (1985).

114. Erhan, S.Z., and M.O. Bagby, *J. Amer. Oil Chem. Soc. 68:* 635 (1991).

115. Erhan, S.Z., M.O. Bagby, and H.W. Cunningham, *J. Amer. Oil Chem. Soc. 69:* 251 (1992).

116. Bagby, M.O., and S.Z. Erhan, U.S. Patent 5,122,188 (1993).

117. Cashau, G.R., in *Soybean Utilization Alternatives, Feb. 16–18*, University of Minnesota, St. Paul, MN, 1988, pp. 295–300.

118. Landell Mills Commodities Studies Ltd, *White Paper*, American Soybean Association, St. Louis, MO, 1988.

119. *Special Report, Vegetable Oil Based Newsinks and Their Printability Properties*, IFRA Publications, Darmstadt, Germany, 1991.

120. Cunningham, H.W., in *New Technologies for Value-Added Products from Protein and Co-Products, Symposium Proceedings*, edited by L.A. Johnson, 80th Annual Meeting and Exposition of the American Oil Chemists' Society, Cincinnati, OH, 1989, pp. 44–47.

121. Barham, H.N, Jr., and H.N. Barham, Sr., U.S. Patent 4,208,433 (1980).

122. *Federal Register*, March 4, 1987, pp. 6493–6497.

123. Marking, S., *Soybean Dig*. Feb. 1987, p. 27.

124. Weigel, J.C., in *New Technologies for Value-Added Products from Protein and Co-Products, Symposium Proceedings*, edited by L.A. Johnson, 80th Annual Meeting and Exposition of the American Oil Chemists' Society, Cincinnati, OH, 1989, pp. 48–51.

125. *ASA Member Letter*, Nov. 1988.

126. Kapusta, G., *J. Amer. Oil Chem. Soc. 62:* 923 (1985).

127. Kapusta, G., in *New Technologies for Value-Added Products from Protein and Co-Products, Symposium Proceedings*, edited by L.A. Johnson, 80th Annual Meeting and Exposition of the American Oil Chemists' Society, Cincinnati, OH, 1989, pp. 54–58.

128. Zoebelein, H., *INFORM 3:* 721 (1992).

129. Pryde, E.H., and K.D. Carlson, *J. Amer. Oil Chem. Soc. 62:* 916 (1985).

Chapter 22

Soy Foods

Lester A. Wilson

Iowa State University
Ames, IA

Introduction

From 2800 B.C. to the present, soybeans have been cultivated in China (1). Ancient texts give detailed information on varieties, cultivation practices, and harvest and preservation procedures. The utilization of soybeans, alone or in combination with other cereals, is well documented. The origin of soybeans is believed to be within the boundaries of modern China. However, some authorities believe that the point of origin is elsewhere within East or Southeast Asia.

Due to their ease of production, long-term storage characteristics and exceptional protein content, soybeans have been highly prized as a nutritional food source. Consequently, soybeans were introduced into Korea and Japan, gaining widespread acceptance throughout those cultures. Soybeans are traditionally consumed in nonfermented and fermented forms (Tables 22.1, 22.2). The nonfermented forms include soymilk, yuba, tofu, and toasted soy protein powders. The fermented forms include miso, natto, soy sauce (shoyu), and tempeh. While the names of these soy foods vary

TABLE 22.1 Nonfermented Soy Food Products and Common Names by Country (2,3)

	Nation of origin			
Product	Japan	China	Korea	Other
Fresh soybeans	Edaname	Mao-dou	Put kong	
Soybean sprouts	Daizu no moyashi	Huang-dou-ya	Kong na moal	
Soy milk	Tonyu	Dou-jiang	Kong kook Doo Yoo	
Soy milk film	Yuba	Dou-fu-pi		Fu chok (Indonesia) Fu chok (Malaysia)
Soybean curd (fresh)	Tofu	Dou-fu	Doo bu	Tahu (Indonesia) Tau-foo (Malaysia)
Frozen-dried	Kori tofu	Dong-dou-fu		Tokua (Philippines)
Deep fried	Aburage Nama-age Atsu-age Gan-modoki Use-age Kara-age			
Toasted soy flour	Kinako	Dou-fen	Kang ka rau	Bubuk kadele (Indonesia)

from country to country within Asia, this chapter always uses the Japanese terms.

These foods play important roles in the diet and culture of the people in Asia. Despite significant changes in the variety of foods consumed over the years, it is estimated that nearly 10% of the Japanese population's protein intake is from these basic soy foods (2). A heavy dietary reliance on rice, which is low in protein, the essential amino acid lysine, and lipid, has led to universal use of soybean and fish products as alternative protein sources throughout Asia (5). Supplemental soybean use, in particular, has made a significant improvement in the nutritional well-being of the population. The relative protein contents of various soy foods are listed in Table 22.3.

Soybeans are relatively high in lysine but low in methionine. Together, these protein sources—corn and soybeans—compose what is known as nutritionally complete "complementary protein." The amino acid balance for soy protein is comparable to animal sources based upon human requirements.

Changes in food consumption and nutritional intake by Japanese and other Asian consumers are evidence of a general "Westernization" of their dietary patterns. The correlation of this trend with increased heart disease and related disorders, hitherto inconspicuous in Japan, has been widely reported. The Japanese government has recently taken action to develop a guideline for citizens to be more aware of dietary health issues (5).

The per capita production and consumption of soybeans in several Asian countries have tended to remain constant or increase (Table 22.4) (3). In recent years, increases in utilization in Indonesia, Malaysia, Philippines, Thailand, Japan, and Korea have resulted from increased consumption of some soy foods (soymilk and tofu). Historically, production of Japanese soybeans decreased until ten years ago; then production gradually increased to 200,000 metric tons (MT) per year, because of Japanese governmental policy designed to encourage farmers to grow soybeans as an alterna-

TABLE 22.2 Fermented Soy Food Products and Common Name by Country (2,3)

	Nation of origin			
Product	Japan	China	Korea	Other
Whole soybeans	Natto			Tempeh (Indonesia and Malaysia)
	Hamanatto			
Soy paste	Miso	Jiang	Doen jan	Tauco (Indonesia and Malaysis) Tao si (Philippines)
Soy sauce	Shoyu	Jiang-you	Kang jang	Kecap (Indonesia and Malaysia) Tayo (Philippines)
Soy curd		Sou-fu		
Soy pulp				Tempeh gembus Oncom ampas tahu (Indonesia)

tive to rice cultivation. This policy resulted from increasing annual rice surpluses.

Present Japanese soybean production constitutes 20% of total national demand for soyfoods. Soybeans from the United States and China are imported into Japan at well over 4,000,000 MT per year, with 80% of the raw tonnage (mainly U.S. soybeans) being processed for oil extraction. Of this, 80% of the soybean meal is used as feed, the remaining 20% being used for human foods (nonfermented and fermented soy foods), amino acid mixtures, and new protein foods such as textured soybean protein. This compares with about 2% in the United States.

Soybean Chemical Composition

Soybeans are a rich source of protein (30 to 40%) and oil (20%) (Table 22.3). They contain 9 to 12% total sugars, of which 4 to 5% is sucrose, 1 to 2% is raffinose, and

TABLE 22.3 Chemical Compositions of Soy Foods (2,3,4,6,7,8,9)

	Moisture	Protein[a]	Fat	Carbohydrate Soluble (percent)	Fiber	Ash
Nonfermented						
Soybean[b]	12.5	35.3	19.0	23.7	4.5	5.0
Soymilk	94.0	3.0	1.0	1.0	0	0.3
Tofu						
Momen	86.8	6.8	5.0	0.8	0	0.6
Kinugoshi	89.4	5.0	3.3	—	—	0.6
Packed	90.0	4.5	3.2	—	—	0.6
Aburage	44.0	18.6	33.1	2.8	0.1	1.4
Kori	10.4	53.4	26.4	7.0	0.2	2.6
Yuba	8.1	50.2	33.4	5.3	0.2	2.8
Soybean sprouts	88.3	5.4	2.2	2.6	0.8	0.7
Kinako	5.0	35.5	23.4	26.4	4.6	5.1
Fermented						
Tempeh	64	18	4	1	—	1.0
Natto	59	17	10	12	2.0	3.0
Miso[c]	45.7	13.1	5.5	19.1	2.0	14.6
Hamanatto	36	26	12	14	3.0	12.0
Soy sauce[d]	72	7	0.5	2	0	18.0
Soy paste	50	14	5	16	2.0	5.0
Ko chu jang	48	9	4	19	4.0	20
Fermented soy curd	60	17	14	0.1	—	9.0
Fermented soy pulp						
Tempeh gembus	81	5	2	11	—	1.0
Oncom ampas tahu	84	4	2	8	—	2.0

Note: Given as % approximate chemical composition as is basis.
[a]Protein = N x 5.71
[b]Obtained in Japan
[c]Dark yellow
[d]Regular; nonspecialty

3.5 to 4.5% is stachyose (see Chapter 2). While all of the sugars are fermentable by microorganisms, raffinose and stachyose (oligosaccharides) are not digestible by humans and other monogastric animals (e.g., pigs and poultry). These sugars produce intestinal gas and discomfort in humans and a loss of feed efficiency in livestock operations. The removal of the oligosaccharides with an increase in sucrose (or other simple sugars) would not only eliminate the oligosaccharide problem, but improve soybeans as an industrial fermentation feedstock. Typically, Chinese and some Japanese soybeans are higher than U.S. soybeans in carbohydrate content, which may make them more desirable for fermented soy food products than U.S. soybeans (15,16). However, care must be taken when interpreting "sugar" levels of Japanese soybeans in the literature and Japanese government reports, because the sugar referred to is a figure obtained by acid hydrolysis, not the free sugar present in the soybean or its total carbohydrate content (3).

Soybeans also contain less than 1% starch, 5% ash, and 4.5% crude fiber (2,14) (Table 22.3). Over half of the crude fiber in soybeans contributes to the physiologically important dietary fiber that is necessary for proper human nutrition. In recent years, the importance of dietary fiber in the diet has received considerable attention in the United States. While the role of fiber in reducing the incidence of colon cancer and heart disease is not well understood, the potential health benefits of increasing fiber in the diet should not be overlooked. Soybean hulls contain about 87% crude fiber, consisting of cellulose, hemicellulose, lignin, and uronic acids.

Linoleic acid comprises 50% of the available fatty acid and is believed to be conducive to decreasing the content of blood cholesterol (see Chapter 23). New varieties containing different amounts of saturated and unsaturated fatty acids have been developed, as noted in Chapter 2. Soybeans also contain vitamins B_1, B_2, B_6, and E (17,18). Additionally, trypsin inhibitors, goitrogenic substances, isoflavones, saponins, and hemagglutinins are present in small or trace amounts.

Certain phytochemicals in fruits, vegetables, and grains may possess cancer-preventive properties that inhibit tumor initiation, prevent oxidative damage, and affect steroid hormones or prostaglandin metabolism to block tumor promotion. Isoflavones are one class of these compounds that are found in soybeans in high amounts. The major soybean isoflavones, genistein and daidzein, have been identified for a considerable period of time. Because these compounds appear to act as anticarcinogens by exerting a biological antioxidant effect, their contents and

TABLE 22.4 Per Capita Annual Consumption of Soybeans (kg) in Selected Asian Countries (3,10,11,12,13)

Country	1968	1978	1988	1994
China	6.7	5.9	4.8	4.4
Indonesia	3.4	5.9	7.8	10.5
Japan	6.3	6.8	7.1	7.4
Korea	7.9	9.5	9.0	9.1[a]
Malaysia	1.8	1.9	3.5	4.7
Philippines	0.0	0.2	0.3	0.4
Thailand	0.4	1.0	2.0	1.8

[a]Additional 5.1 kg in North Korea.

bioavailabilities in foods have been a topic of recent interest. Isoflavone (potential cancer-preventive properties) profiles of traditional soybeans and soy foods showed that soy foods contained 6 to 20% of the isoflavone of soybeans (Wang, H.-J. and P.A. Murphy. *J. Agric. Food Chem.*, in press).

In spite of the fact that soybeans are excellent nutritional staples, they are not consumed raw, because of their hard texture and undesirable flavor. Undaunted by these attributes, the people of Japan and other Asian countries have developed a variety of sophisticated processed soy foods. These products and their culinary presentation have done a great deal to enhance the table aesthetics, palatability, and nutritive value of soy protein.

Traditional soybean foods are classified into two groups: non-fermented foods, including regular, deep-fried, frozen-dried, roasted tofu, soybean protein film (yuba), and soybean sprouts; and fermented soybean foods, such as miso, soy sauce (shoyu), and fermented whole soybeans (natto).

Unfermented Soy Foods

Soymilk

Soymilk is a very popular beverage with the Chinese, though considerably less so for the Japanese consumer. However, soymilk production is very important to Japanese tofu producers, because it is the intermediate product in the manufacture of tofu. Since 1978, Japanese soymilk consumption has increased. To some extent, this is the result of effective marketing campaigns that have advertised soymilk as having physiological benefits, particularly as a healthful pick-me-up "energy drink" for stressed workers and business persons (5).

Many Japanese, like their Western counterparts, find the flavor and odor of soymilk undesirable. This flavor and odor are formed by the oxidation of specific unsaturated fatty acids by lipoxygenase enzymes during the grinding of the seed. Understanding how off-flavors and undesirable flavors interact with soy proteins and similar seed constituents may lead to improved processing systems. Breeding programs at Purdue University (West Lafayette, IN), Iowa State University (Ames, IA), and the Japanese Ministry of Agriculture, Forestry, and Fisheries (Tsukuba, Japan) have developed lipoxygenase-null varieties. Davies (19) reported that the flavor of soymilk was improved by using lipoxygenase-2-null soybeans. Additionally, odor formation may be circumvented by heat inactivation of the enzyme before the beans are ground (although this significantly lowers yields) or by masking the flavor with additives (20).

Soymilk is traditionally made by soaking soybeans in water (1:10) overnight, then grinding the beans in a mill with additional water being added during the grinding step (Fig. 22.1). The resulting slurry is boiled and stirred for 15 to 30 min. This heating step improves the nutritional value of the milk, by inactivating trypsin inhibitors, and improves the flavor, by inactivating lipoxygenase and volatilizing some of the off-flavor compounds that appear during grinding. Heating also increases the shelf life of the milk by reducing its microbial load (critical control point).

The heated slurry is then filtered through a cloth or nylon bag to separate the undispersible fiber residue, *okara*, from the soymilk.

A more recent innovation utilizes MicroSoy® flakes instead of whole soybeans (Nichii Company of America, Jefferson, IA). Soy flakes are dispensed into a paddle mixer (3.5 kg MicroSoy flakes to 40 L of potable water for a 5% solids soymilk), rehydrated and blended at room temperature for 10 min. The resulting slurry can be steam-injected into a cooker and cooked by direct culinary steam or indirect steam injection with continuous agitation. The temperature of the slurry is held for 7 min at 95°C (203°F) and subcooked for 40 s before extraction in a "Takai Automated Soymilk Plant" machine (Takai Tofu and Soymilk Equipment Co., Ishikawa-ken 921, Japan). The cooked slurry is sieved through a 120-mesh roller screen to extract the milk. The insoluble materials are expelled onto a 100-mesh roller drum and roller press to further extract milk before expelling the solid okara. The soymilk at 90°C (194°F) is collected in a coagulation tub, where it is allowed to cool to 85°C (185°F) prior to tofu manufacture or bottling and refrigeration. The resulting soymilk from either whole soybeans or flakes may have flavors added to mask the beany flavor. It may also be homogenized, pasteurized, and sterilized before being bottled, aseptically packaged, or retorted.

Typically, high-protein, clear- or yellow-hilum (see Chapter 2), large-seed soybeans are preferred for soymilk production. Two hundred grams of soybean will yield about 1 L of soymilk (2). The chemical composition of soymilk is given in Table 22.3. Soymilk can be made more shelf-stable by spray-drying or roller-drying it into a dry powder (as is done with cow's milk in the United States). Spray-dried soymilk is often used in confections, meat fillers, and beverages.

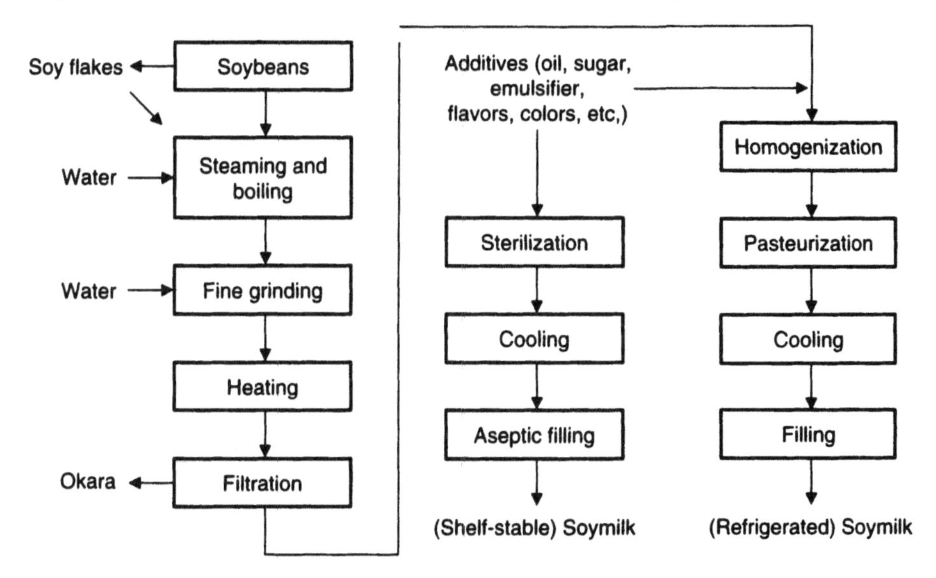

Fig. 22.1. Steps in refrigerated and shelf-stable soymilk production. *Source:* Wilson, L.A., et al., *Japanese Soyfoods: Markets and Processes,* MATRIC, Iowa State University, Ames, IA, 1992. Reprinted with permission of MATRIC.

Fig. 22.2. JAS Seal of Approval. *Source:* Wilson, L.A., et al., *Japanese Soyfoods: Markets and Processes,* MATRIC, Iowa State University, Ames, IA, 1992. Reprinted with permission of MATRIC.

Soymilk is certified by labeling, which includes the Japanese Agricultural Specifications (JAS) Seal of Approval (5) (Fig. 22.2).

Japanese Agricultural Specifications classify soymilk into four groups: regular soymilk, reconstituted soymilk, soft drinks, and soy protein beverages. Although soymilk has a similar composition to cow's milk, the oil-to-protein ratio is lower. For this reason, reconstituted soymilk is supplemented with oil. In addition, soymilk may be supplemented with sugar to enhance its palatability in soft drinks. These soymilk beverages often contain fruit juice, cocoa, sugar, flavors (artificial or natural), stabilizers, and other ingredients to enhance customer acceptance by masking soy flavors.

Tofu (bean curd) has been produced in Japan for over 2,000 years. It was introduced from China along with the agronomic introduction of soybeans (1). Then as now, the production of tofu was largely a small-scale enterprise consuming less than 60 kg of soybeans per day. The shelf life of tofu can be quite variable ranging from 1 to 5 d for fresh tofu; 1 to 3 wk for pasteurized tofu; and six months to two years for tofu processed aseptically. The initial microbial load and storage temperature largely govern the shelf life (22,23,24). Some products (e.g., silken tofu) are more difficult to transport due to temperature abuse and physical damage.

Large-scale factories consume 2 to 3 MT of soybeans per day (Fig. 22.3). These large manufacturers have developed integrated production and marketing systems. From the factory to the supermarket showcase, superior product quality is ensured by sophisticated, timely distribution using refrigerated transport and display systems. Likewise, some supermarket chains are now producing their own tofu "in-house" where it is kept in refrigerated display cases. Aseptically packaged tofu is gaining in popularity. This product has greater utility with respect to preservation, storage and transportation. Consequently, its production scale is much larger, exceeding 6 MT of soybeans per day. However, much of this tofu is for the Japanese export market due to legislation protecting the small tofu producers (3).

The official Japanese sanitary guidelines for soy milk and tofus, first published in 1959, remain relatively unchanged (1):

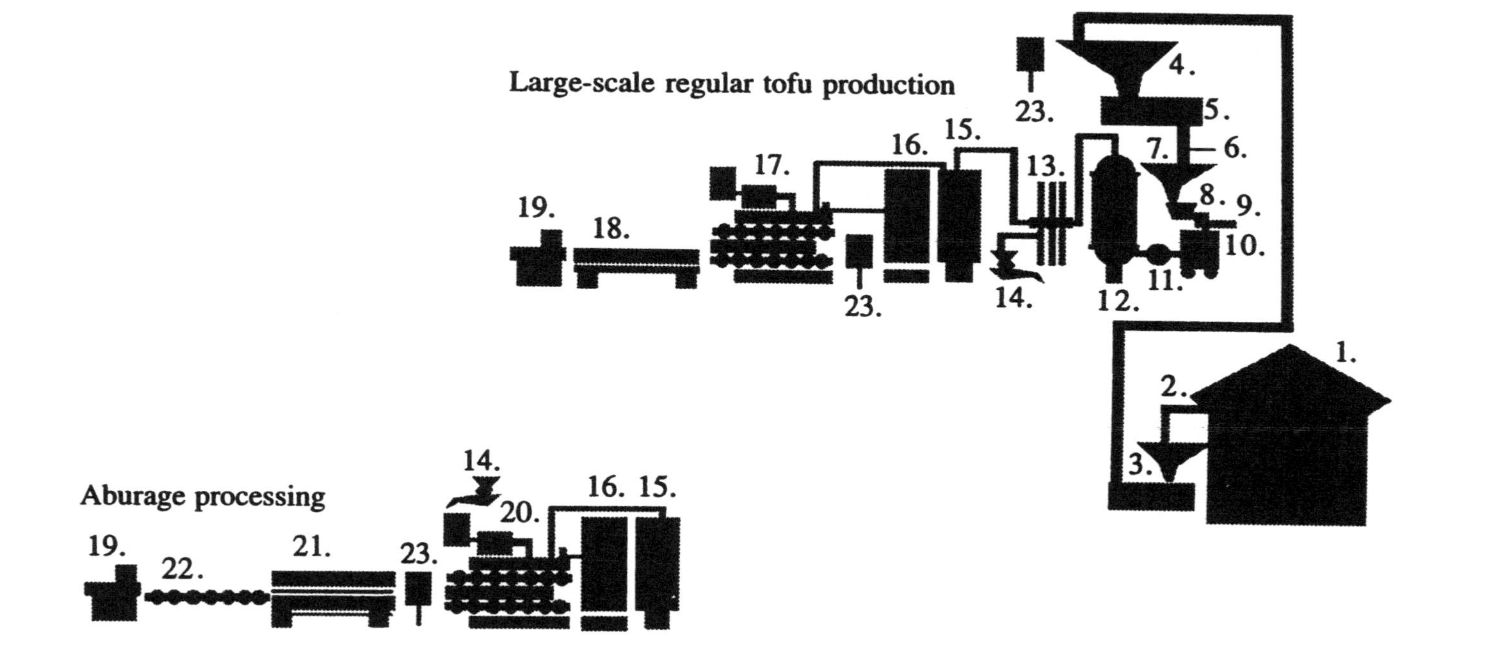

Fig. 22.3. Large-scale manufacturing operation for tofu production: (1) soybean pit (storage) with air conveyor, (2) cyclone, (3) storage and weight measure, (4) divider and storage, (5) soybean tank, (6) washer and conveyor, (7) hopper, (8) grinder, (9) antifoam measure dispenser, (10) slurry tank, (11) pump, (12) cooker, (13) filter, (14) rotary feeder (residue discharge), (15) soymilk storage, (16) coagulant storage, (17) coagulation vessel, tofu molder, and press, (18) tofu cutter and tofu cooler, (19) tofu packer, (20) coagulation vessel, tofu molder, press, and cutter, (21) deep fryer, (22) conveyor for packed products, and (23) control station. *Source:* Wilson, L.A., et al., *Japanese Soyfoods: Markets and Processes*, MATRIC, Iowa State University, Ames, IA, 1992. Reprinted with permission of MATRIC.

Standards of manufacturing of bean curd:

1. Soybean as ingredient shall be of good quality and shall not contain foreign substances.
2. Soybean as ingredient shall be sufficiently washed in water.
3. Soybean juice and soybean milk shall be sterilized by the method of heating at the boiling state for 2 minutes or by a method having the same or superior effect.
4. Soaking of the bean curd shall be performed while continually changing the water.
5. Wrapped bean curd (meaning bean curd prepared by adding a coagulating agent to soybean milk, filling the milk in a package, then heating to coagulate) shall be sterilized by the method of heating at 90°C (194°F) for 40 minutes or by a method having the same or superior effect.
6. Tools used for manufacturing bean curd shall be sufficiently washed and sterilized.
7. Water used for manufacturing bean curd shall be potable water.

Standards of storage of bean curd:

1. Bean curd shall be stored refrigerated or in a sufficiently washed and sterilized water tank with continually changed potable cold water. However, this does not apply to bean curd for itinerant sale and bean curd ordinarily intended for immediate sale after molding and without soaking.
2. Bean curd for itinerant sale shall be sufficiently washed and kept cool using sterilized tools."

Tofu

Momen (Cotton) Tofu. The traditional production process for momen tofu, the most popular kind of tofu in Japan, is shown in Fig. 22.4. After soybeans are soaked in water for 8 to 12 hr (25), they are ground with water into a slurry using a stone-mill or stainless steel centrifugal grinder. Alternately, MicroSoy Flakes (Nichii Company of America, Jefferson, IA) could be used, as noted in the soymilk section of this chapter.

Water and an antifoaming agent are added to the slurry before it is heated. After heating for 5 to 10 min to reduce the beany flavor and antinutritional factors, the slurry is filtered to remove any insoluble soybean solids. This filtration is accomplished by hand, air, hydraulic, or mechanical pressing. The residue (okara) remains in a cotton or nylon cloth filter bag. Traditionally, the okara is often used as animal feed or landfill (buried). Currently, however, some okara is being processed for use in new (proprietary) dietary and medicinal products.

Control of the percent solids and temperature of the soymilk, amount of coagulant, and the stirring of the coagulating soymilk are critical quality control points, because they influence the texture and yield of the tofu. Firm tofu can be produced by using a lower-solids soymilk (5 to 8%), higher coagulation temperature (90 to 95°C, 194 to 203°F), and vigorous mixing during the coagulation step. A softer, larger-yielding tofu can be produced with a high-solids soymilk (10 to 13%), coagulated at a lower temperature (70 to 80°C, 158 to 176°F), with a minimum of stirring (only enough to thoroughly disperse the coagulant in the soymilk). The amount and type of coagulant are also critical. Lack of sufficient coagulant will fail to coagulate the soymilk, but too much coagulant can produce low yields of small, hard curds with uneven texture and bitter (magnesium chloride) or chalky (calcium sulfate) taste (26). It is recommended that the solids content of the hot soymilk be measured and standardized before adding the coagulant (27). The correct amount of coagulant can be determined by plotting the amount of coagulant added to a known solids "hot" soymilk against the transparency of the resulting whey after curd formation (Figs. 22.5 and 22.6). The correct amount of coagulant is determined by the point where the whey transparency plot (line) becomes parallel with the *x*-axis. A similar plot can be made using yield of tofu, firmness of tofu, or dry tofu solids. The optimum amount of coagulant is also influenced by the solids content of the soymilk; the more concentrated the soymilk, the more coagulant required (26).

Calcium sulfate dihydrate ($CaSO_4 \cdot 2H_2O$), not to be confused with anhydrous calcium sulfate or calcium sulfate monohydrate, at a rate of about 2% to 3% of the

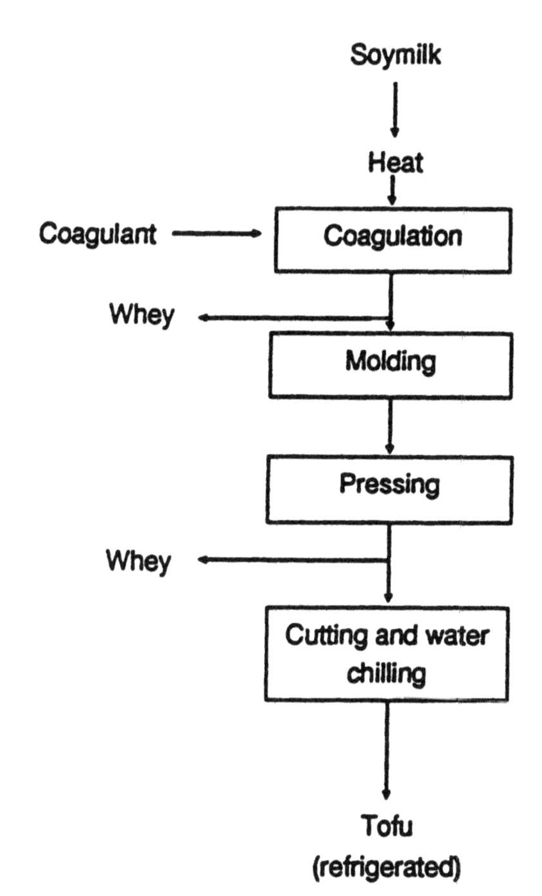

Fig. 22.4. Steps in momen (cotton or regular) tofu production. *Source:* Wilson, L.A., et al., *Japanese Soyfoods: Markets and Processes,* MATRIC, Iowa State University, Ames, IA, 1992. Reprinted with permission of MATRIC.

soybeans used for the batch (0.013 to 0.023 N), is mixed with water and added to 70° to 90°C (158 to 194°F) soymilk, depending on the desired firmness of the tofu (26,28,29,30). Calcium sulfate, produced as a byproduct of the soda industry, is the most extensively used coagulant, although products derived from powdered gypsum and seawater are also available. A concentrated solution of magnesium chloride ($MgCl_2$), called *nigari*, has been commonly used for over 100 years. Recently, its use has been widely replaced by calcium sulfate and glucono-delta-lactone (GDL). Both of these coagulants are GRAS (generally recognized as safe) and contribute to the public's dietary calcium intake. However, nigari's perceived superior contribution to tofu texture and flavor has been a reason for its continued use by some processors.

The coagulant causes the soymilk to gel. Breakage of the gel into curds facilitates separation of the whey from the curds. After whey removal, the curds are transferred to a perforated press-box, covered with a cloth, pressed, and shaped. Unfortunately, bamboo mats instead of perforated metal plates are often used by small processors to help distribute the press weight, potentially increasing the microbial load of the tofu by the end of the production day. The press-boxes are made from aluminum or stainless steel. For sanitary reasons, wooden press-boxes are not advisable; however, some small operations still use them. Pressing is accomplished with hydraulic, air, or manual (ratchet) presses. From a microbiological quality standpoint (critical control point), the tofu temperature should not fall below 60°C (140°F) during the pressing step.

After the pressing weight is removed, the tofu is cut, then taken out of the press-box within a refrigerated water bath. Great care is taken not to damage the fragile blocks. Soaking is continued to cool the blocks and to remove excess coagulant. The tofu is held within these water baths until packaged or is refrigerated at 0 to 10°C (32 to 50°F) prior to packaging and shipment. Tofu is either packed in plastic containers or sold "in bulk," as unpackaged blocks, directly to consumers. To extend the shelf life of finished packaged tofu, the package can be submerged in water and pasteurized in the package. This process is used in California to ship tofu to the midwestern United States. Typically, 10 kg of soybeans will yield 40 to 50 kg of soft momen tofu or 15 to 30 kg of firm momen tofu.

There is considerable interest in reducing the amount of soaking time for whole soybeans. This would result in more economical processing: less water and energy consumption, lower labor costs, less okara production, more uniform tofu, and a faster response time when filling orders. Three approaches are used or are under consideration (3):

1. Dehulling the beans prior to soaking them (or grinding the dehulled beans with water)
2. Flaking the dehulled beans prior to grinding or soaking them
3. Making soy flour from the beans prior to a combined heating and soaking step

The first strategy was observed in use in Japan in 1989. The second strategy, the use of flakes, was proposed in 1989 by one manufacturer (Nichii Company, Japan), who has subsequently built processing facilities of this type in both Japan and the United States (31). Due to the extremely high cost of industrial land in Japan, there is

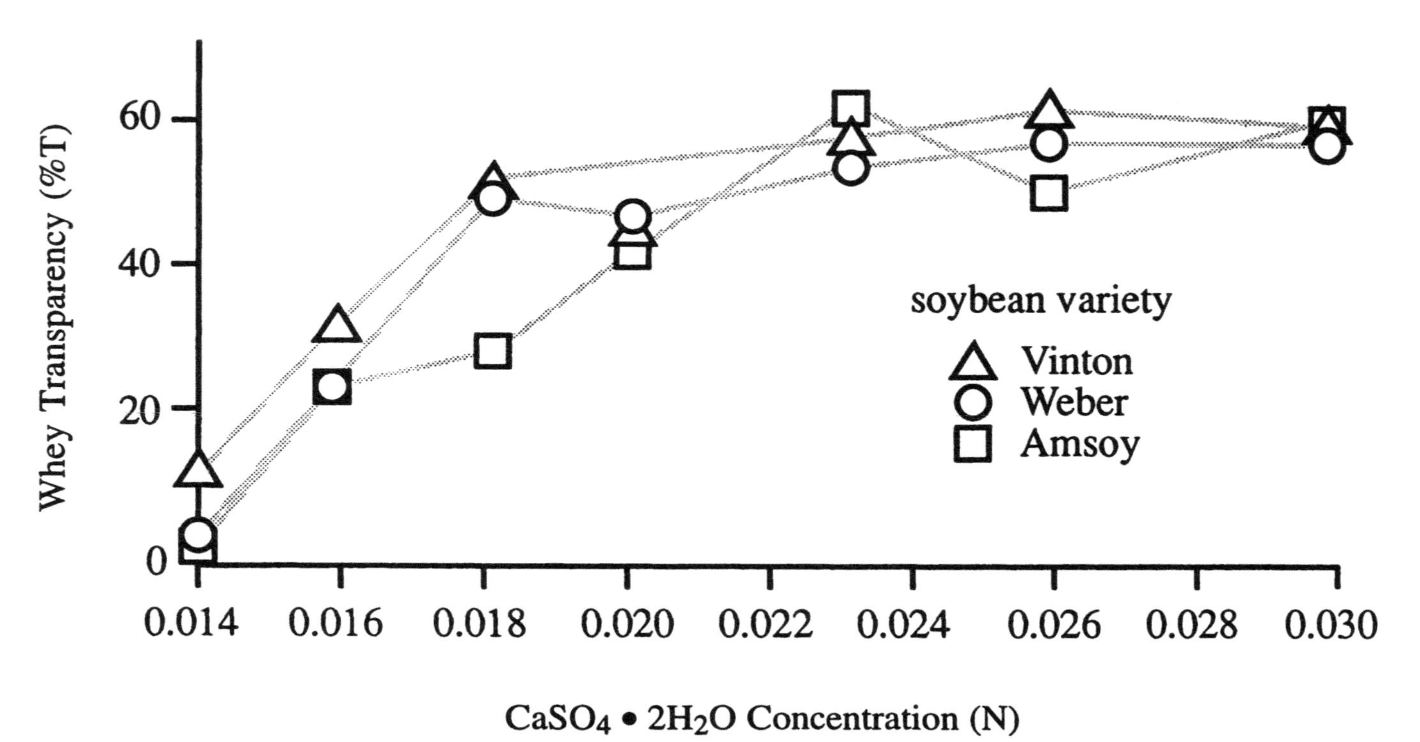

Fig. 22.5. Percent transmittance of whey versus coagulant concentration for soymilks at 6% solids made from Weber, Vinton, and Amsoy soybeans. A concentration of 0.023 N was selected as the optimum coagulant concentration. *Source:* Johnson, L.D., Influence of Soybean Variety and Method of Processing on Tofu Manufacturing, Quality, and Consumer Acceptability. M.S. Thesis, Iowa State University, Ames, IA, 1984, p. 128.

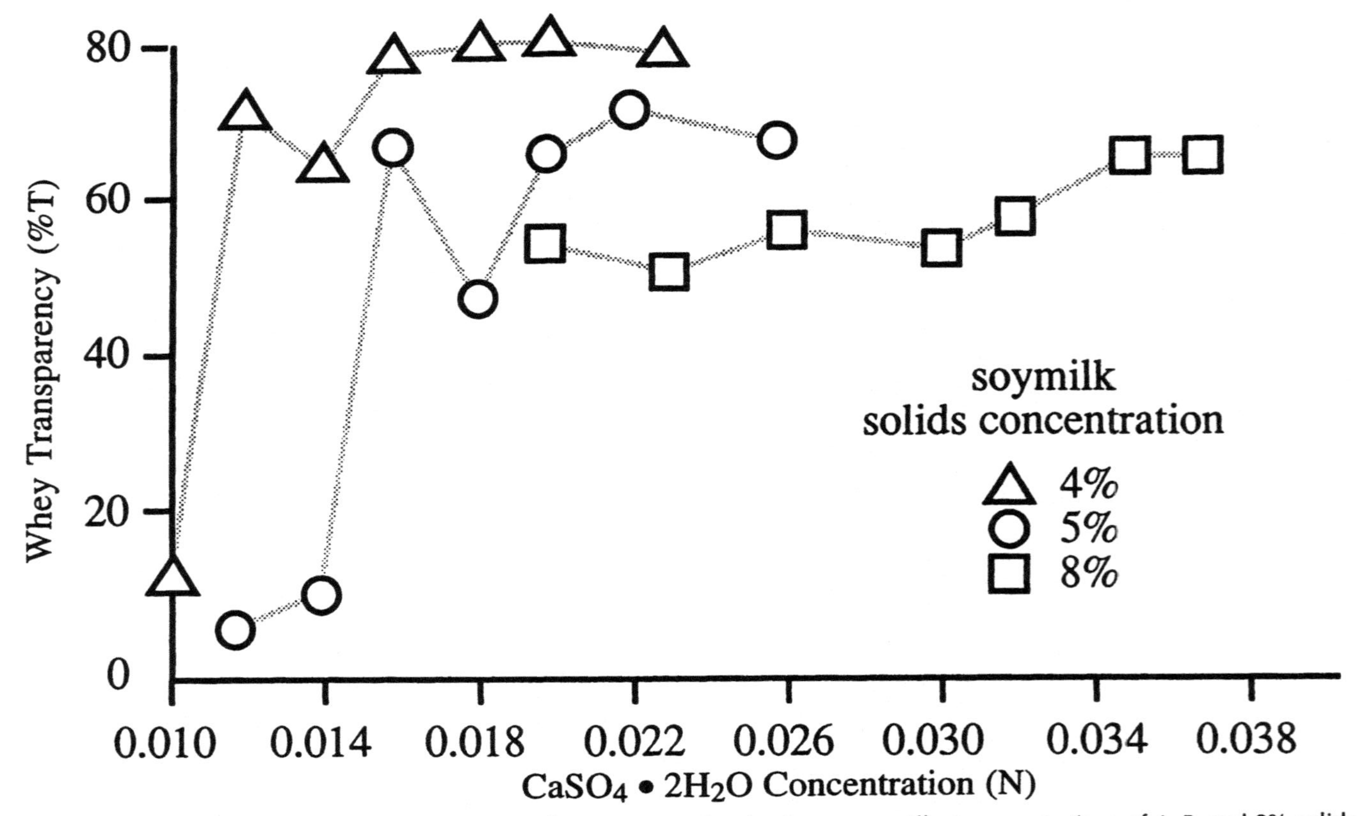

Fig. 22.6. Percent transmittance of whey versus coagulant concentration for Amsoy soymilk at concentrations of 4, 5, and 8% solids. Concentrations of 0.018N, 0.19N, and 0.035N, respectively, were selected as the optimum coagulant concentrations. *Source:* Johnson, L.D., Influence of Soybean Variety and Method of Processing on Tofu Manufacturing, Quality, and Consumer Acceptability. M.S. Thesis, Iowa State University, Ames, IA, 1984, p. 128.

considerable interest in locating soybean-processing plants outside the country. We have found that flaking of soybeans for soymilk and tofu manufacture generates a 20 to 40% cost savings, because of reduced water and electrical usage during the tofu-manufacturing process, and significantly shorter processing (soaking) times (15 min versus 12 h). Tofus produced from flakes were lower in fat than those produced from whole beans. Flake production in the United States is cost-effective for some Japanese producers. The third strategy in the foregoing list is "steam infusion cooking" (32), a technique that has been commercially adopted in parts of the United States and has been test-marketed in Japan.

Kinugoshi (Silken) Tofu. The name "silken" is often misconstrued as referring to the use of a silk cloth instead of a cotton cloth for filtering soymilk. It actually refers to the fine, delicate texture of the tofu. Kinugoshi has a much more homogenous, delicate texture, is softer and has a smoother mouth feel than "regular" momen tofu. This is due to the use of GDL, developed expressly for kinugoshi and packed tofu. GDL coagulates the soymilk slowly due to the slow release of gluconic acid.

Kinugoshi tofu is unique in that it is produced by coagulating whole soymilk without separation and removal of the supernatant (whey) (Fig. 22.7). This results in a tofu with higher nutrient content, softer texture, and lower antinutritional factors; however, whey inclusion usually compromises flavor. More concentrated soymilk is used for kinugoshi (12 to 13% solids) than for momen tofu (5 to 10% solids).

Kinogoshi tofu can also be produced by the rapid mixing of a calcium sulfate suspension (0.5 to 0.6% of the soymilk by volume) and soymilk at about 70°C (158°F). This processing is done within either heated or unheated coagulation vessels. After the milk has coagulated in the vessel, it is carefully removed and irrigated with running water within a holding tank.

Packed Tofu. Packed tofu differs from the others previously discussed in that it is processed within its sealed retail container. It is manufactured by pouring cooked soymilk into a plastic rectangular container or plastic-lined fiberboard box with a coagulant (often a mixed coagulant such as a mixture of calcium sulfate and GDL) (Fig. 22.8).

The container is then heat-sealed with a plastic sheet, or the mixture is sealed within a similar commercial plastic packaging system, then heated in water at 90 to 95°C (194 to 203°F) for 40 to 50 min. The whole soymilk coagulates in the container. Again, whey is incorporated into the product, as it is with kinugoshi tofu. Packed tofu is more sanitary than other types, because all pathogenic bacteria are killed during heating. Additionally, the product is protected from contamination during manufacture, storage, and distribution. Packed tofu is more shelf-stable and easier to transport than either momen or kinugoshi tofu. For these reasons, large continuous-production operations are possible. This product is increasing in popularity with supermarket chains and other large retailers.

Aseptically Packaged Tofu. Aseptically packaged tofus are essentially the packed tofu products mentioned in the previous section, except that all ingredients are commercially sterilized prior to formulation and packaging (Fig. 22.9).

These products were specifically developed and manufactured with an emphasis on their marketing as an export product. However, unanticipated consumer enthusiasm has resulted in advantageous openings into domestic tofu markets. To protect traditional tofu manufacturers (a well-organized small-business lobby) from this threat, production of aseptically packaged tofu for domestic markets has been strictly limited by government regulation to 1% of total tofu sales (3).

Production begins with the heat sterilization of soymilk at 135°C (275°F) for several seconds using a plate heat exchanger. The sterilized soymilk is then mixed with a sterile-filtered GDL solution, packaged within a plastic container, and sealed. All preparation is done under carefully controlled conditions within a "clean room." The packages are then heated in hot-water baths to coagulate the tofu. The resulting product has a shelf life of 6 mo to 2 yr at ambient temperatures.

Deep-Fried Tofu. There are three main types of deep-fried tofu: *namage* (single-fried tofu), *aburage*, and *gan-modoki* (double-fried tofus). Deep-fried tofu also has a longer shelf life and is more transportable than conventionally packed, fresh momen tofu. Additionally, it can be made on a relatively large scale using continuous fryers.

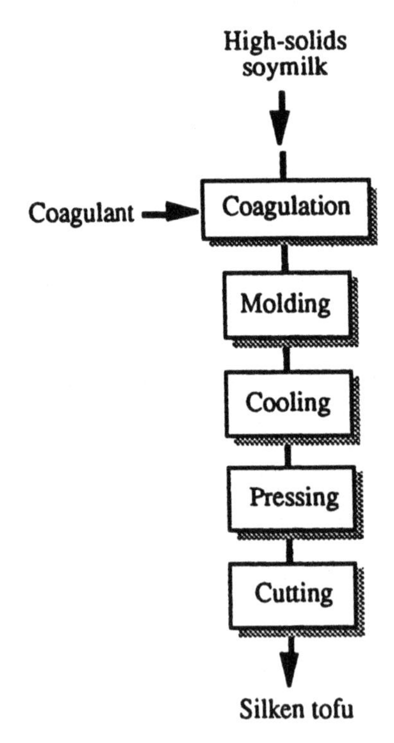

Fig. 22.7. Steps in kinugoshi (silken) tofu production. *Source:* Wilson, L.A., et al., *Japanese Soyfoods: Markets and Processes,* MATRIC, Iowa State University, Ames, IA, 1992. Reprinted with permission of MATRIC.

Namage is a fresh, single-fried tofu that is prepared by deep-frying cut pieces of pressed tofu. While the inside texture is nearly the same as the original tofu, the surface is lightly browned and firm.

Aburage is made swollen and porous by an initial first frying in low-temperature oil (110 to 120°C, 230 to 248°F) (Fig. 22.3). This texture is then fixed by removing additional water by a second frying in high-temperature oil (180 to 200°C, 356 to 392°F). The best tofu of this type is obtained when the tofu expands to three times its original size. This swelling can be negatively influenced by improper heating conditions during the initial stages of soymilk and tofu production. Careful control of heat regimes during these and subsequent stages of production ensure minimal protein denaturation and maximum retention of air bubbles within the product.

Gan-modoki is another kind of double-fried tofu. It is made from dehydrated minced tofu and is a mixture of tofu, Chinese yams, finely chopped carrots, kelp, hemp seeds, sesame seeds, and other ingredients. It is usually molded into a ball, deep-fried, and sold hot to the customer. This product has a more porous texture than aburage.

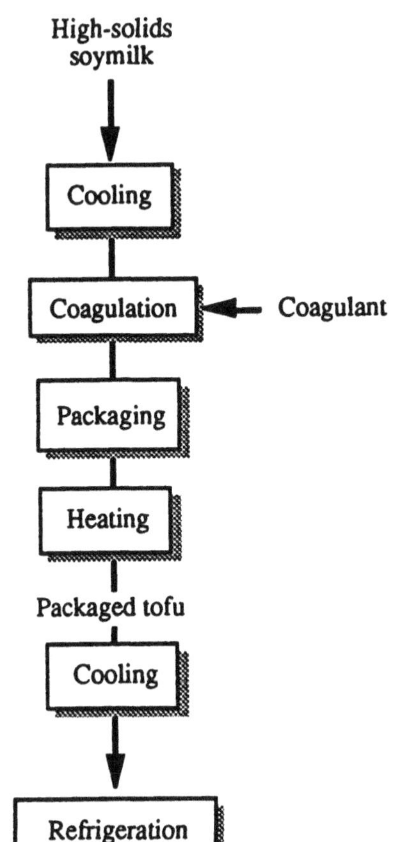

Fig. 22.8. Steps in packaged tofu production. *Source:* Wilson, L.A., et al., *Japanese Soyfoods: Markets and Processes*, MATRIC, Iowa State University, Ames, IA, 1992. Reprinted with permission of MATRIC.

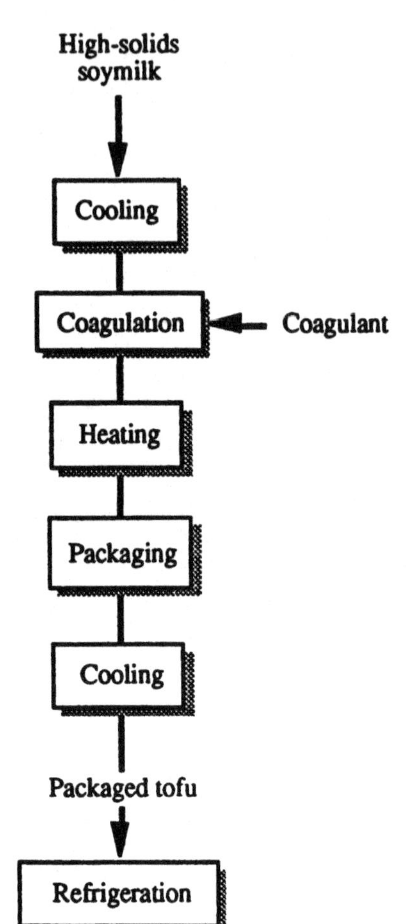

Fig. 22.9. Steps in aseptically packaged tofu production. *Source:* Wilson, L.A., et al., *Japanese Soyfoods: Markets and Processes,* MATRIC, Iowa State University, Ames, IA, 1992. Reprinted with permission of MATRIC.

Kori Tofu (Dried-Frozen Tofu). Kori tofu is a frozen, dehydrated tofu. The dehydration process does not detrimentally affect the nutritional value or digestibility of the material. Upon rehydration, its texture is very different from the original tofu.

The production process for kori tofu begins when hard and sandy-textured tofu is made from soymilk coagulated with calcium chloride. The whey is removed, and the mixture is blended to break down the curds. This releases more whey, which is again removed. The curds are then transferred to wooden boxes for pressing (Figs. 22.10 and 22.11).

The resulting tofu is very firm. This is cut into pieces of 60 × 72 × 25 cm (24 × 28 × 10 in) or similar size. They are then forced-air frozen on metal trays at –10°C (14°F). After freezing they are stored for three weeks at –1 to –2°C (30 to 28°F).

Once thawed, the aged tofu, now very spongy, is easy to dehydrate. This is usually accomplished by mechanical compression, followed by hot-air drying until the product reaches 17 to 18% moisture. Care is taken during the drydown to ensure against cracking of the blocks. Final drying is accomplished in the open air.

Historically, kori tofu was produced during the winter season as a result of the natural freezing of stored tofu. Some tofu in the northeastern part of Japan is still preserved as kori tofu following historical preparation methods. Almost all commercially dehydrated products are made using large-scale freezing units. Kori tofu's long shelf life and its transportation, and storage characteristics make it an attractive product for large-scale manufacturers. Several production facilities currently use over 10 MT of soybeans per day, and the products from three of these companies hold more than a 70% share of the Japanese market (3).

Older Japanese consumers prefer kori tofu, which swells to a large size yet remains soft when cooked, and they appreciate the taste of the product. Younger consumers do not relish the taste and usually purchase other tofus. In the past, to enhance desirable textural characteristics, the dried product was exposed to ammo-

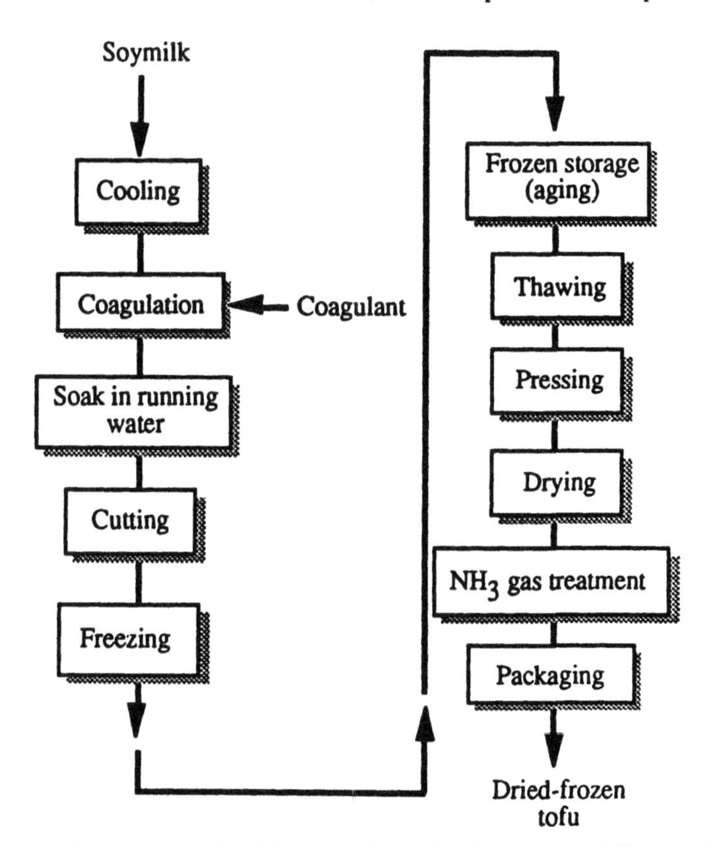

Fig. 22.10. Steps in kori (dried-frozen) tofu production. *Source:* Wilson, L.A., et al., *Japanese Soyfoods: Markets and Processes,* MATRIC, Iowa State University, Ames, IA, 1992. Reprinted with permission of MATRIC.

nia gas. Fortunately, soaking in *kansui* (a solution of sodium or potassium carbonate or potassium phosphate) has replaced the practice of ammonia exposure. Ammonia use is undesirable, particularly with respect to worker safety. In addition, it was known to impart an unacceptable brown color and a characteristic off-odor to the stored product (5).

The yield of kori tofu is about 4.5 to 5.0 kg from 10 kg soybeans, with each dried piece of kori tofu weighing about 20 g. Kori tofu, with 53% protein and 26% fat (Table 22.3), is considered to be highly nutritious and desirable by consumers. It is more commonly used in western Japan.

Other Nonfermented Soy Foods

Yuba (Soybean Protein Film). Some consumers use fresh film for cooking, whereas others prefer dried yuba. This is packaged in sheet form, in small rolls, or in pieces of various shapes and sizes.

To prepare yuba, soymilk of the same solids content as that used for tofu is heated in a flat pan until it nears the boiling point. A protein film then forms on the surface of the milk and is then removed with a bamboo stick (or a sanitary stainless

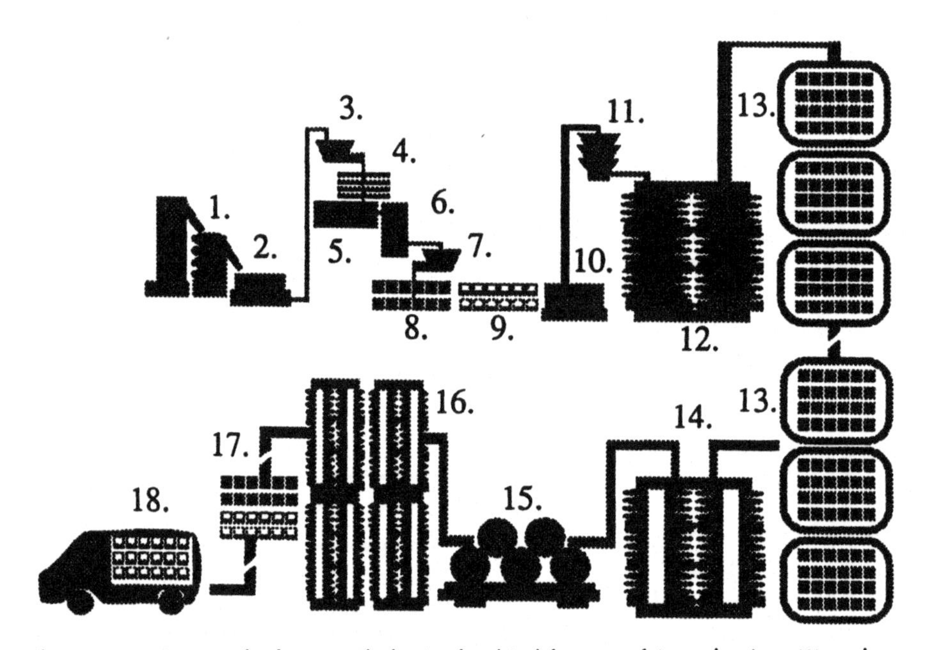

Fig. 22.11. Process for large-scale kori tofu (dried-frozen tofu) production: (1) soaking tank, (2) grinder, (3) cooker, (4) filter, (5) soy milk, (6) coagulation tank, (7) grinder, (8) molding box, (9) dehydration, (10) soaking, (11) cutter, (12) freezing equipment, (13) frozen storage, (14) continuous thawing, (15) press dehydration, (16) drier, (17) trimming, and (18) packaging. *Source:* Wilson, L.A., et al., *Japanese Soyfoods: Markets and Processes*, MATRIC, Iowa State University, Ames, IA, 1992. Reprinted with permission of MATRIC.

steel rod), and carefully dried at ambient temperature. After the first film is lifted, continued heating produces successive films, each of which is in turn removed. As the films are removed, the solids content of the soymilk is gradually reduced to the point where it is impossible to form another film.

Yuba is rich in protein and oil; however, it is an expensive product because of its labor-intensive production. Yuba is primarily a local food in western Japan, although it is increasing in national and international popularity (especially Korea). The product is gaining widespread exposure as an ingredient in vegetarian dishes.

Kinako (Roasted Whole Soybean Flour). To make kinako, whole soybeans are roasted in a pan or rotating roaster, dehulled, cracked to grits, then ground to a fine powder with an impact grinder. The powder is then roasted until it has a desirable toasted flavor. Great care is taken not to exceed 220°C (428°F) within 30 s. This constitutes overheating and is deleterious to the protein, forming a bitter taste and lowering its nutritive value.

The hull is usually excluded from the product, although in some cases it is included to increase the fiber content. The digestibility of kinako is lower than that of tofu. It is often mixed with sugar and salt and is used to coat baked mochi. It is also used as a cake base. Green kinako is made from domestic soybeans, which have green hulls and cotyledons.

Fresh (Edamame) and canned soybeans. Soybeans that are green, soft, and large can be harvested when they are about 80% mature and prepared as fresh beans or peas. They can also be canned using the appropriate times and temperatures for low-acid foods.

Texturized Soy Protein–Based Foods. Texturized soy protein products, made by texturizing whole soybeans with single- and twin-screw extruders (see Chapter 8), are also manufactured and marketed in Japan. They are similar to extruded foods in the West and are primarily marketed as meat analogs, but only to specialty markets (hospital food services, individuals on restricted diets, some canned foods, etc.) (3). Consumers generally prefer "fresh" products to those that are reconstituted or extended. A related product, called "emulsion curd," is a tofu made from soybean protein isolate, soybean oil, and water. When these materials are mixed at a specific ratio, a stable curd resembling tofu is formed; this is then frozen or dried. The curd recovers much of its original shape and texture when rehydrated or thawed, unlike the traditional dried or frozen tofu. For this reason, it often replaces traditional tofu as an ingredient in dried or frozen foods.

Fermented Soy Foods

Fermented soy foods usually contain salt and the byproducts from a desirable fermentation. Both inhibit or slow the spoilage of these products and allow them to have a relatively long shelf life compared with fresh soy products such as traditionally prepared momen tofu (33).

Miso

Miso is made by mixing cooked soybeans with *koji* (starter culture, often fermented rice), and salt water. This material is then fermented for several months (Fig. 22.12). There are several miso products, which differ in the type of koji used for the fermentation (34). Rice koji is used to make rice miso, barley koji to make barley miso, and soybean koji to make soybean miso. A high ratio of rice or barley to soybeans results in a more lightly colored and sweeter miso. In conjunction with the ratio of rice or barley to soybeans, the hydrolysis of starch to maltose and glucose is essential to miso production. The fermentation period for high-wheat or -barley miso is usually shorter than for miso with greater soybean content. "Soybean miso" has the longest fermentation period, taking from 1 to 2 yr to produce an acceptable flavor.

The salt content of miso is usually about 10% or more by weight. Recent research has succeeded in depressing the salt content by adding alcohol or extra yeast to suppress undesirable "wild" fermentations.

To make miso, washed soybeans are soaked in water, then autoclaved (1.5 to 2.0 h at 0.5 atm) until softened (Fig. 22.12). In large plants, continuous soybean cookers are used. The cooked soybeans are cooled to 35 to 40°C (95 to 104°F), then mixed with koji and salt. The type of koji starter and whether rice or barley is used are important components in the biochemical process responsible for the flavor characteristics of the final product (Fig. 22.13). This mixture is then agitated in a semisolid state with water or previously drained cooking liquid. The resulting material is packed within fermentation casks, tightly covered with a thin plate or wax sheet, top-weighted, and allowed to ferment.

In the past, miso was produced by a seasonal, natural fermentation, which ran through the summer months. Temperature-controlled fermentation has made it possible to ripen miso within three months. The lactic acid bacterium *Pediococcus halophilus* and the yeast *Saccharomyces rouxii* are often added to the mash to accelerate the ripening process. The yield from 100 kg of soybeans and 100 kg of rice is about 300 to 400 kg of miso.

For other "specialty misos" the ingredients and fermentation vary. For white miso, rice is the dominant raw material, and the water in which the soybeans soak is removed. This procedure prevents the product from browning during fermentation, so that it retains the desired "white" color. Of some 600,000 MT of soybeans used for miso in Japan, 70% was for regular miso and 30% was used for white miso (34).

Soybean miso is made exclusively from soybeans. Cooked soybeans are ground, molded into balls, then covered with powdered koji starter and incubated in a koji room to promote the growth of *Aspergillus oryzae*. After four days, soybean koji germination is complete. The mash is then mixed with water and salt, and ripened in casks. This miso is aged for one year. The interrelationships between the various ingredients that produce the unique sensory properties of miso (color, texture, flavor, taste, aroma) are complex (see Fig. 22.13). A key factor in the quality of the final product is the enzymatic action of the microorganisms and how they influence the composition of the substrate (rice, barley, soybeans, rice and barley, rice and soybeans, barley and soybeans).

A spicy red "hot" soy paste (*ko chu jang*) is made in Korea by mixing a cooked soybean *koki* with cooked rice and red pepper in brine. The mass is allowed to ferment and ripen for several months (2).

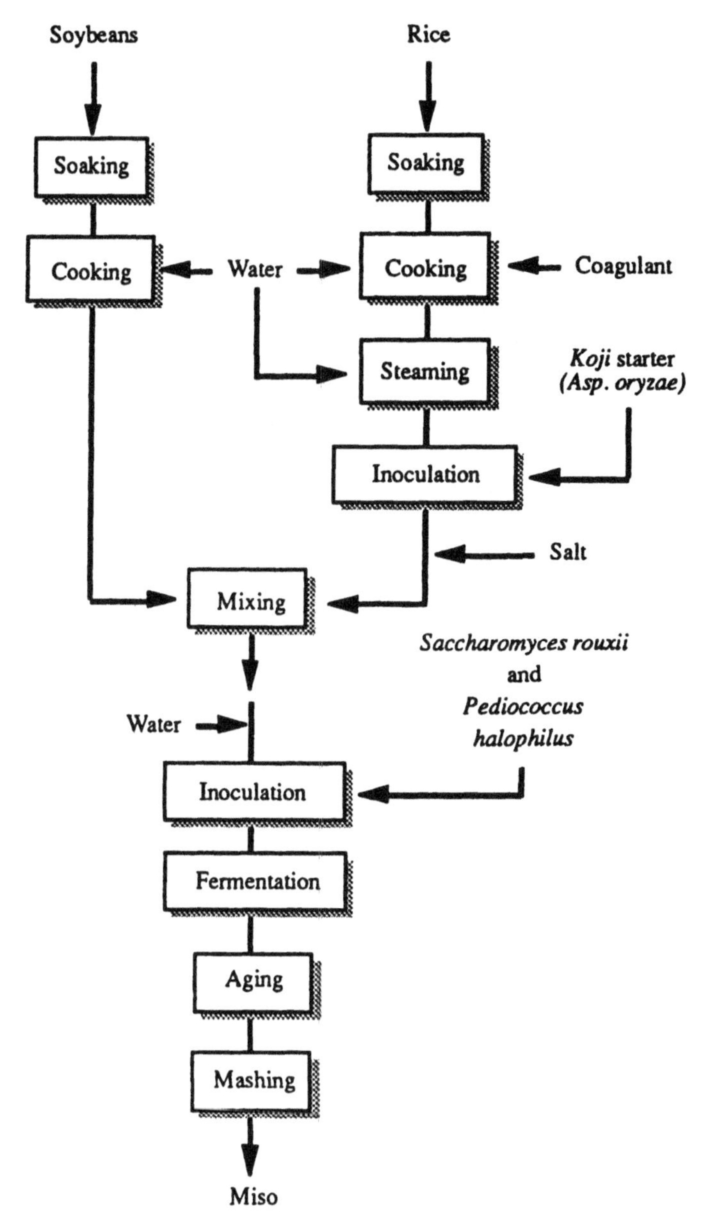

Fig. 22.12. Steps in miso production. *Source:* Wilson, L.A., et al., *Japanese Soyfoods: Markets and Processes*, MATRIC, Iowa State University, Ames, IA, 1992. Reprinted with permission of MATRIC.

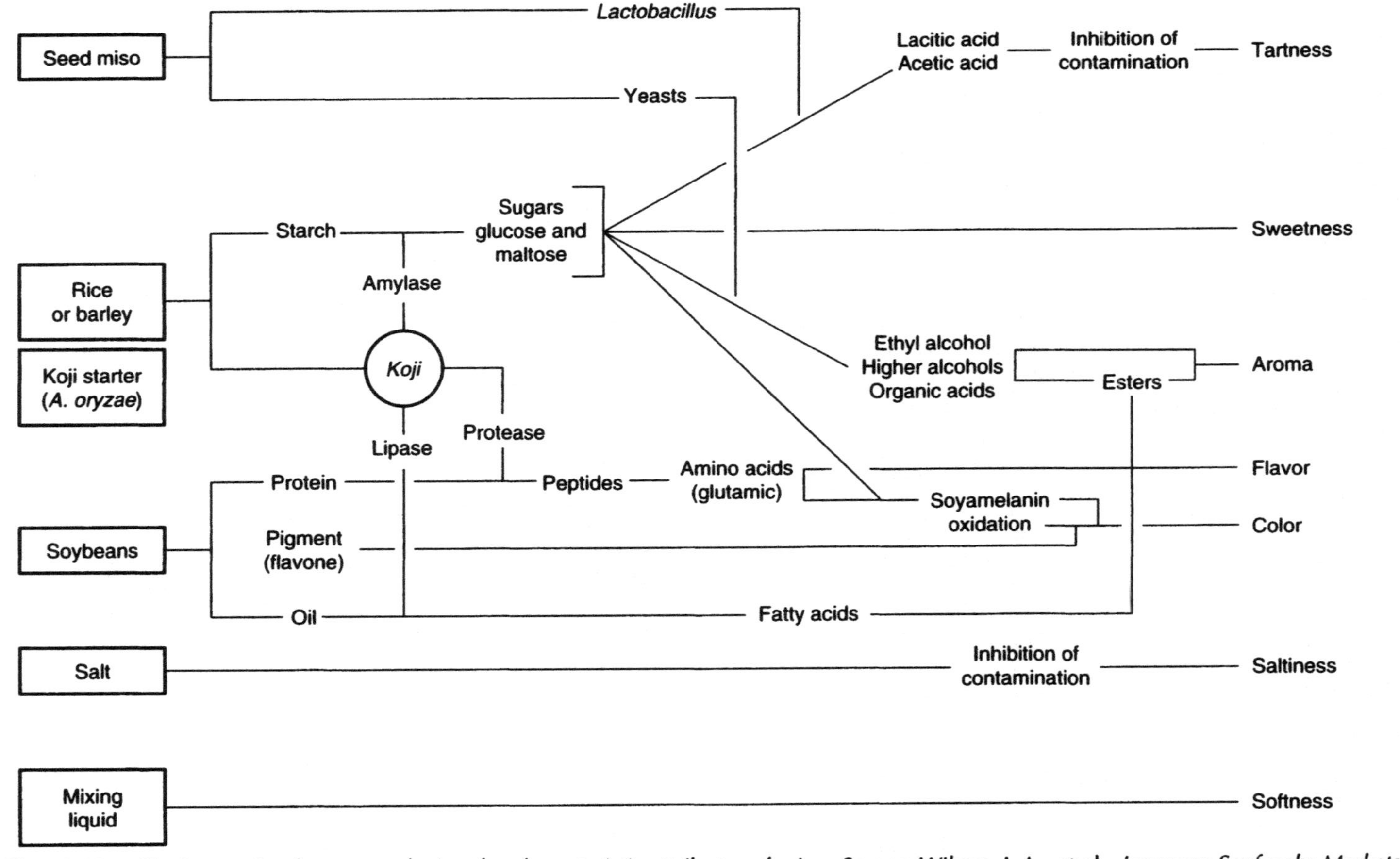

Fig. 22.13. The interactive factors producing the characteristic attributes of miso. *Source:* Wilson, L.A., et al., *Japanese Soyfoods: Markets and Processes*, MATRIC, Iowa State University, Ames, IA, 1992. Reprinted with permission of MATRIC.

The introduction of supermarkets throughout Japan has stimulated the development of innovative packaging strategies. Previously, difficulties were encountered when some plastic packaging swelled and ruptured from carbon dioxide produced by the product once it was sealed in the container. This was especially a problem in supermarket shelf displays. Heating of the miso and incorporation of ethyl alcohol have largely solved these problems. The average Japanese citizen consumes around 20 g of miso a day, often in the form of hot miso soup served with vegetables (leafy vegetables, daikon radish, onions, etc.), seaweeds, or tofu. "Instant" miso is made by drying miso in a vacuum drum drier. Ingredients, such as lyophilized (freeze-dried) vegetables and deep-fried tofu, are often added to complete the product for marketing as a fast-food convenience item. Production drying technology influences the ultimate quality of the finished product.

Shoyu (Soy Sauce)

Soy sauce manufacturing was modernized much earlier than that of other soy foods. Soon after World War II (1945), the introduction of labor-saving machinery and mass production techniques spurred both profitability and growth in the industry. Defatted soybeans, the meal byproduct of soybean oil extraction, were already used prior to World War II as the raw material for soy sauce. When whole soybeans are used, soybean oil separates and collects on the surface of the sauce. For this reason defatted soybeans are preferred (4,5).

To make soy sauce, defatted soybean meal (30 to 60 nitrogen solubility index (NSI) (see Chapter 8) is well moistened with water and then cooked in a kettle or continuous cooker. The meal is then cooled to protect the protein from further denaturation. Although the protein is denatured to some extent during the cooking process, overdenatured meal can have an adverse affect on protein hydrolysis during aging and is rigorously avoided. Concurrent with meal preparation, wheat kernels are roasted and then broken into grits. These are then mixed with soy sauce koji starter, spread on top of the steamed soybean meal, and then mixed together. This material is piled on the floor of the koji room at 30 to 35°C (86 to 95°F) to start the growth of *A. oryzæ*. As the organisms grow, they congeal the mixture. The growing clumps of cultured material are continually separated from the remaining mash. This is done to ensure an adequate supply of fresh air to the growing culture.

Forty-eight hours are necessary for the preparation of soy sauce koji. The koji is greenish-yellow with white spots and has a sharp, volatile flavor. It is then mixed with salt water in a fermentation cask or tank. The material, now called *moromi*, is stirred with a paddle or by compressed air at least once a day to supply oxygen, remove carbon dioxide, and homogenize the mixture. Stirring time is reduced as fermentation progresses.

Enzymes from *A. oryzae* hydrolyze starch and protein, producing the characteristic flavor and aroma of soy sauce. The growth of various added or natural microorganisms assists fermentation and ripening. Historically, this aging took about one year. Today, it can take as little as six months because of technical processing improvements (e.g., continuous-agitation fermenters). Some brands employ accelerated ripening at 40°C (109°F). In these cases the ripened moromi is filtered to remove insoluble material, then heated (pasteurized) for 30 min at 60 to 70°C (140 to 158°F). To produce 50 kg of soy sauce, 10 kg soybeans (corresponding to 3 kg defatted soybean meal) are used.

Soy sauce quality determinations (or the nitrogen utility value) are based on the product's nitrogen content. Standard soy sauce contains 1.5% nitrogen and has a salt content of about 18%. Within the past few years, low-salt soy sauce has entered the market, reflecting the interest in reducing salt intake exhibited by the majority of Japanese and American consumers.

The nitrogen content of soy sauce (nitrogen utility) results from the fermentation conversion of proteins to amino acids. This value is influenced by the temperature and duration at which the soybean meal is steamed and, to a lesser extent, the strain or strains of fermentation microorganisms. Formerly, the nitrogen utility value of most products was 55 to 65%, but production improvements (e.g., fermentation temperature control technology) has raised this to 90% (2,3). Another method for increasing nitrogen utility is to heat the soybeans in the presence of hydrochloric acid to partially hydrolyze the protein in the meal. The acid is then neutralized with caustic soda, and the material is supplemented with salt and wheat bran koji and allowed to ripen at 30°C (86°F) for several months. This process raises the nitrogen utility value to 90%; however, the resulting product has an odor and flavor that differ from those of traditional soy sauce. Soy sauces produced by each production method are currently available as brand name products in both the United States and the Pacific Rim.

Annually in Japan, roughly 1.25 million MT of soy sauce are produced. This processing consumes from 170,000 to 180,000 MT of defatted soybeans and wheat. Five firms account for 50% of total production. Each Japanese citizen consumes a daily average of about 30 g of soy sauce.

Natto

Natto is a cooked whole soybean product fermented with *Bacillus natto*. There are similar products in the Indonesian and Thai markets, but not in China. Originally natto was developed using *B. natto* grown in rice straw. Preparation involved wrapping steamed or cooked whole soybeans within bundles of these prepared rice straws, then incubating them, allowing the *B. natto* in the straw to grow into the soybeans. Today, natto is made using pure-cultured *B. natto*.

Modern production is relatively straightforward. Soybeans are soaked in water overnight, then cooked to soften them. The cooled mash is inoculated with a commercial *B. natto* culture (a liquid suspension or starchlike powder), then mixed in a rotary cask. Fresh packets are then prepared by wrapping 50 to 100 g of mash with a thin piece of perforated polyethylene film or by placing the mixture in shallow polystyrene or wood trays. These are stacked in the fermentation room at 30 to 40°C (86 to 104°F) for 24 hr or until the soybeans are fully covered by a white, sticky glutamic acid polymer. The product is then transferred to a cold room for storage or transported to market.

Within the last few years a number of large-scale natto production plants have been established. Their existence is the direct result of solid-state temperature and humidity control systems that have facilitated automated production techniques. Additionally, the development of active-enzyme treatments to decompose starch and protein has reduced soaking and cooking times, accelerating the breakdown of the cooked soybeans and softening product texture. These additives are believed to have the additional benefit of aiding the digestion of the soy food within the human intestine.

Small soybeans are often preferred for natto production. It is perceived that small beans absorb water more easily, shortening the steaming time. Additionally, the larger surface-to-volume ratio of small beans may be a factor in establishing the proper degree of *B. natto* colonization and growth. Smaller beans are also easier to eat. Hardness problems, frequently related to small bean size in the United States (35; Kim, C.J., L.A. Wilson, and K.H. Hsu, *Cereal Chem.*, in press), have not been reported to be a problem in Japan. In 1984 roughly 1000 natto production plants used 150 kg of soybeans per day. The larger plants used from 2 to 3 MT daily. The total use of soybeans for natto production in Japan is from 70,000 to 80,000 MT annually.

Tempeh

Tempeh is a fermented whole soybean product that originated in Indonesia but is now equally popular in Malaysia. Tempeh is made by soaking soybeans overnight, then boiling them with the hulls (Malaysian) or without the hulls (Indonesian) for 30 min. The excess water is drained off, and the beans are placed on a tray for inoculation with a piece of tempeh or *Rhizopus oligosporus* (2,36). The beans are then allowed to ferment at room temperature for 1 to 2 d or at 30 to 32°C (86 to 90°F) for 20 h (Fig. 22.14). During fermentation, white mold mycelium covers the beans and binds them together into a solid sheet. The tempeh sheet is then cut into smaller pieces and, since it is a perishable product, sold that day. If it is to be stored for future use, it is usually blanched, sun-dried, or frozen. Tempeh is usually cooked before it is eaten. Preparation usually involves frying, deep-fat frying, or baking the product. It is also added to soups and fast foods and is used as a meat replacement in main dishes (37).

Sou-Fu

Fermented tofu (soy curd) is made in China by inoculating small cubes of pasteurized firm tofu (less than 70% moisture) with *Actinomycor elegans*. Other molds, such as *Mucor* and *Rhizopus*, can also be used. Depending upon the mold used, fermentation lasts from 3 to 7 days at 20°C (68°F). After fermentation, the cubes are placed in a 12% salt and rice wine brine for several months. The finished product is packaged with the brine and sterilized prior to marketing. The product has a mild flavor, a salty taste, and a creamy cheese texture (38).

Japanese Agricultural Standards

The Japanese Agricultural Standards (JAS) were enacted in 1951 to improve the quality of processed foods, simplify the Japanese trading system, promote uniform labeling, and to make it easier for the consumer to select high-quality processed foods (21,39). Unlike the U.S. Food and Drug Administration Standards of Identity, these standards are optional. The JAS quality standards define the food product, specify the application range, and define quality judgment criteria. The standards also set labeling formats, such as volume or weight, and similar criteria. These standards are applied to all products before they are shipped.

In 1988 the JAS covered over 70 items of processed foods including shoyu, kori tofu, vegetable proteins and their products, and refined soybean oil. These foods are therefore allowed to carry the JAS seal on their labels. Foods processed outside of Japan can be labeled with this seal if the company producing this product has petitioned for certification and received permission from the Japanese Government.

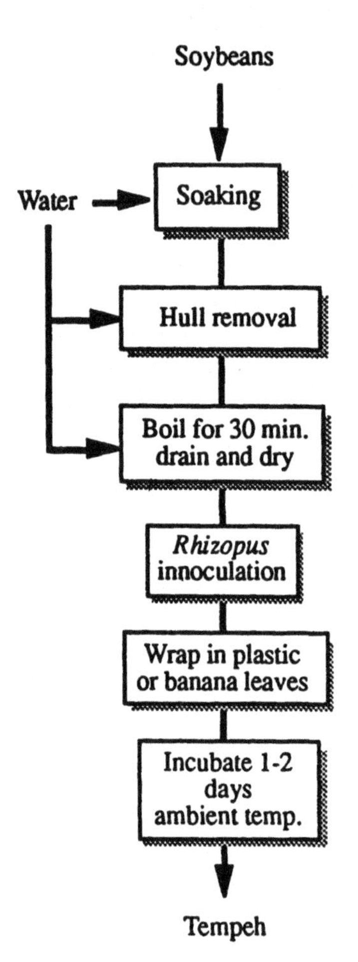

Fig. 22.14. Steps in tempeh production. *Source:* Wilson, L.A., et al., *Japanese Soyfoods: Markets and Processes*, MATRIC, Iowa State University, Ames, IA, 1992. Reprinted with permission of MATRIC.

Identity Preservation and Transportation

There are three main avenues by which U.S. soybeans are exported to Asia for food use. Traditionally, when importers purchase soybeans, they are sized, cleaned, and bagged on arrival. The importer then sells the beans directly to large tofu manufacturers or through various distributors to the smaller tofu shops and other soy food producers. The importation of mixed-variety Indiana, Ohio, and Michigan (IOM) beans is an example of this system.

Soybeans for crushing (resulting in oil and meal) constitute the second route through the distribution system. These soybeans are not specifically grown for the soy foods market and are generally composed of mixed varieties of U.S. No. 2 soybeans. The beans are often imported directly to the crushing company, where they are sorted by size and cleaned. They are then bagged, put into containers for sale to soy food processors, or crushed for oil and meal. Alternatively, the company may sell the beans to other primary or secondary distributors (15). Some of the crushers also extrude meal to produce soy-based meat analogs.

More recently, Japanese companies have contracted with specific U.S. seed companies or directly with American farmers to supply specialty "identity-preserved" soybeans for the production of specific soy foods in Japan. These purchases were initially related to tofu production; however, specialty purchasing of beans for miso and natto production is increasing in frequency (31).

Specialty soybeans are typically cleaned and bagged at the farm or seed company and then loaded into 18 MT (6601 bu) containers. These are then shipped by rail to U.S. west coast seaports, where they are loaded onto container ships bound for Japan. Container shipping maintains segregation (identity) of the individual lots, unlike the mixed cultivar U.S. No. 2 and IOM soybeans routinely bulk-shipped by freighter. The quality of container-shipped beans is usually much higher, since they are not a mixed lot that requires regrading and the multiple handling steps that directly expose the grain to physical degradation.

U.S. farmers, seed companies, and co-ops interested in directly marketing their soybeans to Japan may do so by contacting a local exporter, a Japanese trading company, or a Japanese soy food processor (15).

Soybean Quality Characteristics

While plant breeders, seed companies, brokers, and soy food manufacturers are all interested in identifying measurable soybean characteristics that can be used to characterize the end product quality of soy foods, very little published research is currently available. Likewise, computerized retrospective literature searches yield few references (Shurtleff, W., personal communication). The majority of reports concern the influence of particular varieties or soybean processing methods on the resulting soymilk and tofu rather than on identifying specific, measurable characteristics (40). Worldwide, most soybean producers and soy food processors recognize that in order to make quality tofu, high-quality soybeans are a must; however, actually identifying the "favorable characteristics" of desirable soybeans is not an easy task.

Quality characteristics of soybean-based foods, such as tofu and soymilk, have been defined subjectively (color, flavor, texture, etc.) by approximate analysis and texturally by using instrumentation. However, the relationships of these end-product quality factors to composition of the soybean are largely unknown.

Many variables contribute to soybean quality, including soybean variety, soybean environment (field and storage), phytic acid (phosphorus content), bean water uptake, composition of the protein fraction (glycinin, β-conglycinin, and their subunits), removal of the insoluble solids (okara), soymilk solids concentration, coagulation temperature, type of coagulant used, curd breakage, whey removal, pressing time and force, and plant and personnel sanitation.

Overview

Research at Iowa State University and the body of published articles on soy food quality suggests that the environment under which the beans are grown and the variety of soybean are the greatest initial influences on tofu quality (3). Neither the total amount of soybean protein or its NSI correlated with the yield or the textural qualities of tofu made from those soybeans. Glycinin appears to play a more significant role in forming the texture (elasticity, chewiness, brittleness) of tofu than does β-conglycinin.

The length of soybean storage, even under closely controlled refrigeration, significantly influences the textural properties of the resulting tofus (3). After 15 months of storage, all correlations between glycinin or β-conglycinin and the textural properties of the tofus were lost.

High-temperature and -relative humidity storage conditions of soybeans also caused significant decreases in the quality of soybeans and significant changes in textural and appearance characteristics of tofus. The effects were very variety-dependent, allowing no overall prediction indicator of tofu quality. For example, the NSI of Vinton 81 soybeans was significantly correlated with fracturability and cohesiveness, whereas the NSI of Pioneer 9202 soybeans was correlated only with cohesiveness, and no correlations were found for all other stored varieties between NSI and textural characteristics (40,41). The significant change in extractable lipid content suggests that its role in tofu quality may be more important than previously realized.

The percentage protein of soybean varieties should not be used as the sole criterion for the breeding or purchasing of soybeans for tofu manufacture. However, low β-conglycinin levels in the soybeans may increase tofu yield. Some U.S. soybeans compare favorably in composition, tofu yield, and tofu sensory properties to Japanese soybean varieties.

Judging Quality

The following recommendations apply to soybeans intended for soy foods (3,15).

Tofu. For tofu, the Japanese currently prefer soybeans with a clear to light-colored hilum and large seed size (>180 g/1000 beans, 6.5 to 7.0 mm). (However, some beans with black hila have been observed blended into some lots.) Ideal soybeans should be cream-colored, without cracks in the seed coat or seed, and firm to the touch. While the seed coat should be intact, it should be thin enough to be easily removed (usually self-exclusive properties). The protein content of the beans should be >40% (moisture-free basis). Some processors specify a high NSI or protein dispersibility index (PDI) in the belief that a high value will mean a larger yield of soymilk, a larger yield of tofu, or less degradation from overheating during shipment. Of these perceptions, the second is relatively unlikely, while the first and last may have some validity.

There is also a commonly held belief that soybean color, size, and protein level indicate whether a high-yielding, white to cream-colored tofu with acceptable texture will be produced. However, processors have acknowledged that these characteristics have proven less than reliable for identifying desirable bean varieties. Recent sensory and analytical evaluations of U.S. and Japanese varieties confirm that many U.S. varieties (Vinton 81 is the current standard variety) should be as acceptable as Japanese varieties for use in high-value soy foods.

Often, varieties that pass these screening specifications fail to produce desirable tofus. For that reason, each processor has developed specific processing parameters for different commonly used soybeans (e.g., Enrei, Vinton 81, IOM). Some companies are now setting specifications for not only color, protein, and NSI but also peroxide value (PV) and thiobarbituric acid (TBA) tests (measure of oil oxidation) or acid value, or free fatty acid (FFA) level (enzymatic activity releasing fatty acids). If the beans do not perform under these conditions, they are deemed unreliable and are rejected. Plant breeders, processors, trading company representatives, educators, government agencies all desire an objective testing system that would identify desirable imported beans prior to purchase. Likewise, such a test could be used by processors' quality control departments to monitor supply integrity during transit and storage. Production departments could then use these test results to set the optimal processing parameters for individual varieties.

Miso. Miso processors prefer somewhat different soybeans than do tofu producers. They prefer soybeans with a white to yellow hilum and cotyledon, although brown, black, and purple are acceptable as long as they are large in size (200 to 250 g/1000 beans), preferably with intact thin hulls. These beans should have high total sugar contents, with sugars of relatively high fermentable or hydrolyzable value. Protein contents should be of average value; oil content is seldom considered.

Natto. In the production of natto, round, small(~170 g/1000 beans)-sized (5 mm diameter, although one prefecture prefers 8 to 10 mm diameter) soybeans are preferred. The seedcoat and hilum should be white to a light cream color. It is said that "the brighter the light yellow, the better the taste." The 100,000 MT of soybeans used by producers yields roughly 200,000 MT of natto per year. Approximately 60% of the beans used for natto production are of Chinese origin, while the remaining 15% are Japanese, and 26% are from U.S. sources). Producers commonly experience two problems with U.S. soybeans. They are typically larger in size than is desired and have insufficient "sugar" content. The price of soybeans for natto is five times the cost of normal soybeans (e.g., 2000 to 3000 ¥/60 kg for tofu vs. 15,000 ¥/60 kg for natto). It is possible that a marketing opportunity may exist with respect to the production of specialty small, "high-sugar" soybeans.

References

1. Hapgood, F., *National Geographic 172:* 66 (1987).
2. Snyder, H.E., and T.W. Kwon, *Soybean Utilization*, New York: AVI Books, Van Nostrand Reinhold Co., 1987, pp. 218–241.
3. Wilson, L.A., P.A. Murphy, and P. Gallagher, *Japanese Soyfoods: Markets and Processes*, MATRIC, Iowa State University, Ames, 1992. (CCUR, Iowa State University, CP 1, 1991).
4. Watanabe, T., and A. Kishi, *The Book of Soybeans*, Japan Publications, Inc., New York, 1984.
5. Watanabe, T., *Food Processing in Rural Areas of Japan*, Japan Food and Agriculture Association of the United Nations (FAO), Tokyo, 1988.
6. Watanabe, T., Industrial Production of Soybean Foods in Japan. *United Nations Industrial Development Organization (UNIDO), Expert Group Meeting on Soya Bean Processing and Use, Peoria, IL*, U.S. Dept. of Agriculture, Washington, DC, 1969.
7. Watanabe, D.J., H.O. Ebine, and D.O. Ohida, *Soybean Foods*, Kohrin Shoin, Tokyo, 1971 (in Japanese).
8. Saio, K., *Cereal Foods World 24:* 342 (1979).
9. Mheen, T.I., T.W. Kwon, and C.H. Lee, Traditional Fermented Food Products in Korea. Paper presented at the 8th ASCA Conference, Medan, Indonesia, Aug. 11–18, 1981.
10. Food and Agricultural Organization, *Food Balance Sheets and Per-Capita Food Supplies*, U.N. Food and Agricultural Organization, Rome, 1970.
11. Food and Agricultural Organization, *Food Balance Sheets and Per-Capita Food Supplies*, U.N. Food and Agricultural Organization, Rome, 1980.
12. Food and Agricultural Organization, *Food Balance Sheets and Per-Capita Food Supplies*, U.N. Food and Agricultural Organization, Rome, 1990.
13. Webb, A., and K. Gudmunds, *PS&D '90, '95*, U.S. Department of Agriculture, Economic Resource Service, Rockville, MD (electronic database).
14. Wilson, L.A., V.A. Birmingham, D.P. Moon, and H.E. Snyder, *Cereal Chem. 55:* 661 (1978).
15. Griffis, G., and L. Wiedermann, *Marketing Food-Quality Soybeans in Japan*, American Soybean Association, St. Louis, MO, 1989.
16. American Soybean Association, *Soya Bluebook*, American Soybean Association, St. Louis, MO, 1989.
17. Fernando, S.M., and P.A. Murphy, *J. Agric. Food Chem. 38:* 163 (1990).
18. Guzman, G.J., and P.A. Murphy, *J. Agric. Food Chem. 34:* 791 (1986).
19. Davies, C.S., S.S. Nielsen, and N.C. Nielsen, *J. Amer. Oil Chem. Soc. 64:* 1428 (1987).
20. Johnson, K.W, and H.E. Snyder, *J. Food Sci. 43:* 349 (1978).
21. Japan Hygiene Association. *Specifications and Standards of Food, Additives, Etc. (Food Sanitation Law). Ministry of Health and Welfare Notification No. 370, December 28, 1959.* Japan Hygiene Association, Tokyo, 1987.
22. Rehberger, T.G., L.A. Wilson, and B.A. Glatz, *J. Food Protection 47:* 177 (1984).
23. Tuitemwong, K., and D.Y.C. Fung, *J. Food Protection 54:* 212 (1991).

24. Fouad, K.E., and G. Hegeman, *J. Food Protection 56:* 157 (1993).
25. Hsu, K.H., C.J. Kim, and L.A. Wilson, *Cereal Chem. 60:* 208 (1983).
26. Shurtleff, W., and A. Aoyagi, *Tofu and Soymilk Production*, Soyfoods Center, Lafayette, CA, 1984.
27. Johnson, L.D., and L.A. Wilson, *J. Food Sci. 49:* 202 (1984).
28. Johnson, L.D, *Influence of Soybean Variety and Method of Processing on Tofu Manufacturing, Quality, and Consumer Acceptability*. M.S. Thesis, Iowa State University, Ames, 1984, p. 128.
29. Lim, J.M., L. DeMan, and R.I. Buzzell, *J. Food Sci. 55:* 1088 (1990).
30. Sun, N., and W.M. Breene, *J. Food Sci. 56:* 1604 (1991).
31. Iowa Soybean Association, *Iowa Soybean Review 2:* 17 (1991).
32. Johnson, L.A., C.W. Deyoe, and W.J. Hoover, *J. Food Sci. 46:* 239, 248 (1978).
33. Keith, E., and H. Steinkraus, *Industrialization of Indigenous Fermented Foods*, Marcel Dekker, Inc., New York, 1989.
34. Shurtleff, W., and A. Aoyagi, *The Book of Miso*, Soyfoods Center, Lafayette, CA, 1984.
35. Arechavaleta-Medina, F., and H.E. Snyder, *J. Amer. Oil Chem. Soc. 58:* 976 (1981).
36. Wolf, W.J., and J.C. Cowan, *Soybeans as a Food Source*, CRC Press, Inc., Cleveland, OH, 1975.
37. Shurtleff, W., and A. Aoyagi, *The Book of Tempeh*, Soyfoods Center, Lafayette, CA, 1985.
38. Hesseltine, C.W., and H.L. *Wang, in Soybean: Chemistry and Technology, Vol. 1*, edited by A.K. Smith and S.J. Circle, AVI Publishing Co., Westport, CN, 1972.
39. Intra-Governmental Council on Standards and Certification Systems. *Standards and Certification Systems in Japan: Measures for Improving Market Access*, Tokyo, 1989.
40. Narayan, R., G.S. Chauhan, and N.S. Verma, *Food Chem. 30:* 181 (1988).
41. Chen, H.-P. Effects of Glycinin, β-Conglycinin, Their Subunits and Storage Conditions on Tofu Sensory Characteristics. Ph.D. Thesis, Iowa State University, Ames, 1993, p. 97.

Chapter 23

Nutritional Aspects of Soybean Oil and Soy Proteins

P.J. Huth

Kraft General Foods Technology Center
Kraft General Foods, Inc.
Glenview, IL

Introduction

In the United States, soybean oil (SBO) is the predominant edible oil in the food supply. In 1992, SBO accounted for 78% (5 million MT, 11.1 billion lb) of the total edible fats and oils used in the United States for manufacturing salad and cooking oils, baking and frying fats, and margarine, as shown in Fig. 23.1 (1). Of all the SBO available for consumption in the United States, 54% (2.7 MMT, 6.0 billion lb) is estimated to be partially hydrogenated, which accounts for about 42% of the total visible fat in the U.S. diet, as shown in Fig. 23.2 (1).

Total fat intake in the United States has decreased over the past decade from approximately 36% of energy (% kcal) to 34% kcal (2). Saturated fat intake has also decreased during this period and now contributes about 12% kcal, whereas polyunsaturated fat has increased to about 7% kcal. The proportions of total dietary fat coming from visible and invisible fats are approximately 46.6 and 53.4%, respectively (3). Since SBO accounts for the majority of the visible fat in the diet (~78%), it can be estimated that SBO contributes about 12% kcal to the diet, of which 5.7%

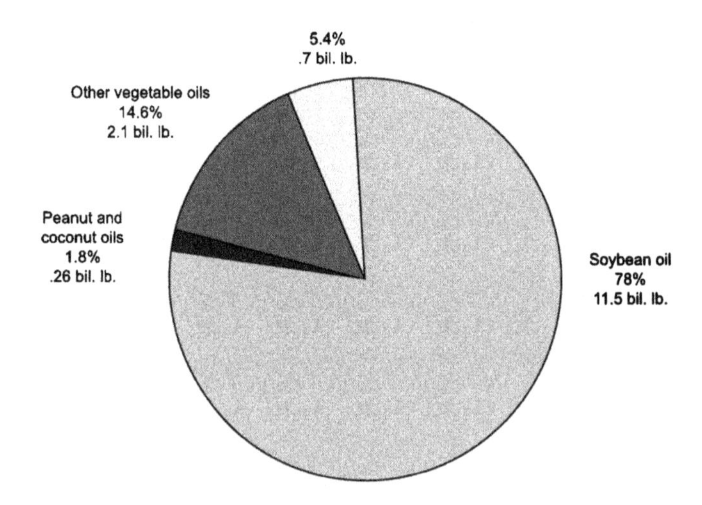

Fig. 23.1. Contribution of soybean oil to total fats and oils used in edible products in the United States during 1992. *Source:* U.S. Department of Agriculture, OCS-41, July 1994.

kcal is in the form of liquid oil and 6.7% kcal is partially hydrogenated (1). SBO supplies about 3.0% kcal as linoleic acid and 1.8% kcal as saturated fat to the diet. The average intake of *trans* isomeric fatty acids in the U.S. diet is around 3% kcal (8 to 10 g per day), of which about 17% is derived from ruminant food sources and 83% from partially hydrogenated vegetable oil (4). Based on edible oil disappearance data, SBO accounts for approximately 87% of all hydrogenated vegetable oil used in edible products in the United States (1). From these data, it is estimated that SBO contributes to the diet approximately 2.1% kcal as *trans* fatty acids and accounts for about 70% of the total *trans* fatty acids.

Fat Absorption

In normal individuals, typical edible fats and oils are digested and absorbed to a high degree. The absorption coefficients in humans of 36 different vegetable oils and 20 animal fats in humans are all in excess of 91% (5). The absorption coefficient of soybean oil is about 95%. In general, natural fats and oils with melting points below 50°C are nearly completely absorbed in humans. The absorption coefficients for high-melting fats and hydrogenated vegetable oils used in manufacturing margarines and shortenings are 79–98% and are related to the melting point of the fat (5). The absorption of hydrogenated corn oil with melting points of 33°C and 50°C were about 95% and 88%, respectively (5). As shown in Table 23.1, fat absorption is clearly related to the melting point of the fat.

Numerous studies have compared the absorption of the *trans* isomers elaidic (*t*-18:1, n-9) and linoelaidic acids (*t*-18:2, n-6) with their *cis* isomer counterparts, oleic and linoleic acid. The results on this topic in virtually all studies conducted in animals and humans indicate that the absorptions of *cis* and *trans* fatty acids are sim-

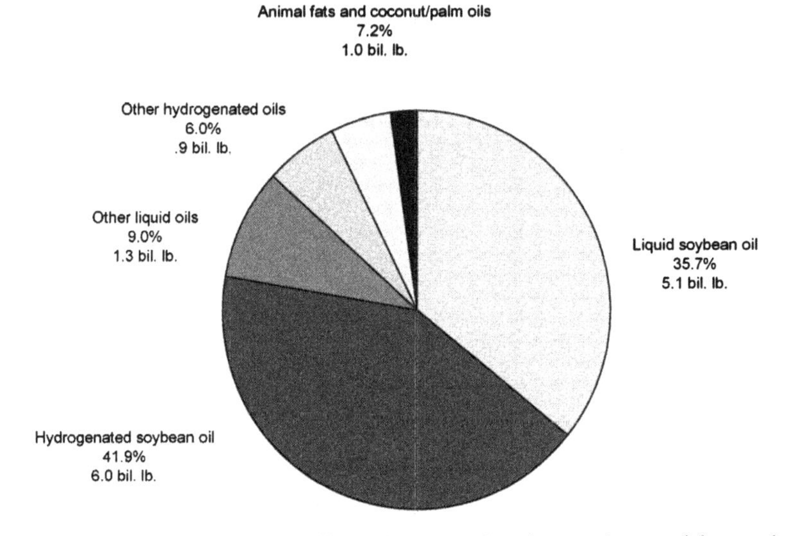

Fig. 23.2. Contribution of partially hydrogenated soybean oil to total fats and oils used in edible products in the United States during 1992. *Source:* U.S. Department of Agriculture, OCS-41, July 1994.

ilar (6,7). In human studies, it was found that the absorption of 9-*trans*, 12-*trans*, 12-*cis*, 13-*trans*, or 13-*cis*-18:1 isomers were not different compared to 9-*cis*-18:1 (8, 9). Moreover, the hydrolysis of triglycerides by pancreatic lipase is not inhibited by *trans* fatty acids contained in hydrogenated vegetable oils (7). These findings are supported by multigenerational studies in animals showing that diets containing high levels of partially hydrogenated vegetable oil in the form of margarine and adequate levels of essential fatty acids (EFA) did not affect weight gain, growth rate, reproductive and lactation performance, and longevity (10, 11).

The digestion and absorption of fats and oils are generally similar in human, dog, and rat, with some notable exceptions. Cocoa butter and 46°C hydrogenated cottonseed oil have markedly lower absorption coefficients in rats than in humans (Table 23.1). Other species, such as the guinea pig, have been found to absorb some high melting vegetable oils and several other fats poorly(5).

The positional distributions of long-chain saturated fatty acid–containing triacylglycerides, as well as the level of divalent cations, also influence fat absorption. Absorption of stearic acid is markedly reduced when it is esterified in the *sn*-1,3 rather than the *sn*-2 position of SOS or OSS triglycerides (12). Recently, de Fouw et al. (13) demonstrated that *sn*-2 palmitic acid–containing fats were better absorbed than fats having palmitic acid in the *sn*-1,3 position. Calcium and magnesium, when present in significant amounts in the diet, also can impair the absorption of stearic acid by forming insoluble fatty acid soaps (13).

Other factors that influence the absorption of high-melting fats include the mode of preparation and emulsification of the fat system. Emken et al. (14) found that, as indicated in Table 23.2, tristearin was poorly absorbed by human subjects (34%) when the triglyceride mixture was emulsified at 70°C in a sugar-casein-water mixture. In contrast, approximately 90% of tristearin was observed when the preparation was emulsified at 80 to 85°C (14). Thus, high-melting fats, such as tristearin, appear

TABLE 23.1 Digestibility of Fats and Oils

		Coefficient of digestibility (%)		
Fat or oil	Melting point, °C	Human	Rat	Guinea pig
Mutton	50	88.0	84.8	—
Hydrogenated cottonseed	54	—	68.7	—
Hydrogenated cottonseed	46	94.9	83.8	—
Hydrogenated peanut	50	92.0	—	—
Hydrogenated peanut	52.4	79.0	—	—
Hydrogenated corn	43	95.4	—	—
Margarine	35	96.7	97.0	—
Soybean	—	97.5	98.5	—
Olive	—	97.8	92.0	77.0
Lard	37	97.0	96.6	75.2
Butter	34.5	—	90.7	—
Cocoa butter	28	94.9	81.6	—
Tripalmitin	66.5	—	27.9	—
Tristearin	70	—	18.9	—

Source: Deval, H.J., Jr., in *The Lipids, Their Chemistry and Biochemistry, Vol. II*, pp. 218–221.

TABLE 23.2 Percent Absorption of Palmitic and Stearic Acid Relative to Linolenic Acid

				Percentage absorbed			
				Mono-acid TG[c]		Mixed-acid TG[c]	
Subject	Diet	Sample (hr)	Temp. (°C)[b]	16:0	18:0	16:0	18:0
1	PUFA	6	70		33.7		
2	SAT	6	75		58.5		
3	SAT	4	85	93.5	87.0		
4	PUFA	4	80	97.2	94.1		
5	SAT	4	80			85.5	96.2
6	SAT	4	80			99.0	93.5
7	PUFA	2	80			99.5	90.9

[a]Percent calculated from deuterated 16:0 to 18:3(n-3) ratios in the samples and the ratios in the fed meal.
[b]Temperature used for preparation of emulsion containing deuterated fat.
[c]Deuterated 16:0 and 18:0 fed as tripalmitin or tristearin (mono-acid TG) to subjects 1-4 or as a randomly esterified triacylglycerol (mixed-acid TG) containing deuterated 16:0, 18:0 and 9*t*-18:1 to subjects 5–7.
Source: Emken, E.A., et al., *Biochim. Biophys. Acta 1170:* 173, 1993.

to be readily absorbed when heated and adequately emulsified to avoid the formation of microcrystalline particles. Although stearic acid and tristearin are generally considered to be poorly absorbed based on animal data (15), these data, as well as other human studies that report good absorption (16,17), indicate that additional work is needed to characterize the relationship between the physical chemistry of dietary hard fats and their gastrointestinal absorption characteristics.

Nutritional Aspects of Soybean Oil

Soybeans have an oil content of about 20% on a dry weight basis. Soybean oil (SBO) is the major edible oil used in the United States for producing salad dressings (90%), baking and frying fats (72%), margarines (88%), and salad and cooking oils (76%) (1). Refined SBO is composed of approximately 98% triglycerides, less than 1% free fatty acids, and less than 1% unsaponifiables. SBO is composed of 60.8% polyunsaturated fatty acids (PUFA), 24.5% monounsaturated fatty acids (MUFA), and 15.1% saturated fatty acids (SAFA) (Table 23.3). The essential fatty acids linoleic acid (18:2, n-6) and α-linolenic acid (18:3, n-3) account for 89 and 11% of the total n-6 and n-3 PUFA fraction of SBO, respectively. These fatty acids cannot be synthesized by the body and must be supplied by the diet. The level of n-6 PUFA in SBO is somewhat lower than in other major vegetable oils such as corn and sunflower oils, and is twice as high as in canola oil. On the other hand, SBO contains over 2 times as much more saturated fat than canola oil. α-Linolenic acid, an n-3 fatty acid that is nearly absent from the other plant oils, is present in SBO and canola oil at about 7.2 and 10.3%, respectively (Table 23.3).

The physiological effects of vegetable have been primarily evaluated on their fatty acid composition and more recently on their unsaponifiable level and composition. The major fatty acids in the diet that are supplied in large part by SBO are the unsaturated fatty acids: n-6 PUFA, n-3 PUFA, *cis*-MUFA and *trans*-MUFA. Each of

TABLE 23.3 Typical Fatty Acid Composition (%) of Soybean, Sunflower, Corn, and Canola Oils

Fatty acid	Soybean	Sunflower	Corn	Canola
16:0	10.9	6.2	11.4	4.1
18:0	4.0	4.7	1.9	2.1
18:1, n-9	24.2	20.5	25.4	56.7
18:2, n-6	54.1	69.0	60.9	26.8
18:3, n-3	7.2	<0.1	0.7	10.3
Total PUFA	60.8	69.0	61.6	37.1
Total MUFA	24.5	20.5	25.4	56.7
Total SFA	15.1	10.8	13.3	6.2

[a]Values are normalized to total 100 ± 0.5%.
Source: Meydani, S.N., et al., *J. Amer. Coll. Nutr. 10:* 406, 1991.

these fatty acid categories have important effects on lipid metabolism that have the potential to alter the outcomes of major chronic diseases. The most predominant chronic diseases in affluent societies are coronary heart disease (CHD) and the malignant diseases (cancer). Current U.S. dietary recommendations for fat and fatty acids are based primarily on the established relationships between dietary total fat and specific fatty acids on blood lipids, which can affect the incidence of such diseases as CHD.

The current U.S. dietary guidelines recommend that diets contain less than 30% kcal from fat; that of this, less than 10% kcal be derived from SAFA and 10 to 15% kcal from MUFA; and that PUFA should be increased from the current level of 7% kcal to no more than 10% kcal. In addition, cholesterol intake should be limited to less than 300 mg/d (19). The role of SBO in the diet is significant because of its fatty acid composition and the preponderance of SBO in the U.S. food supply. It is calculated that SBO may supply as much as 35% of the total fat in the U.S. diet, 43 of the n-6 PUFAs, 70% of the *trans*-MUFA, and 15% of the SAFAs.

It is a concentrated source of energy (calories), is highly digestible, provides essential fatty acids, Vitamin E and is a rich source of polyunsaturated fatty acids. Like most fats and oils, SBO provides 9 kcal/gram or about 120 kcal per tablespoon (14 g) serving. The National Research Council and the Food and Agriculture Organization/World Health Organization recommend about 24% kcal as essential fatty acids. A tablespoon (14g) serving of SBO will satisfy the daily essential fatty acid requirement for a healthy child or adult.

Cardiovascular Effects

Plasma Lipids. The relationship between high levels of plasma cholesterol and CHD has been well established from epidemiological evidence as well as primary and secondary prevention trials (20–26). These studies demonstrated that high levels of dietary SAFA are related to an increased incidence of CHD and that dietary modification can lower plasma cholesterol and, in some cases, the incidence of CHD. These results were achieved primarily by lowering dietary SAFA, increasing n-6 PUFA, or both.

Changes in plasma cholesterol can be predicted based on alterations in the levels of SAFA and PUFA in the diet using equations developed by Keys et al. (27) and Hegsted et al. (23) and recently updated by Hegsted et al. (28):

$$\Delta\text{Cholesterol} = 2.10(\Delta\text{SAFA}) - 1.16(\Delta\text{PUFA}) + 0.067(\Delta\text{Dietary cholesterol})$$

Hence, dietary SAFA is most effective in altering plasma cholesterol, whereas addition of PUFA (i.e., linoleic acid) has a modest independent lowering effect. Somewhat uncertain are the monounsaturated fatty acids, which have generally been viewed as not having an independent effect on plasma lipids (23,27,28).

Numerous animal and human controlled-feeding trials conducted over the past 40 years have demonstrated that increasing PUFA or MUFA intake at the expense of lowering SAFA usually results in a decrease in plasma total cholesterol (TC) and, LDL-cholesterol (LDL-C) (31–34). However, it is less clear whether oleic acid has independent effects on TC and LDL-C. Early studies suggested that linoleic acid lowers TC more than oleic acid, whereas more recent investigations indicate that the TC- and LDL-C–lowering actions of these unsaturated acids were identical (32). On the other hand, recent meta-analysis of human clinical studies published between 1970 and 1993 suggest that MUFA also independently lowers plasma cholesterol but to a lesser degree than PUFA (29,30). Table 23.4 shows results in which young men were fed diets containing high levels of MUFA (oleic acid), provided by olive oil or n-6 PUFA (linoleic acid), provided by SBO without significant cross-contamination of either of the test fatty acids (35). Compared to a diet high in 12:0 to 14:0 SAFA (dairy butter), both the MUFA and n-6 PUFA diets markedly reduced TC and LDL-C. Diets high in stearic acid, provided by cocoa butter, were also hypocholesterolemic compared to the 12:0 to 14:0 SAFA diet, but the MUFA and n-6 PUFA diets were more hypocholesterolemic than the stearic acid diet. These results suggest

TABLE 23.4 Effects of Different Dietary Fats on Plasma Levels of Lipids and Lipoprotein Cholesterol

Diet	TC	LDL	HDL	TG	Apo A-I	Apo B
CB	165 ± 3	103 ± 3	44 ± 2	87 ± 3	98 ± 2	86 ± 2
B	176 ± 3	113 ± 3	45 ± 2	88 ± 3	98 ± 2	88 ± 2
OO	152 ± 3	92 ± 3	48 ± 2	84 ± 3	99 ± 2	81 ± 2
SO	139 ± 3	83 ± 3	45 ± 2	73 ± 3	95 ± 2	69 ± 2
CB v B	−11 ± 4[b]	−10 ± 4[a]	−0.3 ± 3	2 ± 5	−1 ± 3	−2 ± 3
CB v OO	13 ± 4[c]	11 ± 4[b]	−4 ± 3	2 ± 5	−1 ± 3	5 ± 3
CB v SO	26 ± 4[c]	20 ± 4[c]	−1 ± 3	13 ± 5[b]	3 ± 3	17 ± 3[c]
B v OO	24 ± 4[c]	21 ± 4[c]	−3 ± 3	3 ± 5	−1 ± 3	7 ± 3
B v SO	37 ± 4[c]	31 ± 4[c]	−1 ± 3	14 ± 5[c]	3 ± 3	19 ± 3[c]
OO v SO	13 + 4[c]	9 ± 4	2 ± 3	11 ± 5[a]	4 ± 3	12 ± 3[c]

Note: Values are mg/dL (least-square means ± SEM from ANOVA).
Abbreviations: TC, total cholesterol; TG, triglycerides; LDL, low-density lipoprotein cholesterol; Apo A-1, apolipoprotein A-1; Apo B, apolipoprotein B; CB, cocoa butter; B, dairy butter; OO, olive oil; SO, soybean oil.
[a] $P < .1$
[b] $P < .05$
[c] $P < .01$
Source: Kris-Etherton, P.M., et al., *Metabolism 42:* 121, 1993.

that unlike the 12:0 to 14:0 SAFAs, stearic acid does not raise TC. In addition, these results indicate that n-6 PUFAs are more potent than oleic acid in lowering plasma total cholesterol. Although the biochemical mechanisms that explain the effects of these fatty acids on plasma lipids are not clearly understood, recent studies in rodent models suggest that unsaturated and saturated fatty acids can differentially alter the rate of LDL-C clearance from the plasma by modifying the hepatic LDL-receptor binding system (36). Animals fed diets containing 12:0 to 16:0 saturated fatty acid triacylglycerides experienced a down-regulation of hepatic LDL-receptor activity and an increase in hepatic LDL-C production rates, resulting in increased plasma LDL-C levels, whereas the feeding of unsaturated fatty acid triacylglycerides such as triolein resulted in an up-regulation of hepatic LDL-receptor activity and a decrease in hepatic LDL-C production rates, resulting in a decrease in plasma LDL-C levels. On the other hand, the medium-chain saturated fatty acids (8:0 and 10:0), as well as stearic acid (18:0), had no effect on these parameters, which suggests that these fatty acids have a neutral effect on plasma LDL cholesterol.

In some studies, diets high in PUFA have been reported to reduce high-density lipoprotein cholesterol (HDL-C) (37). This is viewed as an undesirable effect, because HDL-C is considered to have cardioprotective effects and has been viewed as a predictor of risk for CHD. This effect has been observed in diets containing high levels of PUFA (>12% kcal) and, in some cases, using liquid formula diet (37). On the other hand, the HDL-C–lowering effects of n-6 PUFAs are generally not observed with whole food diets at moderate (38) or high levels of n-6 PUFAs (35).

Blood Pressure. Some evidence suggests that high levels of n-6 PUFA intake lower blood pressure (39). In general, studies in this area have produced inconsistent results because of the difficulty in ruling out other dietary factors (e.g., total fat, dietary fiber). Others have reported that blood pressure is lower on a low-fat diet without an increase in n-6 PUFA (40).

Immune System Effects

The amount of total fat in the diet and degree of FA unsaturation have a significant effect on normal immune response and on the expression of inflammatory diseases (41). The n-6 PUFAs (EFA) are necessary for normal immune function by virtue of the arachidonic acid metabolites, prostaglandins, leukotrienes and other bioactive eicosenoids which regulate a broad range of cell mediated immune responses (e.g. delayed hypersensitivity, foreign graft rejection, resistance to pathogenic microorganisms). EFA deficiency impairs normal cellmediated immune responses and repletion of EFA in the diet will restore these impairments. In the absence of EFA deficiency, animals fed a moderate level of total fat with a high proportion of n-6 PUFAs had similar immune responses to those fed the same levels of SAFA. On the other hand, lower responses to immunological tests were observed in animals fed high-fat diets (about 45% kcal) containing high n-6 PUFAs compared to those fed the same level of SAFA. The scientific literature suggests that the effect of n-6 PUFAs on the immune response varies depending on: concentration of dietary fat, duration of supplementation, genetic variation, age of the animal, antioxidant status, EFA status, the types of immunological tests used and the type of PUFA tested, i.e. n-6 or n-3 PUFA. Because of the higher levels of 18:3, n-3 found in SBO (about 7%), researchers have speculated that SBO may have different immunological properties than oils containing low levels of 18:3, n-3 (e.g. corn oil). It is well known that 18:3, n-3 can be elongated and desaturated in cells to 20:5, n-3 which can compete with 20:446 for conversion by cycoloxygenase and can result in reduced production of the 2-series prostaglandins (PG) and 4-series leukotrienes (LT). These are highly immunoreactive and inflammatory compounds. When diets containing 20% kcal as corn oil (2.2% 18:3, n-3), SBO (7.7% 18:3, n-3), linseed oil (62% 18:3, n-3), or a mixture of linseed oil and SBO were compared for their effects on tissue fatty acid composition and PG levels, no differences in spleen 20:5, n-3 levels or in spleen or thymus PGE_2 production was observed between corn oil and SBO. Rats fed the high 18:3, n-3 diets (linseed oil or linseed/SBO mixture) had reduced PGE_2 levels in thymus and spleen (42). Thus, despite the higher amount of 18:3, n-3 in SBO, these results suggest that the amount does not appear to be adequate to cause significant changes in spleen and thymus PG levels or functional immunological alterations when compared to corn oil.

Carcinogenic Effects

Cancer is ranked as a leading cause of death in the U.S. Both genetic and environmental risk factors may affect the risk of cancer. Risk factors include a family history of cancer, smoking, alcohol consumption, obesity, cancer-causing chemicals, and dietary factors. Among dietary factors, the strongest positive association has been found between total fat and risk of some types of cancer. There is compelling evidence, although not conclusive, which demonstrates that the total amount of fats, rather than any specific type of fat, is positively associated with cancer risk (43). No strong association, however, has been demonstrated between fat intake and breast cancer based on the totality of retrospective and prospective epidemiological studies (44,45). Because energy intake and fat intake are highly correlated, the association between dietary fat and cancer is confounded by energy intake. A review of the totality of scientific evidence with careful attention to human and animal studies with isocaloric diets concluded that the effect of dietary fat on tumorigenesis is independent of the effect of calories (46).

Animal studies have suggested that when total fat intake is low but adequate in EFA, linoleic acid is more effective than SAFA, MUFA, and n-3 PUFA in the promotion of experimental carcinogenesis (43). One postulated mechanism by which linoleic acid may act to stimulate the carcinogenic process involves its conversion to arachidonic acid and to the highly immunoregulatory eicosenoids of the prostaglandin-series. However, in high fat diets, once the EFA requirement of the animal is met, tumor growth is dependent upon the level of fat in the diet and not the type of fat (47). On the other hand, in studies of rats fed 40% kcal fat diets from either palm oil, corn oil or SBO, determination of DMBAinitiated tumors revealed that rats whose diets were high in linoleic acid developed significantly more mammary tumors than did rats fed the EFAadequate (4.5% kcal) palm oil diet (48).

TABLE 23.5 Summary of Selected Human Studies Reporting Negative Association Between PUFA and Cancer (18)

Author	Type of study	Cancer	Result
McKeown-Eyssen (1984); [50]	International correlation	Colon (mortality)	Vegetable fat $r^a = -0.28$
Kaizer (1989); [51]	International correlation	Breast (mortality)	Fish consumption, $r^a = 0.47$ ($p = 0.0066$), assumed to be due to fatty acid composition of fish
Tuyns (1988); [52]	Case-control	Colorectal (incidence)	Oils of high P:S ratio (corn, soybean, sunflower), $RR^b = 0.48$ ($p < 0.0001$)
Verrault (1988); [53]	Cancer cases only	Breast	PUFA, $OR^c = 0.6$, χ^2 trend = −2.17 ($P = 0.03$)

[a] r = Correlation between dietary factor and mortality
[b] RR = Relative risk; ratio of the number of persons with colon cancer in the third quartile of PUFA consumption compared to the number of colon cancer cases in the lowest quartile of PUFA consumption.
[c] OR = Odds ratio; ratio of the odds of having lymph node involvement at diagnosis in the highest quartile of PUFA intake to the same odds among those in the lowest intake of PUFA.

Twenty-five percent fewer tumors, however, were observed in the SBO group compared with corn oil which may be due to the ratio of n-6 to n-3 PUFA in the two fats, 6.8:1 vs. 87:1, respectively.

Linolenic acid competes more effectively for esaturase than linoleic acid resulting in production of the 3-series prostaglandins and attenuates the conversion of linoleic acid to the 2-series prostaglandins. Most studies evaluating the potential anticarcinogenic properties of the n-3 PUFAs have generally used the longerchain n-3 PUFAs (20:5, n-3 and 22:6, n-3) from marine oils. Preliminary evidence also indicates that linseed oil (61% alinolenic acid) effectively suppressed the growth of transplanted mammary tumors (49). These data support the hypothesis, as yet unproven, that the shorterchain n-3 PUFA of plant origin may modulate the action of linoleic acid on tumor promotion in experimental tumorigenesis.

Slover et al. (54) and others have reported as many as 20 *trans* and *cis* positional isomers of 18:1 fatty acids in dietary fats made from PHSBO. The principal trans fatty acid group in PHSBO is *trans* 18:1 and accounts for the majority of the total trans isomers. *trans* Isomers of 18:2 are also present in PHSBO at low levels (12%) with occasional reports of 45% depending on the type of catalyst and degree of hydrogenation, those isomers of quantitative importance are: *cis*-9, *trans*-13; *cis*-9, *trans*-12; and *tran*-9, *cis*-12 (55).

Despite convincing evidence in experimental animals for the involvement of linoleic acid in carcinogenesis, this has not been found to be the case in humans. Numerous studies reviewed in the literature indicate that PUFA does not increase the risk, incidence, or mortality of breast and colon cancers (41). Indeed, three studies have reported a negative association between vegetable oil PUFA and cancer (Table 23.5).

Partially Hydrogenated Soybean Oil

The nutritive value of partially hydrogenated soybean oil (PHSBO) compared with unhydrogenated soybean oil has been at issue for many years because of the complex array of fatty acid geometric *trans*- and positional isomers formed during hydrogenation. Numerous studies have been conducted over the years to characterize the physical, biochemical, and physiological consequences of altering the geometry and position of double bonds as the result of catalytic hydrogenation.

Slover et al. (54) and others have reported as many as 20 *trans* and *cis* positional isomers of 18:1 fatty acids in dietary fats made from PHSBO. Isomers of quantitative importance were *cis*-9, *trans*-13; *cis*-9, *trans*-12; *trans*-9, *cis*-12; and *cis*-9, *cis*-15 octadecadienoic acid (49).

Figure 23.3 illustrates that the distribution of *cis* and *trans* double bonds found in the 18:1 isomers of hydrogenated vegetable oils ranges from 6 to 16, with the majority found between carbons 9 and 12 (56). Butterfat also has a similar number of isomers, of which *trans*-vaccenic acid (18:1, *trans*-11) is the primary isomer (57). From a nutritional standpoint, it is important to understand the biochemical and physiological similarities and differences of these isomeric fatty acids because of their prominence in the food supply.

One of the primary reasons that *trans* acids pose the potential for unique biological properties stems from the conformational changes that *trans* double bonds confer on the fatty acid structure, which are reflected by the increased melting point. Molecular modeling of 18-carbon fatty acids (Fig. 23.4) illustrates that the isomeric fatty acids of oleic and linoleic acid have a straighter configuration than their *cis* counterparts (58). These models correctly predict what has been found biochemically: that the biological properties of *trans* acids are intermediate between those of saturated fatty acids and *cis*-unsaturated fatty acids.

Metabolism and Biochemistry

The enzymes triacylglycerol acyltransferase, phosphatidylcholine acyltransferase, and lecithin:cholesterol acyltransferase show distinct selectivity for incorporation of *cis* and *trans* acids into the primary lipid pools in the body (59–61). The relative selectivity into plasma lipid fractions is shown in Fig. 23.5, in which humans were fed fats containing various *cis* and *trans* 18:1 isomers as well as other fatty acids that typically occur in partially hydrogenated soybean oil. The data shows that

1. 18:1 *trans* isomers incorporate less into triglycerides than *cis*-18:1.
2. *trans* 18:1 isomers have very low incorporation into cholesterol esters.
3. *trans* 18:1 isomers are preferentially incorporated into the l-acyl position of phospholipids.

Emken et al. (58) have pointed out from these data that although there are some similarities in the tissue distribution patterns for 18:1 *trans* isomers and saturated fatty acids, it is a gross oversimplification to conclude that the *trans* isomers and saturated fatty

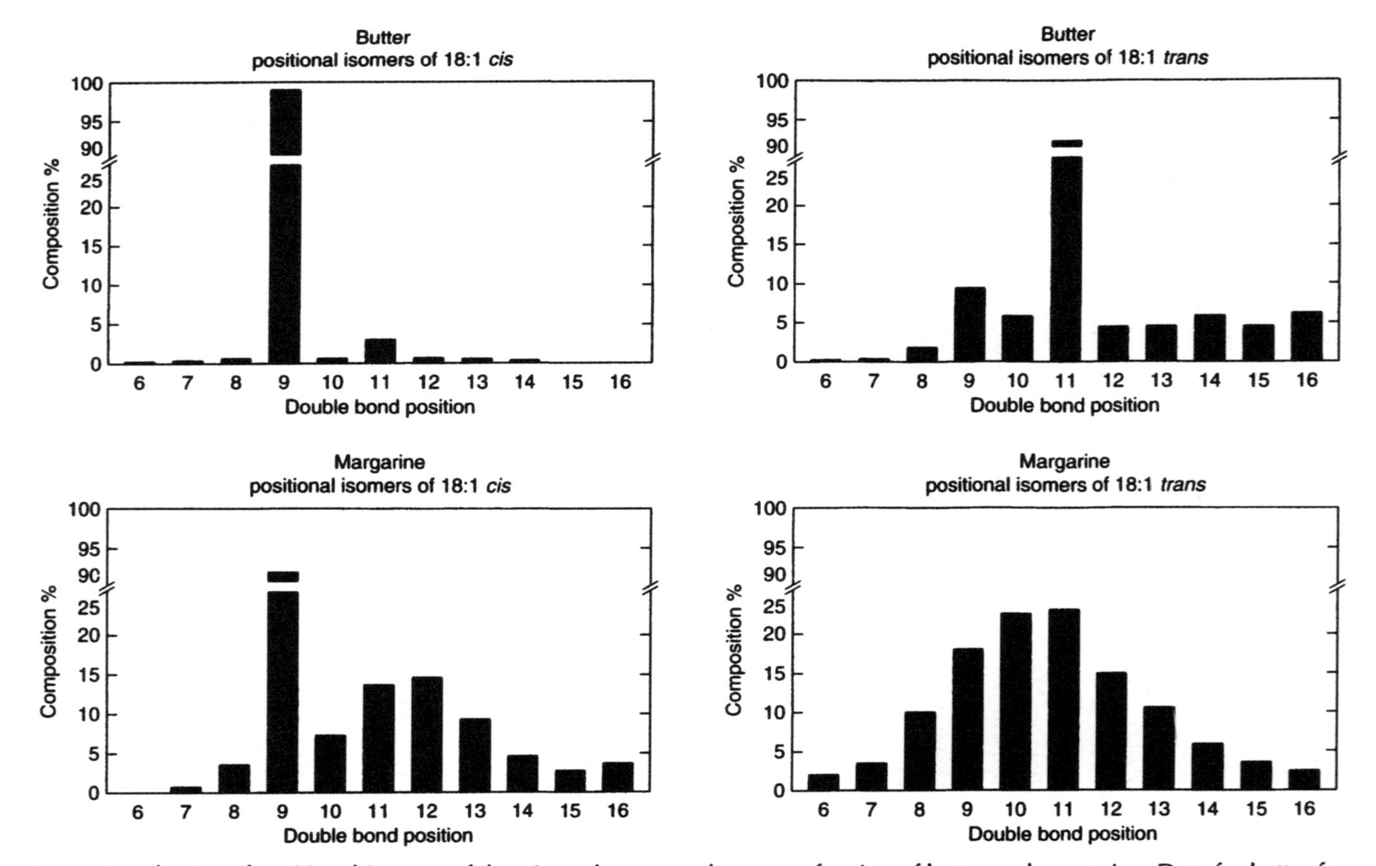

Fig. 23.3. Distribution of positional isomers of the *cis*- and *trans*octadecenoate fraction of butter and margarine. Data for butter from Parodi (57); data for margarine from Sampugna et al. (56).

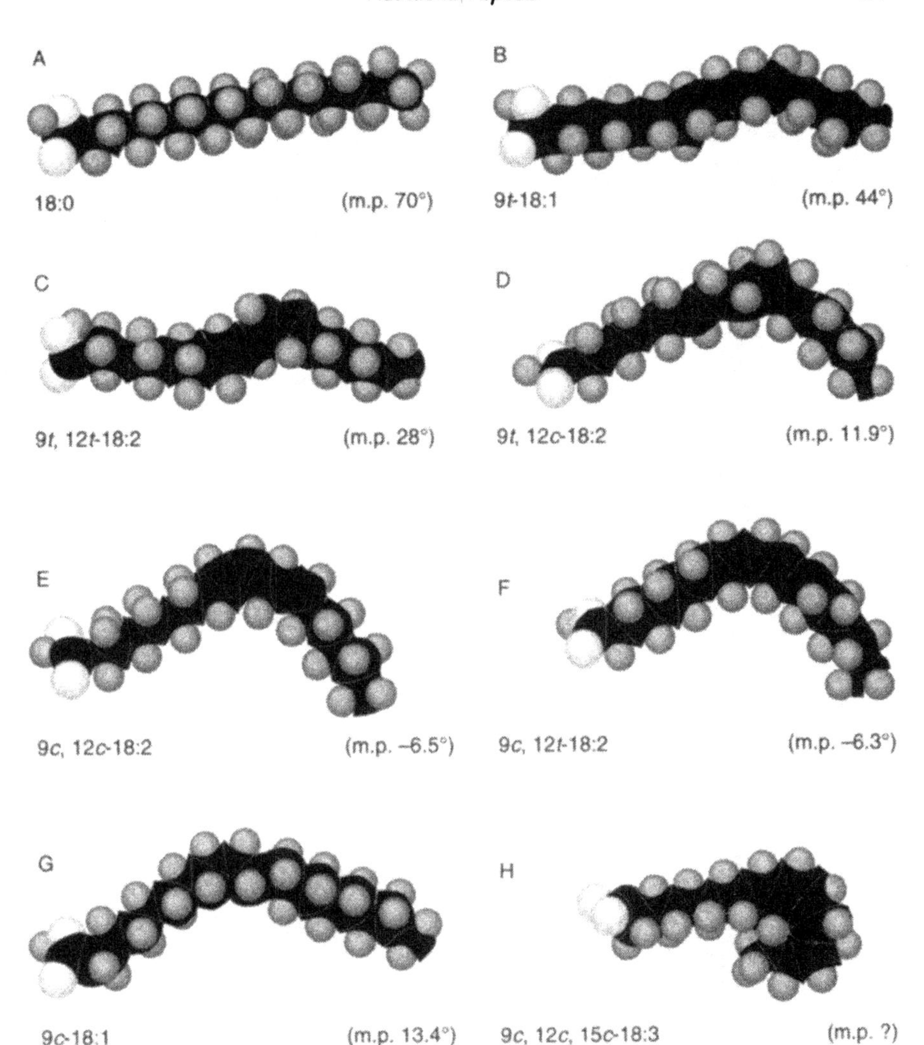

Fig. 23.4. Calculated minimum energy configurations and melting points of some *cis* and *trans* isomers and nonisomeric 18-carbon fatty acids. *Source:* Emken, E.A., in *Health Effects of Dietary Fatty Acids*, edited by G.J. Nelson, American Oil Chemists' Society, Champaign, IL, 1991, pp. 245–260.

acids are metabolically equivalent. Rather, the data suggest that the metabolism of 18:1 *trans* isomers is intermediate between those of saturated fatty acids and *cis* 18:1.

Oxidation

Studies indicate that the fatty acid isomeric distribution pattern in the tissues of humans is very similar to that of PHSBO. The *trans* acid content of various tissues of the body ranges from 0.1 to 3.9%. Because hydrogenated vegetable oil supplies

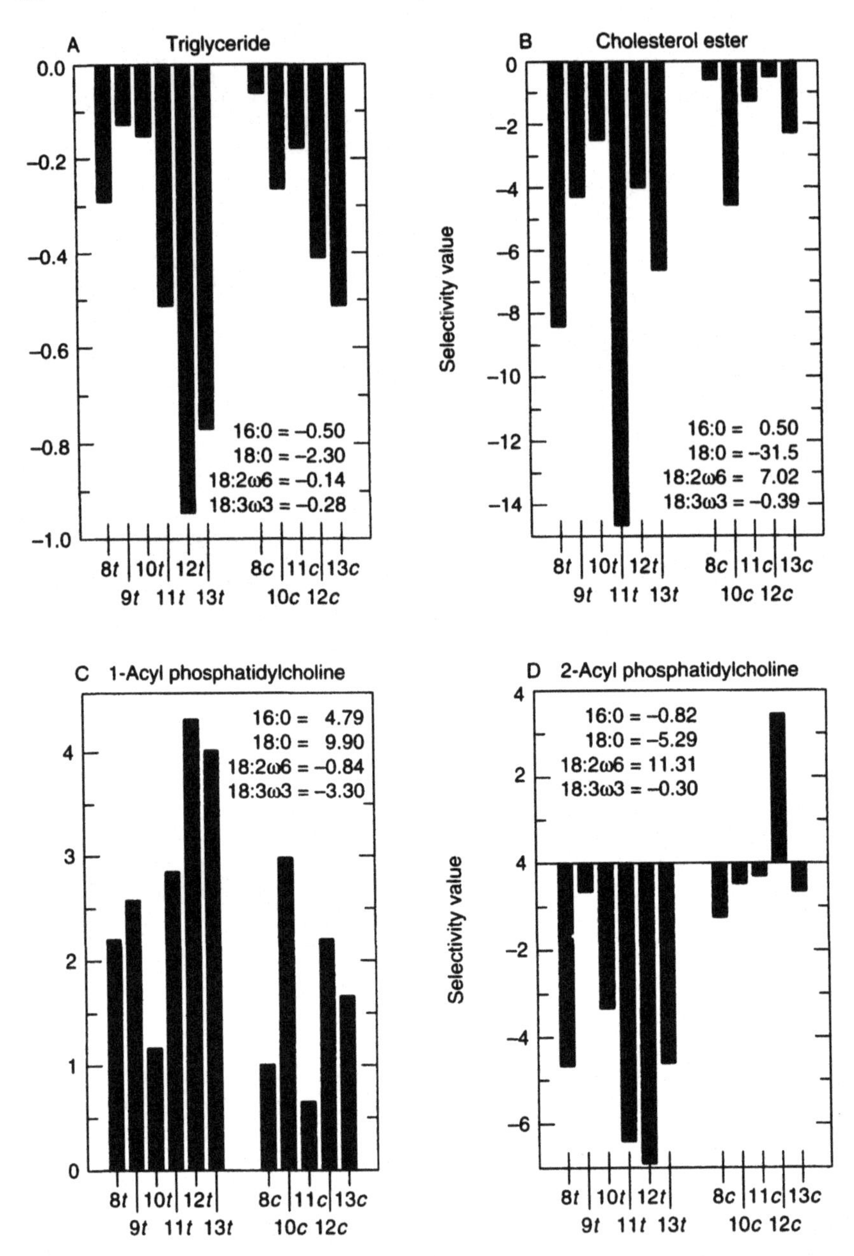

Fig. 23.5. Deuterium isotope tracer data for incorporation of *cis* and *trans* 18:1 positional isomers in human plasma (A) triglyceride, (B) cholesterol ester, (C) l-acyl-phosphatidylcholine, (D) 2-acyl-phosphatidylcholine. Selectivity values for incorporation of individual 18:1 isomers are relative to 9c-18:1. Selectivity values for palmitic (16:0), stearic (18:0), linoleic (18:2, n-6), and linolenic (18:3, n-3) are listed in inserts (58). *Source:* Emken, E.A., in *Health Effects of Dietary Fatty Acids*, edited by G.J. Nelson, American Oil Chemists' Society, Champaign, IL, 1991, pp. 245–260.

TABLE 23.6 Influence of Partially Hydrogenated Vegetable Oils on Serum Lipoprotein Cholesterol

	Dietary fat (%)			Final cholesterol	Change cholesterol[a]	*Trans*/ 18:2, n-6
		18:2, n-6				
	trans	Control	Experi.	(mg/dL)	(mg/dL)	Ratio
Anderson, 1961; [64]	35	58	6	185	+21	5.8
Mensink, 1990; [66]	27	10	11	182	+10	2.5
de Iongh, 1965; [65]	18	57	6	182	+19	3.0
Anderson, 1961; [63]	27	18	13	209	+25	2.1
Zock, 1992; [73]	19	29	10	189	+ 6	1.9
Judd, 1994; [75]	17	16	16	213	+10	1.1
McOsker, 1962; [69]	18	56	22	163	+ 4	0.8
Nestel, 1992; [70]	18	15	18	229	+14	0.8
Judd, 1994; [75]	10	15	15	211	+ 8	0.6
Lichtenstein, 1993; [68]	14	29	26	205	+11	0.5
Erickson, 1964; [66]	11	35	31	193	+ 4	0.4
de Iongh, 1965; [65]	10	57	37	164	+ 1	0.3
Laine, 1982; [67]	8	43	33	161	+ 9	0.2

[a]Difference between control and experimental diets.

about 6% of the total fat in the U.S. diet, the low *trans* content of human tissue lipids strongly suggests that *trans* isomers are preferentially oxidized to CO_2 and water by the body. *In vivo* oxidation studies of 18 carbon *cis* and *trans* acids in rats, as well as human studies, show that *trans* fatty acids: (1) are more rapidly removed from the plasma than oleic acid and (2) do not accumulate in tissue lipids, supporting the notion that *trans* fatty acids are rapidly oxidized (59–62).

Cardiovascular Effects

Plasma Lipids. The cholesterol raising potential of partially hydrogenated vegetable oil (PHVO) and trans fatty acids has been evaluated in numerous studies over the past 40 years because high plasma cholesterol is a recognized risk factor for coronary heart disease (CHD). Early studies on PHVO were conducted before adequate experimental design criteria for human feeding studies were well established and before routine analytical methods were available for determining lipoprotein fractions other than total cholesterol (e.g. LDLC, HDLC, VLDL, Apo B, Apo A1). Regardless, these early studies provided helpful data as well as experimental insight for future work. In general, studies conducted prior to 1990 yielded inconsistent results and a general consensus that trans fatty acids were equivalent to cisMUFA (63,71). In 1990, Mensink and Katan questioned these conclusions in a report which compared the effects of 40% kcal total fat diets containing high levels of SAFA (19% kcal), *trans* (11% kcal), or oleic acid (23% kcal) on plasma lipoprotein cholesterol in young adults (72). The *trans* diet raised TC and LDLC compared to the oleic acid diet but lowered TC and LDLC compared to the SAFA diet which is consistent with others (71). However, the *trans* diet was also found to lower HDLC compared to the oleic and SAFA diet, leading the authors to conclude that *trans* fatty acids are at least

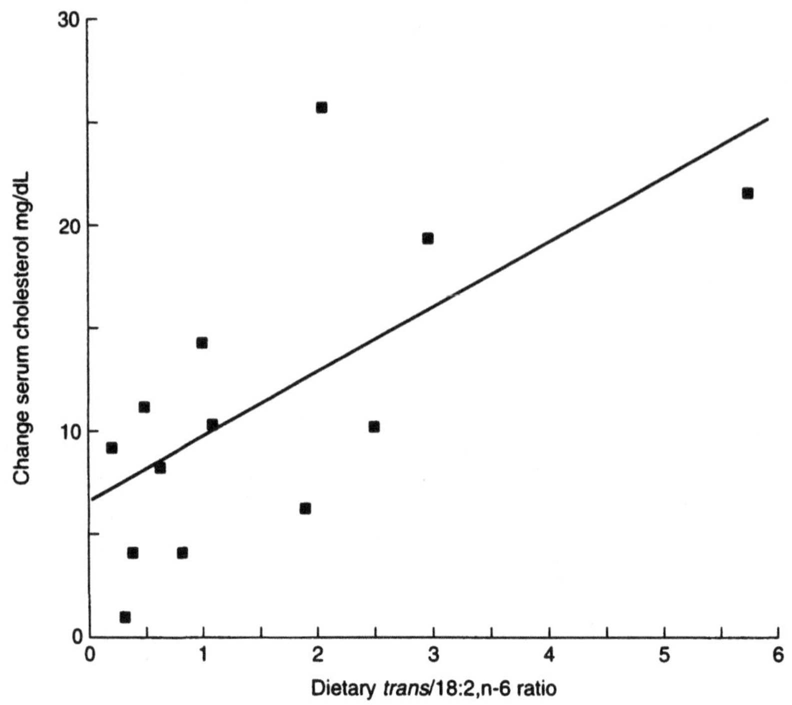

Fig. 23.6. Linear regression analysis of the change in serum cholesterol as a function of the dietary *trans*/18:2, n-6 fatty acid ratio. Data from Table 23.6, refs 58–69. ($r = 0.67$, $P = 0.01$). Change in cholesterol (10 mg/dL) = 61.9 + 30.5 (*trans*/18:2,n-6). *Source:* Adapted from Emken, E., *INFORM 5:* 906 (1994).

as unfavorable as the cholesterol raising saturated fatty acids. Judd et al. (1994) also compared the plasma cholesterol effects of diets high in oleic acid (17% kcal) or 12:0 16:0 SAFA (16% kcal) with diets containing moderate *trans* (4% kcal) or high *trans* (7% kcal) (75). The oleic acid diet produced a lower TC and LDLC when compared to all the diets, whereas both the moderate and high *trans* fatty acid diets produced a lower TC and LDLC than the SAFA diet. The high *trans* fatty acid diet caused a small but significant lowering of HDLC compared to the oleic acid diet whereas the moderate *trans* fatty acid diet had no effect. Finally, Zock and Katan compared diets containing linoleic acid with *trans* fatty acids and stearic acid (73). Linoleic acid produced a lower TC (7 mg/dL) and LDLC (9 mg/dL) than did either *trans* fatty acids or stearic acid, which had equivalent effects. These results, which suggested a cholesterol raising effect for stearic acid, are in contrast to findings by others showing that stearic acid is not hypercholesterolemic (35,74).

Table 23.6 summarizes the results of early as well as more recent controlled human feeding studies showing the observed changes in serum cholesterol as a result of manipulating linoleic acid and *trans* fatty acids in the diet compared with control diets high in oleic or linoleic acid (64–70,72,73,75). Linear regression analysis of these data indicate that the observed changes in serum cholesterol are directly correlated to the ratio of *trans*/18:2, n-6; see Fig. 23.6 (76). These data indicate that *trans* fatty acids and linoleic acid appear to affect serum cholesterol levels in opposing directions. However, because *trans* fatty acids and linoleic acid are highly correlated ($P = 0.78$), it is difficult to determine which has the predominant effect on plasma cholesterol; this problem underscores the need for designing feeding trials in which only one fatty acid is varied.

Taken together, *trans*-18:1 monoene fatty acids from PHVO raise plasma TC and LDL-C relative to their *cis*-18:1 counterpart (oleic acid) and the *cis*-18:2 diene (linoleic acid), whereas *trans* fatty acids are not as potent as the 12:0 to 16:0 SAFA in raising TC and LDL-C. It is well documented in humans, for example, that dairy butter raises plasma TC and LDL-C compared with margarine (77). The independent effects of *trans* fatty acids on plasma lipids, as well as the relationship between *trans* fatty acids and stearic acid remain unclear. Studies in hamsters suggest that individual dietary fatty acids modify the rate of LDL-C clearance from the plasma, and so alter plasma LDL-C levels, by affecting the activity of the hepatic LDL-C receptor system and the rate of LDL-C production (36). These studies have demonstrated that unlike the 12:0 to 16:0 SAFAs, which down-regulate hepatic LDL-C receptors and raise plasma LDL-C, oleic acid actively up-regulates this system, resulting in lower plasma LDL-C. If the unsaturated double bond of oleic acid is isomerized to the *trans*-18:1 form or is fully saturated to 18:0, the up-regulation of the hepatic receptor was lost, resulting in no alterations in plasma LDL cholesterol and suggesting a neutral effect of stearic and *trans* fatty acids (78).

Soy Protein

Hypocholesterolemic Effects of Soy Protein

The hypocholesterolemic and antiatherogenic effects of soybean protein, compared with animal protein or casein, has been recognized in experimental animals for many years (79,80). Animal protein, such as casein, is generally more hypercholesterolemic than plant proteins when fed to a number of animal species such as rabbits, rats, hamsters, guinea pigs, pigs, and monkeys (81). On the other hand, studies conducted in humans have shown variable results. Some clinical trials in which soybean protein was substituted for animal proteins have reported significant reductions in total and LDL-cholesterol, whereas others have found only moderate effects, and in some studies little or no effects were observed. Several factors appear to play a role in the varying responses of plasma cholesterol to soy protein, including degree of hypercholesterolemia, amount and type of dietary protein, and characteristics of soy protein preparations (82,83).

In general, studies conducted in normocholesterolemic humans (≤205 mg/dL) in which soy protein was substituted for animal and dairy proteins in mixed protein diets or provided as a supplement have demonstrated variable results on plasma lipoproteins, suggesting a lack of effect (82). In contrast, greater responses have been obtained in persons with hypercholesterolemia (82). As illustrated in Fig 23.7 (A), however, regression analysis of 12 soy protein feeding experiments conducted in normocholesterolemic subjects with baseline cholesterol levels of ≤205 mg/dL

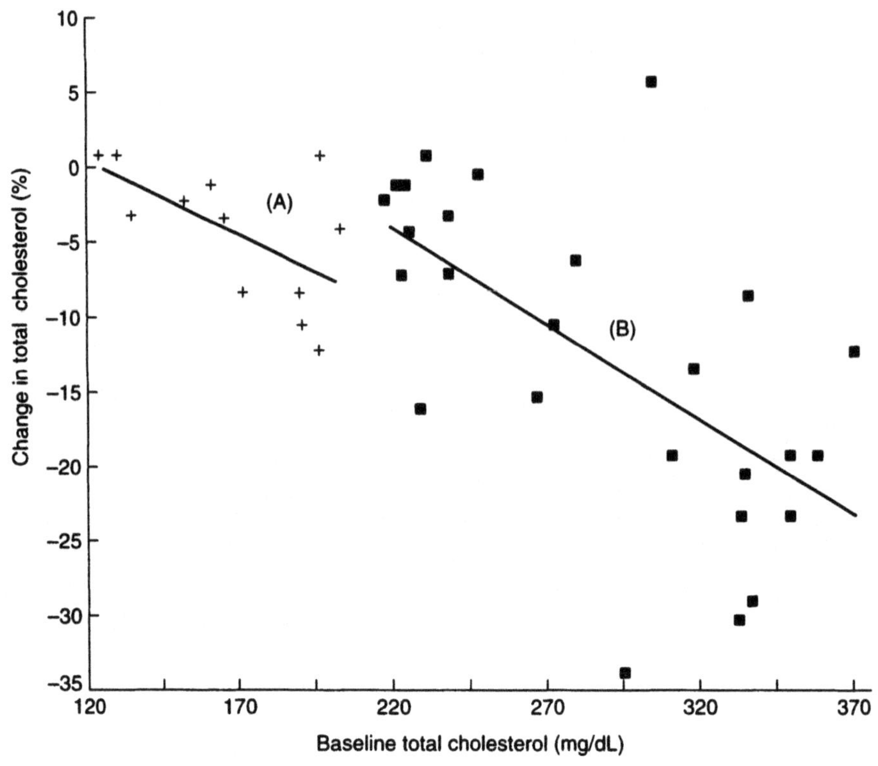

Fig. 23.7. Linear regression analyses of the changes in plasma or serum cholesterol as a function of baseline levels in normocholesterolemic (A) and hypercholesterolemic (B) human subjects. Data from refs 84–90 and 91–109, respectively.

results in a negative linear correlation ($P < 0.04$; $r = 0.59$) between baseline cholesterol and changes in concentration after feeding soy protein. In Fig 23.7 (B), regression analysis of 28 soy protein feeding experiments conducted in hypercholesterolemic subjects with baseline cholesterol levels of ≥220 mg/dL also produced a negative linear correlation ($P < 0.0006$) with a similar correlation coefficient ($r = 0.63$). The slopes of the normocholesterolemic and hypercholesterolemic regression lines are not different ($P < 0.67$). These data suggest that the percentage reduction in plasma cholesterol that results from soy protein feeding depends on the initial (baseline) plasma cholesterol level. For every 10 mg increase in initial cholesterol levels, there is approximately a 0.9 to 1.3 unit decrease in the percentage change in plasma cholesterol caused by feeding soy protein. Bakhit et al. (110) found that in individuals with baseline concentrations ≥220 mg/dL, consumption of 25 g soy protein isolate (about 25% of total protein intake) resulted in a progressive decrease in total cholesterol as baseline values increased. On the other hand, for subjects with baseline cholesterol ≤ 220 mg/dL, soy-protein consumption resulted in no change or, in some cases, a concentration increase.

Studies in animal models indicate that a dose response increase in plasma cholesterol is observed with increasing amounts of animal protein (111), whereas this is

TABLE 23.7 Plasma Cholesterol Concentrations at the End of Each Experimental Period (adapted from ref. 107)

	Baseline	SF	ISP/SCF	ISP/C	NFDM/C
Total-C	228	217[a]	211t[b,c]	210t	235
LDL-C	175	166	160t[b]	161	179
HDL-C	37	37	35	35	36

Note: Values are mean values.
Abbreviations: SF, soy flour; ISP/SCF, isolated soy protein/soy cotyledon fiber; ISP/C, isolated soy protein/cellulose; NFDM/C, nonfat dry milk/ cellulose.
[a,b]Significantly different from NFDM/C: [a]$P < 0.05$, [b]$P < 0.01$.
[c]Significantly different from baseline. $P < 0.05$.

not observed with the feeding of high levels of soy protein (112). Application of the soy protein–induced hypocholesterolemia to humans, however, have, in general, involved total replacement or very high levels of soy protein and/or food items that are of questionable practical use (82,83). Potter et al. (113) reported that acceptable diets can be constructed to replace ~50% (50 g) of dietary protein with isolated soy-protein and soy cotyledon fiber using common baked products. Results of the dietary treatments in mildly hypercholesterolemic men are shown in Table 23.7. Both isolated soy protein diets resulted in significant reductions in total cholesterol and LDL-cholesterol compared to the nonfat dry milk–based diet and baseline. These results agree with the literature, indicating that the cholesterol lowering effects of soy-protein in hypercholesterolemic individuals occur mainly in the LDL fraction and do not alter HDL-cholesterol or triglyceride levels (82,83).

Which components of dietary soy protein modify total and LDL-cholesterol and the mechanism by which they do so are not known. Since animal proteins tend to have higher levels of essential amino acids than plant proteins except for arginine, it has been postulated that the levels of lysine and methionine or the ratio of lysine to arginine could be related to its atherogenic potential (114). Casein and a casein amino acid mixture, for example, produced similar elevations in plasma cholesterol when fed to rabbits, indicating that the amino acid composition of proteins appears to be an important determinant (115,116), whereas similar studies with amino acid mixtures showed little correlation between the lysine/arginine ratio and plasma cholesterol (117). Other characteristics and effects of proteins that have been postulated to influence blood lipid responses include protein digestibility (118), hormonal alterations (e.g., insulin and glucagon) (119), and increased fecal steroid production, and cholesterol excretion (120,121). These factors, however, appear to be weakly associated. More recently, studies have demonstrated that casein-fed rabbits, in addition to having higher plasma cholesterol levels than soy protein–fed animals, also showed impaired hepatic removal of plasma LDL as well as increased LDL production rates (122,124). These results suggest that casein-induced hypercholesterolemia is associated with impaired LDL clearance and is consistent with a down-regulation of hepatic LDL receptors. In hypercholesterolemic humans, soy protein diets increased mononuclear cell LDL degradation 16-fold compared to a casein diet (125). No clear relationship, however, was found between the reduction of total and

LDL-cholesterol and the increased LDL degradation. Taken together, these findings in animals and humans suggest that some component(s) of protein may regulate the expression of lipoprotein receptors, thus controlling plasma LDL-cholesterol.

Summary

Soybean oil is the major contributor of visible fat to the U.S. food supply, providing about 12% of total energy to the diet. The fatty acid composition of SBO is unique among other edible oils. Like other unsaturated vegetable oils, SBO is a rich source of the essential fatty acid linoleic acid, but only SBO provides the other EFA, α-linolenic acid. The linoleic-to-linolenic ratio in SBO is 7:1, which is within the recommended range (i.e., above 4:1) (109). Current dietary guidelines recommend that the intake of polyunsaturated fatty acids (linoleic acid) should be increased from the current level of 7% kcal to no more than 10% kcal and that saturated fat should be reduced from the current amount of 12% kcal to less than 10% kcal. SBO currently supplies about 43% of the total linoleic acid and about 15 of the SAFA to the U.S. diet. Understanding the nutritional role of isomeric fatty acids is of obvious importance because of the wide use of hydrogenated vegetable oils in the food supply. The biochemical knowledge of isomeric fatty acids indicate that 18:1 and 18:2 *trans* fatty acids are metabolized similarly to their *cis* counterparts. Multigenerational studies in experimental animals have demonstrated that consumption of 18:1 *trans* fatty acids from PHVO results in normal growth, weight gain, and reproductive capacity with no indications of histopathological effects. On the other hand, it is clear that the effect of *trans* fatty acids from PHVO on plasma total and LDL-cholesterol is different from oleic acid and the 12:0 to 16:0 saturated fatty acids. Current data indicate that *trans* fatty acids raise TC and LDL-C compared with oleic acid and are less potent than saturated fatty acids in that regard.

Another component of soybeans that appears to have positive nutritional effects is isolated soy protein. The independent cholesterol-lowering effects of soy protein have been known for years. Recently, however, it has been demonstrated that significant reductions in plasma cholesterol can be achieved in moderately hypercholesterolemic individuals with reasonable levels of soy protein and provided in nutritionally acceptable food products.

There are still unanswered questions about the nutritional effects of isomeric fatty acids. In light of studies indicating that *trans* fatty acids do not appear to differ from stearic acid in their blood lipid effects, it is unclear whether *trans* fatty acids have an independent hypercholesterolemic effect or are neutrocholesterolemic. Resolution of this issue has pragmatic implications for communicating accurate and effective nutritional guidance to consumers.

References

1. Commodity Economics Division, Economic Research Service, U.S. Department of Agriculture, *OCS-41*, July 1994.
2. McDowell, M.A., R.R. Briefel, and K. Alaimo, *Phase I, 1988–91*, National Center for Health Statistics, Hyattsville, MD, 1994.

3. *Food, Fats and Oils*, 7th edn., by the Institute of Shortening and Edible Oils, Inc., Washington, DC, 1994, p. 25.
4. Hunter, J.E., and T.H. Applewhite, *Amer. J. Clin. Nutr. 54:* 363 (1991).
5. Deuel, H.J., Jr., in *The Lipids, Their Chemistry and Biochemistry*, Vol. II, Interscience Publishers, Inc., New York, London, 1955, pp. 218–221.
6. Senti, F.R., (Contract no. FDA 223-83-2020), August 1985.
7. Emken, E.A., *Annu. Rev. Nutr. 4:* 339 (1984).
8. Emken, E.A., H.J. Dutton, W.K. Rohwedder, H. Rakoff, R.O. Adlof, R.M. Gulley, and J.J. Canary, *Lipids 15:* 864 (1980).
9. DeJarlais, W.J., R.O. Adlof, J. Mackin, R. Dougherty, and J.M. Iacono, *Metaboliosm 28:* 575 (1979).
10. Alfin-Slater, R.B., A.F. Wells, L. Aftergood, and H.J. Deuel, Jr., *J. Nutr. 63:* 241 (1957).
11. Alfin-Slater, R.B., L. Aftergood, and T. Whitten, *J. Amer. Oil Chem. Soc. 53:* 468A (1976).
12. de Fouw, N.J., G.A.A. Kivits, and W.G.L. van Nielen, *INFORM 3:* 516 (1992).
13. Mattson, F.H., G.A. Nolan, and M.R. Webb, *J. Nutr. 109:* 1682 (1979).
14. Emken, E.A., R.O. Adlof, W.K. Rohwedder and R.M. Gulley, *Biochim. Biophys. Acta 1170:* 173 (1993).
15. Nelson, G.J., and R.G. Ackman, *Lipids 23:* 1005 (1988).
16. Olubajo, O., M. Marshall, J. Judd, and J. Adkins, *Nutr. Res. 6:* 931 (1986).
17. Bonanome, A.G., and S.M. Grundy, *J. Nutr. 119:* 1556 (1989).
18. Meydani, S.N., A.H. Lichtenstein, P.J. White, S.H. Goodnight, C.E. Elson, M. Woods, S.L. Gorbach, and E.J. Schaefer, *J. Amer. Col. Nutr. 10:* 406 (1991).
19. Expert Panel Report, *Arch. Intern. Med. 48:* 36 (1988).
20. Heiss, G., N.J. Johnson, S. Reiland, C.E. Davis, and M.B. Tyroler, *Circulation 62* (Suppl)*:* 116 (1980).
21. Keys, A., *Circulation 41:* 1 (1970).
22. Gordon, T., W. Kagan, M. Garcia-Palmieri, W.B. Kannel, W.J. Zukel, J. Tillotson, P. Sorlie, and M. Hjortland, *Circulation 63:* 500 (1981).
23. Hegsted, D.M., R.B. McGandy, M.L. Myers, and F.J. Stare, *Amer. J. Clin. Nutr. 17:* 281 (1965).
24. Hjermann, I., I. Holme, and P. Leren, *Am. J. Med. 80* (Suppl)*:* 7 (1986).
25. Dayton, S., M.L. Pearce, S. Hashimoto, W.J. Dixon, and U. Tomiyasu, *Circulation 49* (Suppl)*:* 1 (1969).
26. Ginsberg, H.N., S.L. Barr, A. Gilbert, W. Karmally, R. Deckelbaum, K. Kaplan, R. Ramakrishnan, S. Holleran, and R.B. Dell, *N. Engl. J. Med. 322:* 574 (1990).
27. Keys, A., J.T. Anderson, and F. Grande, *Metabolism 14:* 776 (1965).
28. Hegsted, D.M., L.M. Ausman, J.A. Johnson, and G.E. Kallal, *Am. J. Clin. Nutr. 57:* 875 (1993).
29. Shaomei, Y., J. Derr, T.D. Etherton, and P.M. Kris-Etherton, *Amer. J. Clin. Nutr.*, in press.
30. Mensink, R.P., and M.B. Katan, *Arteriscler. Thromb. 12:* 911 (1992).
31. Weisweiller, P., P. Janeschek, and P. Schwandt, *Metabolism 34:* 83 (1985).
32. Mensink, R.P., and M.B. Katan, *N. Engl. J. Med. 321:* 436 (1989).
33. Kromhout, D., A.C. Arntzenius, N. Kemper-Voogd, N.J. Kemper, J.D. Barth, H.A. van der Voort, and E.A. van der Velde, *Atherosclerosis 66:* 99 (1987).
34. Becker, N., D.R. Illingworth, P. Alaupovic, W.E. Connor, and E.E. Sundberg, *Amer. J. Clin. Nutr. 37:* 355 (1983).

35. Kris-Etherton, P.M., J. Derr, D.C. Mitchell, V.A. Mustad, M.E. Russell, E.T. McDonnell, D. Salabsky, and T.A. Pearson, *Metabolism 42:* 121 (1993).
36. Dietschy, J.M., S.D. Turley, and D.K. Spady, *J. Lipid Res. 34:* 1637 (1993).
37. Mattson, F.H., and S.M. Grundy, *J. Lipid Res. 26:* 194 (1985).
38. Iacono, J.M., and R.M. Dougherty, *Amer. J. Clin. Nutr. 53:* 660 (1991).
39. Iacono, J.M., P. Puska, R.M. Dougherty, P. Pietinen, E. Variainen, U. Leino, M. Mutanen, and S. Moisio, *Amer. J. Clin. Nutr. 38:* 860 (1983).
40. Kestin, M., I.L. Rouse, R.A. Correll, and P.J. Nestel, *Amer. J. Clin. Nutr. 50:* 280 (1989).
41. Kinsella, J.E., Lokesh, B., Broughton, S., and J. Whelan, *J. Nutr. (Suppl):* 24–44 (1990).
42. Marshall, L.A., and P.V. Johnson, *Lipids 17:* 905 (1982).
43. Dupont, J., P.J. White, E. Schaefer, S. Meydani, C.E. Elson, M. Woods, S.L. Gorbach, and M.P. Carpenter, *J. Amer. Coll. Nutr. 5:* 438 (1990).
44. Howe, G.R., *Cancer 74:* 1078 (1994).
45. Willett, W.C., *Cancer 74:* 1085 (1994).
46. Dept. of Health and Human Services, Food and Drug Administration. Federal Register 58(3): 2787 (1993).
47. Ip, C., and C.A. Carter, *Cancer Res. 45:* 1997 (1985).
48. Sundram, K., H.T. Khor, A.S.H. Ong, and R. Pathmanathan, *Cancer Res. 49:* 1447 (1989).
49. Fritsche, K.L., and P.V. Johnston, *J. Amer. Oil Chem. Soc. 65:* 509 (abstract) (1988).
50. McKeown-Eyssen, G.E., and E. Bright-see, *Nutr. Cancer 6:* 160 (1984).
51. Kaizer, L., N.F. Boyde, V. Kriukor, and D. Tritchler, *Nutr. 12:* 61 (1989).
52. Tuyns, A.J., R. Kaaks, and M. Haelterman, Nutr. *Cancer 11:* 189 (1988).
53. Verreault, R., J. Brisson, L. Deschenes, F. Naud, F. Meyer, and L. Belanger, *J. Natl. Cancer Inst. 80:* 819 (1988).
54. Slover, H.T., R.H. Thompson, C.S. Davis, and G.V. Merola, *J. Amer. Oil Chem. Soc. 62:* 775 (1985).
55. Ratnayake, W.M., and G. Pelletier, *J. Amer. Oil Chem. Soc. 69:* 95 (1992).
56. Sampugna, J., L.A. Pallansch, M.G. Enig, and M. Keeney, *J. Chromatogr. 249:* 245 (1982).
57. Parodi, P.W., *J. Dairy Sci. 59:* 1870 (1976).
58. Emken, E.A., in *Health Effects of Dietary Fatty Acids*, edited by G.J. Nelson, American Oil Chemists' Society, Champaign, IL, 1991, pp. 245–260.
59. Emken, E.A., W.K. Rohwedder, R.O. Adlof, W.J. Dejariais, and R.M. Gulley, *Biochim. Biophys. Acta. 836:* 233 (1985).
60. Emken, E.A., R.O. Adlof, W.K. Rohwedder, and R.M. Gulley, *Lipids 24:* 61 (1989).
61. Emken, E.A., W.K. Rohwedder, R.O. Adlof, and R.M. Gulley, *Lipids 22:* 495 (1987).
62. Ide, T., and M. Sugano, *Bichim. Biophys. Acta. 794:* 281 (1984).
63. Senti, F.R., *Health Aspects of Dietary Trans Fatty Acids*, Life Sciences Research Office, Federation of American Societies for Experimental Biology, Bethesda, MD (1985).
64. Anderson, J.T., F. Grande, and A. Keys, *J. Nutr 75:* 388 (1961).
65. De Iongh, H., R.K. Beerthuis, C. Den Hartog, L.M. Dalderup, and P.A. Van Der Spek, *Nutritio et Dieta 7:* 137 (1965).
66. Erickson, B.A., R.H. Coots, F.H. Mattson, and A.M. Kligman, *J. Clin. Invest. 43:* 2017 (1964).
67. Laine, D.C., C.M. Snodgrass, E.A. Dawson, M.A. Ener, K. Kuba, and I.D. Frantz, *Amer. J. Clin. Nutr. 35:* 683 (1982).

68. Lichtenstein, A.H., L.M. Ausman, W. Carrasco, J.L. Jenner, J.M. Ordovas, and E.J. Schaefer, *Arterio Thromb. 13:* 154 (1993).
69. McOsker, D.E., F.H. Mattson, H.B. Sweringen, and A.M. Kligman, *JAMA 180:* 380 (1962).
70. Nestel, P.J., M. Noakes, G.B. Belling, R. McArthur, P.M. Clifton, E. Janus, and M. Abbey, *J. Lipid Res. 33:* 1029 (1992).
71. Mattson, F.H., F.J. Hollenbach, and A.M. Kligman, *Amer. J. Clin. Nutr. 28:* 726 (1975).
72. Mensink, R.P., and M.B. Katan, *N. Engl. J. Med. 323:* 439 (1990).
73. Zock, P.L., and M.B. Katan, *J. Lipid Res. 33:* 399 (1992).
74. Bonanome, A., and L.N. Grundy, *N. Engl. J. Med. 318:* 1244 (1988).
75. Judd, J.T., B.A. Clevidence, R.A. Muesing, J. Wittes, M.E. Sunkin, and J. Podczasy, *Amer. J. Clin. Nutr. 59:* 861 (1994).
76. Emken, E., *INFORM 5:* 906 (1994).
77. Wood, R., K. Kubena, B. O'Brien, L. Tseng, and G. Martin, *J. Lipid Res. 34:* 1 (1993).
78. Woollett, L.A., C.M. Daumerie, and J.M. Dietschy, *J. Lipid Res. 35:* 1661 (1994).
79. Ignatowski, A., *Arch. Med. Exp. Anat. Pathol. 20:* 1 (1908).
80. Meeker, D.R., and H.D. Kesten, *Arch. Pathol. 31:* 147 (1941).
81. Van der Meer, R., and A.C. Beynen, *J. Amer. Oil Chem. Soc. 64:* 1172 (1987).
82. Carroll, K.K., *J. Amer. Diet Assoc. 91:* 820 (1991).
83. Erdman, J., and E. Fordyce, *Amer. J. Clin. Nutr. 49:* 725 (1989).
84. Carroll, K.K., P.M. Glovannetti, M.W. Huff, O. Moase, D.C.K. Roberts, and B.M. Wolfe, Amer. *J. Clin. Nutr. 31:* 1312 (1978).
85. Glovannetti, P.M., K.K. Carroll, and B.M. Wolfe, *Nutr. Rep. Int. 6:* 609 (1986).
86. Van Raaj, J.M.A., M.B. Katan, J.G.A.J. Hautvast, and R.J.J. Hermus, *Amer. J. Clin. Nutr. 34:* 1261 (1982).
87. Van Raaj, J.M.A., M.B. Katan, C.E. West, and J.G.A.J. Hautvast, *Amer. J. Clin. Nutr. 35:* 925 (1982)
88. Miyazima, E., S. Takeyama, N. Tada, T. Ishikawa, and H. Nakamura, *Nutr. Sci. Soy Protein 2:* 31 (1981).
89. Goldberg, A.P., A. Lim, J.B. Kolar, I.L. Grundhauser, F.H. Steinke, and G. Schonfeld, *Atherosclerosis 43:* 355 (1982).
90. Sacks, F.M., J.L. Breslow, P.G. Wood, and E.H. Kass, *J. Lipid Res. 24:* 1012 (1983).
91. Sintori, C.R., E. Agradi, F. Conti, O. Mantero, and E. Gatti, *Lancet 1:* 275 (1977).
92. Sintori, C.R., E. Gatti, O. Mantero, F. Conti, E. Agradi, E. Tremoli, M. Sintori, L. Fraterrigo, L. Tavazzi, and D. Kritchevsky, *Amer. J. Clin. Nutr. 32:* 1645 (1979).
93. Sintori, C.R., C. Zucchi-Dentone, N.M. Sintori, E. Gatti, G.C. Descovich, A. Gaddi, L. Cattin, P.G. DaCol, U. Senin, E. Mannarino, G. Awellone, and L. Columbo, *Ann. Nutr. Metab. 29:* 348 (1985).
94. Descovich, G.C., C.G. Ceredi, A. Gaddi, M.S. Benassi, G. Mannino, L. Colombo, L. Cattin, G. Fontana, U. Senin, E. Mannarino, C. Caruzzo, E. Berielle, C. Fragiacomo, G. Noseda, M. Sintori, and C.R. Sintori, *Lancet 2:* 709 (1980).
95. Hodges, R.E., W.A. Krehl, D.B. Stone, and A. Lopez, *Amer. J. Clin. Nutr. 20:* 198 (1967).
96. Verrillo, A., A. de Teresa, P.C. Giarrusso, and S. La Rocca, *Atherosclerosis 54:* 321 (1985).

97. Gaddi, A., G.C. Descovich, G. Noseda, C. Fragiacomo, A. Nicolini, G. Montanari, G. Vanetti, M. Sintori, E. Gatti, and C.R. Sintori, *Arch. Dis. Child. 62:* 274 (1987).

98. Widhalm, K. in *Nutritional Effects on Cholesterol Metabolism*, edited by A.C. Beynen, Voorthuizen, The Netherlands, Transmondial, 1986, pp. 133–140.

99. Miyazima, E., S. Takeyama, K. Kondo, A. Kagami, N. Suzuki, N. Tada, T. Ishikawa, and H. Nakamura, *Nutr. Sci. Soy Protein 3:* 90 (1982).

100. Wolfe, B.M., P.M. Glovannetti, D.C.H. Cheng, D.C.K. Roberts, and K.K. Carroll, *Nutr. Rep. Int. 24:* 1187 (1981).

101. Vessby, B., B. Karlstrom, H. Lithel, I.-B. Gustafssohn, and I. Werner, *Hum. Nutr. Appl. Nutr. 36A:* 179 (1982).

102. Schwandt, P., W.O. Richter, and P. Weisweiler, *Atherosclerosis 40:* 371 (1981).

103. Kolb, S., and D. Sailer, *Nutr. Rep. Int. 30:* 719 (1984).

104. Holmes, W.L., G.B. Rubel, and S.S. Hood, *Atherosclerosis 36:* 379 (1980).

105. Shorey, R.L., B. Brazan, G.S. Lo, and F.H. Steinke, *Amer. J. Clin. Nutr. 34:* 769 (1981).

106. Grundy, S.M., and J.J. Abrams, *Amer. J. Clin. Nutr. 38:* 245 (1983).

107. Huff, M.W., P.M. Giovannetti, and B.M. Wolfe, *Amer. J. Clin. Nutr. 39:* 888 (1984).

108. Wolfe, B.M., and P.M. Giovannetti, *Nutr. Rep. Int. 32:* 1057 (1985).

109. Mercer, N.J.H., K.K. Carroll, P.M. Giovannetti, F.H. Steinke, and B.M. Wolfe, *Nutr. Rep. Int. 35:* 279 (1987).

110. Bakhit, R.M., B.P. Klein, D. Essex-Sorlie, J.O. Ham, J.W. Erdman, and S.M. Potter, *J. Nutr. 124:* 213 (1994).

111. Terpstra, A.H.M., L. Harkes, and F.H. van der Veen, *Lipids 16:* 114 (1981).

112. Huff, M.W., D.C.K. Robers, and K.K. Carroll, *Atherosclerosis 41:* 327 (1982).

113. Potter, S.M., R.M. Bakhit, D.L. Essex-Sorlie, K.E. Weingartner, K.M. Chapman, and R.A. Nelson et al., *Amer. J. Clin. Nutr. 58:* 501 (1993).

114. Kritchevsky, D., S.A. Tepper, and D.M. Klurfeld, *J. Amer. Oil Chem. Soc. 64:* 1167 (1987).

115. Huff, M.W., and K.K. Carroll, *J. Nutr. 110:* 1676 (1980).

116. Katan, M.B., L.H.M. Vroomen, and R.J.J. Hermus, *Atherosclerosis 43:* 381 (1982).

117. Carroll, K.K., *J. Amer. Oil Chem. Soc. 58:* 416 (1981).

118. Kuyvenhoven, M.W., W.F. Roszkowski, C.E. West, R.L.A.P. Hoogenboom, R.M.E. Vos, A.C. Beynen, and R. Van der Meer, *Br. J. Nutr. 62:* 333 (1989).

119. Sanchez, A., M.C. Horning, G.W. Shavlik, D.C. Wingeleth and R.W. Hubbard, *Nutr. Rep. Int. 32:* 1047 (1985).

120. Noseda, G., C. Fragiacomo, G.C. Descovich, R. Fumagalli, F. Bemini, and C.R. Sintori, in R. Fumagalli, D. Kritchevsky, and R. Paoletti, Elsevier/North Holland Biomedical Press, Amsterdam, 1980, pp. 355–362.

121. Fumagalli, R., L. Soleri, R. Farina, R. Musanti, O. Mantero, G. Noseda, E. Gatti, and C.R. Sintori, *Atherosclerosis 19.*

122. Khosla, P., S. Samman, K.K. Carroll, and M.W. Huff, *Biochim. Biophys. Acta. 1002:* 157 (1989).

123. Samman, S., P. Khosla, and K.K. Carroll, *Lipids 24:* 169 (1989).

124. Kurowska, E.M., and K.K. Carroll, *J. Nutr. 120:* 831 (1990).

125. Lovati, M.R., C. Manzoni, A. Canavesi, M. Sintori, V. Vaccarino, M. Marchi, G. Gaddi, and C.R. Sintori, *J. Clin. Invest. 80:* 1498 (1987).

Chapter 24

Soybean Processing Quality Control

Sherman A. Boring

Consultant
The Sea Ranch, CA

Introduction

The high quality of soybean products is an important factor in their dominance of world vegetable oil and protein markets. The soybean processing industry has developed manufacturing systems and procedures, trading rules, comprehensive product quality specifications, and quality control procedures to ensure finished products that consistently meet customer and consumer needs.

This chapter describes sampling; testing; process control points; specifications for raw, in-process, and finished products; analytical procedures; and communication and storage of control laboratory data. Also discussed are automation of analyses, applications of statistical quality control, quality improvement programs, and laboratory management and design.

Quality Specifications

Essential for quality control is a system of specifications defining raw and in-process materials and finished products.

The Trading Rules of the National Oilseed Processors Association (NOPA)(1) govern commerce in soybean meal and oil, including quality requirements for the materials listed in Table 24.1.

Sometimes these quality requirements are supplemented by additional specifications written into sales contracts for customers' special needs.

Manufacturers of finished products, particularly fats and oils, have also developed comprehensive specifications for their in-process materials and finished products based on their own experience, on consumer research, and, in the case of industrial products, on their customers' own specifications.

TABLE 24.1 Quality Requirements included in NOPA Trading Rules

NOPA rule no.	Product(s) covered
2	Soybean flakes and 44% protein soybean meal; soybean flakes and high-protein or solvent-extracted soybean meal
102	Crude soybean oil
103	Crude degummed oil; once-refined soybean oil; fully refined soybean oil
104	Refining byproduct lipid (soapstock from alkali refining); acidulated refining byproduct lipid (tank bottoms)

Sampling

A key part of the control of processing of soybean products is periodic sampling and testing of raw, in-process, and finished products to determine compliance to specifications or need for corrective action. Because of the great variety of situations in which samples are taken for analysis, it is not possible to provide detailed guidelines here for proper sampling procedures and equipment, but processors should carefully think them out. It is axiomatic that samples must be representative of the material sampled; otherwise, analyses are not only useless, but misleading. Sampling ports and valves, for example, in continuous processes must be located at points where material flows are thoroughly mixed and homogeneous and where reactions in the particular process step have gone to completion. Sampling personnel must be trained with regard to proper use, cleanliness, and maintenance of sampling devices such as triers, probes, and automatic samplers and dividers. Samples need to be taken at the proper time. For example, sufficient agitation time must be allowed before sampling tanks; in continuous processes the operator must wait to take a sample until process equilibrium is reestablished after changing operating conditions. Sampling lines and valves should be thoroughly flushed out before samples are taken. Sampling of deodorized oil, in particular, requires thorough cleaning and flushing of sample valves and spouts to avoid false peroxide value analyses.

It should be standard practice to take and analyze a second sample of in-process materials before corrective action when the first sample is out of specification, except for a few lengthy analyses, such as that for phosphorus in oil by AOCS method Ca 12-55, which requires about 3 hr to complete.

NOPA and AOCS Sampling Methods

The NOPA *Yearbook and Trading Rules* (1) specify detailed sampling methods and equipment for soybean meal and for soybean oil in bulk shipments. *Official Methods* of the American Oil Chemists' Society (AOCS) (2) cover sampling of soybeans, meal, soy flour, commercial fats and oils, soapstock, industrial oils, and derivatives. NOPA Trading Rules recommend as a guideline that sampling of soybean oil for export be done by an independent surveyor. References to specific NOPA and AOCS sampling procedures are made in the quality control program tables in this chapter.

Automatic Samplers

Automatic samplers are specified by NOPA trading rules for soybean meal at point of origin, barge transfer facilities, and export vessel loading facilities. Specific types of samplers are prescribed, as well as procedures for certifying proper installation and semiannual checks of their condition and proper operation.

Automatic samplers for soybean oil are permitted by NOPA rules for barge and shipload quantities. These are of the continuous bleed type, conforming to AOCS Method C 1-47. Automatic samplers, properly operated and maintained, are preferable to other sampling methods when permitted by trading rules. See Chapters 4 and 9 and reference (3) for further discussion of sampling.

Analytical Procedures

The *Official Methods* of the American Oil Chemists' Society (2) provide authoritative analytical procedures for almost all needs in the soybean meal and oil industry. AOCS, through its Uniform Methods Committee, maintains an active program of reviewing and updating existing methods and issuing new ones. Changes for 1993 include ten new methods and 45 additions and revisions. Every control laboratory should have a copy of the AOCS Methods and subscribe to the annual update program. The quality control program tables in this chapter state the AOCS Method number, where available, for each analysis.

Other Methods Sources

AOCS methods should be used wherever possible. However, there are many additional sources (4,5,6,7,8,9,10,11) of useful analytical procedures for fats and oils not covered by AOCS Methods.

Test procedures for quality of many packaging materials are provided by the American Society for Testing Materials (ASTM) (12). In addition, manufacturers of various types of packaging materials have formed trade associations, such as the Plastic Bottle Institute and the Closure Manufacturers Association, which have developed standardized test methods and general specifications for their particular products. This information is often available through suppliers of packaging materials.

Statistical quality control texts (13,14,15) provide information on sample sizes, data analysis, and acceptance and lot inspection plans.

Automated Analyses

Control laboratories should consider automated equipment wherever large numbers of repetitive analyses and similar operations are made. The types available include

1. *Autosamplers*, which rack up multiple samples for the same type of analysis and feed them automatically in sequence into autoanalyzers
2. *Autoanalyzers*, which perform the same analysis repetitively and record the results (16)
3. *Robots*, computer-controlled mechanical devices, perform repetitive tasks such as weighing, filtration, titration, or injection of samples and can be programmed to alternate among several kinds of tasks (17)
4. *In-line analyzers*, which are installed in the processing lines or vessels and, in extraction plants and oil refineries, may include nephelometric turbidimeters, spectrophotometers for color and chlorophyll measurement, refractometers, oxygen analyzers, and moisture testers (18)

5. *Computer-aided inspection* by optical scanners, automatic weighers, and other sensors, which can inspect every package on the conveyor belt and trigger the rejection of defective packages

The benefits of equipment such as the foregoing are reduced manpower requirements and faster and often more accurate analyses. The disadvantages are higher capital costs and, in the case of the first three types listed, the fact that they lend themselves best to saving up and running several samples through the equipment, one immediately after another. This is not desirable for in-process analyses, which need to be run immediately for rapid feedback of information. Each possible application should be judged on its individual merits.

Plasma (ICP) spectrometers have recently been adopted by some processors to analyze oils for trace elements such as iron, nickel, copper, sodium, phosphorus, calcium, and magnesium. ICP instruments are expensive but offer great advantages in speed and accuracy. All elements that an ICP is equipped to determine are measured essentially instantaneously (19).

Flavor Methodology

Flavor is probably the single most important quality attribute of edible oils. Ideally, deodorized soybean oil would be completely bland, tasteless, and odorless and would remain flavor-stable over its shelf life. This ideal can be rather closely approached by proper oil processing as discussed in Chapters 5, 9 through 12, 14, and 18 through 20. An important quality control function is determining the acceptability of oil flavor after deodorization and as finished product. This may be done by sensory evaluation, instrumental techniques, or both. Sensory testing is more often used despite its subjectivity.

A complete description of sensory methodology for quality control of oils is beyond the scope of this chapter, but briefly it consists of

1. Creating standard oil samples representing several levels of blandness (flavor intensity), assigning each sample a flavor score on a scale of 1 to 10, with 1 having a very strong flavor and 10 being the perfectly tasteless ideal, under standardized tasting conditions;
2. Training analysts by repeatedly exposing them to these standard samples and their flavor scores
3. Establishing a specification for minimum acceptable level of blandness on this scale, generally determined by consumer research under use conditions of the oil product
4. Routine testing of production samples for flavor by these trained analysts, who frequently refer to the standard samples to maintain "calibration"

Flavor testing is done by the trained analysts under standardized conditions, as individuals or as panels, usually consisting of four to ten analysts. Processing personnel should be trained and included in the panels. The tasters assign a flavor score to each sample; in panels, the average of all panelists' scores is recorded. Because of the large number of samples that are routinely flavored in quality control laboratories, it is not practical to panel-test all of them. Some or even most samples, therefore, are evaluated by individuals. After each sample is flavored, its score is assigned and compared against the minimum product specification, thus determining its acceptability or nonacceptability. For example, assume that standard oil samples have been prepared with scores of 2, 4, 6, 8, and 10, with 2 representing a very strong flavor and 10 no flavor at all. If consumer marketing research has established that an oil with a flavor score of 6 represents a threshold beyond which the amount or intensity of flavor present is just discernible as a negative in various consumer use conditions, then a specification of 6 would be set as the lowest acceptable score, and any oil with a lower score would be unacceptable for shipment. However, because of the subjective nature of flavor testing, an oil scored below 6 should be retested for verification before making the final decision to reject.

Sensory evaluation requires special physical facilities, procedures, and maintenance of a supply of standard flavor samples, kept in frozen storage (20,21,22,23); see also AOCS Method Cg 2-83 (flavor intensity section) and the treatment of laboratory design in this chapter for further discussion.

Instrumental methods have been developed to replace or supplement sensory evaluation of oil flavor. These are generally based on the principle of heating the sample to evolve odorous volatiles, capturing these volatiles by a purge-and-trap technique, and then quantifying them by glass capillary chromatography. Although many kinds of volatiles are released during heating, the sum of the concentrations of pentanal and decadienal has been found to be highly correlated inversely to sensory flavor scores of refined, bleached, and deodorized (RBD) soybean oil (24,25). See also AOCS Method Cg 1-83.

Oil Color Methodology

Color, like flavor, is a very important attribute of soybean oil products. Color control is a key aspect of processing, and oil color is the most frequently run analysis in a refinery laboratory. The predominant method of measurement in the United States is the Wesson Lovibond system, AOCS Method Cc 13b-45, involving visual matching of the sample against combinations of yellow and red glasses in a colorimeter that provides standardized viewing conditions. Visual matching is an acquired skill requiring careful training of analysts, so this is less than an ideal system. Despite its shortcomings, the Wesson Lovibond system is still used in the United States in preference to other methods, which include the AOCS spectrophotometric, Cc 13c-50, which is more objective, and the British Standard Lovibond, Cc 13e-92. The latter is frequently used in international oil commerce (26).

Smalley Check Sample Series

AOCS maintains a program of check samples. The following sample series are of interest to control laboratories in soybean extraction plants or oil refineries:

- Soybeans
- Soy meal
- Crude soybean oil
- National Institute of Oilseed Processors (NIOP) fats and oils
- Gas chromatography
- *Trans* fatty acids
- Edible fats
- Vegetable oil color
- Nuclear magnetic resonance (NMR) for solid fat content

In each of these, several samples are sent out annually to all collaborators, all receiving the same samples. They perform the specified analyses and report them to the Smalley coordinator. A statistical evaluation of the results is made, and all collaborators, identified only by code numbers, are issued a report of their own and all other collaborators' performance (accuracy), measured by their deviations from the average of all results for each type of analysis. This information is very valuable to the control laboratory manager in identifying areas where improvement in analytical work needs to be made.

Another feature of the Smalley program is the availability of samples whose analyses are accurately known. These can serve as standards for reference and training purposes. Information about the Smalley program may be obtained by contacting

AOCS Technical Department
P.O. Box 3489
Champaign, IL 61826-3489 USA
Telephone (217) 359-2344
Fax (217) 351-8091

Quality Control Program

General Considerations

The two main functions of a quality control laboratory are the following:

1. To ensure compliance of finished products to specifications with the goal of consumer/customer satisfaction
2. To support plant operations with prompt analyses of raw and in-process materials, and assist in solving plant technical problems and in improvement projects

Considerably more quality control worker-hours in a well-managed soybean processing plant are expended on in-process control analyses than on finished-product analysis and inspection. This reflects the philosophy that if production processes are properly controlled, the resulting finished-product will, of necessity, be correct and a large amount of finished-product testing will not be needed. Finished-product testing is still necessary, but the emphasis should be on process control, not on after-inspection.

A related point is that processes must be basically stable and capable of consistently meeting quality specifications. If a process relies on frequent adjustments based on conventional in-process analyses, then the cause of instability must be found and corrected. Analyses generally cannot be performed fast enough or frequently enough to keep an unstable process on target. The main function of in-process testing should be to verify that the process at start-up is producing the desired quality, followed by periodic checks to verify that it is holding stable.

The need for rapid and accurate in-process analyses is accentuated by strong trends in soybean processing, as in other industries, toward reduced inventories of raw, intermediate, and finished goods, such as "just-in-time" and "order-based" production programs, "straight-through" processing, and increased adoption of continuous systems. These trends reduce the margin for operating errors in the plant, so rapid feedback of analyses, making them as "real-time" as possible, is more important than ever before.

Quality Control Analysis Schedule

A quality control program covering raw, in-process materials, and finished-products is given in Table 24.2, for a soybean oil extraction plant, and in Table 24.3, for a soybean oil refining plant. These tables describe sampling points, sampling frequency, the analyses performed, and test methods used. Critical control points are identified. Analytical limits are not given, because these are either defined by trading rules or generally depend on finished-product specifications.

The program described is necessarily general in nature and may need to be modified for specific operating systems and products. Most of the analyses are done in the control laboratory, but some of the less complex tests may be done by trained operators in small testing stations set up in the plant.

It should be standard practice in carrying out this program to take and analyze a second sample of in-process materials before corrective action when the first sample is out of specification, except for a few lengthy analyses, such as phosphorus in oil by AOCS Method Ca 12-55, which requires about 3 hr to complete.

Data Management

Communication of Data

As stated earlier, a primary job of quality control is to provide analyses of in-process samples. To be of maximum value, these analyses must be promptly performed and communicated to processing personnel. Good record keeping is essential, not only for immediate process control, but also to make it easy to retrieve past data for studying

TABLE 24.2 Quality Control Program at Soybean Extraction Plant

Analysis	Method*	Sampling frequency	Comments
Soybeans (incoming or to process)			
Sampling	Ac 1-45	*Incoming*: each shipment *To process*: before startup and each change of source of soybeans	
Moisture and volatile	Ac 2-41		
Oil Content	Ac 3-44		
Protein	Ac 4-41		
Free fatty acids	Ac 5-41		
Hexane (incoming)			
Hexane specifications	H 16-56	First five shipments from each new supplier; if within specification, reduce subsequent sampling and analysis to spot-check basis	Method covers six attributes
Extraction process			
Cracking	Screen test	Startup and every two hours	Done by operators
Flaking	Micrometer		
Soybean meal from process			
Sampling	Ba 1-38	Start-up and once per shift on a composite of hourly samples, or as required to maintain control	Add other tests as required by sales contract
Moisture and volatile	Ba 2-38		
Oil content	Ba 3-38 or calibrated IR		
Protein	Ba 4d-90 or calibrated IR		
Crude fiber	Ba 6-84		
Urease	Ba 9-58		
Soybean meal blending			
Moisture and volatiles	Ba 2-38	Startup and every four hours, or as required to maintain control	
Protein	Ba 4d-90 or calibrated IR		
Oil content	Ba 3-38 or calibrated IR		
Soybean meal shipments			
Sampling	NOPA rule 2	Each shipment	Use only AOCS methods listed in NOPA rules
Other tests as for meal from process			

Crude soybean oil from process			
Sampling	C 1-47	As soon as operational stability is achieved after startup and after change of source of soybeans; thereafter, every four hours or as required to maintain control, but analyses for unsaponifiable, green color, bleached color, and neutral oil content may be reduced or deleted unless a problem is suspected	
Flash point	Cc 9b-55		
Moisture and volatile	Ca 2c-25		
Neutral oil content	Ca 9f-57		
Green color	NOPA Tent		
Phosphorus	Ca 19-86		On degummed oil only
Free fatty acids	Ca 5a-40		
Unsaponifiable	Ca 6a-40		
Bleach test	Cc 8e-63 and Cc 8b-52		
Insoluble impurities	Ca 3-46		
Crude soybean oil shipments (official loading sample)			
Sampling per NOPA trading rule; all tests as for crude oil from process	Each shipment[a]	Perform flash point, unsaponifiable, green color, bleached color, and insoluble impurities on (1) oil from new season soybeans or after change in the source of soybeans until experience with the quality of oil from these sources is acquired; (2) when problems are suspected. Thereafter the frequency of these tests may be reduced.	

[a]By mutual agreement between supplier and customer, up to five loading samples of tank truck shipments may be composited for analysis if trucks are loaded from same crude oil storage tank at crude mill.

*AOCS unless specified otherwise

TABLE 24.3 Quality Control Program at Soybean Oil Refinery

Analysis	AOCS or other method	Sampling frequency	Comments
Arriving crude soybean oil (official loading sample)			
All tests as for crude oil from process, Table 24.2		Each shipment; see Table 24.2, note [a]	For settlement purposes and basis for possible segregation of incoming crude oil with respect to quality
Calcium, magnesium	Ca 15-87		Optional; perform if relatively high levels of calcium and magnesium in crude oil are associated with difficulties in achieving low phosphorus in refined oil
Arriving caustic soda			
Sodium hydroxide, %	Titration	First five shipments from each new supplier; if within specification, reduce subsequent sampling and analysis to spot-check basis	
Sodium carbonate, %	Titration		
Arriving bleach clay			
Bleach clay efficiency	Cc 8f-91	First five shipments from each new supplier; if within specification, reduce subsequent sampling and analysis to spot-check basis	
Particle size	Per method in clay supplier's specification		
pH			
Crude oil storage tank feeding refinery			
Free fatty acid	Ca 5a-40	Each tank before startup	Basis for setting plant refining conditions
Neutral oil content	Ca 9f-57		
Bleach test	Cc 8e-52 and Cc 8b-52		
Caustic storage tank			
Sodium hydroxide,%	Titration	After any addition to tank	Basis for dilution to refining strength
Diluted-caustic storage tank			
Sodium hydroxide, %	Titration	After each tank is made up	To verify proper dilution
Feed to primary centrifuge			
Sodium hydroxide, %	Titration to 4.0 pH	As soon as operating equilibrium is established after startup and after any change in operating conditions that might affect oil quality	**Critical**, to verify proper caustic dosage
Oil from water wash centrifuge(s)			
Free fatty acid	Ca 5a-40	As soon as operating equilibrium is established after startup, then every 4 h and after any change in operating conditions that might affect oil quality	
Phosphorus, ppm	Ca 12-55		**Critical**
Soap, ppm	Cc 17-79 or Ca 15b-87		**Critical**

Oil from vacuum dryer			
Moisture Content, %	Ca 2e-84	As soon as operating equilibrium is established after startup, then every 4 h and after any change in operating conditions that might affect oil quality	
Spent bleach clay			
Oil content, % (dry basis)	Soxhlet extraction	Weekly composite of each clay type	
Moisture content, %	Ba 2a-38		
Bleached, filtered oil			
Peroxide Value (PV)	Cd 8-53	As soon as operating equilibrium is established after startup, then every 4 h and after any change in operating conditions that might affect oil quality	**Critical**[a]
Chlorophyll, ppb	Cc 13d-55		
Clarity	Visual and filter test	Each batch	**Critical**
Hydrogenated oil (sampled in batch convertor)			
Refractive index (RI) (soft fats)	Cc 7-25	Each batch	RI by plant operator
Iodine value (hardfats)	Cd 1-25 or Cd 1b-87[b]		
Hydrogenated oil (sampled after final filtration)			
Clarity	Visual	Each batch	**Critical**
Nickel, ppm	Ca 18-79		**Critical**
Peroxide value	Cd 8-53 or Cd 8b-90	Once per shift	**Critical**[a]
Storage tank, hydrogenated stock (each type)			
SFI[c]	Cd 10-57	Each tank before pumping to blend tank or deodorizer	Perform tests required by product specification[c]
SFC[c]	Cd 16-81 or Cd 16b-93		
Iodine value	Cd 1-25 or Cd 1b-87		
Nickel, ppm	Ca 18-79		
Color	Cc 13b-45		
Free fatty acid	Ca 5a-40		
Iron, ppm	Ca 18-79		
Trans fatty acids[c], %	Ce 1c-89		
Fatty acid composition	Ce 1-62		
Peroxide value	Cd 8-53 or Cd 8b-90		**Critical**[a]
Blend tank (hydrogenated stock)			
Same tests and comments as for storage tank, but delete IV, iron, nickel tests			

TABLE 24.3 Quality Control Program at Soybean Oil Refinery—*continued*

Analysis	AOCS or other method	Sampling frequency	Comments
Deodorized product (sampled after deodorizer filter)			
Color	Cc 13b-45	Hourly (semicontinuous) or each batch (batch type)	
Free fatty acid	Ca 5a-40		
Clarity	Visual		
Flavor	Sensory		
Deodorized product in storage tank			
Color**	Cc 13b-45	See footnote [d]	
Free fatty acid	Ca 5a-40		
Clarity**	Filter test		
Flavor**	Sensory		
Peroxide value**	Cd 8-53 or Cd 8b-90		
Oil stability index	Cd 12b-92		
Refractive index[e]	Cc 7-25		
Iron, ppm	Ca 18-79		
Nickel, ppm	Ca 18-79		
Dissolved oxygen, %	Per supplier's instructions		
Dissolved nitrogen, %[f]	Per supplier's instructions		
Storage tank headspace oxygen, %		Daily	If tank is N_2-inerted
All other analyses required by product specification			
Packaged product			
Color	Cc 13b-45	Startup, every 4 h, and after change of feed tanks	
Peroxide value	Cd 8-53 or Cd 8b-90		
Flavor	Sensory		
Clarity	Filter test		
Refractive index (RI)[e]	Cc 7-25		
Dissolved oxygen, %	Per supplier's instructions		On liquid oil only
Dissolved nitrogen, %[f]	Per supplier's instructions		
Occluded gas, %	Gravimetric	Startup and half hourly	For shortenings only; also examine visual appearance of shortening for smooth white texture, freedom from yellow greasy streaks, excessive air holes, foreign material
Visual appearance			
Consistency (shortenings)	Cc 16-60	Startup and every 4 h	
Headspace oxygen, % (bottles)	Per supplier's instructions	Only if bottles N_2-inerted	
Arriving packaging materials			

Inspection as appropriate for the particular material[g]		First five shipments from each new supplier; if within specification, reduce subsequent sampling and analysis to spot-check basis	
Finished package integrity and appearance			
Defect inspection[h]		Startup and half hourly and after correction of defects	Examine 10 packages and 10 cases
Cap torque (bottles)			Test five packages
Seam dimensional check (cans)		One/seaming head/day	See discussion in text concerning packaging materials and can seaming
Package fill control			
Net weight[i]		Start-up and half hourly and after correction of fill	Weigh five packages
Loaded tankcar or tanktruck of finished product			
Color	Cc 13b-45	Each shipment	
Peroxide value	Cd 8-53 or Cd 8b-90		
Clarity	Filter test		
Flavor	Sensory		
Refractive index (RI)[e]	Cc 7-25		
Wastewater analyses			
Biochemical oxygen demand (BOD), mg/L		See footnote [j]	
Fat, oil, and grease (FOG), mg/L			
Suspended Solids (SS), mg/L			
Acidulated soapstock shipment			
Total fatty acid, %	G 3-53	Each shipment	
Moisture, %	Ca 2a-45		
pH	G 7-56		

[a]As a measure of oxygen exposure and heat abuse of oil, perform peroxide value test and possibly others as discussed in AOCS Method Cg 3-91. Every effort should be made with in-process oils to reduce peroxide value to as near zero as possible. See Reference (27).

[b]See references (29,30) for discussion of "3-minute iodine value" method. This modification of Method Cd 1-25 reduces the time required to determine IV but requires use of a toxic reagent (mercuric acetate).

[c]Perform tests required by product specification.

[d]After completion of each deodorization run into tank, analyze tank sample for all specification attributes prior to packaging product or loading shipments from the tank. In addition, when more than 16 hours have elapsed since packaging or loading product from the tank, again perform tests marked ** prior to packaging or loading.

[e]This procedure needed only if plant produces a variety of products at the same time. Perform refractive index (RI) analysis on storage tank sample and corresponding loaded-shipment sample or packaged-product sample and compare RIs as a rough check on possibility of packaging or shipping wrong product. RIs should agree within 0.0002. If RIs disagree more, perform other analyses as necessary to verify correct product.

[f]Perform dissolved nitrogen test only if plant sparges liquid soybean oil with nitrogen after deodorization for packaging in plastic bottles so as to avoid bottle "panelling" (collapse due to partial vacuum).

[g]Perform tests as appropriate to determine that packaging materials are functional and free of appearance defects. See discussion in chapter concerning packaging materials.

[h]Examples of defects: Package integrity: Loose or cracked caps, holes in containers, defective can seams, sharp edges, leakers including defective induction-sealed foil seals on bottles, unsealed cases, or packages nonfunctional for any reason. Package appearance: Oily, rusty, or dirty containers, loose, torn, crooked, stained or wrong labels, incorrect production date code, product in wrong type case. See also Reference (28).

[i]For products whose labels declare volume of liquid contents, fill control is by weight equivalent of volume calculated from liquid density. Use SPC control chart.

[j]There is considerable variation among processors' facilities, local regulatory requirements, and other factors that require different approaches for monitoring plant wastewaters. In some plants the quality control laboratory does the testing, while others contract this work out. Sampling frequency and test method requirements vary among regulatory jurisdictions.

trends, preparing reports, and other uses. Use of a computer for recording, communicating, storing, and studying control laboratory data accomplishes these purposes.

Computerization

Computerization of laboratory records, with direct data entry by analysts, is strongly recommended. The ideal is a "paperless" system, but one having the capability of printing out data on paper when desired. The greatest single advantage of such a system is that with a local area network, control laboratory data are instantly transmitted to processing and packaging personnel using read-only computers located in plant areas and offices. Plant operating personnel can call up all data, present or past.

It is desirable to have several quality control computer stations located throughout the laboratory and the area where quality control work is done in plant packaging operations.

Since computers generally replace old paper-record methods of recording and transmitting laboratory data in a plant, it is usually best to design the computer data format to look as much as possible like that of the paper system it replaces. This makes the transition easier for all personnel. Because each plant's operations and needs are different, the software will need to be custom-designed, and this requires detailed and careful consultation with the software programmer. Quality control and processing personnel should jointly work with the programmer to be sure the program fully meets the plant's needs. The experience of many companies is that once employees have learned to use the computer and seen its advantages, they become enthusiastic about it.

Advantage should be taken of the computer's capabilities to perform such laboratory tasks as

1. Immediately recording and storing analytical data
2. Automatically comparing analysis results with the appropriate specification, and signaling (with flashing digits, for example) when analyses are out of specification
3. Instantly transmitting data to processing personnel
4. Accepting digital output from compatible laboratory instruments, such as chromatographs, and storing the data in the main database
5. Generating routine laboratory reports
6. Summarizing and graphing data when desired, in both standardized and *ad hoc* ways, by incorporating standard, purchased statistical analysis software
7. Maintaining inventories of and reordering laboratory supplies, keeping personnel records, and performing other business administration tasks

These are only a few examples of many ways that a computer can speed up communications and reduce clerical effort and paper handling.

Extensive experience has indicated that ordinary computers are able to stand the physical stresses of a laboratory environment if the keyboard has a clear vinyl protective cover.

Statistical Process Control (SPC) and Target Quality

Statistical control charts can be a useful tool to aid in process control. The benefits are more uniform quality, improved efficiency, and less rework. These charts were developed in the 1920s and have been increasingly used in industry, but only fairly recently in soybean processing applications. The charts, usually referred to as SPC or $\bar{X}$-R charts, are applicable to many continuous or semicontinuous processes. Each chart covers one particular variable, such as moisture content of meal or phosphorus in refined oil. As each analysis result becomes available, it is immediately plotted on the chart. Plotting by the process operators is recommended.

If the result falls within the control limits for average ($\bar{X}$) and range (R), the process is considered to be in control for that variable, and no process correction is needed; only normal, random variations are occurring around the target or desired level. When a value falls outside the control limits, this is strong, statistically based evidence that the process is not in control (i.e., is not operating at the desired level for that variable) and that corrective action is needed.

The control limits are *not* based on the specifications for the particular variable but instead on a simple statistical study of the inherent controllability of the particular process. A target line representing the desired or ideal value of the variable is on the chart, usually at the midpoint between the upper and lower control limits (UCL and LCL, respectively).

Control charts basically distinguish random, normal fluctuations around the desired target level and requiring no process correction (analyses falling within control limits), from abnormal fluctuations (analyses falling outside the limits), which indicate that the process is off-target and in need of corrective action. When a plotted point falls outside the control limits, corrective action will be appropriate, even though the variable may still be within the specification limits. Charting may be done manually or by computer using standard SPC software programs, many of which are available and advertised in periodicals of the American Society for Quality Control (ASQC) (31,32).

An example of a control chart is provided in Fig. 24.1, illustrating some of the most common situations in which need for process adjustment is indicated by the plotted points. Information on SPC principles and applications is widely available in the literature; for a cross-section, see references (13,14,15,33,34). In addition, ASQC has an active program of periodical and textbook publications, educational seminars, and professional certification for applications of statistics in industry (31,32).

Laboratory Management

Generally, the operating hours of the quality control laboratory should be the same as those of the plant processes. However, in many plants the process operators perform some of the less complex analyses in testing stations set up in the plant. This may eliminate the need for laboratory staffing on night shifts if few processes are operating. The control laboratory should be responsible for maintaining standardized reagents and

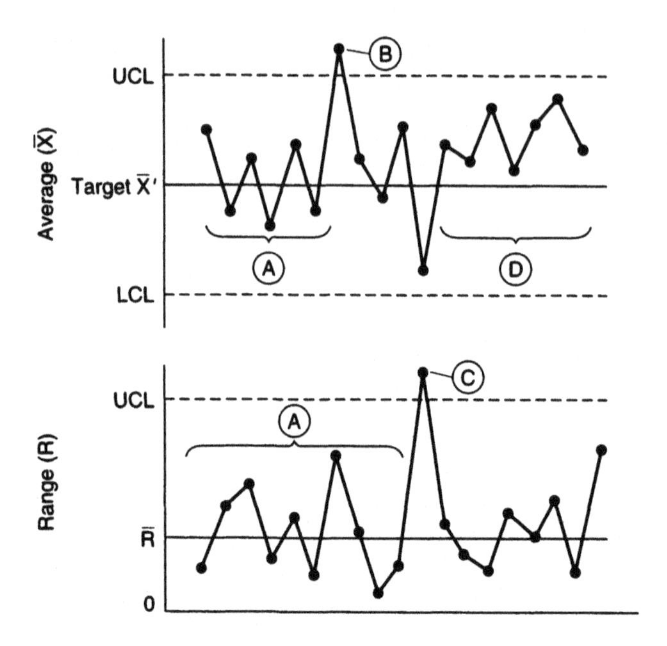

Fig. 24.1. Statististical control chart for average ($\bar{X}$) and range (R). A—These plotted points for average and range represent only normal, random variation around the target (desired) level of the process variable. No process adjustment is needed. B—Process is out of control, not operating at target level. Process adjustment to reduce level of variable is needed. C—Process is out of control for *range*, indicating a significant shift in the average value of the variable. Process adjustment is needed (upward in this case). D—Seven consecutive plotted points are on same side of target line, indicating process is averaging above target. Adjustment (downward) is needed.

calibrating instruments in the testing stations, and there should be an ongoing cross-check analysis program between the control laboratory and testing stations.

The quality control manager should be an experienced chemist with at least a bachelor's degree in chemistry. Other laboratory personnel need to have had basic chemistry and mathematics courses. It is not uncommon to have individuals with science degrees working in plant laboratories. Some companies have a policy of developing plant operators and supervisors either by hiring them as laboratory analysts or by rotating them through laboratory jobs as part of their training.

New laboratory personnel are normally trained in the test methods by working with an experienced analyst, first watching, then doing the work with the trainer observing, then independently doing the tests on production samples and comparing results with the trainer on the same samples. The trainee should be qualified for the various methods one by one, and a record should be kept of progress until the trainee is ready to work independently. Typically the training period in a control laboratory requires 5 to 7 weeks.

It is a good practice to set aside quantities of samples to be used as secondary standards for analytical attributes that do not change appreciably over time. These can be useful for training and reference purposes.

Ready access for all analysts to analytical methods, specifications, and the testing schedule must be provided, and a system must be in place to keep these up to date. A log should be maintained by the quality control manager and checked daily by all analysts at the start of their workday, by which they are kept abreast of changes in standing instructions, specifications, or other information. The computer is useful for this purpose.

It is advisable that the quality control manager should join AOCS and ASQC, both of which offer short courses; local, regional, and national meetings, which develop professionalism and contacts in the industry; and technical education.

Considering the importance of visual color matching of oils, prospective hires need to be given basic color-blindness tests.

Laboratory Design and Safety

The control laboratory should be located so as to facilitate the transport of samples. Exterior doors need to be placed for convenient access with samples and also to allow safe exit from all parts of the laboratory in an emergency. A sample log-in area should be provided near the door through which samples arrive. Some companies have installed pneumatic tube systems for sample delivery to the laboratory. A separate sample preparation room is advisable for an extraction plant laboratory, to reduce dust in the laboratory. The laboratory should be well-lighted, air-conditioned, quiet, and free of excessive vibration. Temperature control is important for regulation of instruments and calibrated glassware. A separate "instrument room," away from heavy foot traffic, is recommended for sensitive and expensive instruments such as gas chromatographs and spectrophotometers. This room should include an area for oil flavor testing, consisting of at least four booths each about 75 cm (2.5 ft) wide, with seats and with short sidewalls to separate panelists so that they do not influence each other when flavoring oils at the same time.

For an extraction plant, laboratory space should be about 140 m^2 (1500 ft^2) including supervisor's office, dishwashing and storage areas for supplies and retained samples.

For a refinery that produces 135,000 MT (300 million lbs)/year and does hydrogenation and packaging, the laboratory space would be ideally about 250 m^2 (2700 ft^2) including supervisor's office, instrument/flavor paneling room, dishwashing room, and storage space for supplies and retained samples. A sketch of such a refinery laboratory is shown in Fig. 24.2.

The kinds of laboratory apparatus, instruments, and reagents required are listed in each AOCS method, so a list of items to equip a laboratory can be assembled once the analyses to be run are known.

Safety equipment should be provided, including ABC fire extinguishers (that is, capable for fighting fires in dry combustibles and liquids as well as electrical fires),

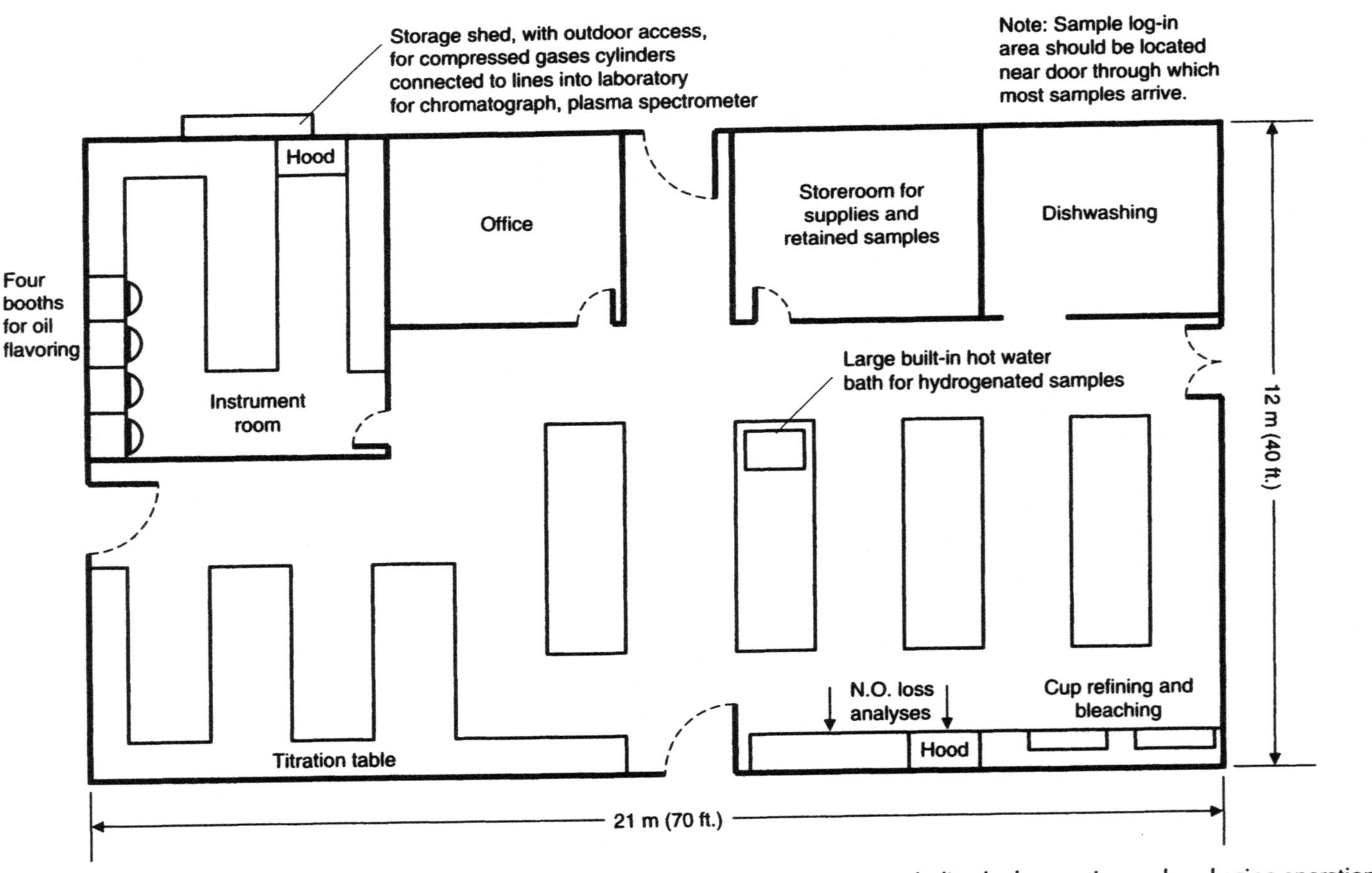

Fig. 24.2. Laboratory suitable for a refinery producing 135,000 MT (300 millions lbs)/year including hydrogenation and packaging operations.

eye/face wash station, safety shower, fire blanket, sprinkler system, and first aid supplies. Fume hoods need to be vented to the outside. Whenever flammable liquids are used, hoods should be equipped with explosion-proof motors, switches, and light fixtures. All personnel must receive safety training and be required to use safety eyewear and other protective gear as appropriate. AOCS (35) and the Association of Official Analytical Chemists (AOAC) (36) provide helpful information on laboratory safety.

For safety reasons, at least two technicians should be on duty at the same time. Where this is not practical, a frequent check-in by a nonlaboratory worker (such as a security guard) is advisable. Strategically located "panic buttons" need to be provided in the laboratory so that an alarm can be easily sounded in the security or processing office if there are periods when only one technician is on duty.

Certain AOCS methods require toxic solvents such as chloroform and carbon tetrachloride. Optional methods using safer solvents have been issued and are listed along with the original methods in Tables 24.2 and 24.3. These are preferred from a safety standpoint, but the literature does contain references to less accurate results in some instances (37).

Laboratory utilities required are hot and cold potable water, filtered/dried compressed air, natural gas, an electrical system with steady voltage, and a source of distilled water (demineralization units are recommended). Approved lockers should be available for storage of flammable or hazardous liquids, and arrangements made for storage and periodic disposal of hazardous wastes. Oil flavor standard samples must be kept in frozen storage until needed. Piping for drains needs to be left exposed and accessible for cleanout, not buried in concrete.

Quality Improvement

Companies in many segments of industry are participating in a movement that stresses continuous improvement of the quality of products and services in order to be more competitive. The movement has been given various names, such as Total Quality Management or the Continuous Improvement Process. A founder and leading spokesman was W. Edwards Deming, who has been credited with setting Japanese industry after World War II on the path to high quality products (see Chapter 27).

The Deming philosophy advocates top management commitment; thorough understanding of customers' needs; the use of scientific methods (stressing SPC); worker training, involvement, and empowerment; and, probably most importantly, a company attitude that improvement of products and services must be never-ending. Commitment of a company to such a program is a top management decision. The quality control manager should play an important role in the team effort of its implementation in the plant. The literature on this subject is extensive. A cross section is given in references (38,39,40,41).

Some companies in the soybean processing industry have begun to adopt formal programs of continuous quality improvement. Specific examples of application by companies that crush oilseeds, extract crude oil, process edible oils and use oils in their products are included in the literature (42,43,44).

References

1. National Oilseed Processors Association (NOPA), *Yearbook and Trading Rules*, reissued annually, Washington, DC.
2. *Official Methods and Recommended Practices of the American Oil Chemists' Society (AOCS)*, edited by D. Firestone, 4th edn., AOCS, Champaign, IL, 1993.
3. Farrer, K.T.H., *Shipment of Edible Oils*, IBC House Byfleet, Kent, U.K., 1992, pp. 23–112.
4. *Official Methods of Analysis of the AOAC*, 14th edition, Association of Official Analytical Chemists (AOAC), Arlington, VA, 1984.
5. Cocks, L.V. and C. van Rede, *Laboratory Handbook for Oil and Fat Analysts*, Academic Press, NY, 1966, pp. 38–43, 120–121.
6. Mehlenbacher, V.C., *The Analysis of Fats and Oils*, Garrard Press, Champaign, IL, 1960.
7. Perkins, E.G., *Analysis of Fats, Oils, and Derivatives*, 1991, American Oil Chemists' Society, Champaign, IL.
8. Rossell, J.B., and J.L.R. Pritchard, *Analysis of Oilseeds, Fats, and Fatty Acids*, Elsevier Applied Press, New York, 1991.
9. Sleeter, R.T., in *Bailey's Industrial Oil and Fat Products, Vol. 3*, 4th edn., edited by T.H. Applewhite, John Wiley & Sons, New York, 1985, pp. 167–242.
10. Sonntag, N.O., in *Bailey's Industrial Oil and Fat Products, Vol. 2*, 3rd edn., edited by D. Swern, John Wiley & Sons, New York, 1982, pp. 407–525.
11. Hamilton, R.J., and J.B. Rossell, *Analysis of Fats and Oils*, Elsevier Press, New York, 1986, pp. 28–35.
12. *Annual Book of Standards*, part 20 (Packaging), 15th edn., American Society for Testing Materials (ASTM), Philadelphia, PA, 1990.
13. Feigenbaum, A.V., *Total Quality Control*, 3rd edn., McGraw-Hill, New York, 1983, pp. 3–40, 171–219.
14. Grant, E.L. and R.S. Leavenworth, *Statistical Quality Control*, 6th edn., McGraw-Hill, New York, 1988, pp. 31–80, 348–360, 293–491.
15. Juran, J.M., *Juran's Quality Control Handbook*, 4th edn., McGraw-Hill, New York, 1988.
16. Sleeter, R.T., in *Bailey's Industrial Oil and Fat Products, Vol. 3*, 4th edn., edited by T.H. Applewhite, John Wiley & Sons, New York, 1985, pp. 167–169.
17. Sleeter, R.T., in *Bailey's Industrial Oil and Fat Products, Vol. 3*, 4th edn., edited by T.H. Applewhite, John Wiley & Sons, New York, 1985, pp. 167–170.
18. Sleeter, R.T., in *Bailey's Industrial Oil and Fat Products, Vol. 3*, 4th edn., edited by T.H. Applewhite, John Wiley & Sons, New York, 1985, pp. 170–171.
19. Sleeter, R.T., in *Bailey's Industrial Oil and Fat Products, Vol. 3*, 4th edn., edited by T.H. Applewhite, John Wiley & Sons, New York, 1985, pp. 221–224.
20. Jackson, H.W., in *Bailey's Industrial Oil and Fat Products, Vol. 3*, 4th edn., edited by T.H. Applewhite, John Wiley & Sons, New York, 1985, pp. 243–272.
21. Evans, C.D., and H. Moser, *J. Amer. Oil Chem. Soc. 48:* 495 (1971).
22. Mounts, T.L., in *Handbook of Soy Oil Processing and Utilization*, edited by D.R. Erickson, et al., American Soybean Association, St. Louis, MO, and American Oil Chemists' Society, Champaign, IL, 1980, pp. 245–266.

23. Weiss, T.J., *Food Oils and Their Uses*, 2nd edn., AVI Press, Westport, CT, 1982, pp. 209–234.
24. Raghavan, S., S. Reeder, and A. Khayat, *J. Amer. Oil Chem. Soc. 66:* 7 pp. 942–947 (1989).
25. Fraser, M., and A. Khayat, *J. Amer. Oil Chem. Soc. 63:* 410 (1986).
26. Bilbin, A., *INFORM 4:* 6 pp. 648–665 (1993).
27. Erickson, D.R., and G.R. List, in *Handbook of Soy Oil Processing and Utilization*, edited by D.R. Erickson, et al., American Soybean Association, St. Louis, MO, and American Oil Chemists' Society, Champaign, IL, 1980, pp. 273–309.
28. Leo, D.A., in *Bailey's Industrial Oil and Fat Products, Vol. 3*, 4th edn., edited by T.H. Applewhite, John Wiley & Sons, New York, 1985, pp. 311–340.
29. Hasheeny-Tonkabony, J.E., *J. Amer. Oil Chem. Soc. 54:* 233 (1977).
30. Cocks, L.V., and C. van Rede, *Laboratory Handbook for Oil and Fat Analysts*, Academic Press, New York, 1966, p. 112.
31. *Journal of Quality Technology*, American Society for Quality Control, Milwaukee, WI, published quarterly.
32. *Quality Progress*, American Society for Quality Control, ASQC, Milwaukee, WI, published monthly.
33. Kivenko, K., *Quality Control for Management*, Prentice Hall, Englewood Cliffs, NJ, 1984, pp. 23–30.
34. Ott, E.R., and E.G. Schilling, *Process Quality Control*, 2nd edn., McGraw-Hill, New York, 1990, pp. 33–42.
35. *Official Methods of the American Oil Chemists' Society*, edited by D. Firestone, 4th edn., 1993, three-page addendum to index discusses laboratory safety.
36. *Official Methods of Analysis*, 14th edn., Association of Official Analytical Chemists, Arlington, VA, 1984, pp. 1010–1015.
37. Steiner, J., *INFORM 4(9):* 955 (1993).
38. Deming, W.E., *The New Economics for Industry, Government and Education*, MIT Center for Advanced Engineering Study, Cambridge, MA, 1993.
39. Deming, W.E., *Out of the Crisis*, MIT Center for Advanced Engineering Study, Cambridge, MA, 1992.
40. Sherkenbach, W., *Deming's Road to Continual Improvement*, SPC Press, Knoxville, TN, 1991.
41. Croy, C., *INFORM 4(8):* 884 (1993).
42. Croy, C., *INFORM 4(9):* 1034 (1993).
43. Croy, C., *INFORM 4(10):* 1120 (1993).
44. Croy, C., *INFORM 4(11):* 1237 (1993).

Chapter 25

Environmental Concerns in Soybean Processing

Norman J. Smallwood

The Core Team, Hammond, LA

Introduction

Issues

An inherent aspect of soybean and soybean oil processing is the normal production and discharge of several waste materials, which must be managed to preclude environmental pollution. Furthermore, there is potential for major environmental pollution problems as the result of human error or catastrophic failure. Plant design, construction, operation, and maintenance must include the capability to comply with the environmental regulatory requirements involving the normal waste material generation, the accidental discharge of material, and the other related matters. The specific environmental issues for soybean and soybean oil processing presented in this chapter are the following (1):

1. Soybean oil and processing reagent spill containment
2. Wastewater collection and treatment
3. Rainwater runoff collection, quality assessment, and treatment if necessary
4. Solid waste disposal
5. Air pollution control
6. Odor control
7. Noise pollution
8. Local appearance standards

Special Focus

Special attention is given to the key aspects of water pollution control. Water pollution is usually the most demanding environmental compliance issue for the industry. Developing a sound strategy and plan for water pollution control is critical in terms of achieving reliable operation, meeting regulatory requirements, and minimizing the long-term financial impact.

Spill Containment

Spill containment is mandated by environmental regulation in most countries. Savings achieved from minimizing the cleanup cost and gaining some economic value (nonfood use) from the recovered oil justify the cost of spill containment in the absence of regulatory requirement. Properly designed spill protection includes dikes around oil and chemical storage tanks, unloading stations, and bulk-oil loading stations (2). The dike walls are designed to handle the total content of each tank in the event of failure.

Wastewater Treatment

System Objectives

The scope and design of wastewater treatment is to accomplish the following (3):

1. Meet the regulatory schedule of compliance and effluent limits
2. Achieve a level of reliability in protecting continuity of plant operations and remaining within effluent limits
3. Provide water treatment capacity for future expansion of production facilities
4. Minimize operating cost
5. Improve the esthetic image of the plant
6. Satisfy spill prevention, control, and countermeasure requirements
7. Accommodate future effluent standards with minimum capital cost

Sources and Characterization of Wastewater Effluent

The following waste streams are present in most soybean and soybean oil processing operations. Appropriate collection, handling, and treatment systems must be provided for each water effluent category.

Process wastewater. This water has been in contact with the product stream and contains some concentration of oil, processing reagents, and sludge material derived from crude soybean oil.

Rainwater runoff. Rainwater (including ice and snow melt) falling on outside surface areas is subject to possible contamination from normal working-area spillage.

Noncontact cooling water. This is cooling water that remains free of contamination while circulated once through a closed piping/coil system to maintain process temperature control.

Sanitary waste. Sanitary waste from employee restrooms and kitchen facilities.

Spent analytical chemical solutions. These result from quality control and research activities.

Hexane. This loss may be incurred in the receipt, storage, and use of hexane for the solvent extraction of soybean oil.

Wastewater Stream Separation, Collection, and Handling

Stream separation and collection is one of the most important aspects of water pollution control (4). Where applicable, *gravity* transport of the various water streams by category to a common site for appropriate disposition is the best approach. The handling of each water-stream category is described as follows:

Process wastewater. Process wastewater from each major source is individually piped to discharge into a process sewer system. Accommodations should be provided for routine, visual inspection, sampling, and analysis of *each* wastewater stream to detect anomalous conditions. From the individual discharge points, wastewater flows through a process sewer line to a pretreatment sump, which is equipped with a mechanical agitator and automatic pH control by the addition of a mineral acid such as sulfuric acid.

Rainwater runoff. Outside areas subject to oil leaks, drips, and spills are provided with concrete mats and dikes for containment. Underground drains are installed for gravity flow of rainwater and/or oil spillage to a collection sump adjacent to the process wastewater pretreatment sump. From the rainwater collection sump, flow can be diverted either to the adjacent process wastewater sump or to a rainwater hold basin. The initial rainwater transports any oil spillage to the rainwater collection sump and on to the process wastewater pretreatment sump. After sufficient rainfall, the water flow into the rainwater sump is essentially pure water; then the flow from the rainwater sump is diverted to a hold basin.

Rainwater collected in the hold basin is subsequently sampled and analyzed. If the rainwater quality is acceptable, it is discharged at a controlled rate to the stormwater drainage system. When the rainwater in the hold basin is found to be unacceptable for direct discharge, it is pumped at a controlled rate to the process wastewater collection sump for subsequent treatment.

Runoff from noncontaminated areas should be isolated for direct discharge to normal storm drainage systems.

Noncontact cooling water. From process operations, noncontact cooling water is usually piped to discharge into a flash tank. For most facilities, water is pumped from the flash tank to a cooling tower for subsequent recirculation and reuse in a closed loop. If used only once through the cooling operation, the water is likely discharged into the stormwater system for disposal. In the event of mechanical failure in a cooling operation, the product stream may enter the cooling water stream and result in subsequent water pollution and significant product loss. The potential problem is avoided by passing the cooling water effluent through a gravity separator prior to final discharge. The use of two oil-water separating decanters connected in series for removal of any entrained oil is the preferred safeguard. Clean water underflows from the second decanter to discharge by gravity through a weir box for flow measurement and on to the final discharge point. Careful inspection of the water surfaces in the decanters every operating shift is needed to detect the presence of oil and for taking immediate corrective action.

Sanitary waste. Sanitary waste is gravity-collected in and transported by a separate sewer system to the appropriate treatment system. If the plant site is served by a sewage treatment facility operated by the local government, connecting to the local sanitary sewer is the appropriate solution. For remote plant sites, treatment of sanitary wastewater is a matter of selecting and utilizing the best option available. Typically, separate treatment of sanitary or domestic wastewater containing human wastes is advisable (4). If mixed with process wastewater, the total flow would have to be treated to sanitary waste standards. For small flows, septic tanks with filters or an absorption field may be acceptable. Compact, preassembled, "package" activated-sludge plants are often used. In some instances, an oxidation pond, an artificial wetlands system, or both may be applicable.

Spent analytical chemical solutions. Disposal of spent reagent solutions from quality control analyses is a complex matter (5). In general, spent reagent solutions fall into three categories for disposal: first, reagents such as caustic soda and mineral acids, which are compatible with process wastewater; second, reagents that are not compatible with process wastewater but are not classified as toxic or hazardous material; and third, reagents that are classified as toxic or hazardous material. Accordingly, all reagents used in a quality control laboratory must be classified and disposed of in the appropriate manner. Incineration may be the best disposal method for hazardous spent reagents (6). Holding in storage tanks for periodic collection and disposal by an approved hazardous waste handler is an option often used.

Hexane. Hexane used for solvent extraction of soybean oil must be contained and isolated from all other wastewater streams, to preclude possible ignition in sewer lines and treatment systems not designed to accommodate low-flash-point material safely. From both economic and process safety perspectives, solvent extraction processes must be designed, constructed, and operated to achieve maximum recovery and reuse of hexane. Provisions must be provided for the complete intercept and safe disposition of any hexane loss from the respective handling and processing facilities.

Process Wastewater Pretreatment Train

The sequence of physical-chemical operations in the pretreatment train consists of pH adjustment, gravity separation of floatable oils, flow equalization, neutralization, and dissolved air flotation (3). Dissolved air flotation is not always utilized in the pretreatment process train. If the pretreated wastewater is discharged to a sewerage system for transport to and treatment in a community wastewater treatment system, inclusion of dissolved air flotation in pretreatment is a matter of economic justification. If the reduction in treatment cost is sufficient to justify the installation and operation of dissolved air flotation, it should be included.

In the cases where biological treatment is carried out as part of the processing plant wastewater treatment, inclusion of dissolved air flotation is usually justified on the basis of financial return and compliance with the quality limits for water discharge.

The sequential operations involved in wastewater pretreatment are described as follows and are illustrated in Fig. 25.1.

pH adjustment. A pH in the range of 2 to 3 is required to achieve optimum gravity separation of floatable oils. pH is adjusted by an instrumented pH control loop, which monitors the wastewater effluent from the pretreatment sump and controls mineral acid addition to the sump for lowering pH as required.

Gravity separation of floatable oils. Wastewater is pumped from the pretreatment sump to a series of decanters for separation of floatable oils. The wastewater stream enters each decanter through a horizontal-plane distributing nozzle located in the center of each vessel; the floatable oil subsequently rises to the top of the liquid column. The accumulation of oil and fatty acid is pumped off as often as appropriate. The water fraction is continually drawn off by gravity underflow to the next decanter connected in series and ultimately to the equalization surge tanks. Some of the solids in the wastewater stream settle to the bottom of the decanters and can be removed periodically.

Flow equalization. Flow equalization for optimum operation of dissolved air flotation is achieved by means of multiple hold tanks. Sufficient hold-tank capacity should be provided for at least 8 h of downtime for maintenance of the dissolved air flotation equipment. Other than for flow equalization, the hold tanks function as gravity separators for further removal of settleable solids and recovery of floatable oil. Each hold tank can be individually valved out of service for removal of accumulated floatable oil and sediment without disrupting normal operation of the system.

Neutralization. Wastewater is pumped from the hold tanks to the dissolved air-flotation process. On the suction side of the transfer pump, caustic soda is injected to neutralize the water stream. Primary neutralization control is provided by an instrumented pH control loop. Manual control is provided for backup operation.

Dissolved air flotation (DAF). There is considerable variation in DAF vessel design; however, the fundamental operating principles are quite common. Neutralized water and cationic-polymer coagulant are discharged into a pressurized flocculation tank (7). In the discharge from the pressure tank, an anionic-polymer coagulant is added to the water stream before it enters the DAF unit. In the round-vessel design, water underflows from the central chamber and rises to a top peripheral collection ring for both recycle pressurization and gravity discharge to biological treatment. Recycled water is fully pressurized and is discharged back into a center coagulation tube. Removal of float is accomplished by a top rotating skimmer. The DAF process is sized to provide a surface loading of less than 3.0 gpm/ft^2 (122 $L/min/m^2$) (3).

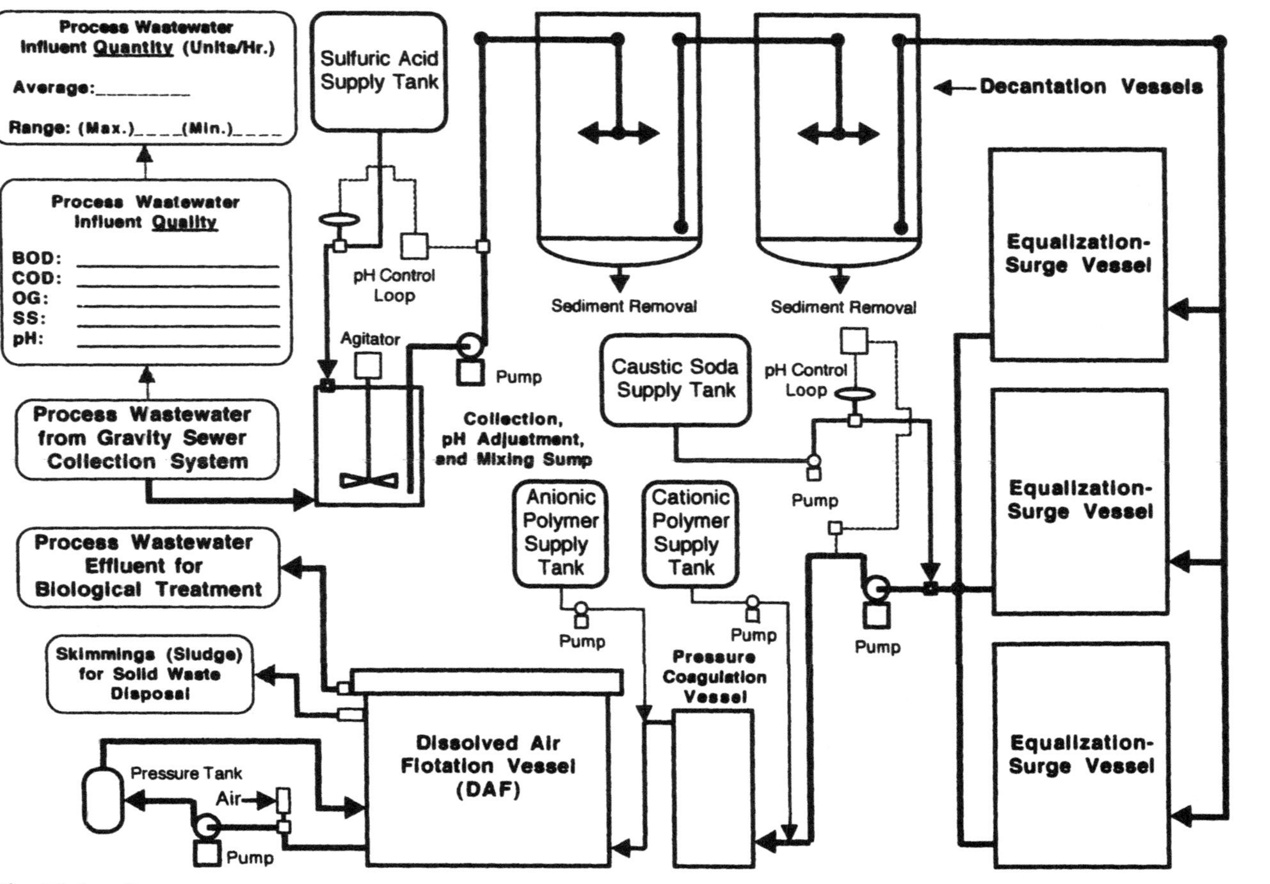

Fig. 25.1. Process wastewater pretreatment system. Reproduced with permission of The Core Team, Hammond, LA.

Biological and Subsequent Treatment of Process Wastewater

After pretreatment, process wastewater is subject to biological treatment to meet the water-quality limits for discharge. The biological treatment and subsequent treatment operations are described as follows:

Biological treatment. Biological treatment can be accomplished by means of an activated-sludge aerated lagoon, an anaerobic system, or a combination of both aerobic and anaerobic treatment (4). An aerated lagoon biological treatment system is illustrated in Fig. 25.2.

Clarification. Treated water is pumped from biological treatment to a clarifier for separating the biological sludge (8). Clarified water flows by gravity to a final monitoring station. The biological sludge either is recycled back to the inlet side of the biological treatment system or is prepared for disposal (see Fig. 25.2).

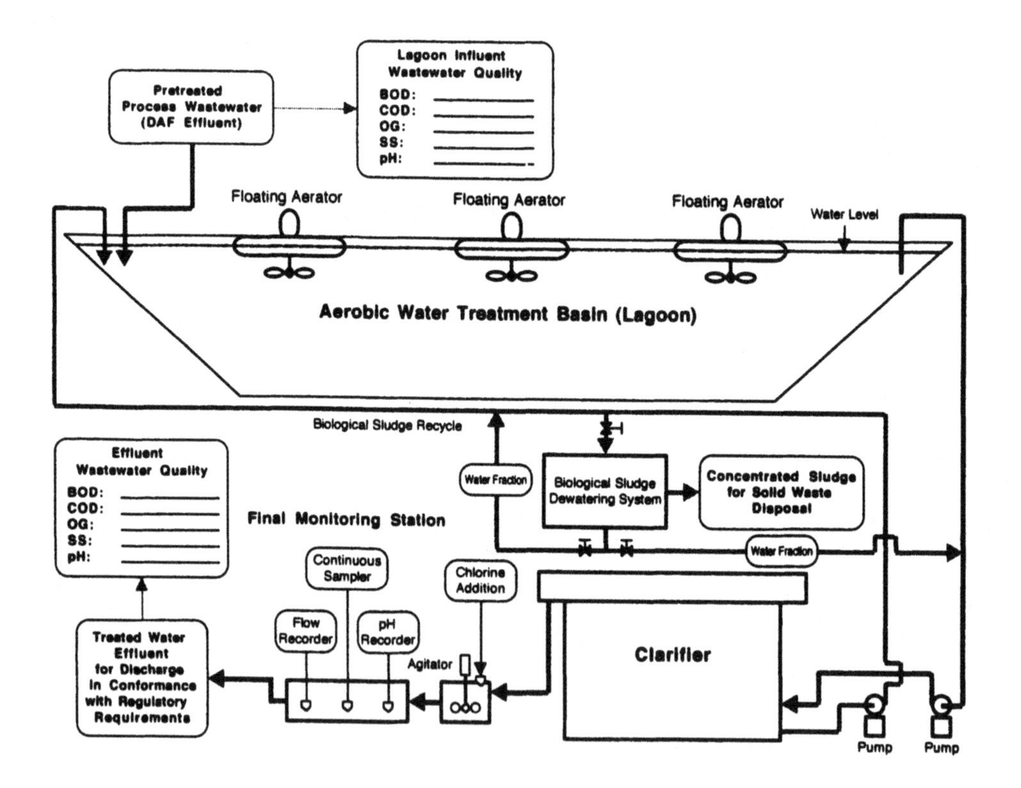

Fig. 25.2. Process wastewater treatment system. Reproduced with permission of The Core Team, Hammond, LA.

Final monitoring. Prior to discharge, the treated water flows through a monitoring station for flow measurement, continuous sampling, and pH recording. Disinfection of the treated water may be required in some countries prior to discharge, particularly when human wastes are mixed with the industrial process wastewater.

Ensuring Reliable Process Wastewater Treatment

To ensure reliable plant performance, the process wastewater treatment system must be properly designed, constructed, operated, and maintained. To meet the reliability requirement, each unit operation of the treatment system must include appropriate redundancy and be sufficiently forgiving to handle the upsets and contingencies normally encountered in processing plant operation (3).

Achieving Compliance with Water Pollution Control Legislation

In the United States, Canada, and many of the industrialized countries, about two decades of experience have been gained in complying with water pollution control legislation. With increasing concern for protecting the environment on a global scale, laws are now being enacted in most of the developing countries. To achieve cost-effective water pollution control, the experience available in the industrialized countries can be of significant help to the developing countries, where technical and financial resources are much more limited. The issues and actions for successfully implementing a processing plant water pollution control project are outlined as follows (9).

Project objectives. The processing-plant water pollution control plan should:

1. Comply with the law
2. Provide a reliable system
3. Gain operational benefits
4. Consider long-term needs
5. Optimize the use of any existing collection and treatment facilities
6. Keep the capital requirements reasonable and justifiable

Process and sanitary wastewater. The design procedure for a system to control process and sanitary wastewater includes the following steps:

1. Identify the present wastewater sources, flow, and disposition
2. Characterize the wastewater sources and streams, quantitatively and qualitatively.
3. Develop strategies and plans for wastewater source reduction, both volume reduction and pollutant reduction.
4. Determine the collection system design.
5. Design the process-wastewater-pretreatment system.
6. Evaluate and select the biological-treatment process.

Spill protection. The design procedure for a spill protection plan includes the following steps:

1. Identify potential sources of spills: overflows, leaks, catastrophic failure.
2. Characterize the potential spills quantitatively and qualitatively.
3. Design the containment system.
4. Integrate the design of spill protection with process wastewater collection and pretreatment.
5. Prepare a spill-response contingency plan consistent with regulatory requirements.

Rainwater runoff from the plant site. The design procedure for a system to handle rainwater runoff includes the following steps:

1. Design the system to accommodate collection and treatment of contaminated-rainwater runoff as required.
2. Determine the detailed requirements specified in the enacted law and published regulations.

Future considerations.

1. Examine the impact of future business expansion or changes on the wastewater collection and treatment system design.
2. Anticipate probable changes in environmental regulations.

The financial consequences of processing plant water pollution control projects are too great to take risks. The project must be done right the first time at the lowest possible cost. The use of reliable and competent technical and managerial resources is mandatory for success.

Solid Waste Disposal

Spent Filter Cake

Spent filter cake (filter aid and bleaching earth) contains from 25 to 100%, dry weight basis, soybean oil. If soybean oil processing is part of a plant complex including soybean processing, it may be permissible to dispose of the spent cake in the protein meal for animal feed (3). The oil content has nutritional benefit, and the inert fraction (clay and diatomaceous earth) in pelletized form can be classified as a processing aid to facilitate free flow of the meal. Other methods of disposal include the use of a solid waste landfill or a landfarm. Spent filter cake must be handled carefully (covered with soil immediately on disposal) to avoid spontaneous combustion caused by rapid oxidation of the thin oil film on the clay and diatomaceous earth particles (see Chapter 17).

Waste Biological Activated Sludge

The excess biomass developed in the aerobic treatment process must be removed from the aeration system routinely or periodically. Usually, a portion of the biomass of settled solids from the final clarifier is diverted as a waste stream containing 2 to 4% dry solids by weight (see Fig. 25.2). This should be dewatered to 20% or more dry solids if disposed of in a landfill site. It may be applied directly to a land farm with or without concentration. The decision to dewater for disposal on a land farm is a matter of handling economics. Probably, concentration of the solids to the range of 8 to 10% by weight will be the least expensive.

Spent Hydrogenation Catalyst

The most cost-effective and environmentally sound method for the disposal of spent catalyst is to ship it to a legitimate business engaged in metal (usually nickel) reclamation. Both the availability and cost of this preferred method of disposal are usually a function of the quantity of desirable metal in the spent catalyst. If the metal content is too low, the reclaimers may refuse to take the spent catalyst. Hydrogenation operating practices involving catalyst usage and handling may be established on the basis of meeting the minimum-metal-content restraint for participating in the reclamation disposal method. The minimum acceptable metal content for reclamation disposal of spent catalyst may vary according to the current market price for the metal involved. The use of open-head steel drums for accumulating and shipping spent catalyst is the most common practice.

If nickel reclamation is not feasible, spent catalyst becomes a solid waste disposal issue. Accordingly, the spent catalyst must be handled under the applicable state and federal regulatory protocol. In some locations, disposal of spent catalyst as a solid waste is quite involved and expensive.

Air Pollution Control

Combustion Products from Boiler Operation

Steam boilers and process heaters fueled with natural gas do not normally pose an air pollution problem if complete combustion is achieved. When fuel oil or coal is used, appropriate stack influent scrubbers are usually required to remove undesirable compounds and particulate matter to yield an acceptable emission (10).

Dust Control

Soybean and soybean oil processing involve unit operations that create dust. In the case of soybean processing, dust is created in bean handling, cleaning, drying, and dehulling. Soybean meal handling and finishing operations result in some dust formation. Soybean oil processing involves the handling and use of processing reagents and aids that generate dust. Each dusty operation must be managed to minimize the quantity of dust that occurs and to provide appropriate dust control facilities necessary to meet air emission regulations consistently (11).

Chemical Vapor

Air pollution from the escape of chemicals such as hexane and ammonia vapor to the atmosphere has recently received regulatory attention in the United States. To comply with these new legislative initiatives, solvent extraction processes using hexane and refrigeration systems using ammonia must be designed, constructed, operated, and maintained to be closed systems. The hexane or ammonia material balance within the system will require rigorous measurement and documentation of any loss for comparison to the mandated threshold limits (11).

Odor Control

For processing plants located adjacent to residential areas, odor can be a troublesome issue. There are several potential sources of odor in soybean and soybean oil processing plants:

1. Soybean drying vapor
2. Meal drying vapor
3. Condensing water cooling tower vapor
4. Acidulation process vapor
5. Anaerobic digestion processes in wastewater treatment
6. Waste sludges and other decomposable solids

The odors generated are usually not particularly objectionable unless anaerobic biological processes are involved. Anaerobic degradation processes generate very foul odors, which will result in immediate and justifiable public complaint. If possible, odor problems are best solved by identifying and eliminating the source. In cases where the source cannot be eliminated, several methods are available to resolve odor problems and are outlined as follows (12):

1. Scrubbing with water alone, with chemical reaction, or by biological degradation
2. Adsorption
3. Incineration
4. Dilution
5. Masking

Incineration to eliminate odors is especially effective when the fouled air can be collected and utilized for combustion in process steam boilers. For this application, the capital cost is limited to structural modifications to isolate the odor, air ductwork, and blowers for moving the fouled air to the air intake of the boilers.

To accomplish odor control by any of these methods, the fouled atmosphere must be confined and conveyed to the control site by carefully designed, constructed, and operated buildings, air collection and conveyance systems, and disposal systems. If there is fouled-air leakage from any aspect of the total system, the efforts for control will be negated.

Noise Pollution

Proper response to the noise level in each operating area of a processing plant will:

1. Provide a productive workplace
2. Safeguard people from hearing loss
3. Meet regulatory requirements
4. Avoid costly litigation of either internal or external origin

A noise level above 45 decibels (dB) indoors and 55 dB outdoors detracts from the concentration and work effectiveness of people (11). Hearing loss can occur at a noise level above 70 dB (11). Exposure to noise is regulated on the basis of time and level of exposure. Because of variation in the regulations by country and the ongoing change in regulations, it is inappropriate to indicate specific exposure data here. The need to know the law and to comply with noise pollution control requirements is the primary issue.

Noise levels around centrifuges, compressors, boiler blowdown, and other operations can be sufficiently high to necessitate ear protection for people working in the immediate vicinity. In some cases where productivity is adversely affected by noise pollution, means of reduction may be the appropriate response. Obtaining comprehensive and accurate data on plant noise levels is the first step. Subsequent noise level measurements can be obtained to discover changes or raise the confidence level in previous data. With the data base, the proper response can be decided for each area of the plant. The response choices are either requiring ear protection consistent with the noise level or taking action to reduce the noise to an acceptable level. Several approaches to reducing noise level are listed as follows (11):

1. Isolate and enclose vibrating equipment
2. Minimize vibration by special mounting devices
3. Install sound-absorbing panels or barriers
4. Improve the balance of rotating parts
5. Reduce frictional resistance
6. Use shock-absorbing devices
7. Eliminate unnecessary noise sources

In considering any of the methods, both the regulatory requirements and the economic benefits must be assessed.

Local Appearance Standards

In the United States, most communities have zoning restrictions and appearance standards for industrial operations such as soybean and soybean oil processing plants. With the practice of good housekeeping, attention to building and facility maintenance, and conformance to the environmental practices presented, complaints from the surrounding community should not be a problem (1).

Material Selection for Equipment and Facilities

Other than design flaws and overall project mismanagement, selection of improper materials for equipment and facilities has and continues to be the most costly mistake in implementing environmental protection projects. In regard to materials, use of concrete or carbon steel, even with corrosion-resistant coatings, for contact with corrosive liquids and vapors is the most common mistake. The occurrence of one failure point—which is inevitable—in the corrosion-resistant coating will result in rapid failure of the total surface, structure, or vessel. Use of fiberglass-reinforced plastic (FRP) with a resin binder appropriate for the service will give years of trouble-free life (3). Grades of stainless steel are widely misunderstood with regard to corrosion resistance. Too often, costly mistakes are made that jeopardize operating reliability of the environmental protection system and add to the operating expense. Careful, knowledgeable selection of material is a must to ensure success.

Operating Practices

The best operating practices to safeguard the environment, to minimize the capital investment, and to minimize the operating expense, including the avoidance of pollution cleanup, are the following (1):

1. Eliminate each potential source if possible
2. Minimize each source that cannot be totally eliminated
3. Operate pollution control systems with equal care and attention to detail as the main-product-stream processes and operations

Recently Enacted Environmental Laws and Regulations

The environmental regulatory climate in the United States is one of constant change involving both new initiatives and refinement of existing environmental law. The following overview of recently enacted environmental law in the United States is from work done by Robert M. Reeves, President of the Institute of Shortening and Edible Oils, Inc. (13).

1. *Clean Air Act of 1990.* Individual states will issue permits to industries in regard to pollutant emissions. The law contains stringent thresholds for pollutant emissions such as ammonia, hexane, and boiler-stack compounds.
2. *Chemical Accident Prevention Regulations.* Under the authority of the Clean Air Act, the U.S. Environmental Protection Agency (EPA) is establishing regulations to prevent chemical spills that result in pollutant emissions to the atmosphere. These regulations appear to be a duplication of some of the existing Occupational Safety and Health Administration (OSHA) requirements.
3. *Emergency Planning and Community Right-to-Know Act (EPCRA).* Administered by the EPA, the regulation requires emergency response plans for and reporting of toxic chemical release during transportation, use, and storage. The list of chemicals is expected to include some of those used in soybean and soybean oil processing.
4. *Stormwater Regulations under EPA.* Stormwater runoff from industrial sites will require permits under one of three possible formats. The edible oil processing industry has elected to file for individual operating site permits.
5. *Off-Site Hazardous Waste Facilities.* Under the authority of the Clean Air Act, the EPA is drafting regulations for emission standards pertaining to off-site points of pollution such as storage facilities, wastewater treatment operations, and solid waste disposal sites.
6. *Oil Pollution Act of 1990.* The legislation requires a plan and the capability for quick response to clean up oil spills of any kind by the industries involved.
7. *Ground and Ground Water Pollution.* Compliance with current environmental regulations should prevent future pollution of industrial site grounds and ground water resources. Uninformed or careless operating practices in the past may have resulted in sites becoming polluted. Particular attention should be given to areas where fuels and solvents have been stored. Clean up can range from *in situ* soil treatment to costly soil removal and replacement. Shallow water-bearing aquifers below pervious soils need to be protected and should be checked in suspect areas. Ground water cleanup is especially costly and often requires long-term procedures.

References

1. Smallwood, N.J., Soybean/Soybean Oil Processing Plant Management, Seminar presentation in Russia, Greece and Turkey sponsored by the American Soybean Association, 1990.
2. Bennett, G.F., F.S. Feates, and I. Wilder, *Hazardous Material Spills Handbook*, McGraw-Hill, Inc., 1982.
3. Smallwood, N.J., in *Proceedings of the Ninth National Symposium on Food Processing Wastes*, U.S. Environmental Protection Agency, EPA-600/2-78-188, August 1978.
4. Metcalf & Eddy, Inc., *Wastewater Engineering*, 3rd edn., revised by George Tchobanoglous and F.L. Burton, McGraw-Hill, Inc., New York, 1991.
5. Karnofsky, B., *Hazardous Waste Management Compliance Handbook*, Van Nostrand Reinhold, New York, 1992.

6. Brunner, C.R., *Hazardous Waste Incineration*, McGraw-Hill, Inc., New York, 1993.
7. Nalco Chemical Company, *The NALCO Water Handbook*, 2nd edn., edited by F.N. Kemmer, McGraw-Hill, Inc., New York, 1988.
8. Weber, W.J., Jr., *Physicochemical Processes for Water Quality Control*, Wiley-Interscience, New York, 1972.
9. Smallwood, N.J., *Managing Industrial Wastewater Treatment Projects*, The Core Team, Hammond, LA, 1993.
10. Bibbero, R.J., and I.R. Young, *Systems Approach to Air Pollution Control*, John Wiley & Sons, New York, 1974.
11. Davis, M.L. and D.A. Cornwell, *Introduction to Environmental Engineering*, 2nd edn., McGraw-Hill, Inc., New York, 1991.
12. Hesketh, H.E., and F.L. Cross, *Odor Control Including Hazardous/Toxic Odors*, Technomic Publishing Company, Inc., Lancaster, PA, 1989.
13. Reeves, R.M., *U.S. Environmental Regulatory Overview*, Institute of Shortening and Edible Oils, Inc., Washington, DC, 1993.

Chapter 26

Cost Estimates for Soybean Processing and Soybean Oil Refining

Richard J. Fiala

Introduction

This chapter will provide guidance for developing the capital cost and operating expense estimates necessary for use in project feasibility studies, evaluation of alternatives, funding requests, and construction budgeting. The intent of this effort is toward handling *major* project activities, which fall outside the domain of plant sustenance, improvements, debottlenecking, and nominal incremental capacity increases. Although these latter projects generally lack the complexity, capital exposure/risk, and demand on resources to follow formalized project development procedures, they do go through the same project stages and thought processes as a major project does.

Among the many critical factors that determine the ultimate profitability of an oilseed-, edible oil–, or soy food–processing plant are its design, location, and capacity. These are also the factors that most affect the capital costs, the subsequent associated operating expenses, and the conversion efficiency and capability (product yield/product mix) of that facility. While this is true of all chemical processing and manufacturing plants, it is of particular importance in the processing of agricultural raw materials and the utilization of their products. These businesses fall into the realm of "commodity businesses," which are typified as market-driven, high-volume, low-margin operations that produce generic (nonunique) products from universally available raw materials. Commodity businesses require low *life cycle costs*—that is, the one-time capital costs plus the "lifetime" value of the direct operating expenses and indirect commercial/administrative expenses—for success.

In order to achieve low life cycle costs for a proposed new facility (whether a greenfield site, a new product addition at an existing site, or a major expansion of an existing facility), it is fundamental for evaluating the business and technology that both the capital cost and the operating expense be estimated appropriately from the *initial project conception to project startup.*

Typical Plant Cost Ranges

Capital costs for greenfield crushing plants will average about $27,500 (U.S.) per metric ton of daily capacity, ranging from $22,500 to $33,000. This is for plants having capacities in the range of 1,000 to 3,000 metric tons per day. Beyond this capacity range the capital cost per ton will be higher for small plants and lower for large plants. Obviously, this large swing in the capital cost range can result in significant cost differences for plants having equivalent capacities. These differences are the result of many factors, such as site conditions, building structure design, automation extent, amount of storage, type of receiving and shipping facilities, availability of utilities and waste treatment, and support facilities included. These same soybean crushing plants will have operating expenses averaging about $16.50 U.S per metric ton (ranging from $14.75 to $22.10). While these operating expense variations primarily depend on capacity, a significant portion of the expenses are functionally related to the process design and the subsequent capital expended.

Likewise, for a greenfield soybean oil refinery, the capital requirements will average about $45,900 U.S. per metric ton of daily refining capacity. The capital range would be from $33,000 to $55,000 per metric ton of daily capacity for a consolidated refining/bleaching/hydrogenation/deodorization facility. The comparable operating expenses for a soybean oil refinery will average about $39.70 U. S. per metric ton of refined product ($35.30 to $48.50 range) with an incremental $22.00 U.S per metric ton for hydrogenated products ($17.60 to $38.50 range). All of these operating expenses are "total" figures, including direct and indirect manufacturing expenses, sales and marketing expenses, and administrative overheads.

In approaching a specific project, therefore, it is essential to estimate correctly both the capital costs and operating expenses, as well as their relationship to each other, for ultimate commercial success.

Levels of Estimating Precision

Different types of estimates are required at different project stages and require different types and levels of information. Engineering and construction companies have established routine estimating procedures (normally computerized), checklists, and techniques for developing, modifying, and updating cost estimates for new facilities, upgrades, and additions to existing facilities. The more precise a project or construction estimate is, the greater the cost for anyone to develop the estimate. It is important, therefore, that estimating techniques be economically appropriate for the type of estimate required.

For example, if a project for a new product with an unknown or marginal probability of success is being evaluated, it does not make sense to develop a detailed, firm cost estimate (having a 10% accuracy and running 3 to 5% of the total projected capital cost) when a budget estimate (having a 25% accuracy and running only 0.3 to 0.5% of the total projected capital cost) can serve the same purpose for a feasibility analysis. This is particularly true during feasibility studies where discounted cash flow (DCF) rate of return is being utilized. With this procedure an error of 15% in the capital cost estimate has much less project impact than a 15% error in operating expenses or sales value for the product(s). Thus, estimating becomes a matter of "dollars and sense."

For the purpose of clarity in the following discussions, the use of the word *cost* will be associated with a *one-time* capital outlay for the construction of facilities, whereas the word *expense* will be associated with the *ongoing* outlay of funds to support operations once a facility is built. The next section of this chapter deals exclusively with costs, whereas the third and final section deals with expenses.

Cost Estimates

Types of Cost Estimates

The capital cost estimates required at each stage during the development of all projects become more detailed and more accurate as a project develops. This increase in detail and accuracy is the result of a progressively improved scope definition and progressively increased engineering and design activities. These progressive cost estimates are normally defined as follows.

1. *Order of magnitude* (OOM) (or *budget*) cost estimates are associated with economic feasibility and early conceptual studies. OOM estimates are normally developed by shortcut methods, which can result in significant deviations from the final project's "true" cost. However, for purposes of decision making and broad-brush development during early project stages, this is generally acceptable. These estimates are inexpensive, relatively quick, and normally within ± 33% accuracy (usually within ± 25% accuracy where historical data is available).
2. *Engineering study* cost estimates are associated with a number of different scenarios that focus on specific issues within either an OOM or a preliminary engineering estimate. Engineering studies can be related to technical analyses (i.e, hot dehulling vs. conventional dehulling) or commercial analyses (i.e., site evaluations, barge vs. rail receipts). The engineering study is normally focused on a particular, narrow-scope issue. The information developed in an engineering study (i.e., the preferred or "optimum process" and the incremental costs associated with it) are transferred to the broader preliminary engineering cost estimate once the optimal solution has been ascertained.
3. *Preliminary engineering* cost estimates are associated with more precise process definition, defined control strategies, and more detailed economic feasibility. Preliminary engineering is specifically intended to focus a project toward a specific goal and answer the "what if" questions normally raised to address business or commercial issues. These estimates are based on a well-defined *process* and *project* scope.
4. *Definitive* (or *construction*) cost estimates are associated with funding requests and are also utilized for project construction budget development. Definitive estimates are based on well-defined and "frozen" scope documents, an in-depth process design package, and enough engineering and design in the civil, structural, mechanical, electrical, and instrumentation disciplines for estimating construction, utilities, and support costs. These estimates are normally within ±10 to 15% accuracy.
5. *Detailed engineering* (*firm bid*) cost estimates are associated with design bid packages—either to provide a contractor's proposal for the work or to provide the construction manager and owner guidance in evaluating competitive contractor bids. Detailed engineering cost estimates are based on specific material takeoffs, standard manpower expenses related to the work, and precise-scope documents, which consist of detailed plans, specifications, time schedules, and

other appropriate, explanatory documentation. These estimates are within 5% accuracy, which is the normal for errors, omissions, changes (EOC) work in a detailed design package.

6. *Final engineering* cost estimates are normally "postcompletion" project recaps to ascertain how accurately the project was estimated, funded, and implemented. The final cost estimate is compared to the definitive estimate to identify those areas where major cost deviations occurred and the extent of EOC work performed. The final engineering cost estimate is a useful tool for engineering or contracting firms for evaluating "in-house" performance and for determining key installation and construction factors for utilization in future project estimates. These details also supply the owner with a documented listing of expenditures for consideration as capitalized versus noncapitalized cost items.

Information Needed for Cost Estimating

Each of the six types of capital cost estimates just described, with the possible exception of OOM estimates, is developed based on specific engineering and design criteria. Enough engineering time and analysis is required to develop these criteria appropriately so that the detail necessary for a particular type or quality estimate is achieved. These design criteria are identified in Table 26.1 as a function of the type of estimate being prepared. To assist in the interpretation of Table 26.1 the following commentary is offered.

1 *The scope statement* describes "what" is included in the estimate. For a Budget Estimate this may be as general as "a 2400 short ton/d (TPD) soybean crushing plant" without further details. However, by the time the project has developed to the definitive estimate stage, all of the process, site, building structure, and auxiliary requirements must be identified for a definitive estimate.

2. *Flow diagrams* describe process and utility flows and relationships. In early project stages this information can be represented by block flow diagrams (BFDs), in which blocks depict major operations and only main streams are defined. However, as the project develops, there is the progressive need for the extended information included in process flow diagrams (PFDs), which show all major equipment pieces, primary product and byproduct streams, major control loops, utility connections, and major line sizes and material of construction. At the definitive estimate stage the details associated with process and instrumentation diagrams (P&IDs) are required; these include not only the information just stated but all process and utility equipment, lines and fittings, instruments and controls, and an identifying numbering system for all items. The reason for this progression is for cost control. A well-thought-out BFD will require from 8 to 16 hrs per sheet, whereas a PFD will require 40 to 60 hrs, and a P&ID 80 to 120 hrs.

3. *Material and energy balances* (MEBs) identify process, utility, ingredient and supply flows (both normal and demand), and storage requirements. Early MEBs associated with budget estimates are generally overview calculations, which later

TABLE 26.1 Estimating Information Required vs. Type of Estimate

Documents	Conceptual stage Budget estimate	Development stage Preliminary estimate	Study stage Study[a]	Approval stage Definitive estimate	Implementation stage Firm estimate
1. Scope statement	General	General/specific	Specific	Specific	Defined bid packages
Historical data	Yes	Yes/no	No	No	No
2. Drawings[b]					
A. Flow diagrams	BFD	BFD/PFD	PFD/P&ID	P&ID	P&ID
B. Site drawing	General	General +	General ++	Specific	Detailed
C. Equipment arrangement	General	General +	General ++	Specific	Detailed
D. Building drawings	Outline	General +		Specific	Detailed bid packages
E. Engr. disciplines	x	General	As required		Detailed bid packages
3. Engineering calcs					
A. Mat'l balance	General	General +	Specific	Specific	
B. Energy balance	Estimated	General	Specific	Specific	Detailed utility packages
C. Process description	General	General +	Specific		
D. Control strategy	General (loop estimate)	General +	Gen'l/specific	Specific	Defined
E. Environmental	General		Gen'l/specific	Specific	Defined
F. Space allocation	General	Gen'l/specific	Specific	Specific	Bid package
G. Equipment list	General	General +	Specific	Specific	Certified drawings/firm
4. Estimating					
A. Equipment	Historical	Experience/verbal	Quoted	Spec'd/quoted	Firm RFQs[a]
B. Installation	Factored	Factored/estimated	Estimated	Designed/estimated	Designed/quoted
C. Buildings	Factored	Factored/estimated	As required	Designed/estimated	Designed/quoted
D. Utility	Historical	Factored/estimated	As required	Calculated	Designed/quoted
5. Accuracy of estimate	±25–33%	±15–20%	N/A	±10%	±5%
6. Cost for estimate (% of total project)					
Project size 0.5 MM–1.0 MM	1.25 ± 0.25	3.25±	N/A	5.56	N/A
1 MM–5 MM	0.80 ± 0.20	1.90	N/A	3.80	N/A
5 MM–25 MM	0.35 ± 0.20	0.80	N/A	1.65	N/A
25 MM–50 MM	0.15 ± 0.10	0.35	N/A	0.80	N/A

[a]For a distinct area within the overall project to answer specific questions

[b]BFD: Block flow diagrams (require 8 to 16 hrs per sheet)

PFD: Process flow diagrams (require 40 to 60 hrs per sheet)

P&ID: Process and instrumentation diagrams (require 80 to 120 hrs per sheet)

[c]Requests for quotations

are extended to each specific line. Early MEBs also identify the environmental emission points, which are addressed in detail later as the project develops.

4. *The equipment list* describes the basic parameters of all of the equipment. This parameter development becomes more detailed and more inclusive through a project. For example, a budget estimate will contain a listing of the major equipment, while a definitive estimate will contain a computer-generated listing, from P&ID and Request for Quotation (RFQ) databases, that contains all of the necessary information for pricing *all equipment* (major specialty items, instruments, electrical items, etc.). The equipment data normally included in equipment lists are tag number, capacity, utility service, size, weight, vendor and model number, motor hp, material of construction, price, and definitive comments. For ease of handling, most contract engineering firms will separate the equipment list into several lists, such as instrument list, specialty item list, motor list, electrical item list, valve list, and others, as required by the project needs.
5. *Site, equipment arrangement, and building drawings* define the arrangement and relationships of buildings and storage structures on the site, equipment within the buildings, and building footprints and profiles. These drawings will allow for space allocation and building cost estimation (initially by factor and later by specific design).

The other line items in Table 26.1 are relatively descriptive and should require little interpretation.

Typical costs for developing the various types of estimates, as a percentage of the total project cost, have been determined. These percentages are shown over a wide range of capital costs in Fig. 26.1 and on the bottom lines in Table 26.1. These cost figures are pointed out because too many otherwise knowledgeable companies will erroneously shortcut or bypass the early estimating stages to "save money" but end up spending considerably more money by not passing through the conceptual and developmental stages, where ideas and alternatives can be assessed both cheaply (because full resources haven't yet been turned loose on a specific design) and quickly (because the documentation formality and detailing hasn't yet come into play). Experience with numerous projects indicates that a good, quality preliminary engineering study and cost estimate will reduce overall project life cycle costs by 5 to 10 times the cost of the study itself through innovative process, layout, building, and energy modifications (from a preconceived starting point).

The information in Table 26.1 is coded to indicate the progressing and sequential degree of technical development during a project. The range of this development is from a broad, general point of view to a specific, comprehensive one. This range is separated into three levels: general, representing the broad conceptual end of the range; intermediate, representing the developing period; and specific, representing the defined end of the range.

For budget estimates in which historical data from past projects are used as a basis for the new estimate, a minimum of two separate cost adjustments must be made.

1. The first adjustment is to bring the historical cost data to current dollars. There are several reliable factors for doing this, depending on whether the historical

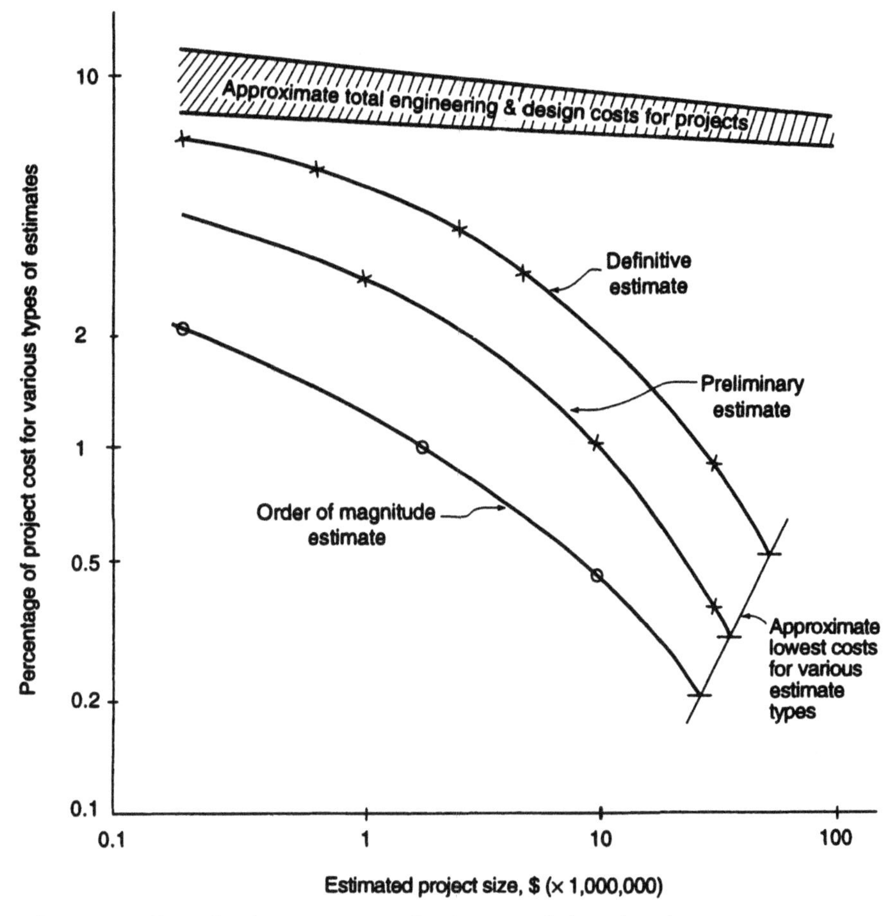

Fig. 26.1. Cost of estimate vs. type of estimate and size of project.

costs are broken down into equipment, supplies, and labor categories or are merely mixed totals.

2. The second adjustment is for capacity. To bring the current-dollar historical cost to the new capacity cost, the formula $C_2 = C_1(Q_2/Q_1)^N$ where C_2 is the new cost, C_1 is the adjusted historical cost, Q_2 is the new capacity, Q_1 is the historical capacity, and the exponent N is related to the type of process and equipment utilized. For soybean plants the exponent N is 0.60 for equipment or 0.75 for the total facility; for oil refineries the exponent is 0.55 for equipment or 0.68 for the total facility. These exponential factors were derived from a limited amount of data and could deviate ±0.05. However, as a first-blush estimate for a new facility the adjustment should be adequate.

This information is plotted in Fig. 26.2 which shows a plot of the plant capacity ratio versus plant capital cost ratio for oilseed-crushing and edible oil–refining plants.

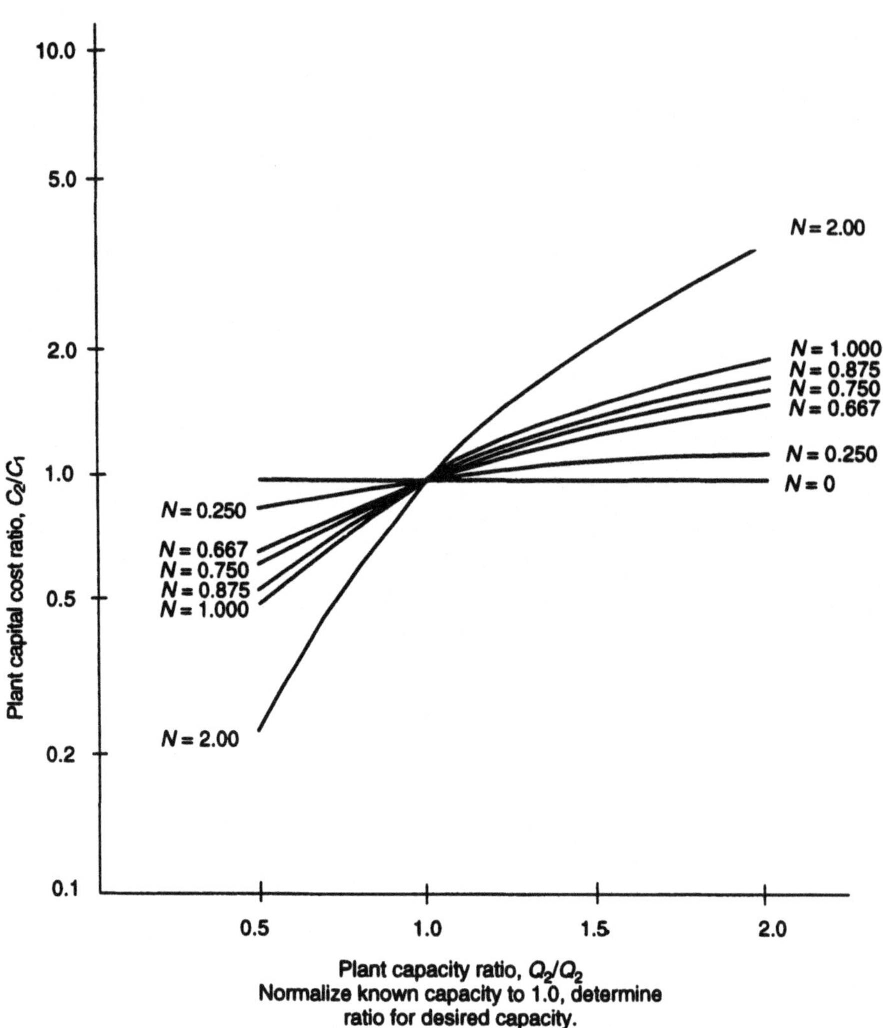

Fig. 26.2. General curve from formula $C_2 = C_1 (Q_2/Q_1)^N$. For soybean plants N = 0.75 (grass roots); for refineries N = 0.68 (grass roots).

3. Obviously, the "other" adjustment to historical cost data that may be required is for scope differences between the projects. It is imperative that the historical project's scope be verified and compared to the new scope. This is of particular importance regarding the extent of utilities, storage facilities, environmental equipment, instrumentation, and other auxiliaries. These facilities can represent 30 to 40% of a project's total cost. This is the area that causes the biggest problem when a new facility is estimated from historical data.

Factors That Affect Capital Cost Estimates

The matrix relationships identified in Table 26.1 deal with the technical information required to develop each type of cost estimate. However, rather than merely to prepare a cost estimate, it is logical to consider the factors that affect capital cost estimates closely so that the company will receive the best information, as early in a project as possible, and at the lowest cost. The sooner that each of these factors can be defined in a project, the sooner an accurate capital cost estimate can be developed. Accurate cost estimates allow management to make better decisions; lower the "surprise factor" and its associated capital risks; and, if approached correctly, allow for developing an improved project. It should also be pointed out that in some cases the best information your management can receive is information that will result in a "no go" project decision, which stops a company from expending further valuable resources (time and capital) on a "dead horse."

As has already been mentioned, three critical factors affect the profitability of *all* projects: design, location, and capacity. Numerous issues related to these and several other factors greatly affect capital cost estimates. These issues should be addressed as early as possible in the project's life cycle, through adequate scope definition and documentation, to *ensure that business, engineering, and operating philosophies are in agreement.* Various sayings express this thought well: "Good, cheap, fast—pick any two;" "Don't order a Cadillac and expect Hyundai prices." Different corporate groups have vastly different wish lists and expect different things.

Following is a brief look at the issues that should be addressed as early in the project as practicable:

Design. Issues in design that should be addressed even at the feasibility stage relate to three general areas: process, support facilities, and mechanical issues.

Process issues include

- *Automation.* Controls and instruments constitute a significant portion of a modern facility's capital cost and also greatly affect operating expenses, quality, and efficiency. These items can be 5 to 25% of the total equipment costs. If an automation strategy has not been adequately defined, an estimate using an "average" cost could be "way off." Initiate a separate study to evaluate a range of automation costs and their associated benefits so that the estimate starts with the correct basis. Most engineering vendor firms are fully capable of supplying this information at a reasonable cost.
- *Product mix.* Can all of the products you plan to produce be handled through a single "line," or are special routings and storage facilities needed to prevent contamination or intermixing? Multiple lines add flexibility, but they also add cost.
- *Energy utilization.* Heat recovery reduces operating expenses but normally adds capital costs. An effort should be made to identify the lowest life cycle costs.
- *Hazardous materials.* Identify special designs and codes. These will sometimes surprise you when it is too late to look for an alternative approach.

Support Facility issues relate to the accessories, such as utility systems, storage of raw materials and products, fire protection and environmental control systems, required to support the business (i.e., operations and commercial aspects). In some instances, these facilities are as expensive as the process systems planned. Thus, definition is necessary. For example, is 10 or 30 days of raw material storage required? The difference may mean multimillion-dollar differences in project costs.

Mechanical issues generally relate to

- *Building design*. Type of structure (preengineered or designed), structure quality, substructure needs, number of operating levels, access to other facilities, building features (i.e., elevators, manlifts, monorails), durability, and expandability are all items that affect cost and should be identified as early as possible to represent the real world.
- *Sanitation features*. These include equipment access for maintenance and operations, special finishes, materials of construction, drainage and process traps, and clean-in-place requirements.
- *Layout*. To enhance communications, shorten conveyances, and reduce operator coverage requires thought.

Capacity. Plant capacity and product mix are normally dictated by marketing requirements. Plant capacity generally has a direct effect on capital costs, increasing as capacity increases (although decreasing on a cost-per-unit-volume basis), and an inverse effect on operating expenses (which decrease as capacity increases). Both of these factors affect the total life cycle cost of the facility. Capacity issues that should be defined are

- *Design capacity*. What are the maximum anticipated flows and what is the turndown requirement? Is full design capacity an immediate need, or can part of the investment be deferred to a later date?
- *Expandability*. Is the design planned for expansion or product mix diversification?

Location. Plant location is both a general business/commercial issue and a capital cost issue. The business factor is normally controlling in terms of site selection. Business issues are related to access to markets and raw materials, access to and expense of utilities and services, availability and skills of labor, and site advantage over a competitor. These location issues ultimately determine the "general area" for the plant location. However, once the "general area" has been defined, the capital cost issues related to a specific site location within the general area become dominant. These issues relate to

- Ground/soil conditions and their effect on substructures
- Fill/leveling requirements
- Utility pickup locations
- Access to transportation (river, rail, truck)
- Sewer and solid waste–handling capabilities of local area

- Land cost
- Local pollution control requirements

Project schedule. An extremely short, optimistic schedule will normally result in greater construction and indirect costs than a comfortable schedule. However, the short schedule may provide a faster revenue recovery. The benefits of a faster start-up should be identified early so that they can be weighed against the costs of a fast-track project effort.

Economic environment. The late 1980s and early 1990s saw a relatively low inflation rate; however, not too long before, this was an important consideration, and "escalation" was a standard line item for all project estimates. This should be considered when a contingency percentage is being determined for a project.

From the foregoing general discussion, it should be obvious that providing good estimates for the various project needs requires an organized approach. The recommended approach toward providing quality estimates is for the estimator to work with management to establish adequate definition of the project scope so as to avoid omissions in the estimate while providing all of the features management expects.

Project checklist and responsibility identification. To achieve adequate project definition, a project checklist is often utilized. A portion of a typical project checklist is shown in a consolidated form in Table 26.2. The project checklist is an independent document, but it is also included in two other documents used during scope development and estimating procedures:

1. An *owner, engineer, contractor responsibility list* identifies and documents the project responsibility for various checklist items or categories. It is mutually prepared by the engineer and client early in a project. For example, if land cost is checked as an owner responsibility, it will not be included in the engineer/contractor supplied cost estimate but will be identified as an exclusion that is to be supplied by the owner. This will keep this item from "falling through the cracks" and being missed in the estimate consolidation.
2. A *cost code system document* categorizes all of the checklist items within general divisions, categories, and subcodes. This allows the proper collection of costs within areas so that estimating factors (for installation, piping and insulation, etc.) are properly applied. For example, an insulation factor would not be applied against land costs or site development but only against equipment, piping, ducting, and so forth.

The most common estimating faults seen in project budgets provided by owners to consultants for confirmation in a preliminary or definitive engineering project are *omission* of auxiliary areas, *misapplication* of cost factors, and *inadequate* construction *assumptions.*

The project checklist is geared for definitive estimates but is also applicable as a guide for lower-quality estimates.

TABLE 26.2 Checklist for Project Cost Estimates

Direct project costs

1.0 Site Work
- Clearing and structure demolition
- Grading, fill
- Railroad trackage: on-site, off-site
- Roadways: on-site
- Underground piping: fire protection loop, utility run-ins, sewer
- Security fencing
- Truck staging and parking
- Stormwater drainage reservoir
- Foundations: piling, caissons

2.0 Buildings
- Process buildings
- Support buildings: shops, part storage, offices, laboratory, guard house, welfare

3.0 Storage, reclamation, and receiving/shipping facilities
- Liquid tank farms
- Solids bins, silos, steel tanks, outside flat storage
- Unloading facilities: marine, truck dumpers, rail pits

4.0 Utilities
- Steam generation facility, including water treatment and reserve fuel handling as required
- Power generation/primary supply
- Cooling water
- Fire fighting water
- Refrigeration
- High-temperature systems

5.0 Instrumentation and controls
- Central control system
- Field-mounted controls

6.0 Environmental controls
- Air pollution control
- Wastewater treatment

7.0 Construction
- Civil/structural
- Piping and insulation
- Mechanical: equipment setting
- Electrical: supply, distribution, MCCs

Indirect project costs

1.0 Engineering and design
- Process, mechanical, civil/structural, electrical, instrumentation
- Programming
- Startup services

2.0 Procurement

3.0 Site surveys, topography and soil studies, environmental permits

4.0 Construction management
- Receiving, warehousing, and control

5.0 Miscellaneous equipment
- Mobile equipment: trackmobiles, diesels, trucks, etc.
- Shop/lab equipment and tools
- Office/welfare furnishings

6.0 Taxes

7.0 Freight

8.0 Land

9.0 Contingencies

Procedure for Developing Project Capital Estimates

Order of Magnitude (or Budget) Cost Estimates. These are normally developed by

1. Updating and/or factoring historical data
2. Applying construction factors to major equipment cost estimates
3. A combination of these two procedures (i.e., applying factors to updated historical equipment costs)

A budget estimate has an accuracy in the range of ±33%, provided that the new facility being estimated has similar process design, auxiliary needs, capacity, and construction materials and a time frame within the past 10 years (maximum). One of the major problems encountered when employing historical data is establishing *good scope information* and an expenditure time line for that historical plant.

Table 26.3 indicates projected *estimating factors* for a soybean plant and oil refinery. All of these estimating factors can be applied to either the major or total equipment costs. It should be pointed out that the equipment costs include both major and minor equipment (i.e., extractors and pumps, diverters, and piping specialties) but not supplies (pipe, fittings, valves, conduit), which are associated with the installation factor. Each engineering firm organizes "factored data" a little differently, so it is sometimes very difficult to compare and reconcile differences reported in the literature. Further, factors can change significantly based on supply of materials and labor from area to area. A rule of thumb is that major equipment cost × 2.5 to 3.5 = Total Project Cost.

TABLE 26.3 Estimating Factors Based on Major Equipment Costs

Capital cost category	Typical%	Range%
Major equipment = *A*	30 %	(27.5–34.5)
Minor equipment = 5% *A*	1.5%	(1.0– 2.0)
Instrumentation hardware = 16.7% *A*	5.0%	(3.0– 7.0)
Freight = 3.3% *A*	1.0%	(0.5– 1.5)
Total equipment = 1.25 *A*	37.5%	(32.0–45.0)
Buildings and storage structures	12.5%	(10.0–25.0)
Utilities	3.5%	(2 – 7.5)
Civil/site work	4.5%	(3.5– 7.5)
Mechanical	6.5%	(4.0– 8.0)
Piping and protective coverings	6.0%	(3.0– 7.0)
Electrical and instrumentation	7.5%	(5.0– 9.0)
Land	2.0%	(0 – 5.0)
Maintenance (parts, tools, mobile eq.)	1.5%	(1.0– 3.0)
Engineering, procurement	10 %	(8.0–12.0)
Expenses (permits, training, etc.)	1 %	(0.5– 1.5)
Contingency and escalation	7.5%	(5.0–10.0)
Total project cost (2.9 to 3.6) x *A*	100 %	

It is again important to note that budget estimates are for a "first-blush" economic evaluation, and the economic evaluation, utilizing whatever method the corporate procedure calls for, should be judged against the high side as well as the "probable" expenditure before the firm commits to the *next* design stage.

Preliminary Engineering Estimates. These are based on better, more thoroughly developed information than budget/study estimates. The following engineering tasks are performed:

1. The *scope statement* is more specific than prior estimates and is firmly based in process details, capacity, building features, automation level, and other decisions.
2. The *flow diagrams* in this case may still be block flow diagrams (BFDs) for the utility systems, but the process is better defined than for budgetary purposes.
3. The *material and energy balance* (MEB) is defined for major product, coproduct, and supply streams. Utilities are calculated from an overview standpoint.
4. The *equipment list* is developed and priced (verbal quotes or recent quotes for similar equipment are used). However, these quotations need not be supported by equipment specs and RFQs.
5. *The overall site plans and equipment arrangement* within buildings are defined, and buildings are estimated by factoring or quick takeoffs.
6. *Installation costs* are factored based on the best information available, utilizing the factor ranges in Table 26.3.

Definitive (Construction Cost) Estimates. These are detailed estimates based on an in-depth process engineering and design package along with the appropriate amount of process, civil/structural, mechanical, and electrical/instrumentation engineering and design to provide suitable information for all process, utility, and auxiliaries equipment and installation takeoffs. Where the design details are relatively straightforward, ratios or factors are still employed to get the cost details. The information normally required for quality definitive estimates is shown in Table 26.1. The documents include (but are not limited to) the following:

1. Scope is defined, frozen, and documented. Any change in scope beyond this point would represent a project "change order" which is directly additive (or subtractive) to (from) the estimate.
2. The process and instrumentation diagrams (P&IDs) are complete.
3. Material and energy balances were developed at the preliminary engineering estimates stage.
4. The equipment list is complete (i.e., not just process equipment but utility and support equipment and specialty items also) and detailed, i.e., supported by formal specs, pricing, and delivery information.
5. Mechanical drawings (including site layout, building profiles and footprints, and equipment arrangements within the processing and major utility buildings,

as well as storage facility/yard layouts) are detailed enough for preliminary concrete, rebar, steel, and siding/roofing takeoffs.

6. A site report (soils, underground, demolition, etc.) is furnished.
7. The control system definition and the instrument and control list are prepared.

Operating Expense Estimates

Operating expense estimates are normally required early in the development of a project for use in economic feasibility calculations. Where multiple plants are operated, historical numbers are usually used for the new facility. Operating expenses consist of several parts:

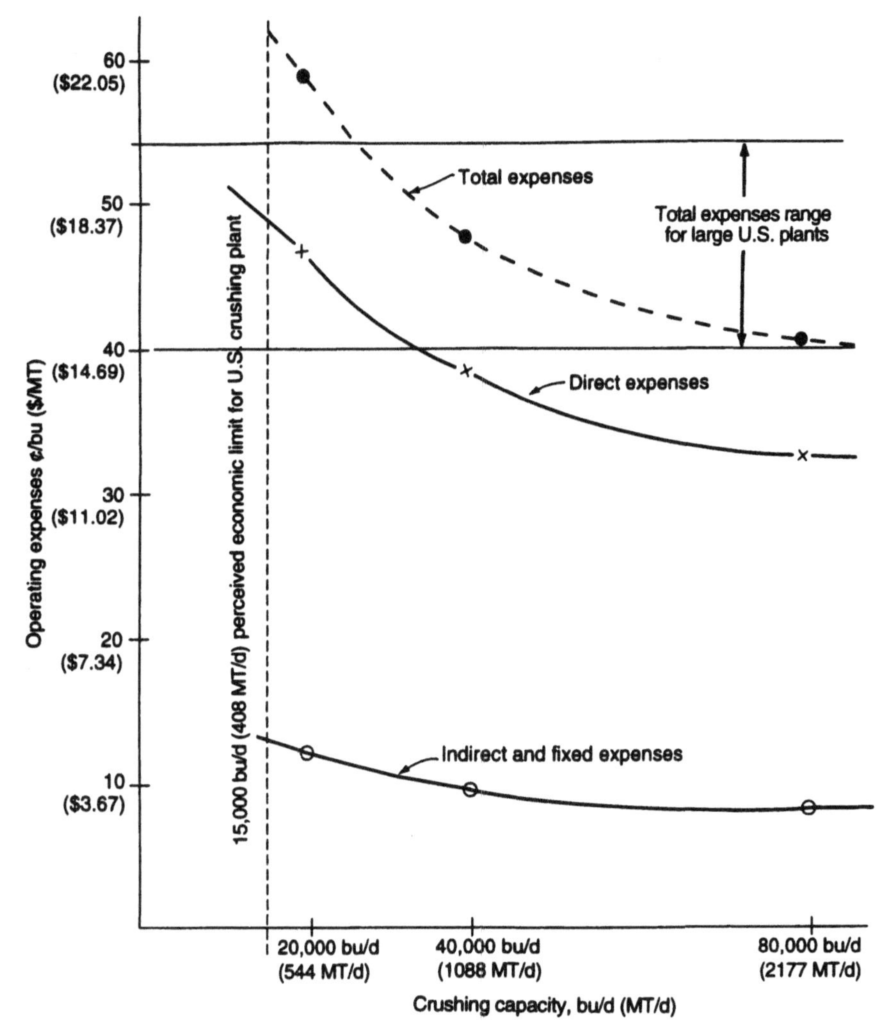

Fig. 26.3. Soybean-crushing plant operating expenses vs. plant size.

1. *Raw material expenses* are related to the quality of the raw materials used (i.e., quality of soybeans/crude oil), and the yields of products derived from the raw materials.
2. *Direct operating expenses* are expenses directly related to the production of the products anticipated. These are relatively constant on a unit basis as long as production levels remain consistent (i.e., hexane, electrical, operating labor).
3. *Fixed operating expenses* are periodic in nature and not necessarily directly related to volume. These will vary considerably with volume. Some of the

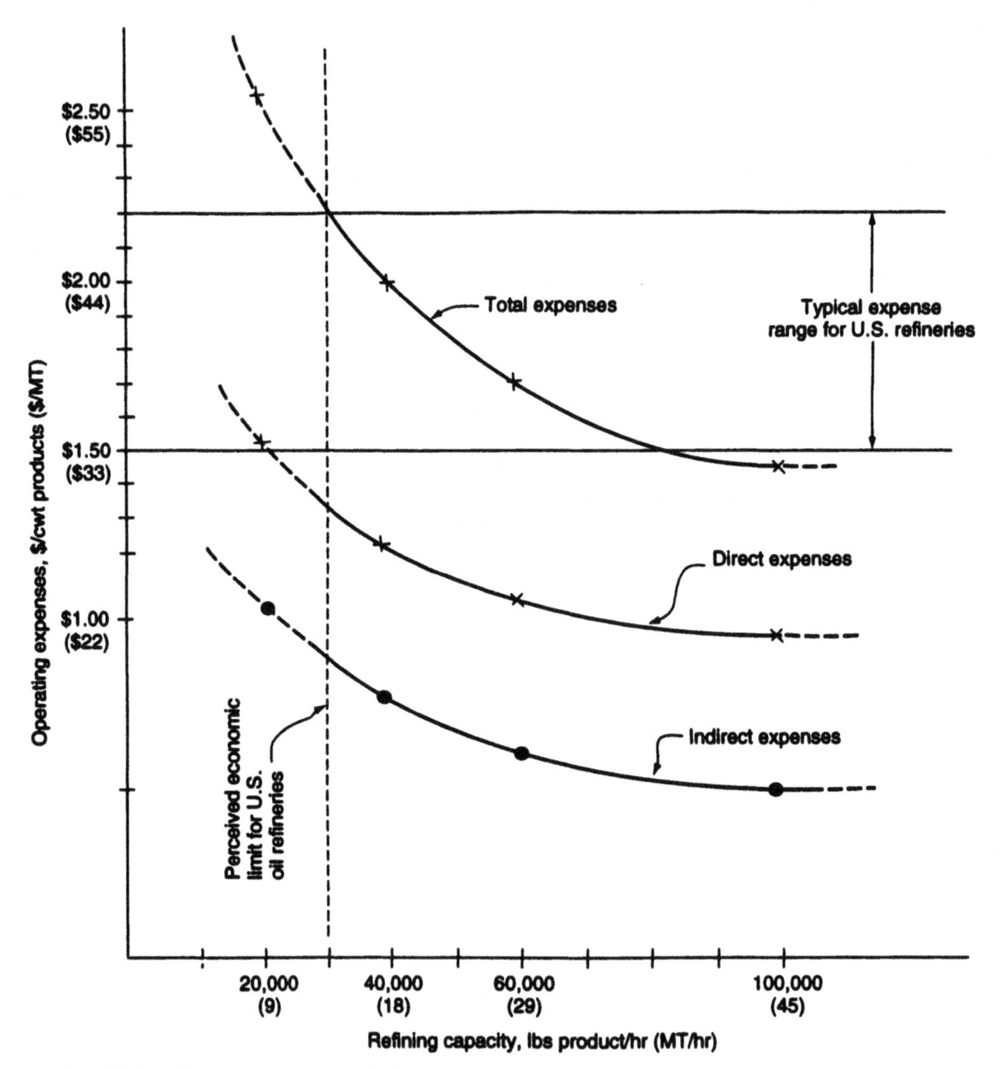

Fig. 26.4. Refining expenses (N.B.D. stages) vs. capacity.

expenses in this category are related to the capital cost and age of the facility (depreciation, taxes and insurance, etc.).

4. *Overhead and commercial expenses* are indirect costs associated with plant, home office administration, and sales/marketing costs. In some cases, R & D expenses are also included in this category.

Typical operating expenses vs. plant sizes are shown for a crushing operation and a refining operation in Figs. 26.3 and 26.4. These statements identify the line items that must be defined for determining the operating expenses for any operation.

When operating expense calculations are prepared based on a specific "running rate," they are then normally increased based on the projected mechanical and process reliability of the operation. Both oilseed (soybean) plants and refineries have relatively high mechanical reliability and will normally operate at 99% of the *scheduled* time, with an operating schedule of 340 to 350 days/yr.

Chapter 27

Plant Management

Norman J. Smallwood

The Core Team, Hammond, LA

Introduction

Responsibility and Results Expected

The responsibility of soybean and soybean oil–processing plant management is to ensure expected performance results over the operating life of the asset. In broad terms, the performance objectives are outlined as follows (1):

1. Provide timely customer service
2. Deliver high-quality products
3. Operate at a competitive cost
4. Produce at operating-capacity volume
5. Give top priority to personnel safety
6. Meet government regulatory requirements

To stay competitive, processing plants must be continually improved. Leadership is required that can achieve excellent plant operating performance by engaging every employee in the process of developing, testing, and implementing better ways of doing things. With every employee involved in identifying, justifying, and initiating incremental change, continuous improvement in operating performance is possible (2).

Resources

To achieve the expected performance results, plant management is concerned with effective utilization of the following resources:

1. Facilities and equipment
2. Material
3. People
4. Finances
5. Information
6. Time

The key concepts for managing the resources involved in soybean and soybean oil–processing plant operation are presented in this chapter (1).

Management System

Traditional, authoritative management practices are most commonly used in processing plants throughout the world. In this chapter, an open, participative management system is presented. It is the most effective method currently being utilized to maximize long-term operating performance. Other than to make some appropriate comparisons, there is little treatment of traditional practices. Traditional management is thoroughly documented in the literature and usually results in mediocre long-term performance due to inherent underutilization of people. During the mid 1960s participative management was introduced to and applied in Procter & Gamble's edible oil–processing plants. In 1980 the A.E. Staley Manufacturing Company utilized open-system, participative management focused on continuous improvement for their new computer-controlled soybean oil–processing plant in Des Moines, IA. Presently, C&T/Quincy Foods' new edible oil–processing plant in Quincy, IL, is operated by an open system of self-managed teams supported by a core resource team. The following presentation of plant management is based on tested, proven strategies and practices that have resulted in performance excellence.

Facilities and Equipment

Production Capacity

Processing plants are designed and constructed to operate at a specified capacity. In establishing the design capacity, assumptions are made in regard to raw material, expected product mix, and operating efficiency. The engineering and design of a plant is usually carried out to err on the conservative side. Accordingly, the resulting plant can have an actual capacity in excess of design. Less frequently, the design assumptions and engineering calculations may be in error so as to result in the actual capacity being less than expected. To confirm the design capacity, make reliable production commitments, assess the actual product-mix impact, and identify unit operation and individual process bottlenecks; determination of actual plant capacity is of key importance. The methodologies for actual-capacity determination and increase are outlined as follows:

Individual Process Capacity Determination.

Analysis:

1. Update process flow diagrams.
2. Calculate theoretical capacity of each unit operation.

Capacity testing:

1. Design the capacity test protocol.
2. Conduct the capacity test.

3. Determine bottlenecks, whether they are equipment rate–limiting or product result–limiting, and if the latter, whether yield, quality, or both if affected.

Capacity documentation:

1. Define the *operating basis*. The *product mix* depends on market forecasts based on recent history. The *feedstock quality*, based on historical data, determines product yield expectation and *operating rate* limitations.
2. Establish the *operating rate* from theoretical analysis, capacity testing results, or both, considering product mix and feedstock quality.
3. Specify the *operating time*, beginning with the total scheduled time available and allowing for scheduled nonoperating time (process startup and process shut down) and process equipment maintenance (scheduled and unscheduled, based on historical data) to determine actual operating time.
4. Calculate capacity.

Total Plant Capacity Determination. Document capacity by process or operation. Identify bottlenecks by process or operation.

Steps Toward Incremental Increases in Production Capacity.

1. Eliminate unit operation and/or process bottleneck.
2. Introduce new or improved process equipment.
3. Upgrade process control.
4. Replace unreliable equipment components.
5. Improve feedstock.
6. Shift product-mix toward higher-processing-rate items.

Major Expansion in Production Capacity. This should be driven by business demand or the rational expectation of operating cost improvement.

Process Operation and Control

Consistency of operation based on optimum procedures and conditions is a fundamental requirement for excellent performance. Operation and control procedures must reflect the appropriate scientific basis, relevant engineering principles, and empirical data. The sources from which operation and control procedures can be established are bench-scale experimentation, pilot plant runs, and carefully controlled plant-scale runs.

The purpose of process control is to maintain selected variables within prescribed limits. A process variable maintained within prescribed limits is defined as a *controlled variable*. A variable that is adjusted to keep a controlled variable within a prescribed range is defined as a *manipulated variable*. For each controlled variable, an optimum control target and a normal operating range around the target are established that will consistently give in-specification quality and expected yield results (see Chapter 24). Outside of the normal operating range, *action range* limits should be established to allow continuation of operation for a maximum specified time to regain proper control without seriously jeopardizing product quality and yield. Finally, the outer limits of the action range are defined as the *shutdown* or *recycle* point. If a controlled variable moves outside the action range, immediate process shutdown or recycle is mandatory. Process recycle is permissible, if technically or economically feasible, when control deviation problems can be readily identified and corrected in a reasonable time. The process variable control target and limits protocol is diagrammed in Fig. 27.1.

Each unit operation of a processing plant involves the control of one or more variables. Effective process control is necessary to achieve expected product quality, yield, and cost results. Control method selection for any process variable involves the following cost-benefit analysis:

1. Cost: initial purchase and installation, ongoing maintenance, expected life
2. Savings potential: yield improvement, energy reduction, reprocessing avoidance, operating personnel requirement reduction
3. Maintenance implications: skill level, spare parts availability, reliability.
4. Product quality specifications
5. Process safety needs

The control methods available are

1. Manual
2. Electromechanical (analog automatic)
3. Microprocessor (digital automatic)

Whether manual or automatic, the basic elements of control are *measurement*, *comparison*, *computation*, and *correction*. A *control mode* is a particular control system response to a change in the controlled variable measurement. The four basic control modes are

1. *On-off (two-position) control*, in which the manipulated variable is either on or off
2. *Proportional control*, a control algorithm (mode) in which change output to the manipulated variable by the controlling device is proportional to the change measured in the controlled variable
3 *Integral control*, a control algorithm that results in resetting the controlling device output to compensate for any offset between the controlled variable and the desired control point
4. *Derivative control*, a control algorithm that anticipates when a controlled variable will reach its desired control point by sensing its rate of change and adjusts the manipulated variable to prevent the controlled variable from overshooting the desired target

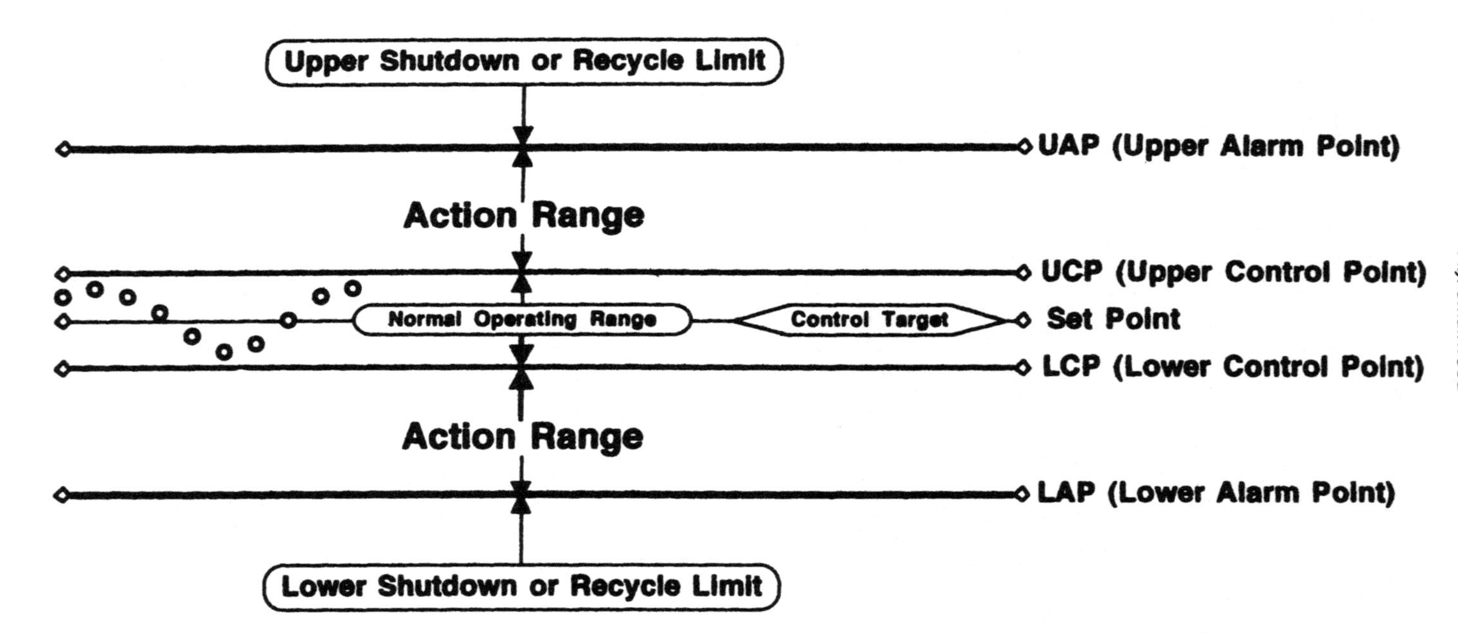

Fig. 27.1. Process variable control protocol. Reproduced with permission of The Core Team, Hammond, LA.

Control mode selection is a matter of carrying out the analysis outlined and matching need with the optimum solution. One or more modes can be utilized to control a given process variable.

Establish reliable data logging of significant process variables. Process log sheets should be formatted to include the control objectives (target, normal operating range, action range, and shutdown/reprocessing limits). Thus, log sheets essentially become control charts. Space should be provided on log sheets to note operator interventions in the action range to regain normal control. Operators need to be thoroughly trained to ensure accurate data recording. For extremely critical process variables, continuous instrument recording of conditions should be provided. With computer control, logging and statistical analysis of process data can be done automatically and continually as programmed.

Maintenance

The broad objectives of plant maintenance are to ensure reliable operation at the lowest possible cost. The starting point for maintenance begins with preparation of the initial equipment specification. Equipment specification should both match present needs and be compatible with anticipated future changes where possible. In preparing specifications, a thorough knowledge of both the operating demands and the environment are critical. Underspecified equipment will prove to be unreliable and result in high maintenance cost. Overspecified equipment has potentially negative implications, including excessive initial purchase and ongoing operating cost.

Selection. Long-term reliability and maintenance implications are significant factors to consider in the actual equipment selection process. Accordingly, the following points should be included in an equipment selection decision:

1. Experience with and reputation of the specific equipment item and manufacturer
2. Availability of replacement parts over the life of the equipment
3. Technical capability to maintain the equipment with either internal or external resources that are readily available
4. Commonality of equipment to facilitate interchange and minimize spare parts inventory requirements
5. Simplicity of design, operation, and maintenance
6. Initial cost compared to the probable total-life cost, which includes the estimated maintenance and operating downtime cost.

Installation and startup. Equipment installation must conform to the conditions specified by the manufacturer, must meet process safety requirements, and must provide appropriate protection from the operating environment. Improper installation will result in poor reliability and high maintenance cost. To ensure that equipment is properly installed, a pre-startup inspection should be conducted before operation is attempted. The pre-startup inspection team should include representation from the engineering, maintenance, and operating functions to be sure that all pertinent issues are covered. A pre-startup checklist, specially prepared for each equipment item, should be followed in the inspection. The general content of a pre-start up checklist is outlined as follows:

1. Assembly
2. Mounting
3. Rotation
4. Process safety devices
5. Personnel safety guards
6. Lubrication: grease application, oil levels
7. Electrical power supply: voltage and amperage needs, wire size, overload protection, switches, code requirements
8. Piping system, including appropriate double valve and bleeder protection
9. Protective startup screen placement in pump inlets
10. Instrumentation and control scheme devices and wiring

Operations. The operating personnel's knowledge of and care in operating the equipment are major determinants of equipment reliability and maintenance cost. To achieve the right level of operating expertise, the following steps are essential:

1. Select personnel who have the right aptitude and potential capability.
2. Train the personnel to operate effectively and efficiently. Ensure that the personnel are familiar with the operating *principles*, *capabilities*, and *limitations*; the *operating procedures* for startup, normal operation, shut down, and emergency and contingency responses; *troubleshooting methods*, including both the trouble-cause-correction charts and diagnostic and problem-solving skills; and *maintenance* procedures for lubrication, repair (within prescribed limits of capability), record keeping, and work order preparation.
3. Ensure that personnel are thoroughly qualified before taking on the actual operating responsibility. Administer a written examination and require personnel to perform an actual demonstration as well as interim operation under observation and guidance.
4. Retrain and requalify operating personnel on a prescribed frequency.

Lubrication. Equipment will self-destruct without lubrication. Accordingly, a thoroughly designed and executed lubrication program is absolutely essential. The elements of a sound lubrication program for each equipment item are outlined as follows:

1. Prepare the format that defines the requirements and will be used to record accomplishment. List all lubrication points, both for grease and for oil. For each point, specify the lubricant and the quantity to be applied. Specify the frequency

of lubrication for each point. For each point specify the frequency for level check and for oil change, and prescribe a sampling of waste oil for analysis (if applicable). The *lubrication record* must show, for each lubrication operation prescribed, the date accomplished and the initials of the person performing the task.
2. Assign responsibility (ownership) for the ongoing implementation of the program.
3. Establish an auditing function to ensure compliance.

Spare parts. Keeping an adequate inventory of spare parts at the plant site is another vital aspect of assuring reliable operation. Considering the cost of a spare parts inventory, careful analysis is required to determine the actual inventory needs. The analysis to determine the spare parts inventory for each equipment item is diagrammed in Fig. 27.2.

Accurate and timely record keeping is needed to guarantee the integrity of the spare parts system. For each equipment item, the spare parts inventory documentation should include part description, inventory level, usage rate, and storage location information. A typical equipment item spare parts record is shown in Fig. 27.3. Record keeping for identical equipment items should be combined. Use of computer systems is the preferred method for spare parts inventory record keeping. Manual systems will do the job but less efficiently. On receipt of spare part orders from vendors, the inventory records need to be immediately updated. A "shoe tag" should be attached to each spare part on receipt. Subsequently, when a part is used, the bottom half of the shoe tag is removed and used for updating the inventory records.

Design and layout of the plant spare parts storeroom is an important consideration. Parts should be easy to locate. Actual storage points should be referenced (numbered) for entry in the spare parts inventory record by individual part.

Maintenance personnel. Readily available personnel with the requisite maintenance skills to match plant needs is another critical link in the total maintenance program. There are a variety of options for providing the skilled personnel:

1. Include an internal maintenance group of the size and with the skill level to meet the plant needs.
2. Develop multiskilled operating/maintenance personnel who have the dual responsibility of operating and maintaining the equipment.
3. Utilize external maintenance services for equipment maintenance.
4. Apply some combination of options 1, 2, and 3.

Careful, ongoing analysis is required to determine which is the best option for any given plant. The best option now is not necessarily the best option in the future.

To provide the required maintenance skills, the first task is to analyze the skill requirements for the particular plant. The skills required are a function of the complexity, size, and condition of the plant. A typical modern processing plant requires the following maintenance skills:

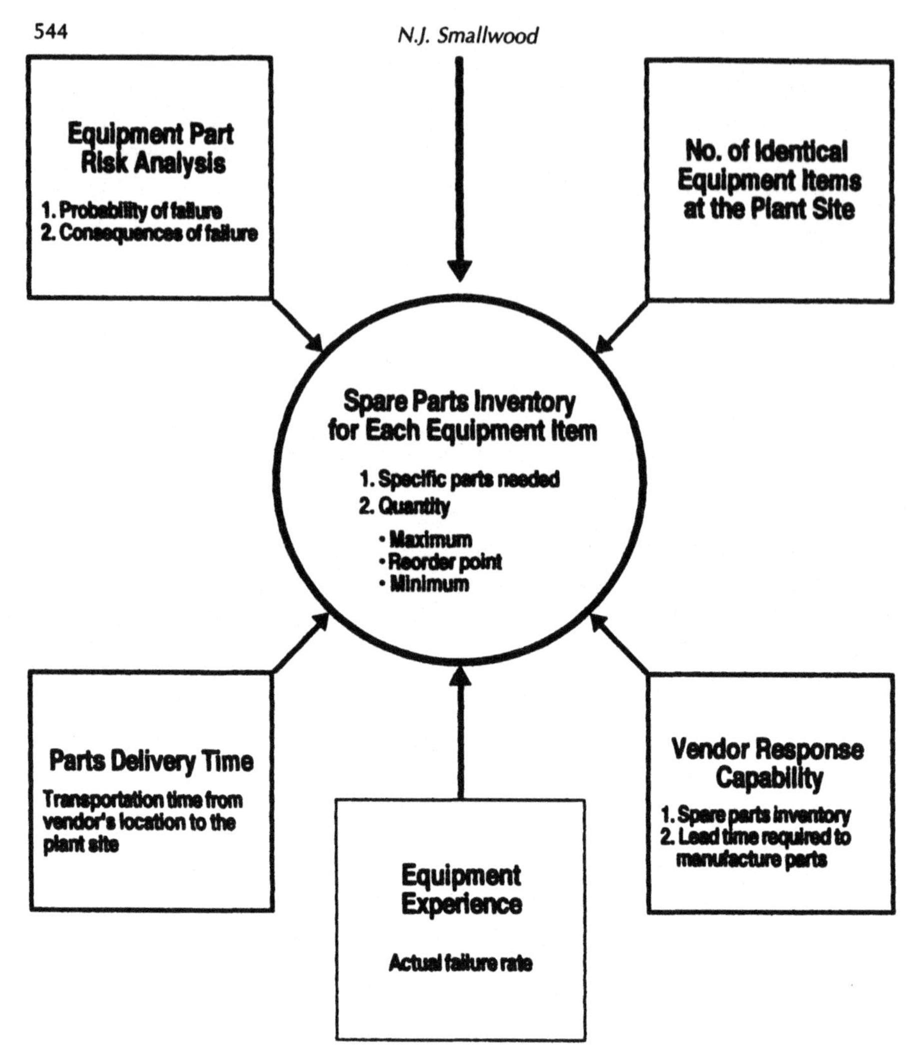

Fig. 27.2. Analysis to determine the spare parts inventory. Reproduced with permission of The Core Team, Hammond, LA.

1. General mechanic
2. Pipe fitter/welder
3. Machinist
4. Electrician
5. Instrument and control technician

Giving some maintenance responsibility to every person in the processing plant within the limits of his or her ability is highly preferred. An individual's role can be limited to that of a helper when called upon or extend to that of a highly-skilled maintenance person with one or more skills.

Skilled maintenance people are universally in short supply. To fill plant maintenance needs, probably the best solution is to identify the people with the interest and aptitude and provide training and skill development opportunities concurrent with their normal work assignments. In other words, "grow your own" maintenance skills. Following are the fundamental steps in skill development (p. 548):

1. Define specifically what is expected (detailed *definition* of the skill).
2. Provide the learning experience (*training*).
3. Verify that the person has mastered the skill (*qualification*).
4. Provide ongoing opportunity for the person to practice and utilize the skill (*practice*).

Preventive maintenance. Preventive maintenance (PM) is highly touted in maintenance literature and by many maintenance professionals. In processing plants, PM is an important part of the maintenance program, but as with everything else, there are practical limits to its application. There must be an economic justification for any PM application. Justification involves risk-benefit analysis from both historical and anticipative perspectives. PM justification includes detailed definition of what is to be done, what it will cost, and what expense it will avoid. One noteworthy argument in considering PM justification is, "If it isn't broke, don't fix it." In many cases, the justified method may be "replace it when it fails." The least expensive and perhaps most beneficial PM application is to look at and listen to equipment and facilities on a regular frequency. Schedule repair based on diagnostic cues before actual failure occurs.

By definition, an equipment item is critical if its breakdown or failure results in rendering the total process, operation, or plant inoperative; if time to make repairs and restore operation could be substantial; and if there is relatively high vulnerability for breakdown or failure. PM is justified for critical equipment items. In a processing plant, the following equipment items are critical:

1. Steam boilers
2. Air compressors (plant and instrument air systems)
3. Centrifuges
4. Vacuum systems
5. Pumps for some services
6. Process safety devices
7. Process control systems in most cases

Pumps that handle abrasive materials are quite vulnerable to seal failure and high rate of internal wear. Accordingly, bleaching earth and catalyst slurry pumps have relatively high failure rates and are typically classified as critical. A PM program for critical equipment items includes the following principal elements:

1. Detailed inspection protocol
2. Schedule to perform the inspection protocol
3. Criteria for replacement of critical parts that have reached the maximum allowable wear limit or operating time

In some cases, it may be possible to justify on-line or on-site critical spare equipment items as part of the PM/reliability program. If a breakdown or failure occurs, the operation can continue by using the standby spare. With the availability of critical spares, no operating downtime is required for regular or preventive maintenance.

Equipment: ______ | Identification No. ______ | No. of Identical Equip. Items: ______

Part Description	Part No.	Vendor No.	Inventory (Quantity)				Annual Usage Rate	Storage Location
			High	Reorder Point	Low	Actual		

Fig. 27.3. Equipment item spare parts record sheet. Reproduced with permission of The Core Team, Hammond, LA.

Individual processes or operations and, in some cases, the total plant may require shutdown for routine cleaning, inspection, or other servicing of certain equipment or facility items. This work is often called *operational maintenance* and is usually performed by operating personnel. Examples of operational maintenance are heat exchanger cleaning, cooling tower cleaning, tank cleaning, centrifuge disk cleaning, vessel internal inspection, and filter cloth replacement. When downtime is taken for the operational maintenance tasks, PM work should be concurrently scheduled and performed.

A vital part of plant maintenance includes attention to buildings and facilities. Properly designed and constructed buildings and facilities will have an indefinite life if sound preventive maintenance is applied. Conducting a comprehensive building and facility maintenance requirement survey with subsequent timely follow-up action is the essence of the program.

Organization. Organization options for managing and administering a plant maintenance program are shown in Figs. 27.4 and 27.5. The more traditional organization

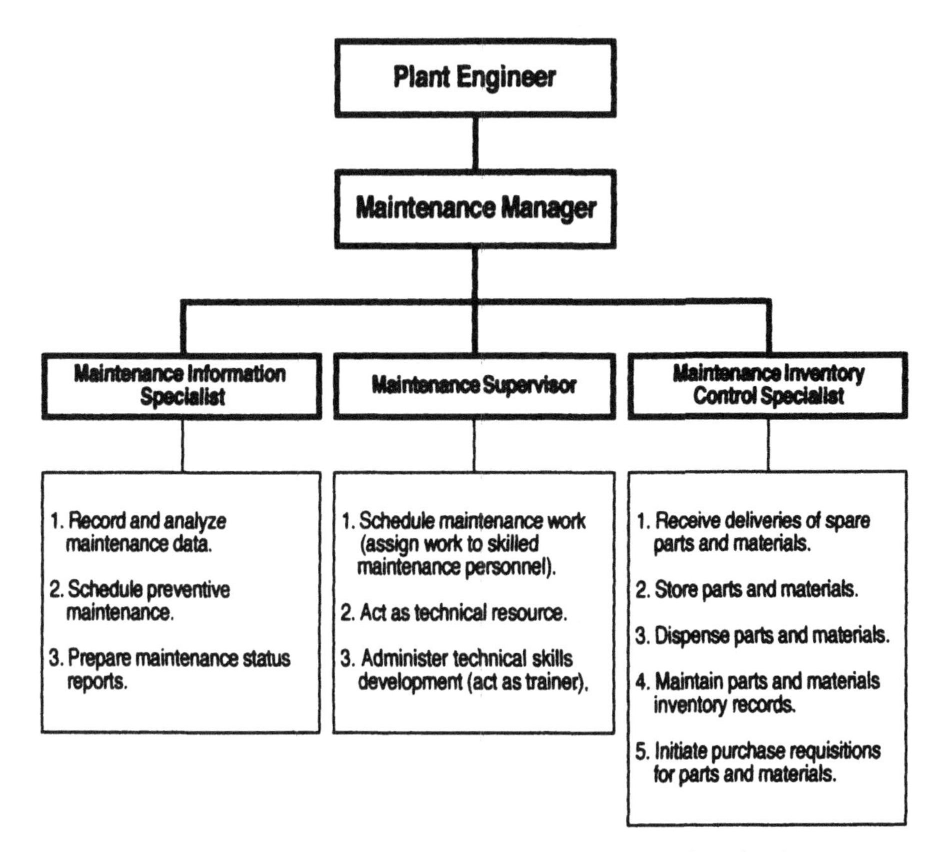

Fig. 27.4. Traditional plant maintenance organization. Reproduced with permission of The Core Team, Hammond, LA.

is diagrammed in Fig. 27.4. It represents an autonomous maintenance organization under the management of the plant engineer. Figure 27.5 reflects a shared approach to the maintenance responsibility in which the plant engineering function supports the operations function, which actually performs the maintenance work. The latter organization is often utilized by management that emphasizes a high level of personnel participation and teamwork.

A written work order is an absolute imperative for an effective maintenance system. A work order is the vehicle/tool for accomplishing the following:

1. Identifying the maintenance need
2. Communicating with the maintenance resources (message)
3. Scheduling
4. Planning and preparing

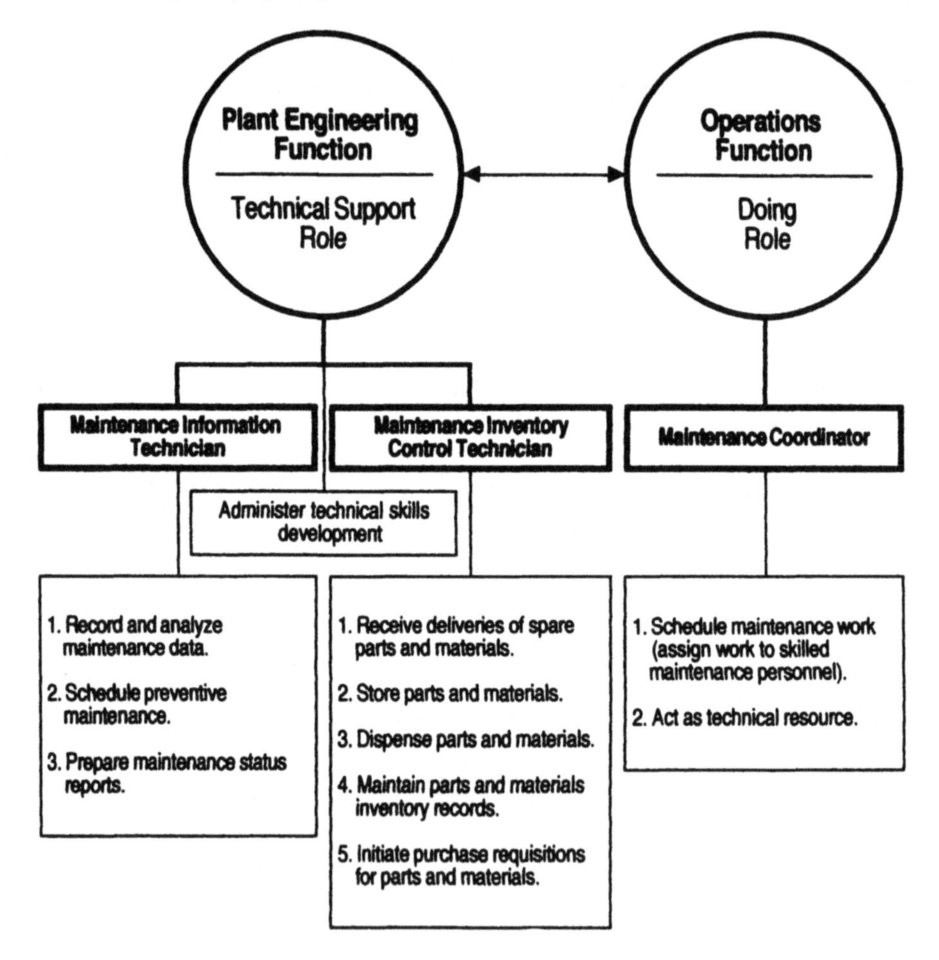

Fig. 27.5. Fast-response plant maintenance organization. Reproduced with permission of The Core Team, Hammond, LA.

5. Documenting what was wrong with what (input for equipment history log); what was done: scope of repair (input for equipment history log); equipment downtime and maintenance time expended; and cost (input for cost accounting records and equipment data log)

A typical maintenance work order is illustrated in Fig. 27.6.

Process and Operation Safety

The purpose of process safety management is to establish codes and practices to ensure safe machinery, equipment, vessels, tanks, buildings, structures, and electrical power supply by means of proper design, operation, and maintenance. With consistent adherence to safety codes and practices, the risk of accidental loss, damage and injury to physical assets and personnel can be minimized. Establishing viable process safety involves the following actions:

Date:	**Maintenance Work Order**	No.	**Priority**
Description (What is wrong?):			Expense Code
Location:			Start Job Site Condition
Availability for Repair: Date:	Time:		End Job Site Condition
Safety Requirements:			Job Cost Labor:
Scheduled for Repair: Date:	Time:		
Mechanic Assigned:	Start: Complete:	Total:	Material:
Probable or Identified Cause of Problem:			Total:

(4-Part Form)

Determination of Work Order Priority		
Priority Designation	Classification	Definition/Application
1	Emergency Immediate Attention Required	a. Personnel safety matter of immediate serious risk b. Process safety matter of immediate significant risk c. Process and or vital equipment inoperability d. Product quality and/or loss of immediate significant consequence
2	Short Notice Completion needed within one day	a. Personnel safety matter of moderate possible risk b. Process safety matter of moderate possible risk c. Process and or vital equipment operability in vulnerable state d. Product quality and/or loss of moderate consequence
3	Routine-A Completion needed within 10 days	Maintenance matters which can be done on a routine basis but should not be delayed for a long period because of some degree of vunerability.
4	Routine-B Completion needed within 30 days	Maintenance matters which can be done on a routine basis and can be delayed for a longer period without serious consequence.

Fig. 27.6. Maintenance work order format. Reproduced with permission of The Core Team, Hammond, LA.

1. Understand the kind of risks: overpressure; underpressure (vacuum); overheating, which risks equipment damage, product degradation, and fire hazard; explosive or flammable material and gases; corrosive or otherwise hazardous material.
2. Identify the actual potential risks by means of process/operation flow diagram study, actual process/operation audit, and literature search for process safety hazards in related operations.
3. Conduct a process safety study. Qualitatively and quantitatively understand each risk/hazard identified. Determine the process safety requirements: standards; regulations; engineering analysis and calculations.
4. Install and maintain process safety devices.
5. Ensure the integrity of the process safety program. Provide effective personnel training; make routine audits; conduct annual inspection and testing as prescribed.

Material Resources

Introduction

Timely delivery, acceptable quality, and proper quantity of materials is the lifeblood of a processing plant. The principal functions and elements to achieve the material resource objectives are presented in outline form with some supporting commentary.

Procurement

Procurement (the purchasing function) involves all the actions necessary to ensure timely delivery of acceptable quality and proper quantity of material resources to the plant site. The salient aspects of procurement are outlined as follows:

Specification.

1. Description
2. Quality requirement (state maximum acceptable defect/variance level if applicable)

Usage rate.

1. Maximum
2. Minimum
3. Average (normal)
4. Predictable trends

Sources (Vendors/Suppliers).

1. Identification
2. Evaluation as to customer service (timely delivery, technical support.); quality performance reputation; location (with freight cost and timely delivery implica-

tions); reliability (known or potential vulnerability); and cost

3. Qualification
4. Utilization (primary or backup supplier)

Delivery (Transportation). To support plant operation, transportation is needed for both incoming and outgoing material. Selection of the transportation means should include an assessment ranging from plant-owned transportation to common carrier (public) transportation. The transportation assessment considerations are outlined as follows:

Options Available.

1. Railroad
2. Truck
3. Barge
4. Ship
5. Pipeline
6. Aircraft

Evaluation.

1. Customer service
2. Quality (product/material protection)
3. Capacity
4. Cost

Rating. The particular transportation options can be rated as primary, secondary, and so forth.

Quality Assurance

Quality assurance is needed for both incoming materials and outgoing products. In-plant capability is needed to handle the input/output material quality assurance function. The major considerations in the quality assurance program are outlined as follows (also see Chapter 24):

1. Inspection: material/product condition; transportation vessel/vehicle cleanliness
2. Sampling: frequency; protocol/methodology
3. Analysis/testing
4. Supplier documentation requirements: certificate of analysis; statistical quality control charts
5. Comparison of inspection, analytical, or testing results with the material quality specification

6. Rejection or complaint protocol
7. Documentation protocol: retained samples; analytical/test result records; performance reports.

Storage

In the ideal case, no material storage capacity would be required. Material input deliveries and product output shipments would be scheduled just in time (JIT) to maintain continuity of operation. With highly reliable material supply and transportation resources, significant operating cost reduction can be gained by JIT delivering and shipping. Furthermore, with JIT delivering and shipping, time-related material and product degradation are minimized, improving both quality and cost.

For each material or product, provide the minimum storage necessary to ensure continuity of operation at the lowest cost. Determination of the optimum storage capacity involves risk analysis and material life consideration coupled with the savings potential from purchasing and shipping material in larger quantities.

Inventory Control

Inventory control is a never-ending balancing act. Too much inventory is cost-prohibitive. To run out of inventory and jeopardize plant operation is intolerable. Inventory control methodology and key considerations are outlined as follows:

Data format for Each Material Item.

1. Code/stock number
2. Description
3. Inventory: maximum; reorder point (normal); priority-expedited delivery point (daily tracking required); actual
4. Normal order quantity
5. Vendors (suppliers): primary; secondary; tertiary
6. Last order (active): quantity; date; supplier; expected delivery date; transportation (delivery) service.

Data collection. Daily input:

1. Receipts
2. Usages
3. Shipments
4. Production
5. Known losses

Physical inventory:

1. Daily: active product stream; critical short inventory items
2. Monthly: all material except that on the annual inventory schedule
3. Annual: spare parts; mobile equipment

Data Processing. Options are manual vs. computer (personal, mini, or mainframe)
Computer Network Capability (systems integration):

1. Internal functional operations
2. External: vendors (suppliers); transportation; customers

People Resources

Organization

To practice participative management successfully within an open system, changes are required in both organization structure and individual roles. The rigid hierarchical and compartmentalized structure of traditional organizations is replaced with a flexible and interactive system capable of bringing together the optimum resources to meet both external and internal needs (see Fig. 27.7).

Individual Roles

The role of each person in the empowered organization is based on the individual ownership of value-added responsibilities and a clear understanding of all customer-supplier relationships, both internal and external (2). Every person is a manager of some unique unit of the business and is held concurrently accountable for the assigned unit and the total business. The statement, "It's not my job," common to traditional organizations, is *not* part of the culture (2).

Personnel Recruiting

Vacancies in the organization are filled by a dynamic process designed to assess the strengths and weakness of each candidate for employment. The capability to function in an interdependent work culture is given careful consideration. The criteria for personnel selection include the following (3):

1. *Personal standards (integrity)*: A person with integrity has much to offer, but a person without integrity is a considerable liability.
2. *Drive and initiative*: Every member of the team must be a self-starter and take appropriate action to get the job done.
3. *Learning capacity*: The ability and desire to learn are essential qualities.
4. *Problem-solving ability*: Recognizing, analyzing, and solving problems are necessary in every job.

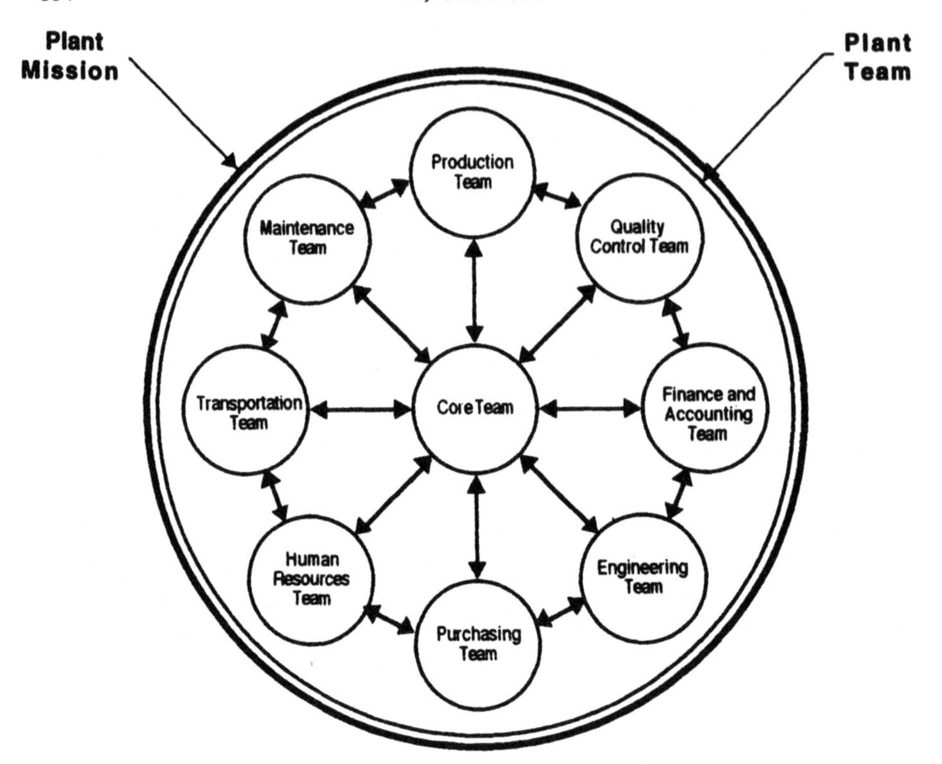

Fig. 27.7. Open system organization structure. Reproduced with permission of The Core Team, Hammond, LA.

5. *Sound relationships*: The capability to develop and maintain sound relationships with others in the work environment must not be seriously impaired.
6. *Communication skills*: The ability to communicate clearly and accurately is essential.
7. *Education, knowledge, and experience*: Existing knowledge and skills are a major asset only for those who possess the attributes listed above.

The personnel recruiting process must be sufficiently thorough to ensure that the criteria listed here are met. Use of a multidiscipline employee task force to assess candidates for employment both yields excellent results and minimizes the possibility of bias. To enhance the organization motivational climate, job openings should be first available to and filled by interested and qualified internal candidates.

Training

Performance excellence can be achieved by organizations that are proficient in training and developing people. For processing plants using computer-based process control, effective training becomes even more essential to achieve and sustain reliable operation. Individual training should be directed, first, to the skills necessary to perform competently the present job responsibilities and, second, to the long-term career goals of the trainee that are compatible with the organization's interests. The scope of training should include both the skill requirements of the specific job and the "big picture" knowledge to enable each person to interact synergistically with the total organization. The following categories of training are recommended for all processing plant employees (3):

1. *People skills.* the basis for, development and maintenance of effective relationships
2. *Team development skills.* the means for achieving effective teamwork through unity of purpose and positive synergistic relationships
3. *Technical skills.* objective (what), theory (why), and application (how) of every aspect of the job
4. *Learning and teaching skills.* understanding the learning process and how to teach others
5. *Creativity skills.* discovering, developing, and applying the inherent creativity of every person
6. *Problem-solving and decision-making skills.* methods and applications
7. *Administrative skills.* knowledge of how the business functions and how each individual is expected to take action to get the job done right

In Fig. 27.8, the categories of learning are given in greater detail for an edible oil processing plant's personnel.

Motivational Climate

The belief that motivation can be imposed by supervision is the basis for frequent misunderstanding and failure. From the investigations of post–World War II behavioral scientists like Frederick Herzberg and Abraham Maslow, motivation occurs only from within each person (4). People are motivated when the proper climate is established to satisfy their particular needs. The key factors to establish a motivational climate are outlined as follows (3):

Responsibility (Ownership). Define and clearly articulate individual and team responsibilities to convey a sense of ownership.

Performance standards. Identify the specific job performance standards and their importance. What are the expectations? What does it take to win?

Accountability. Establish the protocol for determining and posting the score. You can't play to win if you don't know the score.

Critique. Recognize and properly respond to the results achieved.

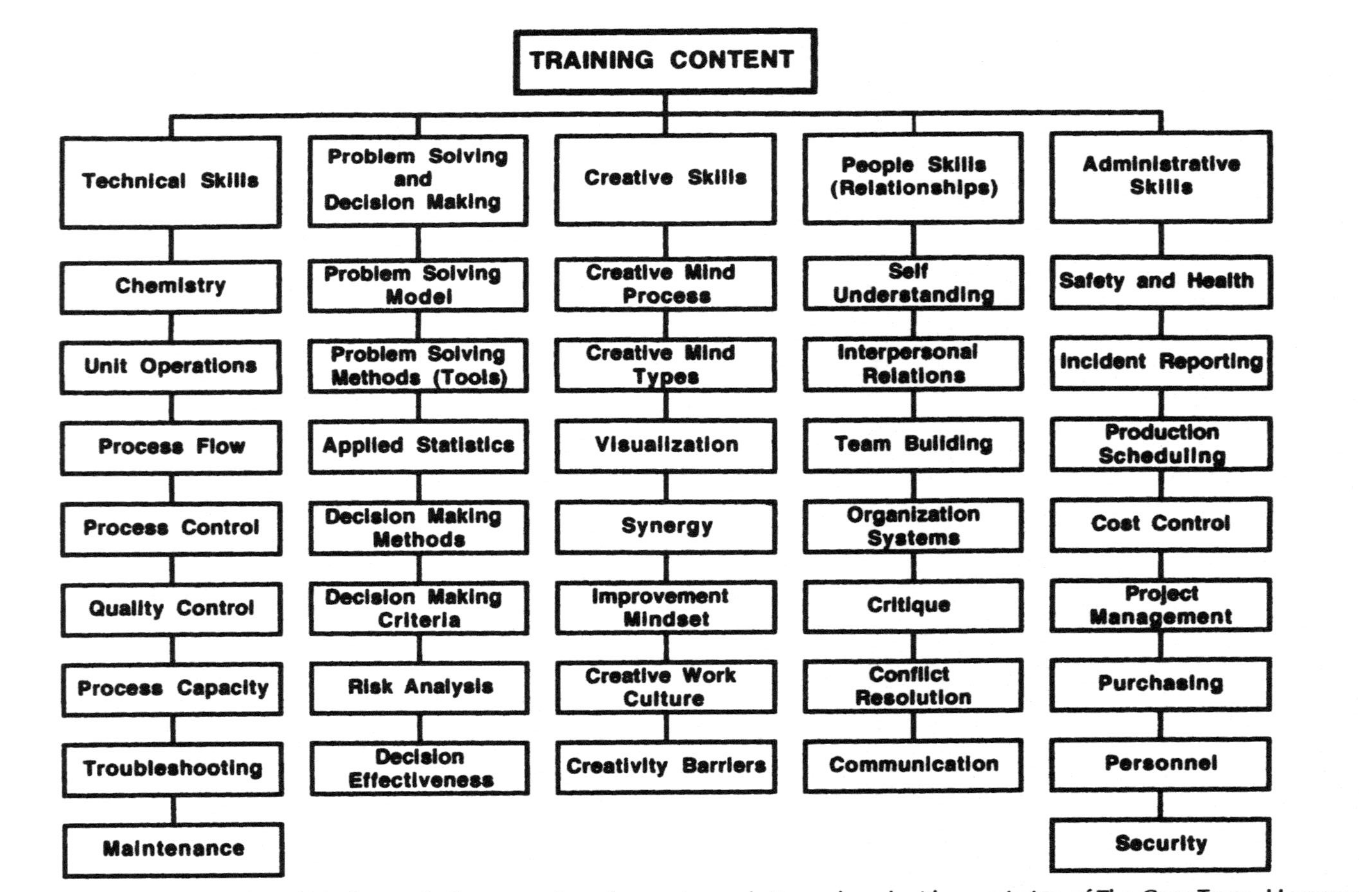

Fig. 27.8. Training for edible fats and oils processing plant personnel. Reproduced with permission of The Core Team, Hammond, LA.

- Positive reinforcement for meeting or exceeding expectations(winning)
- Constructive/productive criticism (coaching) for failing to meet expectations
- Performance assessment to reflect the interdependence of the team and achieve mutual growth

Reward. Reward for the right things: meeting or exceeding expectations.

Team Development

Rarely do plant personnel work in isolation. To be effective, people must learn how to work in cohesive teams (5). Finally, the many work teams making up a processing plant must be able to interact constructively and efficiently to accomplish the mission and objectives. Effective teamwork occurs as the result of a deliberate strategy and careful implementation process. To achieve the necessary teamwork throughout a plant, the process must start at the hierarchical top and work down to include everyone (6). If teamwork is not consistently displayed by the top leadership, it will not occur in the remainder of the organization. The team development process that has been successfully applied in the leading processing plants is diagrammed in Fig. 27.9.

Safety and Health

Recognizing the importance of people and considering the ethical implications, every person's safety and health is the top priority in operating a processing plant. The key elements of the personnel safety and health program to ensure top-priority attention are:

1. Identify the hazards.
2. Provide physical measures and written practices and procedures with high visibility to protect people from the hazards.
3. Do not tolerate violation of safe practices by anyone.
4. Keep the plant clean and orderly.
5. Conduct ongoing inspections to identify safety deficiencies.
6. Maintain on-site first aid capability.
7. Have in place backup procedures and facilities for dealing with serious injuries: ambulance service and a hospital with trauma care capability.
8. Investigate, document and report all injuries. Determine the following: type of injury; degree of injury; cause of injury; implications of injury in terms of lost work time, personal impact, and cost; and for the corrective action required, its description, scheduled completion, actual completion, and cost.
9. Involve every person in the total plant safety and health program. Ensure that every person understands his or her personal responsibility. Give every person a specific role in the safety and health program.

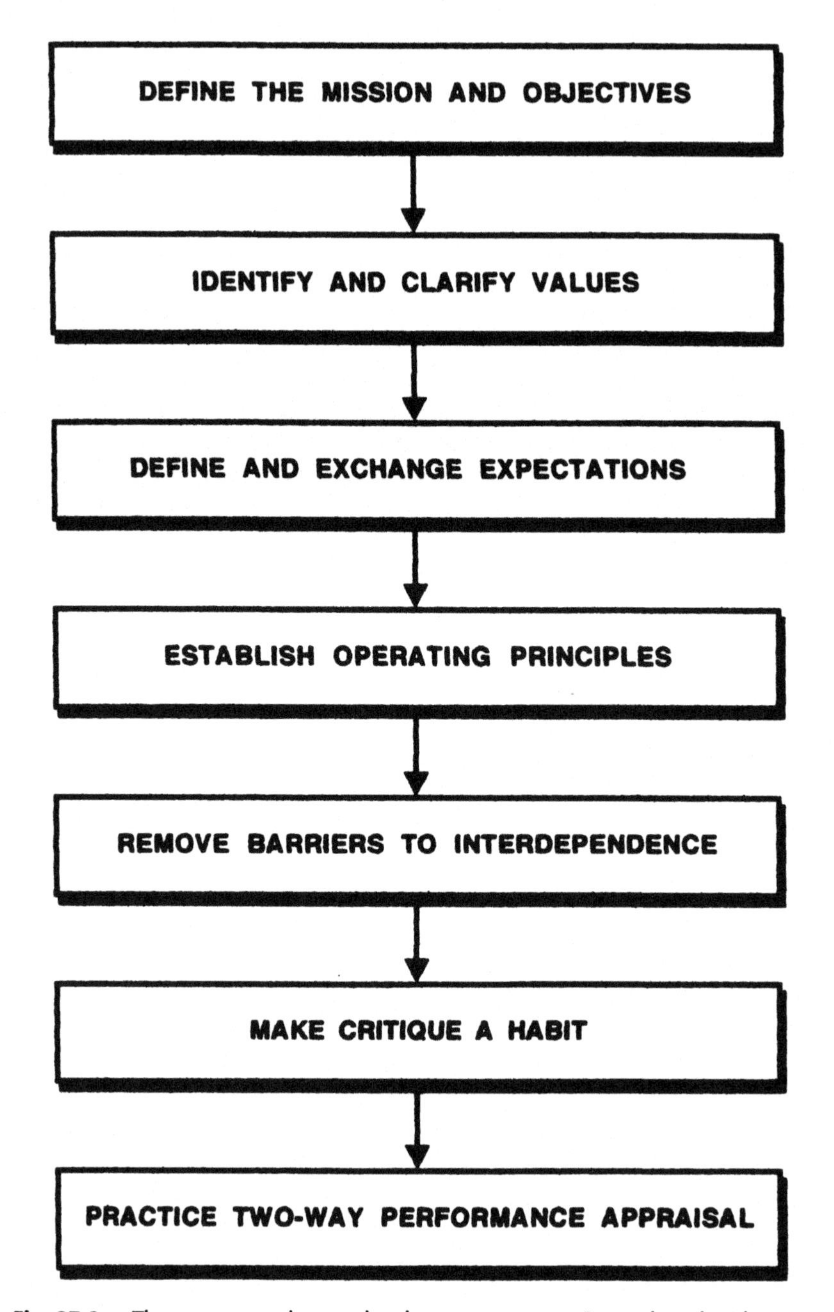

Fig. 27.9. The empowered team development process. Reproduced with permission of The Core Team, Hammond, LA.

10. Schedule monthly safety meetings with mandatory attendance requirement for every person on an operating shift basis. Educate; review the safety record identify problems; plan improvements.
11. Require and provide for thorough physical exams for everyone at the proper frequency.
12. Set safety achievement goals and recognize achievement (celebrate).

Managing with Unions

There is no doubt that managing with union representation is difficult, particularly when an adversarial relationship exists (7). Unions exist because at some point in time, management action or inaction created an adversarial relationship with employees. In reaction, employees turned to union representation as a means to protect their interests and address their concerns.

Sound management works diligently to treat all employees with respect, to be sensitive to their needs, and to respond appropriately whether union representation exists or not. Unfortunately, in some companies, adversarial relationships have existed for decades, and the process of changing the quality of employee-management relations takes time. Change to a climate of trust and cooperation is possible if every manager will consistently engage in open, honest, and respectful interactions across the business.

In cases where the union leadership promotes an adversarial climate as a fear tactic to maintain member allegiance, transformation to a win-win relationship is an even greater challenge. Management must not fall into the trap of responding with equally imprudent behavior. Never assume that the hostility of a few employees represents the feelings of the majority. Doing what is right prevails in the long term.

Not understanding the terms of the contract and not consistently following them are the most common failure of management in working with union-represented employees. When members of management try to function in ignorance, the contract becomes a one-way street. The union is viewed by both management and employees as having all of the power. Management complacency in contract administration can lead to serious erosion of the right to manage, subsequent decline in operating performance, and ultimate failure of the business.

Properly negotiated labor contracts are double-edged swords. Management makes specific concessions to the union-represented employees as documented in the contract. The right to manage the business is retained by management. Furthermore, management has an equal right to file grievances for union or union member breaches of the contract.

To keep the business healthy with union-represented employees, it is vital for every member of management to understand, follow, and skillfully administer the contract for the benefit of everyone. Rigorous training and qualification are required to ensure that every manager is capable of administering the contract correctly and consistently. Preferably, the training and qualification should take place before taking over job responsibilities in a union environment. If deficiencies exist in contract competence for any current managers assigned, they need to be properly addressed immediately.

Financial Resources

Introduction

For products to be purchased, the price must be both affordable to the customers and competitive with respect to other potential suppliers. To survive, a plant cannot operate at a financial deficit. Control of product cost is an essential requirement in operating a processing plant or any venture. Proven methods for controlling product cost are presented under the following headings:

1. Expense categories
2. Budgeting
3. Accounting
4. Reporting
5. Assigning ownership and accountability
6. Developing better methods for cost reduction

Expense Categories

Expense categories are usually established on the basis of functional involvement and accountability. The focus here is limited to the expenses normally incurred by a soybean oil–processing plant. The general expense categories are

1. Material expense (raw material and ingredients from which the finished product is made, including packaging material)
2. Operating expense
3. Finished product transportation expense (if paid by the plant)

The operating expense general category is usually subdivided as follows:

1. Personnel compensation: base pay, bonus pay, benefits
2. Energy, including electrical power and fuel (natural gas, fuel oil, other)
3. Material: operating (reagents, processing aids, supplies); maintenance (replacement parts, supplies); quality control (reagents, supplies): office supplies
4. Water
5. Sanitary sewer charge
6. Wastewater treatment charge
7. Taxes
8. Insurance
9. Depreciation
10. Communication (telephone)
11. Equipment rental

12. Delivery and courier service
13. Solid waste disposal

Further subdivision of expense categories may be needed to define and assign accountability specifically. Each expense category, subcategory, subsubcategory, and so forth is assigned an account number for use in the cost accounting function.

Budgeting

Budgeting is the foundation of cost control. The following question-and-answer presentation is designed to teach basic budgeting concepts and methods.

What is a budget? A budget is the expected cost of taking a specific action or acquiring a specific product or service.

Why go to the trouble of budgeting? Budgeting is done to provide a sound financial basis for a specific business decision; to establish the authority for making expenditures to engage in a specific business activity; and to set the standard for evaluating the actual cost of a business activity.

What are the methods of budget preparation?

1. Historical, based on experience
2. Detailed analysis based on the expected functional methodology and economic circumstances (zero-base budgeting)
3. Combination of historical and zero-base methods

For the advantages and disadvantages of each budgeting method, see Table 27.1.

How accurate should a budget be? There are different outlooks on budgeting accuracy. The positions range from an almost guarantee that actual spending will not exceed the budget to a somewhat flexible policy depending on the circumstances. The most accepted practices on budgeting accuracy reflect the following points:

1. The accuracy of a carefully prepared budget is primarily a function of predicting future events, such as sales volume, product mix, and price escalation, and avoiding catastrophic events such as explosion, fire, or flood, which would seriously impair production capability.
2. Meeting a budget should not be a sure thing. There should be a "stretch factor" in the budget. In terms of probability, the chance of meeting the budget should be about 75%. The value of the "stretch factor" is to encourage everyone to contribute their best effort to controlling and reducing operating expense. If the budget is too generous (a sure thing), there is a tendency to become complacent. It is much better to *slightly* overspend a tight budget than to slightly underspend a fat budget.

Who should prepare the budget, and what is the approval process? Presentation and defense are essential parts of the budgeting process. Each unit manager should prepare a detailed budget proposal for his or her area of responsibility. The origina-

TABLE 27.1 Budget Preparation Methods

Advantages	Disadvantages
Historical Budgeting	
Quick and efficient	Higher-cost results
Reliable predictor for static situations	Questionable value as a standard for evaluating actual cost
Requires little skill or expertise	Not applicable to a changing environment
	Weak argument in defending budget request and receiving approval
	Fosters status quo mentality
Zero-Base Budgeting	
Lower-cost results; forces detailed questioning and analysis of the existing operating environment	Takes considerable time and some expertise
High value as a standard for evaluating actual cost	Effort could exceed benefit if carried to the extreme
Strong argument in defending budget request and receiving approval	
Establishes a climate for mastering the details of the operation and seeking improvement	
Combination of Historical and Zero-Base Budgeting	
Some reduction in cost results	Value largely a function of the judgement of the people involved (when to use historical and when to use zero-base budgeting)
Moderate value as a standard for evaluating actual cost	Confidence level compromised as a standard for evaluating actual cost and in budget proposal defense
Reasonable time requirement	

Source: Reproduced with permission of The Core Team, Hammond, LA.

tor should be prepared to present clearly the budget proposal and defend it thoroughly. The authority level receiving the budget proposal should both challenge and understand the rationale and specific details. For each authority level review, the objectives are to strengthen the budget, ensure understanding of the budget, and obtain approval of the budget. With the review process described, everyone in the management chain of authority is prepared to carry out the budget mission with a high degree of commitment and confidence when final approval is obtained.

Accounting

The role of cost accounting is to document every expense incurred by expense category, to prepare financial reports, and to calculate actual product cost. Expenses are typically allocated to each specific product by means of expense distribution factors. The expense distribution factor process requires rigorous treatment to ensure that the actual product cost calculated is accurate.

Inventory accounting is an equally important task. The key responsibilities are

1. Prepare accurate inventory reports on a daily basis for unit manager information and decision making.
2. Determine product yields, losses, and degradings for each processing system.

3. Report aberrant inventory conditions for immediate corrective action.
4. Participate in and be a resource for physical inventories.

The product accounting points for a typical soybean oil–processing plant are shown in Fig. 27.10.

Reporting

Timely, accurately prepared budget and inventory status reports are critical plant management tools. Use of computer systems enhances report preparation significantly in terms of accuracy, timeliness, and flexibility.

Assigning Ownership and Accountability

A crucial element to success in operating a processing plant is to assign ownership and accountability for everything and involve everyone. Having people empowered, with ownership and accountability, will result in each job being done with excellence and lead to consistent achievement of outstanding performance for the total plant.

Continuous Improvement Resulting in Cost Reduction

With the motivational climate established on the basis of ownership, accountability, and recognition of accomplishment, every employee is consistently engaged in continuous improvement actions, which can result in significant reduction of operating cost. Any continuous improvement that yields financial benefits focuses on the following goals (2):

1. Reduce losses
2. Increase productivity
3. Improve products and services
4. Expand capacity

Capital (Major Improvement) Projects

By definition, a capital project is an investment to add a major new equipment item; extend the useful life of a capitalized asset; expand an operating plant; or build a new plant or facility (see also Chapter 26). The steps required to implement a capital project successfully from start to finish are outlined as follows:

Idea. All improvement projects start with some person's creative idea. The idea could originate with any person on the plant team—operator, mechanic, supervisor, or plant manager.

Definition and Scope. If the idea is considered to have merit, time is allocated to make a preliminary determination of how to do it. The elements of preparing a project scope could include any or all of the following:

1. Draw a simple process flow diagram to illustrate the changes made.
2. Calculate the impact of the proposed changes in terms of improved efficiency, better yield, lower operating cost, and so forth.
3. Decide where to locate the new equipment items required.
4. Determine how the changes will affect personnel safety.

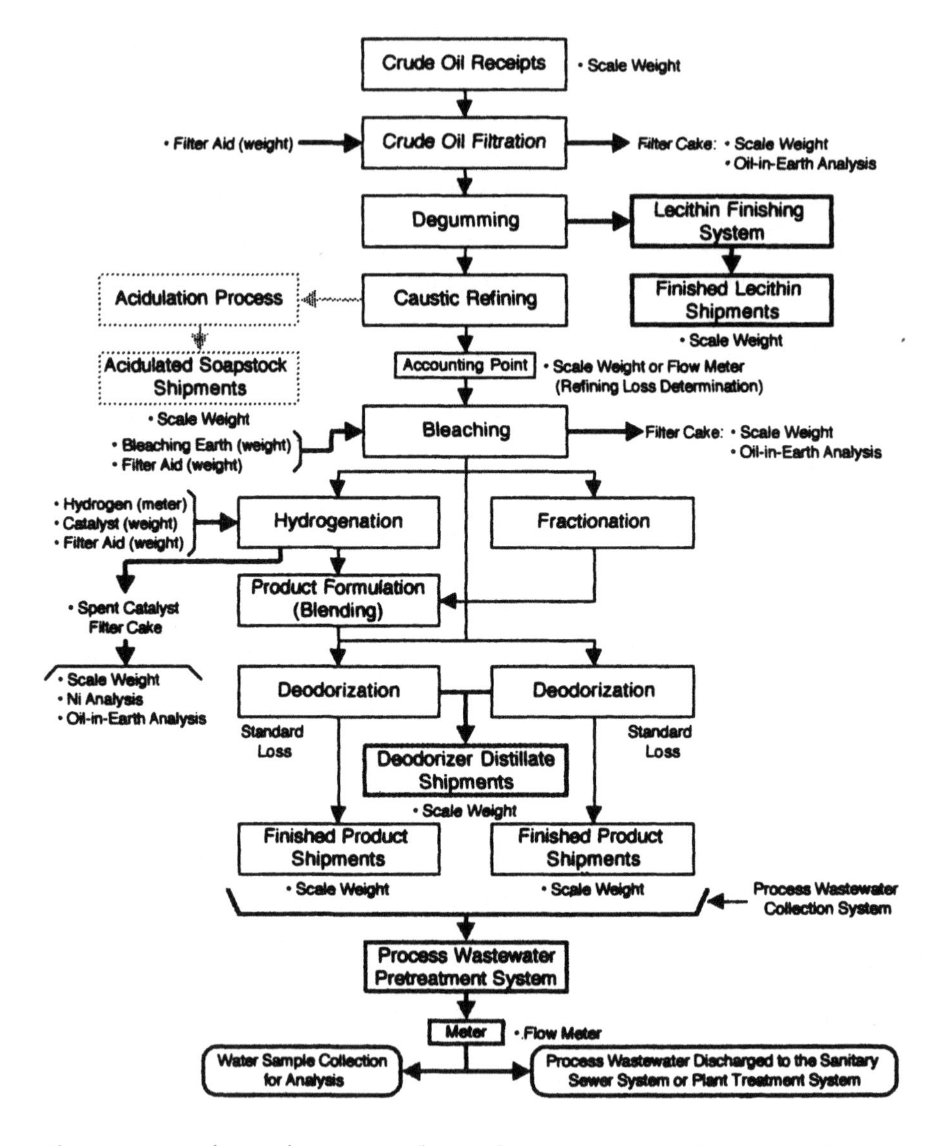

Fig. 27.10. Soybean oil processing flow and accounting points for an operating plant. Reproduced with permission of The Core Team, Hammond, LA.

Justification. After reviewing the preliminary project definition and scope, if there is agreement that further development is warranted, project justification is the next step. Project justification involves the following actions:

1. Prepare a feasibility estimate. Determine the total project cost in material, labor, engineering, and other categories.
2. Determine the benefits to be gained: financial return (amount of savings/yr) from yield improvement, energy cost reduction, operating rate increase, or other effects; personnel safety improvement; environmental protection improvement; or other benefits.

If the benefits are purely financial, the return on investment must be sufficiently attractive to receive authorization. However, when the benefits are limited to personnel safety or environmental protection, financial justification may not be required if the investment will alleviate a high-risk problem.

Engineering. On acceptance of the justification by the plant manager or person(s) designated to approve capital projects, the engineering work is initiated. The engineering work consists of the following (see also Chapter 26):

1. Prepare a detailed project scope, including: detailed process flow diagrams, P & IDs (process and instrumentation diagrams), and detailed layout drawings.
2. Make the engineering study and calculations to specify equipment.
3. Estimate the project based on the detailed scope preparation, engineering calculations, and equipment specification.
4. Verify the project justification.
5. Present the project package for final review by those who are affected by the project.

Production engineering. Prepare the complete engineering package to construct the facility.

1. Procure specified equipment and materials.
2. Make the engineering drawings and sketches detailing the work to be performed.

Construction. When enough of the engineering package is completed, the construction phase is initiated.

1. Plan and schedule the actual construction work.
2. Obtain the necessary construction people resources using internal personnel, outside contractors, or a combination of both.
3. Prepare the site.
4. Carry out the construction work and complete the project.

Pre-startup inspection (PSI). The pre-startup inspection (PSI) precedes the actual startup and consists of the following steps.

1. Make a thorough, detailed inspection of the project to ensure that construction has been completed according to the engineering design and specifications.
2. Check equipment readiness: motor rotation; lubrication; coupling alignment; control loop functioning.
3. Insert a protective screen in the suction-side flange of pumps, to prevent welding slag and metal scraps from entering and damaging the pump on startup.
4. Prepare a *punch list* of any deficiencies found for immediate corrective action.
5. Acquaint the operating, maintenance, and supervisory personnel with the project as part of their final training and qualification to operate, troubleshoot, and maintain the equipment.
6. Ensure that personnel safety measures are in place.
7. Conduct a dry-run operating test.
8. Confirm that the project is complete and that the system is ready to operate.

Actual startup. Actual startup should be performed by the normal operating personnel with the careful guidance of the technical and supervisory support team. As normal operation is achieved, the special support team can phase out of the special role. The startup phase includes the following actions:

1. Compare actual performance results to design (expected) results.
2. Determine the cause of any performance gaps and take corrective action.
3. Identify any installation or construction deficiencies for prompt corrective action.
4. Conduct a critique of the project after expected performance has been achieved, to identify what was done right and identify what can be done better on the next project.
5. Prepare a final report on the outcome of the project for review by those who have a stake.

Information

Introduction

Chris Argyris (4) demonstrated that valid information is an essential resource for operating a business; for success, decisions must be made on the basis of sound information. Furthermore, Argyris concluded that employees who have a sense of ownership in the organization's performance have a higher commitment to provide valid information (4). The wisdom in Argyris's conclusions have been borne out by the results of high-performance management systems where each employee is the unit manager of his or her respective job responsibility. Accordingly, to operate a profitable business, information systems need to be designed to serve the needs of the internal and external customers and suppliers: every player needs to know the score (1).

Information Categories and Their Implications

Information systems pertaining to processing plant operations can be categorized into historical information, real-time information, and future (strategic) information. A sports analogy is useful in understanding the terms and illustrating the significance (8). Real-time data or information is like the score posted while the game is being played. Knowing the score while the game is in progress can have a profound impact on the final outcome (winning or losing). Receiving the score after the game is over is interesting and can have some value in preparing for the next game, but has no impact on the game completed. The game plan (future strategy) prepared for the next game has significant value. Joel Barker points out that the greatest leverage to succeed is in the future (9).

Traditional organizations are skewed toward historical information systems. Occurring in the form of financial and accounting documents, information often lags the actual time of occurrence by up to seven weeks. Consequently, the root cause of problems are slow to be recognized and corrected. Too much effort is directed to dealing with problem symptoms, and problems can be completely overlooked. The lack of real-time information is one of the principal causes of mediocre results or failure.

In high-performance organizations, at the operating level, information is skewed toward real-time systems (8). All players are knowledgeable of the real-time score on the vital aspects of their responsibility. Timely action is taken to win. Managers keep informed on the real-time score from the individual unit managers' scoreboards. Information about future direction (strategic plans) is given significant attention by the top leadership with input and feedback from every member of the organization. Accordingly, all participants in the business know where they are going and how they are going to get there. The information categories and their implications with respect to time are shown in Fig. 27.11.

Real-Time Information as a Means for Providing Timely Critique

Winning (meeting or exceeding the functional-unit performance objectives) is enhanced by the critique process (8). When a unit manager is winning, business managers note the posted score at the job site and give timely, positive recognition. The unit manager will be encouraged and is likely to keep on winning.

If a unit manager is losing, then timely inquiry should be made to understand why. The right empowering questions should be asked to ensure that proper corrective action is being taken.

Information Systems in Computer-Controlled Operations

With computer-based process control, integrated, real-time information systems are a matter of system design, programming, and utilization. Information reliability is a function of the accuracy and reliability of the measuring devices. Information for historical reference can be maintained in an archival data system as programmed.

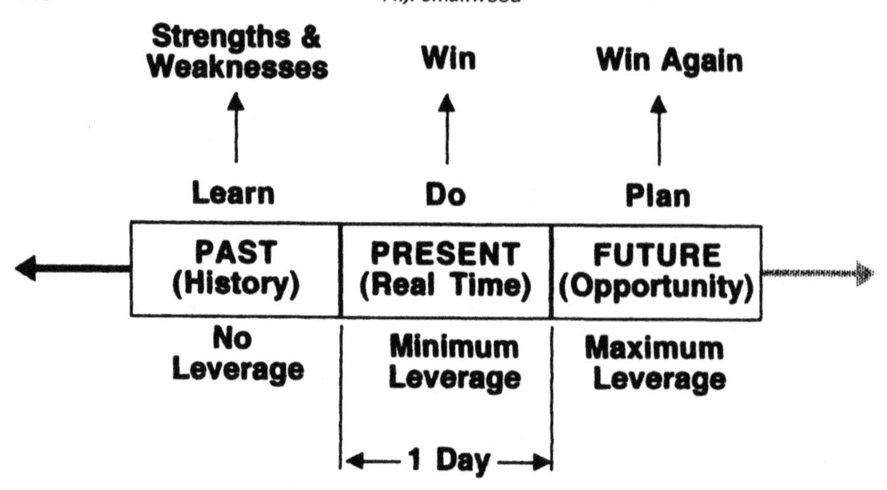

Fig. 27.11. Information categories by time and their implication. Reproduced with permission of The Core Team, Hammond, LA.

Information Systems Content

The kind of information systems applicable to all business activities are shown in the following general outline (8):

1. Customer service, as reflected by number of late and missed commitments
2. Product or service quality, expressed as percent of meeting all specifications or expectations
3. Operating cost, shown in terms of $/unit by category of product or service
4. Production or service volume, determined as a percent of actual capability
5. Personnel safety, measured in terms of total number of injuries and lost-time accidents

Information Format for Tracking Individual Performance Measures

A method to track individual performance measures that gives perspective for all time dimensions (past, present, and future) is illustrated in Fig. 27.12. The method can be performed either manually or by computer and is applicable to every individual job or functional team responsibility.

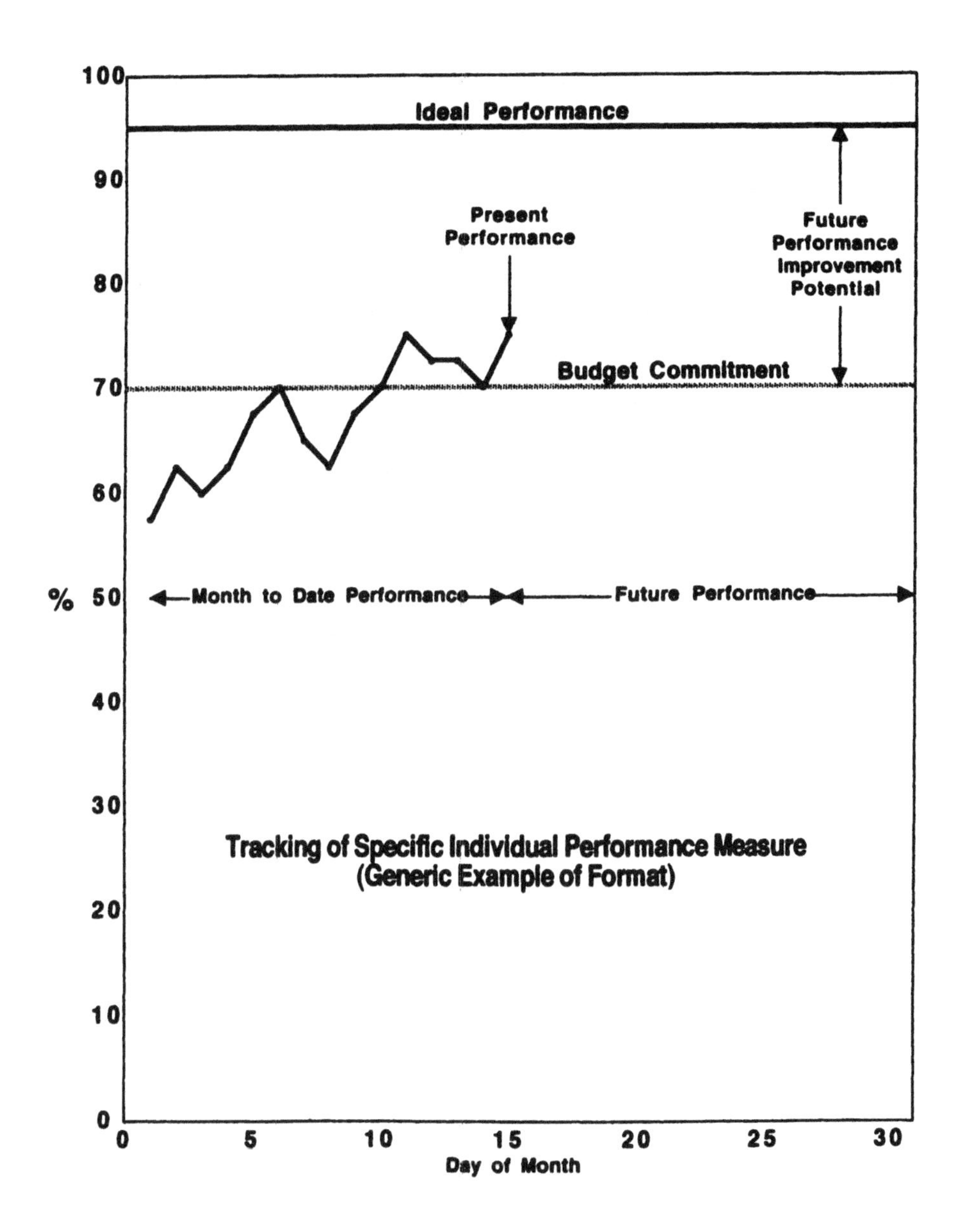

Fig. 27.12. Information format for tracking individual performance measures. Reproduced with permission of The Core Team, Hammond, LA.

Time

Introduction

Time is another vital resource to consider in managing processing plants. The success of a plant is ultimately determined by how each person involved uses his or her time. Is each person's time spent in adding value, or is it wasted?

Work Sampling

Work sampling is an effective technique to emphasize the need for and measure the levels of value-added work performed by individuals, work teams, and the total organization (10). Each member of the organization should routinely engage in self-work sampling for self-measurement and self-critique. Every functional team member should be actively involved in the process of group work sampling. Observations are made within the functional team to measure group value-added work and maintain individual anonymity. Work sampling for the total organization (processing plant, for example) is the composite result of all functional team results. An example work sampling form, with a classification list for the time use observed, is shown in Fig. 27.13.

Management System

Another argument supporting self-managed people and functional teams is effective use of time. Employees with a sense of ownership and accountability will make the most of time. In systems where action is taken on the basis of being told, considerable time is inherently wasted.

Production Scheduling

Effective production scheduling is critical in making the most of time. For a given product mix, the actual plant capacity must be known. Time allocated to produce a given product volume and mix should be on the basis of the actual plant capacity. Production efficiency should always be calculated and reported on the basis of product mix-sensitive actual capacity. Accordingly, people will be encouraged to make the most of the plant's capability and their time. Furthermore, the sales organization and customers will benefit from confidence in the plant's capability for timely delivery of products. During periods of less-than-capacity sales, the operating mode is to run at capacity rate and use the downtime for maintenance, training, or other value-added tasks.

Personnel Training and Development

The impact of time is especially noteworthy in personal training and development. In organizations with ineffective training systems, the prevailing belief is that several

WORK SAMPLING DATA COLLECTION					
Observation Classification			No. Occurrences		
Value-Added Work					
• Operate					
• Sample					
• Analyze					
• Fabricate					
• Erect					
• Repair					
• Train (Skill Development)					
Non-Value-Added Work					
• Receive Instructions					
• Clean					
• Movement with Obvious Work-Related Purpose					
• Procure Tools, Materials and Supplies					
Delay (Time Waste)					
• Personal Matters					
• Movement without Purpose					
• Waiting					
• Re-Work of Any Kind					
Observer:	Date:	Time			
		Stop	Start	Total	
________	________	____	____	____	

Fig. 27.13. Work sampling form. Reproduced with permission of The Core Team, Hammond, LA.

years are required to learn the skills for and become competent in most jobs (11). Organizations with robust self-directed training systems have a sharply contrasting perspective. For handling a new job, the expectation and practice are to learn and demonstrate basic competence in two months, achieve full competence in six months, and reach a level of peak performance in one to two years. The competitive advantage gained by the latter paradigm is most significant, as reflected by the professional contribution and growth of each person.

References

1. Smallwood, N.J., Soybean/Soybean Oil Processing Plant Management, Seminar presentation in Russia, Greece and Turkey, sponsored by the American Soybean Association, 1990.
2. Smallwood, N.J., *The Empowerment Vision*, The Core Team, Hammond, LA, 1993.
3. Smallwood, N.J., in *Proceedings of the World Conference on Oilseed Technology and Utilization*, edited by T.W. Applewhite, American Oil Chemists' Society, Champaign, IL, 1993, pp. 87–91.
4. Dowling, W., *Effective Management and the Behavioral Sciences*, AMACOM, New York, 1978.
5. Blake, R.R., J.S. Mouton, and R.L. Allen, *Spectacular Teamwork*, John Wiley & Sons, Inc., New York, 1987.
6. Blake, R.R., J.S. Mouton, and A.A. McCanse, *Change by Design*, Addison-Wesley Publishing Company, Reading, MA, 1989.
7. Montgomery, S.J., and N.J. Smallwood, Managing with Unions, Empowered Management Skills Seminar presentation, Tela, Honduras, 1994.
8. Smallwood, N.J., Information Systems, Empowered Management Skills Seminar Manual, The Core Team, Hammond, LA, 1993.
9. Barker, J.A., *Paradigms*, Harper-Collins, New York, 1992.
10. Smallwood, N.J., Work Sampling, Empowered Management Skills Seminar Manual, The Core Team, Hammond, LA, 1993.
11. Smallwood, N.J., Self-Managed Training System, Empowered Management Skills Seminar Manual, The Core Team, Hammond, LA, 1993.

Index

Printed in the United States
By Bookmasters